Flora of Eastern Ghats

Volume 5

Nyctaginaceae - Ceratophyllaceae

The Authors

Prof. T. Pullaiah was Professorof Botany at Sri Krishnadevaraya University and retired from active service in the year 2014. He held several positions in the University as Dean, Faculty of Life Sciences; Head, Department of Botany; Chairman, Board of Studies in Botany. He has published 75 books, 335 research papers. Under his guidance, 54 students obtained their Ph. D. degrees and 34 students their M. Phil. Degrees. He is the recipient of Best Teacher Award from the Government of Andhra Pradesh, Prof. P. Maheshwari Gold Medal and Dr. G. Panigrahi Memorial Award of Indian Botanical Society and Prof. Y. D. Tiagi Gold Medal of the Indian Association for Angiosperm Taxonomy. He was President of Indian Association for Angiosperm Taxonomy (2013) and President of the Indian Botanical Society (2014). He was a member of Species Survival Commission of IUCN.s

Dr S. Sandhya Rani is Assistant Professor in the Department of Botany, Sri Krishnadevaraya University. She has 9 years of teaching experience and 20 years of Research experience. She was PI and Co-PI of two major Research Projects funded by DAE. She authored 7 books and published 60 research papers. She has attended 20 seminars/ conferences organized at national and international level and presented papers. She is a Fellow of IAAT and member of IBS and NESA.

Dr S. Karuppusamy is Assistant Professor in the Department of Botany, The Madura College, Madurai, Tamil Nadu. He has published more than 120 research papers and co-authored five books. He is the recipient of Prof. Y. S. Murthy Gold Medal from Indian Botanical Society and Rolla S. Rao award from Indian Association for Angiosperm Taxonomy. He is a Fellow of Indian Association for Angiosperm Taxonomy and Indian Botanical Society. He is the Principal-Investigator of 3 major Research Projects funded by UGC, SERB and NIF. Under his guidance, 4 students obtained their Ph. D. degrees.

Flora of Eastern Ghats

Volume 5

Nyctaginaceae - Ceratophyllaceae

T. Pullaiah

S. Sandhya Rani

S. Karuppusamy

2019

Regency Publications

A Division of

Astral International Pvt. Ltd.

New Delhi – 110 002

ISBN: 978-93-86071-99-6 (International Edition)

Published by : **Regency Publications**
A Division of
Astral International Pvt. Ltd.
– ISO 9001:2015 Certified Company –
4736/23, Ansari Road, Darya Ganj
New Delhi-110 002
Ph. 011-43549197, 23278134
E-mail: info@astralint.com
Website: www.astralint.com

Digitally Printed at : **Replika Press Pvt. Ltd.**

Preface

Flora of Eastern Ghats volume 4 came in 2011 and we took much longer time to bring out the present volume. This is because of our preoccupation with other responsibilities and writing other books. We assure the readers the next two volumes will come in the next year itself.

The first author would like to extend his gratitude to the authorities of Library of Missouri Botanical Garden, USA for the permission to consult the Library. We are thankful to the authorities of Herbaria CAL, MH, BSID, DD, RHT and SKU for permission to consult the Herbaria. They also thank Professor R.R.Venkata Raju and B.Ravi Prasad Rao for their support and help. Photographs for this volume have been provided by A. Lalithamba, Lecturer in Botany, Nellore and authors thank her for the same. Many line diagrams for this book were drawn by Late P. Surya Prakash Babu. We thank Scientists and Scholars at CAL, MH and BSID for extending their support and help during the preparation of this book.

We thank many taxonomists like Dr. K.N.Gandhi, Dr. Sanjappa and Dr.P.V.Prsanna who gave inputs for improving the quality of this book.

T. Pullaiah

S. Sandhya Rani

S. Karuppusamy

Contents

Abbreviations Used in General Text and Enumeration

BLAT: Blatter Herbarium, St. Xavier College, Mumbai

BSID: Botanical Survey of India, Deccan Circle, Hyderabad

CAL: Central National Herbarium, Howrah, Calcutta (nowcalled as Kolkata)

cm: centimeter

DD: Herbarium, BSI, Dehradun

et al.: *et alia* (and others, of people)

FBI: Hooker's Flora of British India

Fig.: Figure

Fl.: Flowering season

Fr.: Fruiting season

Gamble: Gamble's Flora of Presidency of Madras

m: meter

MH: Madras Herbarium, BSI, Southern Circle, TNAU campus, Coimbatore mm: millimeter(s)

Ori.: Oriya

p.p.: *pro-parte*

SACON: Salim Ali Centre for Ornithology and Natural History, Coimbatore

sine. coll.: Without collector's name

s.n.: Without number

SGH: Sri Ganesan Herbarium, The Madura College, Madurai

SKU: Sri Krishnadevaraya University, Anantapur

Sq. km: Square kilo-meter(s)

SVU: Sri VenkateswaraUniversity, Tirupati

Tam.: Tamil

Tel.: Telugu

var.: varities (variety)

Vern.: Vernacular name (s)

Abbreviations – Districts

ATP: Anantapur (now called Anantapuramu)

BGR: Bolangir

CDP: Cuddapah (Now called as Kadapa)

CGP: Chengalput

CMT: Coimbatore

CTR: Chittoor

DGL: Dindigul

DKL: Dhenkanal

DMP: Dharmapuri

EG: East Godavari

GJM: Ganjam

GJT: Gajapathi

GNT: Guntur

KHD: Kalahandi

KRN: Krishna

KRT: Koraput

KMM: Khammam

KNL: Kurnool
MBJ: Mayurbhanj
MDU: Madurai
MBNR: Mahaboobnagar
NA: NorthArcot
NLR: Nellore
NMK: Namakkal
PLN: Pulbhani
PRK: Prakasam
RGD: Rayagadh
SA: SouthArcot
SBP: Sambalpur
SKLM: Srikakulam
SM: Salem
TRP: Tiruchirapalli
TRV: Tiruvannamalai
VJN: Vizianagaram
VLR: Vellore
VSKP: Visakhapatnam
WG: West Godavari

Abbreviations – Plant Explorers

AAA: A.A.Ansari
AL: A. Laxmaiah
AM: A.Mohan
AMK: A.Mukherji
AMR: A.Madhusudana Reddy
ANH: A.N.Henry
ANS: A. Narayana Swamy
APD and *SP*: A.P.Das and Sauris Panda
AU: A.Ugraiah
AVNR: A.V.N.Rao
AWL: A.W.Lushington

BR: B.Ravi Prasad Rao
BSS: B.Sadasivaiah
BSN: B.Suryanarayana
CAB: C.A.Barber
CECF: C.E.C.Fischer
CJP: C.Jayaprada
CPSM: C.P.Sreemadhavan
CS: C.Subbarayudu
DAM: D.Ali Moulali
DBD: D.B.Deb
DCSR: D.C.S.Raju
DN: D.Narasimhan
DRC: D.Rangachari
EC: E.Chennaiah
EKK: E.K.Krishnan
EV: E.Vajravelu
GO: G.Obulesu
GPG: G.Panigrahi
GVS: G.V.SubbaRao
HFM: H.F.Mooney
HHH: H.H.Haines
HOS: H.O.Saxena
HS: H.Santapau
JLE: J.L.Ellis
JS: J.Joseph
JSG: J.S.Gamble
KCJ: K.C.Jacob
KH: K.Hanumanthappa
KMM: K.M.Matthew
KMS: K.M.Sebastine
KNS: K.N.Subramanian
KRM: K.Ramamurthy
KS: K.Subramanyam

KSM: K.Sri Rama Murthy
KT: K.Thothathri
MB: M.Brahmam
MBV: M.B.Viswanatham
MCB: M.Chandrabose
MCKL: M.Chennakesavulu
MM: M.Mohanan
MRR: M.Rama Rao
MSR: M.S.Ramaswami
NPBK: N.P.Balakrishnan
NRR: N.Rama Rao
NV: N.Venkatappa
NVG: N.Venugopal
NY: N.Yesoda
PMR: P.M.Rajeswaramma
PP: P.Perumal
PVK: P.Venkanna
PVP: P.V.Prasanna
RHB: R.H.Beddome
RKM: R.Krishnamohan
RSR: Rolla S.Rao
RVK: R.VijayaKumar
RVR: R.R.Venkata Raju
R.V. Reddy: R.VenkateswaraReddy
SKK: S.Karthikeyan
SKS: S.Karuppusamy
SKW: S.K.Wagh
SNR: S.Nazeeruddin
SPG: Sai Prasad Goud
SRSR: S.R.S.Reddy
SS: S.Sunitha
SSR: S.Sandhya Rani

SX: Saxena
TDCK: T.Dharma Chandra Kumar
TP: T.Pullaiah
VNS: V.Narayanaswami
VRK: V.Ramakrishnaiah
VS: V. Srinivasulu

NYCTAGINACEAE

1. Annual or perennial, prostrate or erect herbs, rarely shrubs, up to 1.5 m high
   2. Flowers showy, with an involucre or prominent bracts.................. **Mirabilis**
   2. Flowers small, bracts or involucel of bracts not prominent........... **Boerhavia**
1. Trees or shrubs, climbing or often straggling, up to 12 m or more long:
   3. Flowers with 3 conspicuously coloured bracts; anthocarp glabrous or pubescent, never viscous .... **Bougainvillea**
   3. Flowers without conspicuous bracts; anthocarp viscous, glandular... **Pisonia**

BOERHAVIA Linnaeus

1. Anthocarp 10-ribbed with wart like appendages ....**B. chinensis**
1. Anthocarp 5-ribbed, glabrous:
   2. Diffuse herb, anthocarp glandular-hairy, top rounded, club-shaped ....**B. diffusa**
   2. Erect herb, anthocarp ribs undulate, top truncate, obconic shaped .... **B. erecta**

Boerhavia chinensis (L.) Rottb., Pl. Hort. Univ. Rar. Prog. 4. 1773; Stemmerik in Steenis, Fl. Malesiana ser. 1, 6: 455. 1964. *Valerina chinensis* L., Sp. Pl. 33. 1753. *Commicarpus chinensis* (L.) Heimerl, Nat. Pflanzenfam. ed. 2. 16c: 117. 1934. *Boerhavia repanda* Willd., Sp. Pl. 1: 22. 1797; FBI 4: 709. 1885; Gamble 2: 1163. 1925.

A creeper with ascending branches, branches glabrous or nearly so, terete. Leaves triangular-ovate, 5-10 × 3-7 cm, base cordate to rounded, margin repand-sinuate or entire, apex acuminate, glandular dotted above, pubescent beneath; petiole 0.5-2 cm long. Flowers pink, in axillary and terminal, long-peduncled umbels; pedicel 1-1.5 cm long; bracts 0; bracteole 1, up to 0.3 cm long, lanceolate; peduncle up to 6 cm long, filiform, glabrous; pedicels up to 0.2 cm long, filiform, glabrous; perianth campanulate, up to1.2 cm long, constricted below the middle; stamens 3-4; much exserted; ovary ovoid, glabrous, style dilated at apex. Anthocarp oblong-fusiform, 7 × 1.5 mm, black, 10-ribbed, viscid-glandular with wart like appendages; seed black, testa shiny, smooth.

Common in most districts in hedges and in forests. Fl. & Fr.: Throughout the year. Vern.: Tel.: *Teega mamidi*.

Pennahobilam (ATP), *TP* & *NY* 368; Batrepalli (ATP), *NY* 853; Amagondapalem RF (ATP), *NY* 1035 (MH); Gulemallebad RF (KNL), *RVR* 1678; Veldurthi RF (KNL), *RVR* 1733; Kodur (CDP), *BR* & *DV* 43480; Chelama (KNL), *JLE* 17982 (MH); Kavalur (NLR), *MVS* & *VSR* 30936; Kondaveedu fort (GNT), *CAB* 4651 (MH); Madgole

(VSKP), *GVS* 29575 (MH); way to Javapuram Konda (VSKP), *GVS* 4259 (MH); Anaparthy (VSKP), *JSG* 41911 (MH); Mahendragiri (GJM), *Srivastava & Party* 38497 (LWG); Pennagaram RF (SM), *EV* 20748 (MH); Threerthamalai RF (DMP), *EV* 51813 (MH).

INDIA: Upper Gangetic plains, Peninsular India.

WORLD: Old World Tropics.

Note: This taxon is sometimes named *Commicarpus chinensis* (L.) Heimerl (1934). Stemmerik (*loc. cit.*) does not accept this generic segregation based on number of calyx ribs.

Boerhavia diffusa L., Sp. Pl. 3. 1753; Gamble 2: 1162. 1925. *B. repens* L., Sp. Pl. 3. 1753; FBI 4: 709. 1885.

Perennial diffuse herb, hirsute, up to 1 m long; branches slender, pubescent. Leaves opposite, sub-equal, ovate, lanceolate, sub-orbicular, thick, 1–4 × 0.5–3 cm, base rounded or sub-cordate, margin somewhat undulate, apex obtuse, apiculate; greenish pubescent above, white beneath; lateral veins 4–5 pairs; petiole slender, up to 1.8 cm long, pubescent. Flowers in axillary and terminal umbels, glandular hairy; peduncle to 5 cm long, secondary peduncles 5, to 1.5 cm long; pedicel < 1 mm long; bracts 1 mm long; bracteole similar to bract; perianth tube 4 mm long; lower part prominently 5-ribbed; upper part deep pink, 4 mm across; 5-lobed; stamens 2 or 3; filaments 3 mm long; anthers 0.5 mm; ovary 0.6 mm long; style 3 mm long; stigma discoid. Anthocarp club-shaped, glandular-hairy, ellipsoid, 3 × 1 mm, rounded at apex.

Common weed of waste places, fields and road sides all over the State. Fl. & Fr.: Throughout the year. Vern.: Tel.: *Atikamamidi;* Ori.: *Kharkharia, Parni, Kechna, Atikapudisag, Guodapurunni;* Tam.: *Mookirattai.*

Yadiki RF (ATP), *TP* & *NY* 1114; Racherla RF, Erramalais (KNL), *RVR* 1371; Gulemallebad RF (KNL), *RVR* 1395; Gani RF (KNL), *RVR* 1504; Palakonda hills (CDP), *CS* 7719; Obulapuram thanda (PKM), *BR* & *BSS* 32791; Chinnamantanala (PKM), *RVK* 15863; Macherla (GNT), *VRK* 3269; Dongabhavigonda (KNL), *JLE* 17984 (MH); Dongalacheruvu (CDP) *KS* 6308 (MH); Tirumala, Papanasananm (CTR), *DRC* 460 (MH); Kondaveedu fort (GNT), *CAB* 4652 (MH); Koruturu (WG), *DN* 84267 (BSID); Maredumilli (EG), *MM* 105002 (MH); Simhachalam (VSKP), *GVS* 19338 (MH); Anantagiri (VSKP), *GVS* 19501 (MH); Karaka (VSKP), *CAB* 1639 (MH); Hogainakkal (SM), *EV* 20507 (MH); Chinakalrayan (SM), *RA* 49966 (MH); Karakambadi, Yercaud (SM), *SKK* 26865 (MH); Pakamalai RF- Gingee (SA), *KRM* 53526 (MH); Shanikulam (SA), *CAB* 850 (MH); Tippukadu RF (NA), *KRM* 16642 (MH); Kottur RF (NA), *MBV* 667 (MH); Rukeekonda (GJM), *VNS* s.n. (MH); Raika to Digi (GJM), *CAB* 1294 (MH).

INDIA: Through out India.

WORLD: Pantropical, Australia.

Boerhavia erecta L., Sp. Pl. 3. 1753. *B. punarnava* Saha & Krishn., J. Sci. Industr. Res. 21c: 149. 1962; Rao & Hemadri, Sci. & Cult. 36: 218-1970.

Annual erect herb, up to 70 cm high; stem pubescent, terete. Leaves opposite, sub-equal, ovate, 2–5 × 1.5–2 cm, glabrous above, pubescent below, base truncate-obtuse, margin undulate, apex apiculate-acute; lateral veins 5–6 pairs; petiole 0.5–1.5 cm long. Flowers white, to 3 mm long, in axillary and terminal, 10–20 cm long, paniculate umbels, zigzag, slender, forked; pedicel to 0.6 cm long; bract and bracteole 1 mm long, lanceolate; perianth tube 2.5 mm long; lower parts with 5 faint ribs; upper part white, with pink tinge, 3 mm across; stamens 1 or 3; filaments 2.5 mm long; anthers 0.4 mm across; ovary 0.5 mm long; style 2.5 mm long. Anthocarp obconical, 2.5 mm long, glabrous, top truncate, the grooves between the ribs somewhat undulate

Rarely found in dry areas. Fl. & Fr.: September –January. Vern.: Tel.: *Punarnava, Tella atikamamidi.*

Kalasamudram (ATP), *KRKS* 39127; Panyam RF (KNL), *RVR* 1576; Pacherla (KNL), *BR* & *BSS* 34235; Thummalapalli (CDP), *KRKS* & KRS 33686 (BSID); Japali (CTR), *AJR* & *K.Raja* 20621; Balapalli (CDP), *JLE* 14221 (MH); Vankatagiri (NLR), *BSN* & *AMR*, 6169 (BSID); Sirumalai hills (DGL), *SKS* 715 (SGH).

INDIA: Andhra Pradesh, Tamil Nadu, Kerala, Maharashtra, Punjab, Rajasthan, Uttarakhand, West Bengal.

WORLD: Pantropical.

BOUGAINVILLEA Comm. ex Jussieu

1. Perianth-tube more or less pubescent; leaves and stems subglabrous; bracts magenta coloured .... **B. glabra**
2. Perianth-tube densely pubescent; leaves and stems pubescent; bracts deep rose, but varying to purple or greenish....................... **B. spectabilis**

Bougainvillea glabra Choisy in DC., Prodr. 13, 2: 437. 1849; Haines, Bot. Bihar & Orissa 2: 756. 1924; Gamble 2: 1164. 1925.

Large, thorny, climbing shrub, spines axillary, up to 1.5 cm long. Leaves ovate, 3–8 × 3–10 cm, base truncate-acute, margin entire, apex acute; lateral veins 5–7 pairs. Flowers axillary in dichotomous cyme; bracts 3, magenta, thin, reticulately veined, 4 × 3 cm; pedicels 3–4 cm long; calyx-tube 2 cm long, pink; lobes 5, pink below, white above, 0.5 cm long; corolla white, 5-lobed, each 2-fid, 0.3 cm across; stamens 8, in different heights, 1.5–2 cm long; anthers yellow, 0.3 mm across; ovary oblong, to 0.5 cm long; style hairy, curved; stigma linear, forked.

Cultivated in gardens. A native of Brazil. Fl. & Fr.: January – April. Vern.: Tel.: *Kagitha pulu, Kagithal chettu;* Tam.: *Kagithapoo.*

INDIA: Cultivated in gardens through out India.

WORLD: Native of Brazil, Widely cultivated in Tropics and subtropics.

Bougainvillea spectabilis Willd., Sp. Pl. 2: 348. 1799; Haines, Bot. Bihar & Orissa 2: 756. 1924; Gamble 2: 1164. 1925.

Large scandent, thorny shrubs; spines stout with recurved tip. Leaves ovate, acuminate, pubescent on both the surfaces, veins 4-5 in alternate pairs. Flowers purple, deep rose or greenish supported by an involucre of 3 brilliantly coloured large bracts exceeding the flowers. Perianth tubular, 5-lobed, limb wider, basal part looking like a sympetalous corolla; stamens 8, filaments unequal; ovary unilocular, ovule 1, basal. Fruit achene, seed erect.

Cultivated in gardens and habitations as an ornamental plant. Fl. & Fr.: August – January. Vern.: Tel.: *Kagithala puvvu;* Tam.: *Kagithapoo.*

Chinnapalli (ATP), *KRKS* 40389.

INDIA: Cultivated in gardens through out India.

WORLD: Native of Brazil, widely cultivated in tropics and subtropics.

MIRABILIS Linnaeus

Mirabilis jalapa L., Sp. Pl. 177. 1753; Gamble 2: 1164. 1925.

Herb, 30-75 cm high, with large tuberous root, stem fleshy. Leaves opposite, triangular-ovate, 4-10 × 3-6 cm, base truncate to shortly decurrent, margin undulate, apex acute to acuminate, hairy on midrib and veins on both sides; petiole to 3 cm. Flowers variously coloured, usually purple, in terminal, sessile corymbs, subtended by involucre of 5, ovate-elliptic, shortly connate bracts, accrescent in fruit; perianth funnel-shaped, 5-lobed, 5 cm long; stamens 3-6, exserted, filaments to 6.5 cm long, anthers 1.5 cm; ovary glabrous, style filiform, stigma capitate. Anthocarp globose, to 6 cm across, black.

Cultivated in gardens and often as an escape. Fl. & Fr.: Throughout the year. Ori.: *Rangbano.*

Dhone (KNL), *RVR* 2512; Maredumilli (EG), *MM* 102699 (MH, BSID); Modugu mamisala (KNL), *JLE* 42210 (MH); Near Tadepalli (EG), *NRR* and *DN* 84334 (BSID); Anantagiri (VSKP), *GVS* 19468 (MH); Sirumalai hills (DGL), *SKS* 644 (SGH).

INDIA: Cultivated throughout.

WORLD: Native of Peru, now cultivated and occasionally escape in tropics.

PISONIA Linnaeus

1. Veins prominent and hairy in lower surface of leaf; perianth lobes not keeled on the inside; anthocarp with 5 rows of mono-serial prickles .... **P.grandis**

1. Veins not so prominent and nearly glabrous in lower surface of leaf; perianth limb with 5 great lobes alternating with 5 smaller lobes, anthocarp with 5 rows of bi-serial prickles .... **P. aculeata**

Pisonia aculeata L., Sp. Pl. 1026. 1753; FBI 4: 711. 1885; Gamble 2: 1163. 1925.

Scandent shrub with axillary spines, spines recurved, 7 mm long; stem terete, finely pubescent. Leaves opposite, elliptic to ovate, 2-7 × 1.5-4.5 cm, base rounded, margin entire, apex obtuse to acute, glabrous; petiole up to 1.4 cm long, pilose. Flowers in axillary and terminal, pedunculate, pubescent cymes, pedicels pubescent; bracts and bracteoles up to 0.1 cm across, pubescent on the lower surface, glabrous above. Female flowers: perianth up to 0.3 cm long, tubular, densely pubescent without, lobes 5, triangular; ovary elongate, glabrous, style thick, stigma fimbriate. Anthocarp oblong; seed oblong, hyaline, testa shiny.

Occasional in open forests and rare in waste lands in most districts. Fl. & Fr.: September-January. Vern.: Tel.: *Embuddi;* Ori.: *Hathi ankuso;* Tam.: *Marugalli.*

Yadiki RF (ATP), *TP* 488, *TP* & *NY* 1110; Macherla RF (GNT), *VRK* 3891; Palakonda hills (CDP), *CS* 8277; Way to Komativaricheruvu (CTR), *GVS* 46808 (MH); Anantagiri (VSKP), *JSG* 41939 (MH); Sirivaka (EG), *CAB* 5266 (MH); Devipatnam (WG), *CAB* 4922 (MH); Rambha (GJM), *SX* & *MB* 2506, 3148 (RRL-B); Surangi (GJM), *SX, MB* & *Panigrahi* 2418 (RRL-B); Gingee RF (SA), *KMS* 12335 (MH); Kadaladi forest (NA), *EV* 54530 (MH); Gingee RF (SA), *S.R. Rajan* 77391 (MH); Hogainakkal (SM), *EV* 23539 (MH); Padavadi (TRV), *VSR* & *MVS* 33244; Alagar hills, near Nupuragangai (MDU), *SKS* 68 (SGH).

INDIA: Deccan Peninsula, Odisha.

WORLD: (Sub)tropical America, Africa, SE Asia.

Pisonia grandis R.Br., Prodr. 422. 1810. *P. alba* Span., Linnaea 15: 342. 1841; FBI 4: 711. 1885 p.p. *P. morindifolia* R.Br. ex Wight, Icon Pl. Ind. Orient t. 1765. 1852; Gamble 2: 1164. 1925.

Evergreen tree, up to 12 m high. Leaves ovate, ca 11 × 5.5 cm, petiole 3 cm, base cuneate, entire, acute, glabrous, veins hairy below. Inflorescence terminal, corymbose cymes; male flower tubular, 5-lobed; stamens 8-9; male flower much smaller than female flower. Anthocarp clavate, 5-ribbed, angles with only 1 row of prickles.

Cultivated in gardens. Fl. & Fr.: September – April. Vern.: Tam.: *Lachakattakeerai.*

Eluru (WG), *KS* 5157 (MH).

INDIA: Planted in Gardens.

WORLD: Indo-Pacific Tropical Tree.

AMARANTHACEAE

1. Bracteole wings entirely adnate to midrib, staminodes fimbriate, prominently appendaged within:
    2. Aquatic; bracteoles not spinescent .... **Centrostachys**
    2. Terrestrial; bracteoles spinescent .... **Achyranthes**

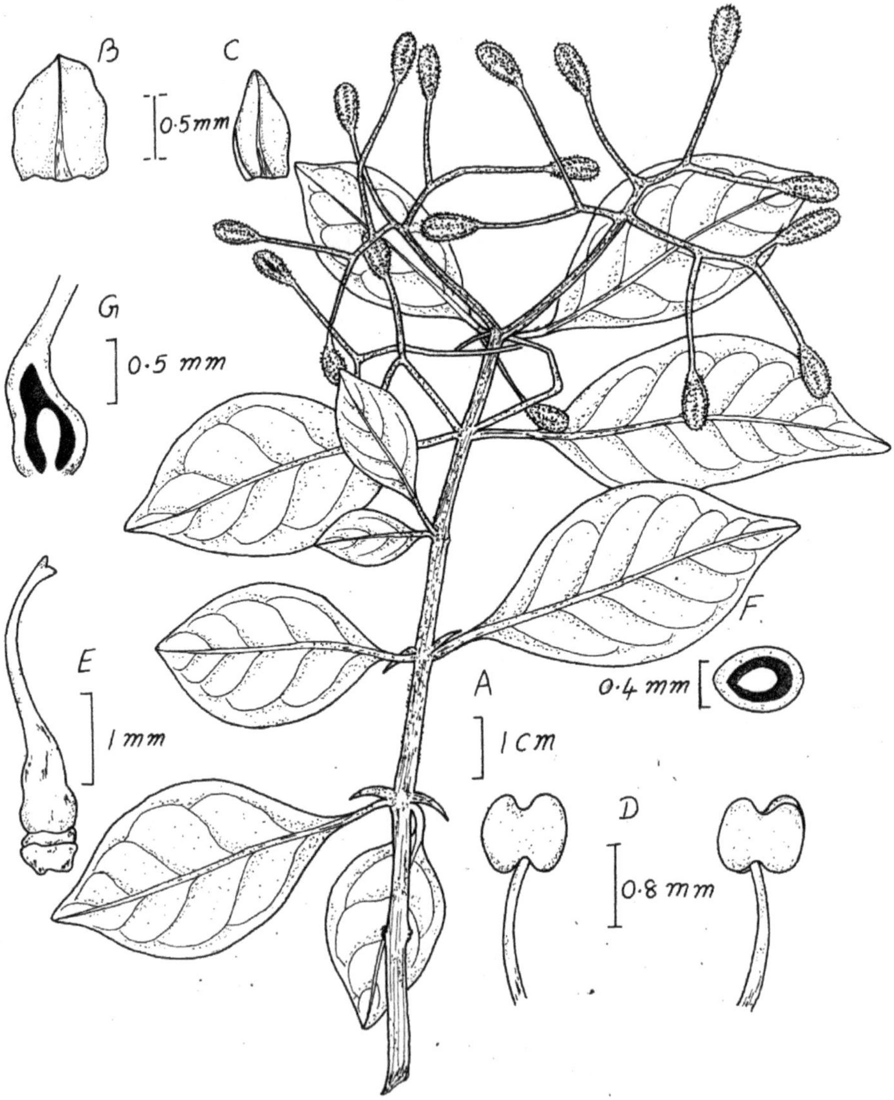

Figure 1. **Pisonia aculeata** L.
A. Twig, B. Bract, C. Bracteole, D. Stamens, E. Pistil, F. T.s. of ovary, G. L.s. of ovary.

1. Bracteoles not winged, staminodes not fimbriate:
   3. Stamens 2 .... **Nothosaerva**
   3. Stamens more than 2:
      4. Ovary containing 2- many ovules, style persistent, embedded .... **Celosia**
      4. Ovary containing single ovule, style not persistent:
         5. Stems rooting at nodes; flowers spinescent; anthers 1-celled .... **Alternanthera**
         5. Stems lacking roots at the nodes; flowers not spinescent; anthers 2-celled:
            6. Spikes tomentose:
               7. Perianth lobes 5 .... **Aerva**
               7. Perianth lobes 4 .... **Trichuriella**
            6. Spikes not tomentose:
               8. Flowers unisexual or polygamous .... **Amaranthus**
               8. Flowers bisexual:
                  9. Leave alternate:
                     10. Flowers capitate .... **Allmania**
                     10. Flowers spicate:
                         11. Fertile flowers each subtended by 2 lateral sterile modified flowers consisting of branched hooks .... **Digera**
                         11. Fertile flowers not subtended by modified sterile flowers .... **Indobanalia**
                  9. Leaves opposite:
                     12. Stigma bilobed .... **Gomphrena**
                     12. Stigma capitate:
                         13. Filaments alternating with peseudostaminodes .... **Cyathula**
                         13. Pseudostaminodes absent .... **Pupalia**

ACHYRANTHES Linnaeus

1. Basal wing of the spinescent bracteole entirely adnate to the spine; pseudostaminodes fimbriate; stigma capitate .... **A. aspera**
1. Basal wing of the spinescent bracteole free from the spine; pseudostaminodes not fimbriate; stigma slightly capitate .... **A. bidentata**

Achyranthes aspera L., Sp. Pl. 204. 1753; FBI 4: 730. 1885; Gamble 2:1176. 1925.

Key to Varieties

1. Leaves silverily silky beneath .... var. **sicula**
1. Leaves other than silky:
   2. Leaf apex acuminate:
      3. Leaves glabrous or sparsely pubescent..................... var. **porphyristachya**
      3. Leaves pubescent above and below .... var. **pubescens**
   2. Leaf apex acute:
      3. Leaves hairy, brown when dry .... var. **rubro-fusca**
      3. Leaves softly pubescent on both sides or tomentose below; dark brown on drying... var. **aspera**

Achyranthes aspera L. var. **aspera.**

Annual erect herb, up to 1.5 m tall; branches erect to sub-scandent or straggling, much branched, 4-angular, sulcate, densely tomentose. Leaves opposite, elliptic-ovate or obovate, 2–6 × 2–4 cm, sparingly pubescent above, densely so below with white hairs, base rounded or acute, margin entire, apex obtusely acute; lateral veins 5–7 pairs; petiole 0.5–1.5 cm long. Flowers in terminal spike, 19–22 flowers per 5 cm length of spike axis, sessile, light pinkish-green; bracts lanceolate or narrowly deltoid-lanceolate, 3 × 1–1.5 mm, glabrous; bracteoles deltoid-lanceolate, 2–3 × 1–1.5 mm, glabrous, basal wings 0.2–0.3 mm long; spines 1 mm long; perianth segments 5; outer one longer than remaining, 3 × 1–1.5 mm, spiny acuminate, midrib distinct, lateral veins 2, distinct, margin pale; inner ones lanceolate, 2–3 × 1 mm, long acuminate, midrib distinct, margin pale; stamens 5, 1 mm long, alternating with subquadrate pseudo-staminodes, staminal sheath fimbriate at apex, prominently appendaged within; anthers 0.5 mm long; ovary obconical, 0.7 mm across; style 1 mm long; stigma capitellate. Utricle oblong to ovoid, 1.5 mm long; operculum cup-shaped. Seed oblong, testa coriaceous, white, 1–2 mm long.

Common weed on road sides, waste places in all districts. Fl. & Fr.: July-December. Vern.: Tel.: *Uttareni;* Ori.: *Apamaranga;* Tam.: *Nayuruvi.*

Tadipatri – Bugga (ATP), *TP* 448; Jolapuram (KNL), *TP* & *AA* 1922; Rachakuntapalle (CDP), *SSR* & *TP* 33603; Balapalle (CDP), JLE 14969 (MH); Kovuru (NLR), *SSY* 16316; Marripalem (PKM), *RVK* 15765; Diguvametta (KNL), *JLE* 3240 (MH); Balapalle (CDP), *JLE* 14969 (MH); Gobanapalem (WG), *KS* 5144 (MH); Samarlakota (EG), *CAB* 4499 (MH); Maredumilli (EG), *MM* 102548 (MH, BSID); Anantagiri (VSKP), *GVS* 19462 (MH); Adativalasa (VSKP), *GVS* 21586 (MH); Yercaud (SM), *DB Deb* 31284 (MH), *SKK* 23096 (MH).

INDIA: Through out India.

WORLD: Throughout the World in Tropical and warmer regions.

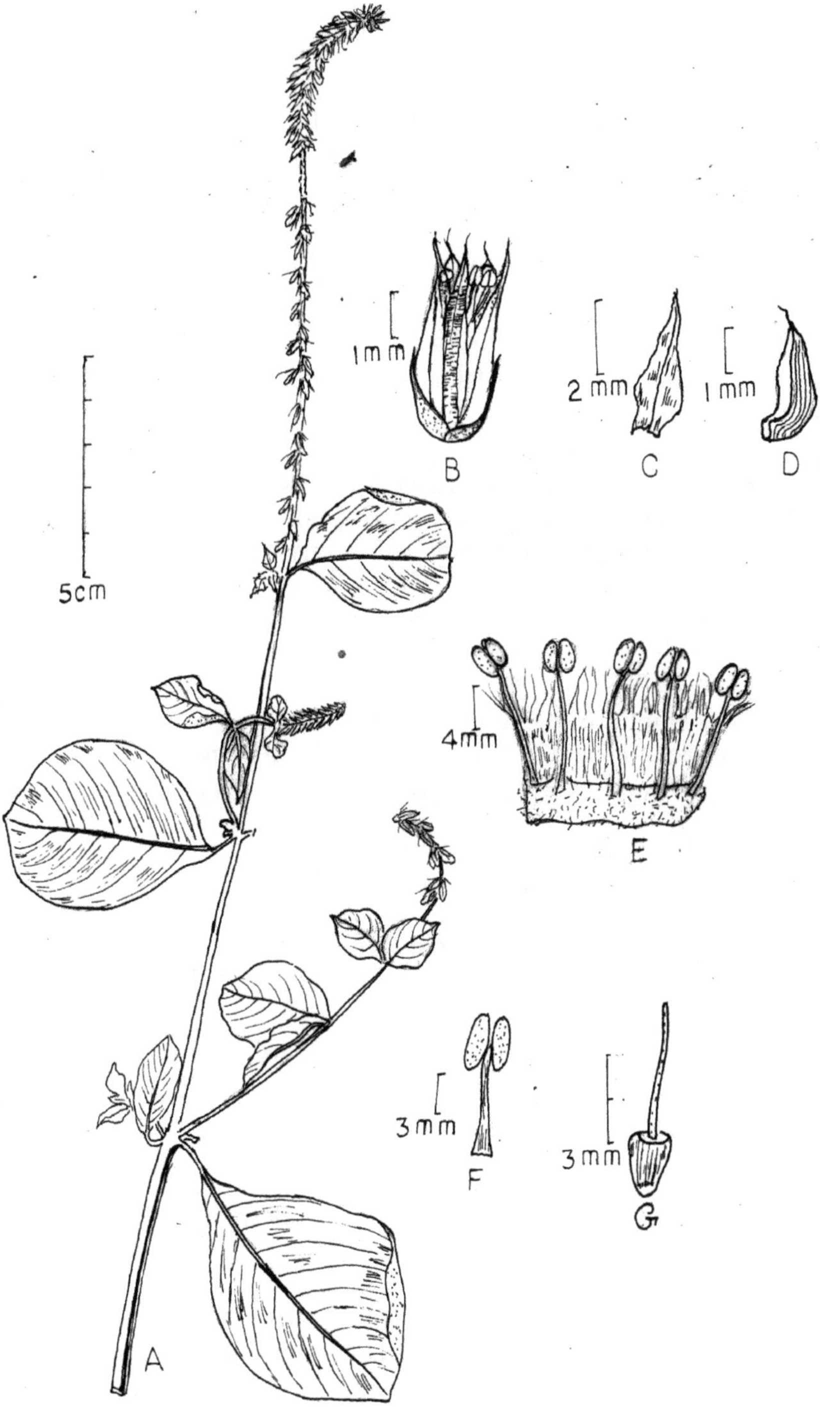

Figure 2. **Achyranthes aspera** L. var. **aspera**
A. Twig, B. Flower, C. Bract, D. Bracteole,
E. Staminal tube opened, F. Stamen, G. Pistil.

Achyranthes aspera L. var. **porphyristachya** (Wall. ex Moq.) Hook.f., Fl. Brit. India 4: 730. 1885; Gamble 2: 1176. 1925. *A. porphyristachya* Wall. ex Moq. in DC., Prodr. 13 (2): 316. 1849; H Hara, Fl. E. Himal. 76. 1966.

Diffusely branched herb. Leaves elliptic, 6-14 × 5-10 cm, glabrous or sparsely pubescent, acuminate. Flowers in slender spikes.

Rare. Fl. & Fr.: June – November.

Pennahobilam (ATP), *TP* 683; Rampa hills (EG), *K. Sampath Kumar* 23232; Simhachalam (VSKP), *NVP* 6615; Machkund (VSKP), *SKW* 4776-77 (BLAT); Waltair-Jeypure road (VSKP), *SKW* 4694 and *STP* 21376 (BLAT); Donubai (SKLM), *NRR* 83789 (MH); Yercaud (SM), *G Bidie s.n* (MH); Yelagiri hills (NA), *MBV* 1354 (MH).

INDIA: Peninsular India, sub Himalayan tracts, Upper Gangetic Plains.

WORLD: Sri Lanka.

Achyranthes aspera L. var. **pubescens** Moq., Kew Bull. 29: 473. 1974. *A. bidentata* sensu Baker and Clarke in Fl.Trop.Afr. 6, 1: 64. 1909; Cavaco, Mém. Mus. Hist. Nat. Paris Sér. B, 13: 125. 1962. *A. aspera* forma *excelsa* Cavaco, Mém. Mus. Nat. Paris Sér. B, 13: 118. 1962 *nom. Illegit. A. fruticosa* var. *pubescens* Moq. DC., Prodr. 13, 2: 314. 1849. *A. aspera* var. *nigro-olivacea* Suesseng. Fedde, Repert. 51: 194. 1942. *A. aspera* forma *robustiformis* Suesseng. Mitt. Bot. Staatss. München 1: 70.1950.

Leaf blades elliptic, ovate, or broadly ovate, 4-20 × 2-5 cm, base rounded or narrowed, apex acuminate. Tepals 6-7 mm. Utricles 3-4 mm.

Occasional in southern Eastern Ghats. Fl. & Fr.: June to December.

INDIA: Andhra Pradesh, Tamil Nadu.

WORLD: South Africa, Madagascar, Zimbabwae, Tanzania.

Achyranthes aspera L. var. **rubro-fusca** (Wight) Hook. f., Fl. Brit. India 4: 730. 1885; Gamble, 2: 1176. 1925. A. *rubro-fusca* Wight, Icon. Pl. Ind. Orient. t. 1778. 1852.

Leaves elliptic-ovate to cordate, acute, entire, hairy, reddish white stems when fresh, brown when dry.

Rare. Fl. & Fr: July – December.

Adativalasa (VSKP), *GVS* 21584 (MH).

INDIA: Maharashtra, Tamil Nadu, Andhra Pradesh.

WORLD: China.

Achyranthes aspera L. var. **sicula** L., Sp. Pl. 204. 1753. *Achyranthes aspera* L. var. *argentea* Hook. f., Fl. Brit. India 4: 730. 1885. *A. argentea* Thw., Enum. Pl. Zeyl. 249. 1861. non Lam. 1785.

Erect herbs, stem hairy. Leaves ovate, 6-13 × 2-4 cm, base rounded, entire, acuminate, dense hairy with gland dots. Flowers in terminal, long spikes; bracts linear, 3 x 1mm. spinescent; gynoecium 3 mm, style 1.5 cm, stigma simple.

Rare in Anantapuram and Kurnool districts. Fl. & Fr.: June – December.

Thalupula (ATP), *KRKS* & *BSS* 41940; S.K.University campus (ATP), *BR* & *V.S.Acharya* 36847; Ramanapenta, (KNL), *S. India flora* 14964 (MH)

INDIA: Andhra Pradesh, Gujarat, Rajasthan, Himalayas.

WORLD: A naturalized weed of uncertain origin now widespread in the tropics and subtropics of the world

Achyranthes bidentata Blume, Bijdr, 545. 1826; FBI 4: 730. 1885; Gamble 2: 1176. 1925.

Erect herb, up to 1.5 m high, stems quadrangular, ribbed; covered with fine hairs. Leaves elliptic-oblong or ovate-lanceolate, 5-20 × 1-8 cm, base acute or obtuse, margin entire, apex acuminate, sparsely hairy above, densely hairy beneath; petioles up to 2 cm long. Flowers in terminal and in the higher leaf axils, in 20 cm long spikes; flowers bisexual; bract narrowly lanceolate brownish, membranous; bracteoles 2, linear; tepals 5, up to 0.55 × 0.15 cm, subequal, pink, ovate-lanceolate; stamens 5, filaments dilated and connate at base to form a cup, filaments above the staminal-cup slender, stamens alternating with pseudostaminodes, pseudostaminodes truncate, irregularly serrate with a dorsal appendage; ovary glabrous, style slender, stigma capitate. Utricle cylindrical, 1 mm; seed annular, smooth, black.

Rare in open forests. Fl. & Fr.: July – November. Tam.: *Sennayuruvi.*

Forest near Sunkarametta (VSKP), *NPBK* 10934, *GVS* 21696 (MH), 21654 (MH); Galikonda (VSKP), *GVS* 21654 (MH); Similipahar (MBJ), *SX* & *MB* 4166 (RRL-B); Kolli hills (NMK), *SKS* 1020, 1021 (SGH); Sirumalai hills (DGL), *SKS* 462 (SGH).

INDIA: Through out India including Temperate and Subtropical Himalayas.

WORLD: Asia from Pakistan, India, Sri Lanka, east to China, Japan, Indonesia and Solomon Islands.

AERVA Forsskal *nom. cons.*

1. Leaves opposite-decussate, lanceolate, seed reniform .... **A. sanguinolenta**
1. Leaves alternate to subopposite, other than lanceolate; seed annular:
   2. Leaves variable, reniform to elliptic-ovate; inflorescence axillary dense, greenish-white spike; flowers bisexual; stigma capitate .... **A. lanata**
   2. Leaves oblanceolate-spathulate; inflorescence terminal, white woolly spikes; flowers dioecious; stigma deeply forked, as long as the style .... **A. javanica**

Aerva javanica (Burm. f.) Juss. ex Schult., Ann. Mus. Hist. Nat. 11: 131. 1808; FBI 4: 727. 1885. var. **javanica.** *Iresine javanica* Burm. f., Fl. Ind. 212. t. 65. f. 2. 1768. *Aerva tomentosa* Forssk., Fl. Aegypt.-Arab. 122.: 170. 1775; Gamble 2: 1178. 1925.

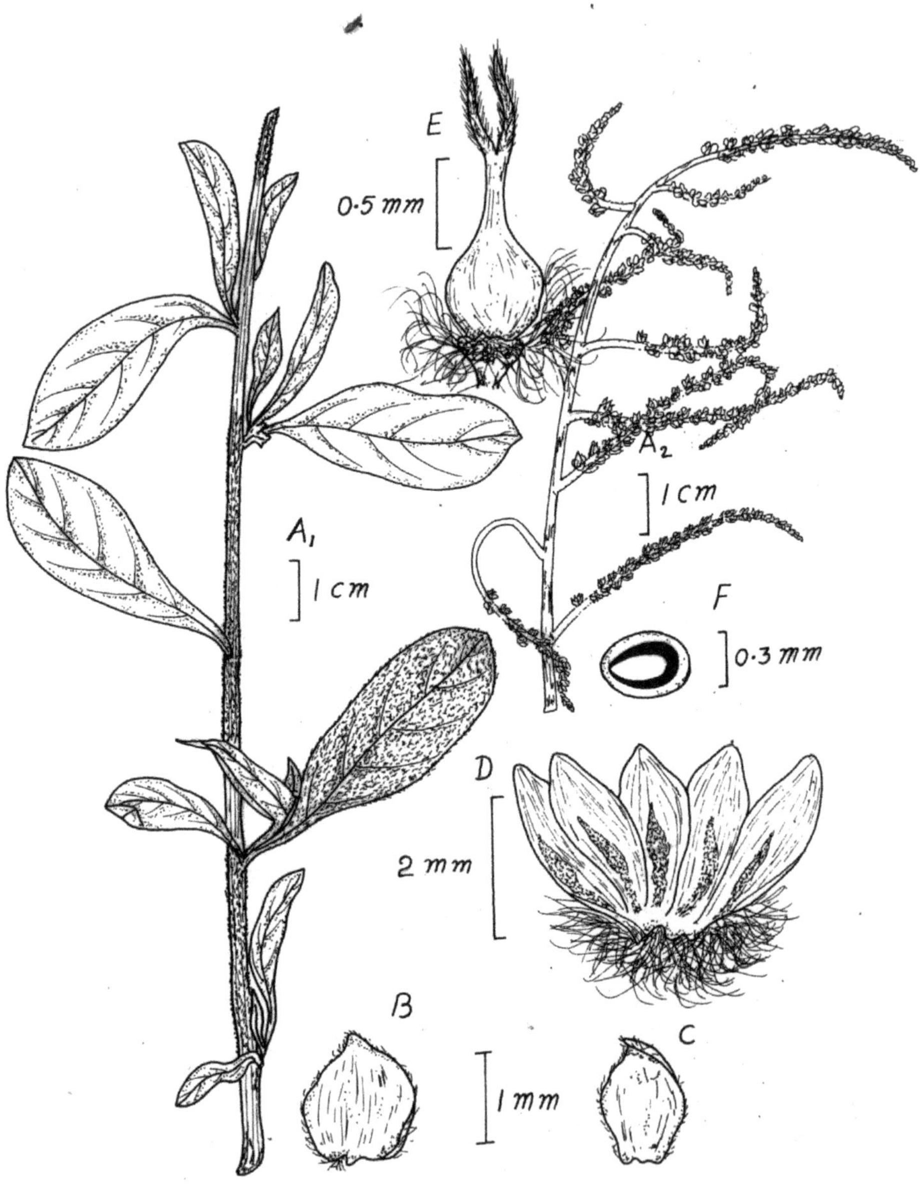

Figure 3. **Aerva javanica** (Burm.f.) Juss. ex Schult.
A. Twig, B. Bract, C. Bracteole, D. Perianth open, E. Pistil, F. Inflorescence, G. Ovary T.s.

Annual or biannual, erect herb, to 1.5 m tall; frequently woody; branches terete, striate, densely whitish tomentose. Leaves opposite at base, alternate towards apex, oblanceolate - spathulate, 5–9 × 1–2.5 cm, densely whitish tomentose, base cuneate to attenuate, margin entire, apex obtusely apiculate; lateral veins 5–6 pairs; petiole 2–3 mm long, tomentose. Flowers unisexual, in terminal, paniculate, 18–25 cm long spikes. Male flower: bracts and bracteoles deltoid, ovate, 0.75–2.25 × 0.5 mm, hyaline, acute or obtuse, midrib obscure, 3 mm long; stamens 5, 1.5 mm long, pseudostaminodes alternating the fertile stamens, gynoecium rudimentary. Female flower: bracts and bracteoles lanceolate, 1–2.5 × 0.5 mm; tepals 5, obovate-oblanceolate to spathulate, 2.5 mm long, tomentose without, obtuse; outer 2 tepals with faint midrib, 2–3 × 0.7–1 mm; inner 3 tepals with thick green midrib, 1–1.5 mm; ovary 0.5 mm across; style 1.8 mm long; stigma deeply forked. Seed round, slightly compressed, black, shining, very faintly reticulate; testa coriaceous, 0.5–1 mm across.

Common in waste lands, open forests and fields in southern districts. Fl. & Fr.: July – December. Vern.: Tel.: *Magabeera, Pindikonda, Pedda pindikonda;* Tam.: *Perumpeelai.*

Garladinne (ATP), *NY* 322; Yadiki RF (ATP), *TP* 468; Gulemallebad RF (KNL), *RVR* 1401; Panyam RF (KNL), *PVP* 1586; Pacherla (KNL), *BR* & *BSS* 34244; Velpula (CDP), *KRKS* & *KRS* 36996; Tirupati (CTR), *BR* & *AL* 28661; Marripadu (NLR), *PMR* 16508; Satrasala (GNT), *VRK* 6780; Cuddapah forests (CDP), *CBDM* s.n (MH); Woddapatti RF (SM), *EV* 24132 (MH). Hogainakkal (SM), *EV* 20639 (MH).

INDIA: Throughout India.

WORLD: Widespread in the drier parts of the tropics and subtropics of the Old World, introduced in Australia and elsewhere.

Aerva lanata (L.) Juss., Ann. Mus. Nat. Hist. Paris 2: 131. 1808; FBI 4: 728. 1885; Gamble 2: 1178. 1925. *Achyranthes lanata* L., Sp. Pl. 204. 1753.

Erect or prostrate, sometimes straggling herb, to 2 m tall; branches slender, 1–3 mm across, weak, white lanate or villous. Leaves alternate, variable, reniform to elliptic-ovate, basally large, to 2.5 × 2 cm and small towards apex, to 0.4 × 0.3 cm, pubescent above, tomentose below, base acute-attenuate, margin entire, apex orbicular or acute-apiculate; lateral veins 5–7 pairs; petiole 1–2 cm long, sessile in apical leaves. Flowers in axillary spikes, bisexual; bracts deltoid-ovate to oblong-ovate, 0.7–1 × 0.5 mm; membranous with a short arista formed by the excurrent midrib, pilose, persistent; bracteoles similar to bracts; perianth lobes 5; outer 2 tepals ovate-oblong, 1.5–2 × 0.5–0.7 mm, hyaline, abruptly connate at apex to a distinct mucro formed by the excurrent mid vein; inner 3 tepals slightly shorter and narrow with a broad central green vitta along the midrib; stamens 5, 1 mm long, pseudostaminodes present at anthesis, attaining about half the length of the style; anthers 0.2 mm across; ovary 0.4 mm long; style 0.2 mm long; stigma 2-fid. Utricle rotund, 1 mm across, compressed. Seed reniform, 0.6–0.8 mm across, black, shining, testa smooth in the centre, faintly reticulate around the margin, coriaceous.

Common weed in all districts. Fl. & Fr.: Throughout the year. Vern.: Tel.: *Telaga pindi konda;* Ori.: *Lopong arak, Paunsia;* Tam.: *Sirupeelai, Koolapoo, Kannupeelai.*

Pennahobilam (ATP), *NY* & *TP* 348 (MH); Penukonda RF (ATP), *NY* 1103; Madhavaram RF (KNL), *RVR* 1642; Rachakuntapalle (CDP), *TP* & *SSR* 33612; Papinenipalli (PKM), *BSS* & *SKB* 31021; Macherla RF (GNT), *VRK* 3259; Chelama (KNL), *JLE* 22045 (MH); way to Amudala kona (CTR), *GVS* 46757 (MH); Tirumala hills (CTR), *DRC* 222 (MH); Somasila project area (NLR), *PV* 109867 (BSID); Kondapalli fort (GNT), *CAB* 4658 (MH); Kondapalli RF (KSN), *P.Venkanna* 5305 (MH); Maredumilli (EG), *MM* 105057 (MH, BSID); Dolphin's nose hill (VSKP), *TAR* 7147 (BSID); Marandahalli RF (DMP), *EV* 57854 (CAL); Kurnubarum RF (SA), *KRM* 90446 (CAL); Gingee RF (SA), *KMS* 12193 (MH); Woddaparthi RF (SM), *EV* 20686 (MH); Tippukadu RF (NA), *KRM* 16586 (MH).

INDIA: Throughout India.

WORLD: Widespread in the drier parts of the Tropics and subtropics of the Old World.

Aerva sanguinolenta (L.) Blume, Bijdr. 547. 18265. *Achyranthes sanguinolenta* L., Sp. Pl. ed. 2. 294. 1762. *Aerva scandens* Wall. ex Moq. in DC., Prodr. 13: 302. 1849; FBI 4: 727. 1885; Gamble 2: 1178. 1925.

Erect or scandent herb, to 1 m high, stems terete, hairy, becoming glabrescent. Leaves opposite-decussate, lanceolate, 3-6 × 2-3 cm, base attenuate, margin entire, apex acute, glabrous above, pubescent beneath; petioles up to 1 cm long. Flowers in axillay, cylindric, white-woolly, to 15 cm long spikes; flowers polygamous; bracts and bracteoles up to 0.2 cm long, concave, white-hairy without; tepals 5, concave, elliptic, hairy without, outer 2 tepals mucronate, inner 3 tepals subobtuse; stamens 5, alternating with subulate pseudostaminodes, filaments dilating and connate at base to form a cup, filaments above the cup slender; ovary glabrous, compressed, style slender, stigma 2-lobed. Utricle circumscissile, glabrous; seed reniform, testa shiny.

Common along forest paths. Fl. & Fr.: July – April. Vern.: Ori.: *Chauladhua.*

Gundlabrahmeshwaram RF (KNL), *RVR* 2394, *BR* & *TSS* 29059; Upper Ahobilam (KNL), *BR* & *BSS* 29203, *JLE* 32582 (MH); Mandapadu (KSN), *PV* 116544 (BSID); Papikonda (WG), *SKW* (1878) (BLAT); Marripakala (EG), *SKW* 1314 (BLAT); Maredumilli (EG), *MM* 100879 (MH); Kuntada RF, Maredumilli (EG), *MM* 101240 (MH, BSID); Sirivaka (EG), *CAB s.n.* (MH); Simhachalam hills (VSKP), *GVS* 19337 (MH); Sapparla (VSKP), *GVS* 29716 (MH); Adativalasa (VSKP), *GVS* 21621 (MH); Shevaroys (SM), *M. A. Lawson s.n.* (MH).

INDIA: Throughout India.

WORLD: E and SE Asia, Malesia.

ALLMANIA R. Brown ex Wight

Allmania nodiflora (L.) R. Br. ex Wight, J. Bot. (Hooker) 1: 226. 1834. *Celosia nodiflora* L., Sp. Pl. 205. 1753. *A. nodiflora* (L.) R. Br. ex. Wight var. *aspera* (Roth) Hook. f., Fl. Brit. India 4: 717. 1885; Gamble 2: 1167. 1925. *Celosia aspera* Roth, Nov. Pl. Sp. 173. 1821. *Allmania nodiflora* (L.) R. Br. ex Wight var. *dichotoma* (Heyne ex Roth) Hook. f., Fl. Brit. India 4: 717. 1885; Gamble 2: 1167. 1925. *Celosia dichotoma* Heyne ex Roth, Nov. Pl. Sp. 172. 1821. *Allmania nodiflora* (L.)

R. Br. ex Wight var. *longipedunculata* Trimen, Fl. Ceylon 3: 394. 1895. *Allmania longipedunculata* (Trimen) Gamble, Fl. Madras 2: 1168. 1925. *A. nodiflora* (L.) R. Br. ex Wight var. *nodiflora;* Hook.f., Fl. Brit. India 4: 716. 1885; Gamble 2: 1167. 1925. *A. albida* R. Br. ex Hook. f., Fl. Brit. India 4: 717. 1885. *A. nodiflora* (L.) R. Br. ex Wight var. *angustifolia* Hook. f., Fl. Brit. India 4: 717. 1885; Gamble 2: 1167. 1925. *A. nodiflora* (L.) R. Br. ex Wight var. *procumbens* Hook. f., Fl. Brit. India 4: 717. 1885; Gamble 2: 1167. 1925. *A. nodiflora* (L.) R. Br. ex Wight var. *roxburghii* Hook. f., Fl. Brit. India 4: 717. 1885.

Annual spreading, prostrate or ascending herb, up to 30 cm long; branches reddish, flexuose below, angular, sulcate, moderately pilose. Leaves alternate, variable in shape and size, linear to elliptic, oblong, obovate, 1.2–5 × 0.5–2 cm, glabrous above, pilose below, base alternate, margin entire, apex rounded or widely retuse to apiculate or shortly acuminate; lateral veins 4–5 pairs; petiole 1–2 mm long, pilose. Flowers pale reddish to stramineous, in terminal or axillary globose spike; bracts deltoid-ovate, 2–3 × 1–1.5 mm, membranous with excurrent dark midrib; bracteoles linear, 3–4 × 1–1.5 mm, pubescent along the keel; perianth lobes 5, elliptic-lanceolate, 4–5 × 1.5–2 mm, silvery membranous with reddish or greenish midrib, excurrent in very short mucro, glabrous or less pilose without; stamens 5, 3.5–4 mm long, subulate; anther 0.5 mm across; ovary depressed-globose, 1 mm across; style slender, 2.5–3 mm long; stigma capitate. Urticle ovoid, 3–4 mm. Seed lenticular, 0.5–1 mm, black, reticulate, shining dull and areolate, testa crustaceous.

Common in waste lands, along road sides, and fields. Fl. & Fr.: July – December. Vern.: Tel.: *Pendibadde aku, Erra maddi aku;* Tam.: *Magulikeerai,Kumuttikeerai.*

Chintakunta (ATP), *TP* 460; Gooty (ATP), *JSG* 16398 (DD); Gulemallebad RF (KNL), *RVR* 1692; Simhachalam hill (VSKP), *NVP* 6616; Chelama (KNL), *JLE* 16714 (MH); Gundlabrahmeswaram (KNL), *JLE* 22226 (MH); way to Humlipass (CDP), *JSG* 15224 (MH); Balapalli (CDP), *JLE* 14264 (MH); Guvvalacheruvu RF (CDP), *RVR* 7809; Way to Duggammakonda, Lankamalais (CDP), *KS* 6334 (MH); Road to Thirumalai (CTR), *KS* 6879 (MH); Satyavedu (CTR), *MC* 45256 (CAL); towards Kailasakona (CTR), *GVS* 46042 (MH); Papavinasanam (CTR), *DRC* 2167 (MH); Nagapatla RF (CTR), *KS* 6858 (MH); Veligonda (NLR), *CECF* 4110 (DD); Nagarjunakonda (GNT), *KT* 9677 (DD); Rampachodavaram (EG), *MSR* 1462 (CAL); Ankanna palem (EG), *DCSR* 244 (DD); Annavaram (EG), *GVS* 68683 (MH); Rampachodavaram (EG), *MM* 105070 (MH); Chintapalli (VSKP), *GVS* 28229 (MH); Galikonda (VSKP), *GVS* 44379 (MH); Anantagiri (VSKP), *NPBK* 10974 (MH); Ganti Konda (VSKP), *GVS* 44244 (MH); Lambasingi hill (VSKP), *HS* 20843 (BLAT); Lakshmipuram (VSKP), *CAB* 1721 (MH); Salur (VZN), *NPBK* 906 (DD); Kilasakona hill (SKLM), *GVS* 46988 (MH); Gingee RF (SA), *KRM* 13061 (CAL); Tippukadu RF (NA), *KRM* 17626 (MH); Alagar hills (Madurai), *KS* 3453 (MH); Pennagaram RF (SM), *EV* 20750 (MH); Rukeekonda (GJM), *VNS s.n.* (MH); Sidevi RF (SA), *KRM* 51160 (MH); Swamimalai RF (SA), *SR Raju & Naganathan* 17913 (MH); Sirumalai hills (DGL), *SKS* 618 (SGH); Kolli hills (NMK), *SKS* 926 (SGH); Alagar hills (MDU), *SKS* 86 (SGH).

INDIA: Throughout India.

WORLD: Tropical Asia from India and Sri Lanka east to Thailand, Malaysia, S.China, Indonesia and Philippines.

Note: *Allmania nodiflora* and *A. longepedunculata* have often been treated as two species, on the basis of sessile heads and pedunculate heads respectively, but this is not a constant character. Pedunculate and sessile heads are sometimes found on the same specimen. The varieties run into one another. So it is preferred to treat them all as *A. nodiflora* (L.) R.Br. ex Wight (see Subba Rao and Kumari, 2008).

ALTERNANTHERA Forsskal

1. Bracts and bracteoles glabrous:
   2. Tepals dissimilar, with a tuft of glochidiate bristles on the keel; fertile anthers 5:
      3. Perianth tip spinous; ovary sessile .... **A. pungens**
      3. Perianth tip mere mucronate, not spinous; ovary stalked .... **A. philoxeroides**
   2. Tepals similar, without glochidiate bristles; fertile anthers 2-3 .... **A. sessilis**
1. Bracts and bracteoles hairy:
   4. Flowers in numerous heads; utricle orbicular .... **A. paronychioides**
   4. Flowers in spikes; utricle oblong .... **A. ficoidea**

Alternanthera ficoidea (L.) Sm. Cycl. 39 (2. Suppl.): n. 8. 1819. *A. tenella* Colla, Mem. Reale Accad. Sci. Torino 33: 131. t.9. 1829. *A. tenella* Colla var. *versicolor* (Lem.) Veldkamp, Taxon 30: 208. 1981. *Telanthera ficoidea* var. *versicolor* Lem., Ill. Hortic. 12, t. 440. 1865.

Annual, decumbent or erect herb, to 1.5 m tall; stem slightly angular, with appressed acroscopic hairs. Leaves opposite, elliptic-oblong or obovate-oblong 1.5–6.5 × 0.3–3.25 cm densely clothed with appressed long hairs below, less above, base attenuate, margin entire, apex acute-apiculate, dotted on both sides; lateral veins 5–7 pairs; petiole to 1 cm long. Flowers silvery or light yellowish in terminal spikes and near the top of the branches, 2–3 spikes together on common peduncle; bracteole lanceolate, 2–2.5 × 0.5–0.7 mm, very acute, densely hairy on dorsal side; perianth 5-lobed, oblong or ovate-oblong, 3–3.5 × 0.8–1 mm, 1-nerved, in lower half rather densely hairy on dorsal side; stamens 5, 1.5–2 mm long, alternating with pseudo-staminodes, strap-shaped, about as long as stamens, at top divided into a few narrow strips; ovary oblong, 2 mm across; style broader than long, 0.25–0.5 mm long; stigma capitellate. Utricle oblong, 1–1.5 mm, compressed, lenticular, testa crustaceous.

Rare in moist waste lands and roadsides. Fl. & Fr.: Throughout the year. Vern.: Tam.: *Ponnankanni*.

Thimmammamarimanu (ATP), *KRKS* and *BSS* 41939; Amarapuram (ATP), *BR* & *ANS* 35596; Balapalli (CDP), *MHR* 6679; Buchireddypalem (NLR), *PMR* & *BR* 19572; Chitrakonda, *SX* & *MB* 6840 (RRL-B).

INDIA: Introduced in several parts of India.

WORLD: Indigenous to Tropical America, Philippines.

Alternanthera paronychioides A. St. Hil., Voy. Distr. Diam. 2: 43. 1833; Veldk., Blumea 19: 167. 1971.

Prostrate creeping herb, stem with long crisped hairs on the younger parts, rooting at nodes. Leaves oblanceolate-elliptic, 1-1.5 × 0.5-0.7 cm, sessile, hairy, apex acute. Flowers in numerous heads; bracts and bracteoles similar, 1.5 mm long, hyaline, spiculate; tepals 5, outer 3 larger, 2.5 × 1 mm, ovate, acute, inner smaller, pilose on back, 3-nerved; stamens 3-5, alternating with sublacerate pseudostaminodes; ovary obovoid, compressed, stigma bilobed. Utricle orbicular; seed discoid, 1 mm across, faintly reticulate.

Rare on bunds of paddy fields and along road sides. Native of Tropical America, naturalized in India. Fl. & Fr.: February – July.

Velgode to GBM (KNL), *BR* & *BSS* 34112; Hatyaraly (CDP), *BR* & *DV* 43464; Papavinasanam (CTR), *AMR* 26466; Pudivalasa (SKLM), *MCK* 18885; Kondapalli village (KSN), *P.Venkanna* 5311 (MH); way to Kolleru lake (WG), *KS* 5103 (MH); Hogainakkal (SM), *EV* 23514 (MH); Sirumalai hills, Kadamankulam (DGL), *SKS* 412 (SGH).

INDIA: Introduced and more or less naturalized in several parts of India.

WORLD: Native of Tropical America, naturalized in the USA, also in Java and other parts of the Old World.

Alternanthera philoxeroides (Mart.) Griseb. Abh. Königl. Ges. Wiss. Göttingen 24: 36. 1879. *Bucholzia philoxeroides* Mart. Beitr. Amarantac. 107. 1825. *Achyranthes paludosa* Bunbury, Proc. Linn. Soc. London 1: 109. 1841. *A. philoxeroides* (Mart.) Standl., J. Wash. Acad. Sci. 5: 74. 1915. *Alternanthera philoxerina* Suess. Repert. Spec. Nov. Regni Veg. 35: 303. 1934.

Perennial herbs with prostrate, fistular, striate, hollow stem, rooting at the nodes and eventually become erect and aerial, often much branched forming dense masses in marshy fields. Leaves opposite, 5-10 x 0.5-2.5 cm, elliptic to obovate-lanceolate with acute base, with midrib prominent on the lower surface; petiole 1-6 mm long. Inflorescence axillary, occasionally terminal, pedunculate, 1-5 mm long, white, globose heads, 10-18 mm across; bracts ovate-lanceolate, glabrous, acute, 1-veined, faintly keeled; perianth dorsally compressed, white glabrous; tepals 5, subequal oblong-lanceolate with mucronate tip 1-5 nerved; fertile stamens 5, pseudo-staminodes longer than the filaments and lacerate at tips; ovary shortly stalked, ovoid, style slender, capitate; stigma densely papillose.

It is an invasive weed, common clumps along the marshy fields in Kolli hills. Fl. & Fr.: August – January. English: Alligator weed.

Kolli hills, way to Seekuparai (NMK), *SKS* 967 (SGH).

INDIA: Assam, Manipur, Kerala, Karnataka, Tamil Nadu.

WORLD: Argentina, Tropical America, Brazil, Java and Australia.

Alternathera pungens Kunth, Nov. Gen. Sp. 2: 206. 1818. *Achyranthes repens* L., Sp. Pl. 205. 1753. *Alternanthera repens* (L.) Link, Enum. Pl. Hort. Berol. 1: 154. 1821; (non Gmelin 1791).

Prostrate-creeper, herb to 50 cm long; rooting at nodes; branches villous with white hairs. Leaves opposite, unequel, obliquely elliptic to orbicular, 1.5–3 × 1–2 cm, distinctly villous above, more below, base attenuate, margin entire, apex mucronate; lateral veins 6–7 pairs; petiole to 0.5 cm long. Flowers white, in axillary clustered heads; bracts oblanceolate, to 5 mm long, midnerve shortly awned; margin scarious; bracteoles 3 mm long, similar to bract; tepals 5; outer 2 lobes oblanceolate, 4–5 × 1.5–2 mm, very rigid, 5-nerved, mid nerve and outer pair stronger, jointed above into a pungent awn; inner one faintly visible, lobes oblong, 3–3.5 × 1–1.5 mm, 3-nerved, margin serrate, obtusely apiculate, with a tuft of glochidiate, barbellate bristles at base; inner 2 lobes linear, curved with a tuft of glochidiate, barbellate bristles at middle, apex acute; stamens 5; filaments 0.3 mm long; anthers oblong, 0.3 mm long; staminodes membranous, alternating the stamens, dentate; ovary depressed-globose, 2 mm long; style broader than long, 0.1 mm. Seed discoid, brown, shining, faintly reticulate; testa crustaceous.

Common weed in waste lands, along road sides and fields in most districts. Fl. & Fr.: Throughout the year. Vern.: Tel.: *Mullu ponnaganti*; Tam.: *Kanthimullu*.

S.K.University (ATP), *TP* & *D.Narahara* 1268; Batrepalli (ATP), *KRKS* & *BSS* 37974; Kalasamudram (ATP), *KRKS* 40038; Veldurthi East (KNL), *RVR* 1712; Palakonda hills (CDP), *CS* 7703; Gandipalem (NLR), *PMR* 16543; Bukkapatnam (ATP), *SKW* 6123 (BLAT); Bogada, Nallamalais (KNL), *JLE* 3253 (MH); Karchuli (GJM), *MB* 2815 (RRL-B); Mangalam RF, Yelagiri hills (NA), *MBV* 639 (MH); Melagiri hills (SM), *S.India Flora* 9771 (MH); Woddaparthi RF (SM), *EV* 20683 (MH); Yercaud (SM), *SKK* 26838 (MH); Sirumalai hills, Kadamankulam (DGL), *SKS* 716 (SGH).

INDIA: Introduced, in many parts of India.

WORLD: Native of the American Continent, now widespread in the tropics and subtropics of the Old World.

Alternanthera sessilis (L.) R. Br. ex DC., Cat. Horti. Monsp. 4: 77. 1813; FBI 4: 731. 1885. *Gomphrena sessilis* L., Sp. Pl. 225 1853. *Alternanthera triandra* Lam., Encycl. 1: 95. 1783; Gamble 2: 1179. 1925. *A. nodiflora* R.Br., Prodr. 417. 1810; FBI 4: 732. 1885.

Annual, prostrate to erect herb, to 50 cm tall, sometimes rooting at basal nodes; branches 5–6-angular, glabrescent, villous at nodes or sparsely hairy. Leaves opposite, variable in shape and size, linear-lanceolate or ovate-oblong or oblanceolate to spathulate, 1–9 × 0.2–2 cm, glabrous or thinly pilose below, base cuneate to attenuate, margin distinctly serrate, apex obtuse or acute; lateral veins 5–7 pairs; petiole to 2 mm long or sessile. Flowers white, in axillary spike; bracts deltoid-ovate, 1 × 0.4 mm, scarious, white, mucronate with excurrent midrib, glabrous; bracteoles similar to bracts, 1–1.5 × 0.5–0.8 mm, persistent; perianth segments 5, ovate-elliptic, 2.5–3 × 0.5–1 mm, 1-nerved, acuminate to rather blunt, white, glabrous, apex shortly mucronate with midrib, margins obscurely denticulate; stamens 3; anthers fertile, staminodes 2, 1–1.5 mm long, alternating with pseudostaminodes resembling the filaments, shorter than fertile ones; ovary depressed-globose, 1 mm across; style 0.2 mm long; stigma simple, fimbriate. Utricle cordiform, strongly compressed, 1.2 × 2 mm; seed discoid-orbicular, 0.5–0.75 mm, brown shining, faintly reticulate, testa crustaceous

Common weed of damp places along ponds and streams. Fl. & Fr.: Throughout the year. Vern.: Tel.: *Ponnagantikura;* Ori.: *Kolamsago;* Tam.: *Ponnankanni.*

Garugudu kona (ATP), *TP* 945; Kalasamudram (ATP), *KRKS* 40435; Gulemallebad RF (KNL), *RVR* 1454; Gani RF (KNL), TP & RVR 1791; Velpula (CDP), *KRKS* & *KRS* 37678; Penchelakona (NLR), *NVP* 6641; Diguvametta (PKM), *AJR* & *PANR* 20408; Ethipothala (GNT), *VRK* 5802; L. Kota reservoir (VZN), *MCK* 18820; Balapalli (CDP), *JLE* 15787 (MH); Mamandur valley (CTR), *KS* 6916 (MH); Nerabylu (CTR), *GVS* 46983 (MH), 46025 (MH); Kondaveedu fort (GNT), *sine coll.* 4659 (MH); Maredumilli (EG), *GVS* 24231 (MH); Addatigala (EG), *GVS* 68576 (MH); Galikonda (VSKP), *GVS* 19594 (MH); Araku valley (VSKP), *NPBK* 10789 (MH); Simhachalam (VSKP), *GVS* 19374 (MH); Singapuram (SKLM), *GVS* 62345 (MH); Javadi hills (NA), *EV* 51967 (MH), *MBV* 848 (MH); Tippukadu RF (NA), *KRM* 17665 (MH); Yercaud (SM), *D.B.Deb* 31399 (MH); Daringabadi (GJM), *CAB* 1382 (MH); Aska (GJM), *s. coll. s.n.* (MH).

INDIA: Throughout hotter parts of India.

WORLD: Widespread in tropics and subtropics of the World.

AMARANTHUS Linnaeus

1. Stamens and tepals 5 each:
    2. Leaf axils with paired spines; styles 2 .... **A. spinosus**
    2. Leaf axils without spines; styles 3:
        3. Flowers in axillary and terminal paniculate spikes:
            4. Fully developed inflorescence large and robust, usually brightly coloured:
                5. Plants glabrous or nearly so:
                    6. Complex thyrsoid structures pendulous; bracts and bracteoles inconspicuously pointed at apex .... **A. caudatus**
                    6. Complex thyrsoid structures erect, bracts and bracteoles conspicuously pointed at apex:

7. Tepals as long as or longer than fruit; bracts twice as long as perianth............. **A. hypochondriacus**

7. Tepals distinctly shorter than fruit; bracts one and half times as long as perianth......................... **A. cruentus**

5. Plants hairy:

8. Fruit longer than perianth .... **A. hybridus**

8. Fruit enveloped in persistent perianth .... **A. retroflexus**

4. Inflorescence moderately large, green; bracts shorter than tepals .... **A. dubius**

3. Flowers in axillary clusters .... **A. polygonoides**

1. Stamens and tepals 3 each:

9. Flowers in axillary clusters:

10. Prostrate herb; utricle 2 mm across .... **A. polygamus**

10. Erect herb; utricle 1 mm across, subglobose.................. **A. roxburghianus**

9. Flowers in axillary and terminal paniculate spikes:

11. Fruit dehiscent with a circumscissile lid:

12. Tepals conspicuously longer than fruit:

13. Bracteoles exceeding or equaling the tepals, awned....... **A. tricolor**

13. Bracteoles shorter than the tepals, acute..................... **A. graecizans**

12. Tepals as long as or shorter than fruit....................... **A. roxburghianus**

11. Fruit indehiscent:

14. Stem erect, somewhat branched; utricle very rugose............ **A. viridis**

14. Stem ascending, much branched; utricles smooth or slightly rugose... **A. blitum**

Amaranthus blitum L., Sp. Pl. 990. 1753. var. **blitum.** *A. lividus* L., Sp. Pl. 990. 1753. subsp. *lividus* var. *lividus*

Annual herb, erect, ascending or prostrate, 6-60 cm long. Leaves ovate-rhomboid ovate, 1-8 × 0.6-6 cm, base cuneate, margin entire or slightly undulate, apex emarginate; petiole 1-3.5 cm long. Flowers green, in slender to stout terminal spikes and axillary cymose clusters, male and female flowers intermixed; bracts and bracteoles deltoid-ovate to lanceolate; perianth segments 3, membranous-margined; stamens shorter than perianth; stigmas 2-3. Capsule subrotund to shortly pyriform, compressed, exceeding perianth; seeds round, compressed, dark brown to black.

Cultivated for leafy vegetable. Fl. & Fr.: August – September.

INDIA: E. Himalyas.

WORLD: Sri Lanka.

Amaranthus caudatus L., Sp. Pl. 990. 1753; FBI 4: 719. 1885; Gamble 2: 1170. 1925.

Erect herb, to 1.5 m high, stem hairy when young, glabrous when mature. Leaves alternate, green or red, rhomboid-ovate or ovate-elliptic, 4-10 × 1-5 cm, base cuneate, margin entire, apex acute, glabrous; petiole 1-15 cm long. Flowers in axillary and terminal, thyrsoid spikes, formed of increasingly approximated cymose clusters, apex pendulous; flowers unisexual, pink; bracts and bracteoles spathulate, elongate, awned; tepals red, shorter than fruit, 2-2.5 mm, transparent; stamens 5; stigmas 3. Utricle circumscissile; seed annular, 0.5 mm across.

Cultivated as leafy vegetable. Fl. & Fr. September – February. Vern.: Tel.: *Netakura*; Tam.: *Thandukeerai*.

Sunkarametta (VSKP), *NVP* 6627 (SKU).

INDIA: Cultivated in various parts of India.

WORLD: Native to Neotropic.

Amaranthus cruentus L., Syst. Nat. ed. 10. 2: 1269. 1759. *A. paniculatus* L., Sp. Pl. ed. 2: 1406. 1763; FBI 4: 718. 1885; Gamble 2: 1170. 1925.

Erect herb, to 80 cm high, glabrous or slightly pubescent distally, especially when young; stem (sub) succulent, green or reddish purple. Leaves alternate, lanceolate to rhomboid or ovate, 3-6 × 0.5-4 cm, petiole length is gradually decreasing from bottom to top, green, base cuneate, margin entire, apex acute to obtuse, glabrous. Flowers in axillary clusters, yellowish green, reddish or purple, unisexual, intermixed. Male flowers, tepals 5, shorter than fruit, stamens 5. Female flowers: Tepals 5, oblong to lanceolate, 1.5-3 mm. Utricle 3-4 mm diam., circumscissile; seed annular, 1 mm across, testa shiny, black.

Commonly cultivated in gardens. Fl. & Fr.: Throughout the year. Vern.: Tel.: *Dantu kura*.

INDIA: Cultivated throughout India.

WORLD: Sri Lanka, Malaysia.

Amaranthus dubius Mart. ex Thell., Fl. Adv. Montpellier 203. 1912.

Annual erect herb, up to 1 m high, glabrous or sparsely pubescent in distal parts, stem green. Leaves alternate, glabrous, ovate-rhomboid, 2-8 × 0.8-4.5 cm, base cuneate, margin entire, apex retuse; petiole of proximal leaves equaling or longer than blade, becoming shorter distally. Flowers dioecious in axillary clusters in terminal panicles and axillary spikes; bracts lanceolate, apex spinescent. Male flowers: clustered at tips of inflorescence branches, tepals 5; stamens 5. Female flowers: tepals 5, oblong-spathulate to oblong, 1.5-2 mm, style branches spreading, stigmas 3. Utricle ovoid – subglobose, circumscissile; seed ovoid, black, compressed.

Introduced and naturalized. Occasional weed in waste places. Fl. & Fr.: September – March.

Banaganipalli (KNL), *MHR* 9706 b.

INDIA: Gujarat, Maharashtra, Andhra Pradesh.

WORLD: Native to South America, Central America. Introduced in the Tropics and Subtropics.

Aamaranthus graecizans L., Sp. Pl. 990. 1753 subsp. **cruentus** C. Towns. in Dassan. & Fosb. Rev. Handb. Fl. Ceylon 1: 14. 1982.

Annual erect herbs, up to 50 cm high. Leaves opposite, glabrous, ovate-rhomboid or linear-lanceolate, 4.5 × 1-3 cm, base cuneate, margin entire, apex acute. Flowers dioecious in axillary cymose clusters; male and female intermixed, males commonest in upper whorls, bracts and bracteoles narrowly lanceolate-oblong. Male flowers: petals 3, 1.5-2 mm, lanceolate-oblong, stamens 3. Female flowers: Petals 3, stigmas 3, slender, flexuose. Utricles sub-globose to ovoid; seeds black, shining compressed.

Common weed in crop fields. Fl. & Fr.: July – December.

Madhavaram RF (KNL), *RVR* 1663; Kambagirikonda (KNL), *SS* & *AMR* 25837; Venkatagiri (NLR), *PMR* 16549.

INDIA: Widely distributed in India.

WORLD: Sri Lanka.

Amaranthus graecizans L., Sp. Pl. 990. 1753. subsp. **thellungianus** (Nevski ex Vassilcz.) Guseve, Bot. Zhurn 57: 462. 1972. *A. thellungianus* Nevski ex Vassilcz in Komarov Fl. U.R.S.S. 6: 365. 1936. *A. polygonoides auct. non* L.; Roxb. Fl. Ind. 3: 602. 1832; Nair, Bull. Bot. Surv. India 8: 88. 1966. *A. polygamus* sensu: Hook. f. in Fl. Brit. India 4: 721. 1885. *p.p.;* Gamble 2: 1171. 1925.

Diffuse, prostrate herb, stem red. Leaves alternate, elliptic-obovate, 1-2 x 0.5-1.5 cm, base attenuate, entire, mucronate, glabrous. Inflorescence axillary fascicles; flowers unisexual. Tepals 3, equal, spreading, oblong-lanceolate. Stamens 3, filaments 1 mm. Utricle ovoid, rugose; seed orbicular, 1 mm across.

Common weed of cultivated fields and waste places. Fl. & Fr.: Jul. – Dec. Vern.: Vern.: Tel.: *Chirikura;* Tam.: *Sirukeerai.*

Tippukadu RF (NA), *KRM* 17640 (MH); Pennagaram (SM), *EV* 22415 (MH); Pennagaram RF (SM), *EV* 20724 (MH).

INDIA: Widely distributed.

WORLD: Sri Lanka.

Amaranthus hybridus L., Sp. Pl. 990. 1753. subsp. **hybridus;** Towns. in Dassan. & Fosb., Rev. Handb. Fl. Ceylon 1: 13. 1980.

Annual, erect herb, up to 2 m tall; branches stout, angular, glabrous or thinly to moderately furnished with short or long muticellular hairs. Leaves alternate, broadly lanceolate to ovate, 3–20 × 1.5–10 cm, gradually narrowed, thinly pilose below on veins, base cordate, margin entire, apex blunt to mucronulate; lateral veins 5–15 pairs; petiole up to 15 cm long. Flowers red or purple, in axillary and terminal spike, varying from a single spike to a broad, much-branched panicle, 10–30 × 3–5 cm; bracts and bracteoles unequal in size, deltoid-ovate 2.5–3 × 0.5–1 mm, pale membranous, scarious, acuminate with a long tip, arista erect and formed by the stout excurrent midrib, exceeding the perianth; flowers androgynous formed of cymose clusters, increasing upwards; perianth segments 5, lanceolate or oblong, 3–3.5 × 1–1.5 mm, acute-aristate or the inner sometimes blunt in the female flower, 3-nerved; stamens 5, 1.5 mm long, dorsifixed, X-shaped; ovary conic or pyriform; styles 2, fimbriate, 1 mm long, curved. Utricle subglobose to ovoid, 2–3 mm, capsule lid swollen with a distinct neck; seed ovoid, 0.75–1.25 mm, compressed, almost smooth, faintly reticulate around the margin, testa crustaceous.

Commonly cultivated in gardens in most districts. Fl. & Fr.: Throughout the year. Vern.: Tel.: *Dantu kura;* Tam.: *Thandukeerai.*

Batrepalli (ATP), *KRKS* 42051; Galikonda RF (VSKP), *NVP* 6625; Dolphin's nose (VSKP), *SKW* 4589 (BLAT); Sunkarimetta (VSKP), *NPBK* 768 (MH).

INDIA: Cultivated throughout India.

WORLD: Sri Lanka, Malaysia.

Amaranthus hypochondriacus L., Sp. Pl. 991. 1753. *A. hybridus* L. var. *hypochondriacus* (L.) H. Rob., Rhodora 10: 32. 1908. *A. hybridus* ssp. *hypochondriacus* (L.) Thell., Adv. Montpellier 204. 1912.

Annual erect herb, up to 70 cm high, glabrous or slightly pilose; stems green or reddish-purple. Leaves rhombic-ovate or oblong-lanceolate, 3-8 x 1.5-3.5 cm, glabrous, base cuneate, margin entire or undulate, apex acute or acuminate; petiole 1- 6 cm long, petiole of distal leaves equaling or slightly shorter than blade, becoming longer proximally. Flowers green in terminal, thyrsoid spikes; bracts green or purple, ovate subulate; Male flowers clustered at tips of inflorescence, tepals 3-5, green or purple, stamens 3-5. Female flowers: tepals 5, style branches spreading, stigmas 2 or 3. Utricle exceeding perianth; seeds white, subglobose.

Cultivated as leafy vegetable. Fl. & Fr.: July – November.

INDIA: Cultivated.

WORLD: Native of North America.

Amaranthus polygamus L., Amoen. Acad. Schreb. ed. 4. 294. 1788; Suryanarayana & Rao, Fl. Nellore Distr. 452. 2002. *A. polygonoides* Roxb., Fl. Ind. ed. 1832. 3: 721. 1832 *illegitimate.*

Annual, prostrate herb. Leaves obovate, 1.5-3 × 0.6-1.8 cm, notched at the tip, base tapering. Flowers 4-5 mm long, pale green, 5-8 clustered in leaf axils; tepals 3, stamens 3. Utricles 1.5-2 mm across; seeds minute, lenticular.

Common weed in moist fields.

Dolphin's nose (VSKP), *TAR* 7143 (BSID).

INDIA: Maharashtra, Karnataka, Andhra Pradesh, Tamil Nadu.

WORLD: East Asia.

Amaranthus polygonoides L., Pl. Jamaica Pug. 27. 1759; Suryanarayana & Rao, Fl. Nellore Distr. 452. 2002.

Annual herb, 15-40 cm tall. Leaves obovate-rhomboid, 1-3 × 0.5-1.5 cm, oblong -rhomboid, apex obtuse with conspicuous mucro, base cuneate and decurrent. Flowers 5-6 mm long, pale green, 6-13 clustered in leaf axils. Male flowers: tepals 5; stamens 5. Female flower: Tepals 5, connate proximally, style branches spreading, stgmas 3. Utricle 2-2.5 mm across, subglobose, enclosed by hyaline, oblong tepals; seeds minute, orbicular.

Occasional on shady hill slopes in Udayagiri hills. Fl. & Fr.: December. Vern.: Tam.: *Pulisirukeeerai*.

Udayagiri (NLR), *BSN* 4846.

INDIA: Kerala, Karnataka, Maharashtra, Andhra Pradesh, Assam.

WORLD: South America, introduced in Europe.

Amaranthus retroflexus L., Sp. Pl. 2: 991. 1753.

Annual, erect herb, up to 70 cm high, densely pubescent. Leaves ovate-rhombic or elliptic, 5-10 × 2-5 cm, shortly hairy on both surfaces, base cuneate, margin entire and undulate, apex acute or notched, with a mucro. Flowers in axillary and terminal thyrsoid spikes; bracts and bracteoles white, subulate; tepals white, oblong or oblong-obovate, membranous, with a green midvein; stamens slightly longer than perianth; stigmas 3. Utricle light green, ovoid, compressed, shorter than perianth, circumscissile; seeds brown or black.

Occasional in waste lands, road sides and hedges of fields. Fl. & Fr.: August – November.

INDIA: Maharashtra, Andhra Pradesh.

WORLD: Native of North America, introduced in the Old World.

Amaranthus roxburghianus H.W.Kung, Fl. Ill. Nord. Chine 4: 19. 1935.

Eerect herb, 30-65 cm high, stem light green, much branched, glabrous. Leaves ovate-rhombic, obovate or oblong, 2-5 × 1-2.5 cm, base cuneate, margin undulate, apex notched, with a mucro; petiole 1-2.5 cm, slender;. Flowers few, sparsely clustered at axils; bracts and bracteoles subulate, ca. 2 mm, abaxially with a distinct midvein, apex long pointed; tepals lanceolate, ca. 2.5 mm, apex acuminate, long pointed; stamens shorter than perianth; stigmas 3. Utricles ovoid, subequal to perianth, ca. 3 mm, circumscissile; seeds brownish black, subglobose, ca. 1 mm in diam.

Occasional in waste lands. Fl. & Fr.: July – October.

Madhavaram RF (KNL), *RVR* 1663; Palakonda hills (CDP), *CS* 7758.

INDIA: Telangana, Maharashtra, Andhra Pradesh.

WORLD: Sri Lanka, China.

Amaranthus spinosus L., Sp. Pl. 991. 1753; FBI 4: 718. 1885; Gamble 2: 1170. 1925.

Armed, annual, erect herb, up to 1 m tall; stem stout, angular, reddish, glabrous basally, multicellular long hairy apically. Spines axillary, 1.5 cm long, paired or clustered, subtending inflorescence. Leaves alternate, ovate to rhomboid-ovate, rarely oblong-lanceolate, 1.5–7 × 0.8–4 cm, glabrous or thinly pilose on the veins on lower side, base cuneate or attenuate, margin entire, apex obtuse-apiculate with fine mucro; lateral veins 5–7 pairs; petiole up to 9 cm long. Flowers pale greenish-white, in axillary or terminal spikes, spikes up to 7 cm long, 0.8 cm diameter; lower clusters entirely female; upper flowers male and bisexual; bracts deltoid-ovate, 1–1.5 × 0.5–0.75 mm, pale scarious margin, pale or reddish awn formed by the excurrent green midrib, to 0.5 cm; bracteoles deltoid-ovate, 2–2.5 × 1–1.5 mm, sub-equaling or shorter than perianth lobes; perianth segments 5; those of female flowers narrowly oblong or spathulate-oblong, 1.5–2.5 × 0.5–1 mm, obtuse or acute, mucronulate, frequently with a greenish dorsal midvein; those of male flowers broader than the female, oblong-lanceolate, 1.5–3 × 0.75–1 mm, acute or acuminate, midrib excurrent; stamens 5; filaments 1–1.5 mm long; anthers oblong, 0.2 mm long, sagittate; ovary 0.5 mm across; style 0.5 mm long; stigmas 3, flexuose or reflexed, 1–1.5 mm. Utricle ovoid, urceolate, 1.5–2 mm, with a short inflated neck below the style base, circumscissile, rarely indehiscent; seed globular, 0.75-1 mm, compressed, black, shining, very faintly reticulate, testa crustaceous.

Common in waste lands, fields, and hedges near villages in most districts. Fl. & Fr.: Throughout the year. Vern.: Tel.: *Mulluthota kura;* Ori.: *Kanta khara, Kantaneutia.*

Gulemallebad RF (KNL), *RVR* 1443; Palakonda hills (CDP), *CS* 9371; Bhumayyagaripalle (CDP), *KRKS* & *KRS* 37618; Talakona (CTR), *AMR* 26045; Buranpur (NLR), *PMR* 16527; Nallapadu (GNT), *VRK* 6770; Gokavaram (EG), *SKW* 1722 and 1773 (BLAT); Maredumilli (EG), *MM* 100807 (BSID); Anantagiri (VSKP), *GVS* 19453 (MH); Salur (VZN), *NPBK* 976 (CAL); Kottayar (NA), *MBV* 610 (CAL); Ettiyampatti RF(DMP), *EV* 51852 (CAL); Malkangiri (KRT), *DCSR* 648 (CAL); Mahendragiri hills (GJM), *CECF* 192 (CAL); Mulakura (KPT), *TKPaul* 1405 (CAL); Govindapur (GJM), *V.P.Samudar* & *SKDas* 26252 (CAL); Budhukhola (GJM), *JKDas* & *Party* 22640 (CAL); Tippukadu RF (NA), *KRM* 16639 (CAL).

INDIA: Throughout India.

WORLD: Presumed American origin, cosmopolitan in warmer regions of the world.

Amaranthus tricolor L., Sp. Pl. 989. 1753. var. **tricolor.** *A. mangostanus* L., Cent. Pl. 1: 32 1755; FBI 4: 720. 1886; Gamble 2: 1171. 1925. *A. gangeticus* L., Syst. Nat. ed. 10. 1268. 1759 incl. var. *tristis;* FBI 4: 719; Gamble 2: 1171. 1925.

Annual erect or decumbent herb, to 50 cm tall; stem angular, glabrescent, upper parts pilose with less crisped hairs. Leaves alternate, rhomboid-ovate, broadly elliptic, 1–4 × 0.5–3 cm, glabrous, white dotted below, base attenuate, margin entire, apex emarginate to apiculate; lateral veins 4–7 pairs; petiole to 7 cm long, longer than lamina. Flowers green to crimson, in globose clusters on axillary and terminal spikes; male and female flowers intermixed; bracts deltoid ovate, 1–1.5 × 0.5–0.7 mm, bracteoles broadly ovate, 2.5–3 × 0.5–1 mm, sub-equal to the perianth, membranous, lanceolate, apex apiculate, midrib green; perianth segments 3, elliptic or oblong-elliptic, 3–3.5 × 1–1.5 mm, narrowed above, pale membranous, midrib excurrent into a long, pale-tipped awn, perianth segments slightly accrescent in fruit; stamens 3; filaments 2–2.5 mm long; anthers 0.3 mm long; ovary compressed, 1 mm across; style 3-fid, minute. Utricle ovoid-urceolate, 2.25–2.75 mm, with a short neck below, circumscissile, membranous; seed ovoid, 1–1.5 mm, black or brown, shining, very faintly reticulate, lenticular, testa crustaceous.

Occasional weed in waste places, fields, and along road sides in most districts. Fl. & Fr.: Throughout the year. Vern.: Tel.: *Perugu thotakura*

Gulemallebad North (KNL), *RVR* 1441; Thummalapalle (CDP), *KRKS* & *KRS* 36415; K.K. Kottala (CDP), *KRKS* & *KRS* 36417; Chennampalli (CTR), *BR* & *DV* 41093; Macherla RF (GNT), *VRK* 3264; Udayagiri (NLR), *PV* 111119 (BSID); Tadepalli (EG), *NRR* & *DN* 84347 (BSID); Anantagiri (VSKP), *HS* 21514 (BLAT).

INDIA: Cultivated throughout India.

WORLD: Tropical Asia. Introduced and cultivated in Africa, West Indies.

Amaranthus viridis L., Sp. Pl. ed. 2. 1405. 1763; FBI 4: 720. 1886; Gamble 2: 1171. 1925.

Annual, erect herb, up to 75 cm tall; stem glabrous when mature, pilose when young, with floccose multicellular hairs, angular. Leaves alternate, deltoid-ovate to rhomboid-oblong, 2–7 × 1.5–5.5 cm, glabrous or shortly to fairly long-pilose when young and on the veins of lower surface, base shortly cuneate or sub-truncate, margin entire, apex obtuse or emarginate-apiculate; lateral veins 7–9 pairs; petiole to 2.5-12 cm long. Flowers in axillary and terminal spikes, often paniculate; bracts and bracteoles deltoid-ovate to ovate-lanceolate, 1–2 × 0.5–0.7 mm, whitish membranous with a very short, pale awn formed by excurrent midrib; perianth lobes 3, very rarely 4; male flower segments ovate-oblong, 2–2.5 × 0.5–1 mm, margin scarious, apex mucronate, midrib green; stamens 3, filaments 1.5 mm long; anthers 0.3 mm long; ovary 0.7 mm across; style 0.2 mm, stigmas 2–3, erect. Utricle subglobose, 1.25–1.5 mm, not exceeding the perianth, indehiscent; seed ovoid, 1–1.25 mm, slightly compressed, dark brown to black, reticulate and verrucose, testa crustaceous.

Very common weed in waste lands, fields and along road sides. Fl. & Fr.: Throughout the year. Vern.: Tel.: *Chilakathota kura*; Tam.: *Kuppaikeerai.*

Kalasamudram, (ATP), *KRKS* & *BSS* 39619; Gulemallebad north (KNL), *RVR* 1442; Thummalapalle (CDP), *KRKS* 36414; Palem (NLR), *SSY* 16323; Garnipenta (PKM), *BSS* & *SKB* 33312; Chintalathanda (GNT), *VRK* 3225; Balapalli (CDP), *JLE* 14973 (MH); Rapur (NLR), *PV* 113645 (BSID); Alapadu – Kolletikota (KSN), *PV* 116513 (BSID); Maredumilli (EG), *GVS* 24256 (MH), *MM* 105058 (MH, BSID); Araku valley (VSKP), *GVS* 19717; Gollapudi (VSKP), *PV* 5270 (MH); Kottaiyur RF- Yelagiri hills (NA), *MBV* 587 (CAL); Pennagaram RF (SM), *EV* 20755 (MH); Keeraikadu – Yercaud (SM), *SK* 26837 (MH); Periyayur RF (MDU), *KS* 4342 (MH).

INDIA: Throughout India.

WORLD: Cosmopolitan in the tropical and subtropical region.

CELOSIA Linnaeus

1. Capsule urceolate, exserted from perianth, persistent style embedded; seed annular; inflorescence lax, thyrsiform spikes .... **C. polygonoides**
1. Capsule pyriform or globose, included in perianth, persistent style apical; inflorescence slender or dense spikes:
    2. Leaves deltoid-ovate; tepals 2-3 mm long; capsule pyriform .... **C. pulchella**
    2. Leaves elliptic-ovate, lanceolate, oblong-lanceolate, linear; tepals 6-8 mm long; capsule globose .... **C. argentea**

Celosia argentea L., Sp. Pl. 205. 1753; FBI 4: 714. 1885; Gamble 2: 1166. 1925.

Key to Varieties

1. Inflorescence dissected, cocks comb like spike .... var. **cristata**
1. Inflorescence dense, cylindrical spike .... var. **argentea**

Celosia argentea L. var. **argentea**

Annual, erect herb, up to 2 m tall; branches ascending, strongly ridged and sulcate, glabrous, reddish or greenish. Leaves alternate, oblong-lanceolate to narrowly linear, 2–15 × 1–3.5 cm, glabrous, sometimes reddish, base attenuate, margin entire, apex acuminate or obtuse, apiculate; lateral veins 7-9 pairs; petiole 2–10 mm. Flowers silvery to pink, in terminal, dense spike, 2.5–10 × 1.5–3 cm, conical at first but becoming cylindrical; peduncle sulcate, up to 20 cm long; bracts and bracteoles lanceolate, the lower deltoid, 3–4 × 2–3 mm, much shorter than tepals, hyaline, aristate with excurrent midrib, margins scarious, persistent; perianth segments 5, elliptic-oblong, 6–8 × 2–3 mm, acute to rather blunt, mucronate, lateral veins 2-4, margin widely hyaline; stamens 5, 6–7 mm long; filaments delicate; anthers creamy; ovary ovoid, 2–3 mm long; style filiform, 5–6 mm; stigmas 2–3. Urticle ovoid, 3–4 mm across; seeds lenticular, 1–1.5 mm diameter, black, shining; testa very finely reticulate, crustaceous.

Common weed in post harvested fields and waste lands. Fl. & Fr.: August – December. Vern.: Tel.: *Gurugu aku;* Ori.: *Sirali, Manjur, Chulia;* Tam.: *Pannaikeerai.*

Nayakunipalle (ATP), *TP* 465; Chinnapalli (ATP), *KRKS* 37804; Veldurthi RF (KNL), *RVR* 1749; Palakonda hills (CDP), *CS* 8243; Nandipadu (NLR), *PMR* 16468; Macherla RF (GNT), *VRK* 3276; Bogada (KNL), *JLE* 32522 (MH); Gundlabrahmeswaram (KNL), *JLE* 22214 (MH); Balaparallaguttakanama (CTR), *GVS* 46784 (MH); Chandragiri (CTR), *S.India Flora* 10061 (MH); Gobanapalem (WG), *KS* 5150 (MH); Araku vally (VSKP), *NPBK* 10734 (MH); S. Kota towards Punyagiri (VZN), *GVS* 32757 (MH); Yelagiri hills, Mangalam RF (NA), *MBV* 648

(MH); Tindivanam to Gingee (SA), *KRM* 60291 (MH); Gingee RF (SA), *KMS* 12178 (MH); Gomukhi, Vellimalai Ghat (SA), *KRM* 52819 (MH); Vepampathi RF (DMP), *EV* 51846 (MH); Hogainakkal forest (SM), *EV* 20675 (MH), *EV* 21911 (MH); Kollimalai Namakkal (NMK), *South Indian Flora* 11377 (MH); Rangaihunda (GJM), *G.K.Das* & Party 22070 (CAL); Malkangiri (GJM), *DCSR* 698 (CAL); Gunpur (KHD), *B.Safui* & Party 13616 (CAL), 13504 (CAL); Khondgunta (GJM), *JSG*13845 (CAL).

INDIA: Throughout India.

WORLD: Pantropical weed.

Celosia argentea L. var. **cristata** (L.) Kuntze, Revis. Gen. Pl. 1: 541. 1891. *C. cristata* L., Sp. Pl. 205. 1753; FBI 4: 715. 1885; Gamble 2: 1166. 1925.

Annual sub-shrubs, to 1.5 m high. Leaves alternate, lanceolate, 4-12 × 0.5-4 cm, base acuminate, margin entire, apex acute, glabrous. Flowers in terminal dense spikes, cockscomb like, variously branched with sterile shoot, red or reddish-orange. Utricle ovoid, enveloped in persistent perianth; seeds compressed-reniform.

Commonly cultivated in gardens. Fl. & Fr.: November – March. Vern.: Tel.: *Kodijuttu chettu;* Tam.: *Kozhikondai.* English: Cocks comb.

Nellore (NLR), *NVP* 6632.

INDIA: Throughout India.

WORLD: Tropical Asia, Africa and America.

Celosia polygonoides Retz., Observ. Bot. 2: 12. 1781; FBI 4: 715. 1885; Gamble 2: 1166. 1925.

Biannual, erect herb, to 40 cm tall; root stock woody; branches angular, villose with yellowish hairs. Leaves alternate, broadly ovate to deltoid-ovate, 1–5 × 1.5–3 cm, pilose on lower side, glabrescent above, base cordate-truncate or cuneate, margin entire, apex obtuse to acute; lateral veins 4–7 pairs; petiole to 20 cm long. Inflorescence terminal raceme; peduncle densely patent-hairy; bracts and bracteoles glabrous, ovate to deltoid-ovate, 15–2 × 1–1.5 mm, whitish, acute, mucronate with the excurrent midrib. Flowers distant to approximate, 2-few in sessile cluster, to 4 mm across; perianth segments 5, elliptic-oblong, 2.5–3 × 1.5–2 mm, mucronate with the shortly excurrent midrib, midrib green; stamens 5, 2.5 mm long; anthers 0.2 mm long; ovary ovoid, 1 mm long; style 0.5 mm, embedded; stigmas 2–3, circinately recurved, sub-equalling or longer than the style, 0.5–0.75 mm. Utricle urceolate, 2.5–3 mm long; seeds reniform, compressed, 0.6 mm across, black, uniformly verruculose, testa crustaceous.

Occasional in waste places and fields, in dry localities of Southern Eastern Ghats. Fl. & Fr.: September – February. Vern.: Tel.: *Eluka-uttareni;* Tam.: *Puliyansirikeerai.*

Pennahobilam (ATP), *TP* & *NY* 366; Jambugumbala RF (ATP), *TP* & *NY* 573 (MH); Kalasamudram (ATP), *KRKS* & *KRS* 37696; Veldurthi RF (KNL), *RVR* 1729; Lankamalai RF (CDP), *SRSR* 13187; Gandipalem (NLR), *PMR* 16542; Kondaveedu (GNT), *SSR* & *KSM* 21464; Srisailam (KNL), *JLE* 16864 (MH); Balapalli (CDP),

JLE 14971 (MH); Madanikonda RF (CTR), *SKW* 6303 (BLAT); Chandragiri fort (CTR), *DRC* 1601 (MH), *DR* 117 (MH); Gingee RF (SA), *KRM* 13050 (CAL, MH); Tippukadu RF (NA), *KRM* 16637 (CAL); Woddapatti FR (SM), *EV* 20699 (MH); Thirukazhukundram, Vedha Gireswar Hill (CPT), *DN* 358 (MH).

INDIA: South India.

WORLD: Sri Lanka.

Celosia pulchella Moq. in DC., Prodr. 13 (2): 238. 1849; FBI 4: 715. 1885; Gamble 2: 1166. 1925.

Erect, glabrous herb, up to 50 cm high, stems and branches striate and often angled. Leaves deltoid-ovate to lanceolate, 2-11 × 1-5 cm, base cuneate, margin wavy, apex acuminate, pubescent, petiole 2-4 cm long. Flowers in terminal, unbranched, 5-15 cm long, slender spikes, 2-4 flowers in clusters, axis sparingly hairy; bracts and bracteoles similar, 1 mm long, ovate, acute, hyaline; tepals 5, 2.5 × 1 mm, ovate, acute, hyaline; stamens 5, filaments united at base; ovary globose, style absent, stigmas 2, curved, pilose. Utricle pyriform, the apex slightly incrassate, circumscissile; seed 0.8 mm, roundish-reniform, compressed, black, muricate.

Rare in Horsley hills. Fl. & Fr.: July – December.

Kukkaladoddi (CTR), *DV* & *MCN* 45923; Horsley konda (CTR), *JSG* 15785, 21141 (MH), *CECF* 4354 (BLAT).

INDIA: Karala, Tamil Nadu, Andhra Pradesh.

WORLD: Sri Lanka.

CENTROSTACHYS Wallich

Centrostachys aquatica (R. Br.) Wall. ex Moq. in DC., Prodr. 13(2): 321. 1849. *Achyranthes aquatica* R. Br., Prodr. 417. 1810; FBI 4: 730. 1885; Gamble 2: 1176. 1925.

Perennial aquatic or sub-aquatic, erect herb, up to 1.5 m high, usually much branched, considerably rooting at the nodes. Leaves oblong-lanceolate to ovate-oblong, 7-15 × 2-5 cm, apex acute to acuminate, base attenuate. Flowers pale, in terminal spikes; bracts deltoid-lanceolate, bracteoles orbicular; tepals 5, the outer segments longer than the rest; stamens 5, filaments shortly connate at the base, alternating with spathulate pseudo-staminodes having fimbriate dorsal scales, style filiform, stigma capitate. Utricle oblong or ovoid, shed together with the perianth and bracteoles; seeds dark brown.

Gamble reported from Northern Eastern Ghats. Fl. & Fr.: July – December.

Berhampur (GJM), *JSG* 42230 (MH).

INDIA: Throughout India.

WORLD: Africa, Bangladesh, Sri Lanka, China, Russia, Thailand, Indonesia, Japan.

CYATHULA Loureiro *nom. cons.*

Cyathula prostrata (L.) Blume, Bijdr. 549. 1826; FBI 4: 723. 1885; Gamble 2: 1172. 1925. *Achyranthes prostrata* L., Sp. Pl. ed. 2.1: 296. 1762.

Erect herbs, up to 50 cm high, stem obtusely quandrangular above the nodes, pubescent. Leaves opposite, ovoid-obovate or rhomboid-oblong, 2-15 × 1-6.5 cm, base acute or rounded, margin entire, ciliate, apex acute or subacute, densely hairy on both sides; petioles up to 0.2 cm long, canaliculated on the upper side. Flowers in terminal, up to 30 cm long recemed spikes, rachis densely pubescent; bracts up to 0.2 cm long, ovate, reflexed after anthesis; flower clusters far apart at the base of the rachis and crowded at apex, shortly stalked. Perfect flowers: Tepals 5, up to 0.3 cm long, subequal, elliptic-oblong, outer 2 tepals more hairy without than the inner ones, 5-nerved, inner tepals 3-nerved; stamens 5, alternating with pseudostaminodes, filaments slender, connate at base to form a cup, pseudostaminodes much shorter than the staminodes, rectangular, cuneate, fimbriate; ovary rounded at top, style filiform, stigma capitate. Utricle glabrous, 1.5 mm, sessile, red.

Rare weed in dry fields in Northern Eastern Ghats. Fl. & Fr.: September – November.

Suganekota (VSKP), *SKW* 4449 (BLAT); Kapilas (Dhenkenal), *SX* 2085(RRL-B); Serango (GJM), *SX* & *MB* 3851 (RRL-B).

INDIA: Peninsular, E. and NE. India, Andaman and Nicobar Islands

WORLD: Widespread in Tropics.

DIGERA Forsskal

Digera muricata (L.) Mart., Nova Acta Phys.- Med. Acad. Caes. Leop.-Carol. Nat. Cur. 13: 285. 1826. *Achyranthes muricata* L., Sp. Pl. ed. 2. 1: 295. 1762. *Digera arvensis* Forssk., Fl. Aegypt.-Arab. 65. 1775; FBI 4: 717. 1885; Gamble 2: 1169. 1925. *Desmochaeta muricata* (L.) DC. Cat. Pl. Horti Monsp. 103. 1813.

Annual, prostrate or decumbent, sometimes erect herb, to 40 cm high; stem glabrous or pilose, angular. Leaves alternate, broadly ovate or rarely sub-elliptic, 2–5 × 0.5–3 cm, glabrous or hairy below, base attenuate or narrowed, margin entire, apex acute or acuminate-apiculate; lateral veins 7–10 pairs; petiole slender, 4–5 cm long. Flowers whitish-pink, in axillary, slender, 10 cm long spike, distantly arranged; fertile flowers each subtended by 2 lateral, sterile, modified flowers consisting of scales; bract obovate, 2.5–3 × 0.75–1 mm, midrib prominent; bracteole lanceolate, 2–2.5 × 0.5–0.75 mm, glabrous, membranous with a green or brownish percurrent midrib; perianth segments 5, ovate-oblong, 3–4.5 × 1–1.5 mm, acute, the inner perianth lobes slightly shorter, more delicate, blunt or erose, 1–3-nerved, margin hyaline with a darker central midrib; stamens 5, 3 mm long; ovary globose, 1 mm across; style slender, to 2 mm long; stigma bifid, finally recurved. Fruit subglobose, 2–2.5 mm, indehiscent, slightly compressed, bluntly keeld, furnished throughout with verrucose, testa crustaceous.

Common in irrigated dry fields. Fl. & Fr.: Throughout the year. Vern.: Tel.: *Chenchali kura*; Tam.: *Kadakeerai, Thottakeerai, Chunnampukeerai.*

Gulemallebad RF (KNL) *RVR* 1414; Thummalapalle (CDP), *KRKS* & *KRS* 33685; Macherla RF (GNT), *VRK* 5831; Forest rangers bungalow (CDP), *KS* 6329 (MH); Nerabylu (CTR), *GVS* 46015 (MH); Mekalavani kunta (CTR), *GVS* 46847 (MH); Penchalakona (NLR), *ASR* 2048 (BSID); Kondapalle RF (KSN), *P.Venkanna* 5307 (MH); South Bank Opposite to Ippur (WG), *CAB* 5332 (MH); Devipatnam (EG), *JSG* 15788 (MH); Javapuramkonda (VSKP), *GVS* 42518 (MH); Sitarampuram (SKLM), *JSG* 12574 (MH); Forest near Gomukhi Dam (SA), *KRM* 77315 (MH); Tippukadu RF (NA), *KRM* 17930 (MH); Goodamalai (SM), *R. Ansari* 50025 (MH); Near Kallandari-Alagar hills (MDU), *KS* 4297 (MH); Sirumalai hills (DGL), *SKS* 103 (SGH).

INDIA: Throughout India.

WORLD: Asia, South, Central and East Tropical Africa.

GOMPHRENA Linnaeus

1. Heads globose or depressed-globose; dorsal crest of the bracteoles well-developed; flowers purplish or crimson .... **G. globosa**
1. Heads ovoid or spiciform; dorsal crest of the bracteoles very narrow; flowers white .... **G. serrata**

Gomphrena globosa L., Sp. Pl. 224. 1753; FBI 4: 732. 1885; Gamble 2: 1179. 1925.

Erect herbs, up to 40 cm, hairy. Leaves elliptic to obovate-oblong, 2-6 × 1.5-3 cm, base attenuate, margin entire, apex acute; petiole to 1cm. Flowers sessile, above the uppermost pair of leaves, in globose or depressed-globose heads, purplish or crimson, bisexual; bracts deltoid, mucronate; bracteoles strongly laterally compressed; tepals 6-6.5 mm long, narrowly lanceolate. Utricle oblong-ovoid, ca 2.5 mm long; seed annular, black, testa shiny.

Commonly cultivated and ornamental plant. Fl. & Fr.: Throughout the year. Vern.: Tel.: *Erra pagada banti, Rudrakshi, Bandumalli*; Tam.: *Vaadamalli*. English: Globe Amaranth.

Kurnool (KNL), *NVP* 6645; Kakinada (EG), *sine coll.* 7265 (BSID).

INDIA: Cultivated throughout India, sometimes as escape.

WORLD: Native of Tropical America, cultivated in the warmer regions.

Gomphrena serrata L., Sp. Pl. 224. 1753; Mears, Taxon 29: 86. 1980. *G. decumbens* Jacq. Hort. Schoenbr. t. 482. 1805; Gamble 2: 1179. 1925. *G. celosioides* auct. non Mart.: Bennet Fl. Howrah 142. 1979.

Annual, ascending or erect herb, to 20 cm high; much branched, spreading from base, densely furnished with long, white lanate hairs, villose. Leaves opposite, elliptic-oblong to oblanceolate 1.5–5 × 0.5–1.5 cm, pilose with lanate hairs, base attenuate, margin entire, apex obtuse-mucronate; lateral veins 5–7 pairs; petiole to 0.5 cm long, pilose. Flowers silvery white, in terminal, to 2.5 cm long spike; peduncle to 6 cm long with pilose hairs and having terminal pair of leaves. basal 1 cm long spike sterile; bracts deltoid-ovate, 2.5–3 × 0.5–1 mm, shortly mucronate with excurrent midrib; bracteoles lanceolate, 3–3.5 × 0.7–1 mm, laterally compressed, hyaline, villose; perianth segments 5, narrowly lanceolate, 4.5–5 × 1–1.5 mm, 1-nerved; the outer 3 more or less flat, lanate at the base, apex mucronate; the inner ones densely lanate almost to the tip, slightly longer than the outer ones; stamens 5, staminal tube subequalling the perianth; anthers yellow; ovary compressed-globose, 1.3 mm; style and stigma together 1 mm, stigma divergent. Capsule shortly compressed, pyriform, to 1.75 mm long; seed ovoid, compressed, brown, faintly reticulate, shining, testa coriaceous.

Common along road sides, waste lands and open forest. Fl. & Fr.: Throughout the year. Vern.: Tel.: *Tella pagada banthi*; Tam.: *Vellaivadamalli*.

Kalasamudram (ATP), *KRKS* & *KRS* 36478; Gulemallebad north (KNL), *RVR* 1432; Kalava, near Hussenapuram (KNL), *RVR* 3140; Palakonda hills (CDP), *CS* 9380; On the way to Duggama konda temple (CDP), *KS* 6330 (MH); Mypadu (NLR), *TAR* & *ARKS* 7491 (BSID); Narasimhulukonda (NLR), *ASR* 2082 (BSID); Maredumilli (EG), *MM* 102700 (BSID); Anantagiri (VSKP), *GVS* 19542 (MH); Saluru (VZN), *NPBK* 1168 (MH); On the way Champakhari (KPT), *DCSR* 1474 (CAL); Alagar hills (MDU), *SKS* 89 (SGH).

INDIA: Assam, Bihar, Gujarat, Kerala, Karnataka, Andhra Pradesh, Odisha, Tamil Nadu.

WORLD: Native to South America.

Note: The speices, to which the name *G. serrata* L. was given, was renamed by Jacquin (1805) as *G. decumbens* without reference to Linnaeus' species name. *G. dispersa* Standl. is a name applied to some phenotypes of the same variable species. *G. serrata* L. is the most common and widely distributed non cultivated species of *Gomphrena*. It is frequently misidentified as *G. celosioides* Mart., a closely related species native to South America except in cultivation.

INDOBANALIA A.N.Henry & B.Roy

Indobanalia thyrsiflora (Moq.) A.N. Henry & B. Roy, Bull. Bot. Surv. India 10: 274. 1968. *Banalia thyrsiflora* Moq. in DC., Prodr. 13: 278. 1849; FBI 4: 716. 1885.

Perennial erect, woody herbs, stems striate, glabrous or sparsely hairy. Leaves alternate, subglabrous to moderately pilose on both surfaces, ovate to ovate-lanceolate, 10-16 × 4-6 cm, base shortly cuneate to cuneate-attenuate, apex acuminate. Flowers white in axillary or terminal spike, pedunculate, thyrsoid panicle of spikes; bracts deltoid-ovate, bracteoles lanceolate; perianth segments 5, whitish or pale green to red or purple, free, equal, 3× 1 mm, 3-nerved, glabrous; stamens 5, filaments jointed into a tube, ovary ellipsoid, compressed, 1-ovuled, style 1, slender, stigmas 2, linear, papillose. Utricle ovoid, 2-2.5 mm, indehiscent; seed ovoid, shining, brown.

Rare in Nallamalais. Fl. & Fr.: July – November.

Nallamalais (KNL), *MHR* 8572.

INDIA: South India. Endemic.

Note: The generic name *Banalia* Moq. (1849) is illegitimate being a later homonymn of *Banalia* Rafin. (1840)-now included in *Croton* L. (Euphorbiaceae) *Indo-banalia* is proposed by Henry and Roy (1948) as an avowed substitute for *Banalia* Moq.

NOTHOSAERVA Wight

Nothosaerva brachiata (L.) Wight, Icon. Pl. Ind. Orient. t. 1776. 6: 1. 1853; FBI 4: 726. 1885; Gamble 2: 1175. 1925. *Acyranthes brachiata* L., Mant. Pl. 1: 50. 1767.

Annual, erect herb, up to 50 cm high, branches spreading from base upwards; stem subterete, striate, sparsely hairy. Leaves decussate, lower leaves on main stem ovoid-elliptic, 1–3 × 0.5–2 cm, upper leaves on main stem and leaves of branches becoming shorter and narrower, reduced in flowering branches, glabrous to thinly hairy above, pubescent below, more on nerves, base acute to attenuate, margin entire, apex acutely apiculate; petiole about half the length of the lamina, to 1.5 cm long on main stem, shorter on branches, reduced towards apex and finally sessile. Flowers pale whitish woolly, bisexual, in axillary, dense paniculate spikes, subtended by reduced leaves; peduncles short, 3 mm long, axis thinly to rather densely pilose; bracts ovate, 0.5 mm, hyaline, acute or shortly acuminate, glabrous or very thinly hairy, nerveless; bracteoles 2, broadly ovate, concave, 1–1.25 mm, hyaline, acuminate, persistent; perianth segments 5, sub equal, oblong, deltoid-ovate, 1.25–1.5 × 0.5–0.75 mm, hyaline, membranous, 1-nerved, subacute to shortly acuminate, villous on the outer suface, with a dark greenish midrib, extends up to $^{2}/_{3}$ of the segment, truncate, acute; stamens 2; filaments 1–2.5 mm long; anthers 2-celled, hemispheric, 0.1 mm across; ovary obconic to depressed-globose, 0.3 mm across, 1-locular; ovule 1 on short, erect funicle; style short, 0.5 mm long; stigma capitellate. Utricle pyriform, oblong, 0.5–0.75 mm, compressed, epicarp membranous, irregularly rupturing, falling off with the perianth-lobes; seed lenticular, 0.5 mm, brown, smooth and shining, testa crustaceous.

Rare, in moist low-lying grounds, along river banks. Fl. & Fr.: September – March. Vern.: Ori.: *Moduranga.*

Pulapally (ATP), *BR & DV* 40586; Lankamalai RF (CDP), *SRSR* 13196; Buranpur (NLR), *PMR* 16522; Velgoda RF (KNL), *SKW* 4886 (BLAT); Gajulapalli RF (KNL), *JLE* 18073 (MH); Mamandur RF (CDP), *AMRao* 984 (BSID); Mamandur (NA), *South India Flora* 1016 (MH); Gingee RF (SA), *KMS* 12356 (MH); Kambakkam hills (CPT), *S. India Flora* 16462 (MH); Vandalur RF (CPT), *ANH* 47121 (MH).

INDIA: N. and Peninsular India, Upper Gangetic plains.

WORLD: SE Asia, Tropical Africa.

PSILOTRICHUM Blume

1. Apex mucronate, glabrous; tepals equal; seed vertical .... **P. ferrugineum**
1. Apex acuminate, hairy; tepals unequal; seed inverse:
    2. Inflorescence elongated spikes; tepals outer 2 membranous, inner 3 densely long-pilose .... **P. elliotii**
    2. Inflorescence lax panicle; tepals outer 2 pungent, inner 3 silky .... **P. sericeum**

Psilotrichum elliotii Baker, Fl. Trop. Afr. 6 (1): 58. 1909. *P. calceolatum* Moq. ex Hook. f., Fl. Brit. India 4: 725. 1885. *nom. illeg.*; Gamble 2: 1174. 1925.

Erect herb, up to 75 cm high; stem and branches in the older parts terete, striate, finally glabrescent at the base. Leaves ovate to broadly oblong, 2-4.5 × 1.5-3 cm, base truncate, margin entire, apex shortly mucronate, sparsely appressed hairy on both sides, sometimes gabrescent above with age; petiole to 5 mm. Flowers in terminal or axillary elongated spikes; bracts up to 0.2 cm long, ovate or ovate-acuminate, white woolly without, reflexed in fruit; bracteoles up to 0.15 cm long, broadly ovate-cordate; tepals 5, up to 0.35 cm long, 3-nerved, outer 2 tepals broader than the inner 3 tepals, margins involute; stamens 5, filaments filiform, slightly connate at base; ovary ovate-oblong, glabrous; style slender, stigma capitate. Utricle compressed; seed ovoid, 1 mm, compressed, brown, testa shiny, coriaceous.

Rare in swampy ground on the edges of forest. Fl. & Fr.: October – January.

Circuit house (VSKP), *GVS* 21473 (MH); Nellore (NLR), *ASR* 2768 (VV); Shanikulam (SA), *CAB* 810 (MH); Betta Mogilalam (SM), *South India Flora* 9825 (MH); Parigam RF (SA), *SR Raju* & *Naganatham* 17896 (MH); Sirumalai puthur (DGL), *SKS* 628 (SGH).

INDIA: South India.

WORLD: Tropical Africa.

Psilotrichum ferrugineum (Roxb.) Moq. in DC., Prodr. 13(2): 279. 1849; FBI 4: 725. 1885. *Achyranthes ferruginea* Roxb., Fl. Ind. 1: 675. 1820 & 2: 502. 1824. *Psilotrichum trichotomum* Blume, Bijdr. 545. 1826; FBI 4: 725. 1885.

Annual, erect, glabrous herb, up to 50 cm high. Leaves variable in shape, lanceolate, oblong-elliptic or obovate, 4-6 × 2 cm, base attenuate, margin entire, apex mucronate, glabrous. Flowers in axillary and terminal spikes, dull red, usually peduncled, sometimes sessile; bracts subulate, lanceolate; perianth segments to 2.5 mm, deeply grooved, outer obtuse. Utricle ellipsoid, 1-1.5 mm; seed vertical, black, testa shiny.

Rare in Northern Eastern Ghats. Fl.: October.

Sonkardih, Bhuinya hills, Keonjor, *HFM* 2816 (DD); Simlipal reserve forest (MBJ), *ARKSastry* & *GPSingh* 12300 (CAL); Podaboia to Bhanjabasa, Simlipal (MBJ), *Bahali* & *Agrawall* 445 (CAL); Girishahardrapur (Sambalpur), *SP* 985 (CAL).

INDIA: Bengal, Odisha, Andhra Pradesh.

WORLD: SE Asia.

PUPALIA A.L. Jussieu nom. cons.

Pupalia lappacea (L.) Juss., Ann. Mus. Natl. Hist. Nat. 2: 132. 1803. FBI 4: 724. 1885; Gamble 2: 1173. 1925. *Achyranthes lappacea* L., Sp. Pl. 204. 1753. *Pupalia atropurpurea* (Lam.) Moq. in DC., Prodr. 13(2): 331. 1849; FBI 4: 723. 1885; Gamble 2: 1173. 1925. *Achyranthes atropurpurea* Lam., Encycl. 1: 546. 1785.

Key to Varieties

1. Leaves orbicular .... var. **orbiculata**
1. Leave ovate-elliptic, oblong:
   2. Leaves glabrous or thinly pilose; spike unbranched.................var. **lappacea**
   2. Leaves tomentose beneath when young, hairy when mature other than above; spike branched ....var. **velutina**

Pupalia lappacea (L.) Juss. var. **lappacea.**

Annual or perennial, erect herb, up to 2 m high; stem much branched and swollen at nodes; branches opposite, slender, obtusely 4-angled to almost terete, thinly pilose to tomentose. Leaves opposite, variable in shape and size, ovate-elliptic to oblong, 2–7 × 1–4 cm, sparsely tomentose or pilose on the veins in lower side, base attenuate gradually, margin entire, apex acuminate to apiculate; lateral veins 7–9 pairs; petiole to 1 cm long, pilose. Flowers in terminal, unbranched, to 50 cm long spike; peduncle tomentose; flowers 3 in a cluster, central one 1-bracteate, lateral flowers 2, bracteolate; sterile flowers modified into hooked bristles in fruit; bracts lanceolate, 1.5–2.5 × 0.5–0.7 mm, persistent, more or less deflexed after the fall of the fruit, sub-glabrous or pilose, sharply awned with percurrent midrib; bracteoles broad, subcordate-ovate, 3–5 × 0.5–1 mm, abruptly narrowed, dark arista formed by the excurrent midrib, membranous with pale margin, densely hairy; bracteoles of sterile flowers ovate-lanceolate, pilose; perianth segments oblong-ovate to lanceolate, 3–6 × 1–1.5 mm, narrowed to obtuse-mucronate at apex, aristate; outer tomentose, dorsally 3-nerved; inner 2-nerved, mucronate at apex with arista; inner most 3 slightly shorter and more densely pilose; stamens 5, 2.5–3 mm long, anthers yellowish; ovary depressed-globose, basally narrowed, 1 mm across; style erect, slender, 0.8–2 mm long; stigma capitate. Capsule depressed globose, circumscissile, 2.5 mm long, top areolate; seed ovoid-oblong, 1.5–2 mm, lenticular, dark brown, shining; testa at first faintly reticulate but finally smooth or punctulate, thinly crustaceous.

Common along road sides, out skirts of the forests and waste lands. Fl. & Fr.: Throughout the year. Vern.: Tel.: *Erra uttareni, Tella uttareni;* Tam.: *Vaelikeerai.*

Kalasamudram RF (ATP), *TP* & *EC* 2840; Somandepalli RF (ATP), *NY* 1086; Gani RF (KNL), *TP* & *RVR* 1776; Kalava, near Hussenapuram (KNL), *RVR* 3143; Thummalapalle (CDP), *TP* & *SSR* 33608; R.K. Palem (GNT), *VRK* 3253; Anantagiri (VSKP), *NVP* 6620; Chelama (KNL), *JLE* 22057 (MH); Balapalle (CDP), *JLE* 14968 (MH); Guvvalacheruvu (CDP), *KS* 6395 (MH); Balaparallagutta (CTR), *GVS* 46804 (MH); Udayagiri (NLR), *SKW* 6737 (BLAT); Kondaveedu fort (GNT), *CAB* 4636 (MH); Boda vaila (KSN), *PV* 6007 (MH); Maredumilli (EG), *MM* 100726 (MH); Sitaram (EG), *GVS* 67572 (MH); Simhachalam (VSKP), *GVS* 21496 (MH) Anantagiri (VSKP), *GVS* 21708 (MH); Salur (VZN), *NPBK* 914 (MH); Woddapathi RF (SM), *EV* 20701 (MH); Kavalur, Javadi hills (NA), *MBV* 746 (MH); 859 (MH). Vepampatti RF, Harur (DMP), EV 51831 (MH) Pakamalai RF, Gingee (SA), *KRM* 5285 (MH); Thippukadu RF (NA), *KRM* 17637 (MH), *SKW* 6850 (BLAT); Pakamalai RF (SA),

KRM 52885 (CAL); Jangigash Road (KHD), *B.Safui* & Party (CAL); Sirumalai hills (DGL), *SKS* 941 (SGH).

INDIA: Throughout India.

WORLD: Sri Lanka, Malaysia, Africa.

Pupalia lappacea (L.) Juss. var. **orbiculata** (Heyne ex Wall.) C. Towns., Kew Bull. 29: 469. 1974. *Achyranthes orbiculata* Heyne ex Wall. in Roxb. Fl. Ind. 2: 507. 1824. *Pupalia orbiculata* (Heyne ex Wall.) Wight Ic. t. 1783. 1852; FBI 4: 724. 1885; Gamble 2: 1173. 1925.

Prostrate herb, hispidulous. Leaves orbicular, 2.5-3 x 4 cm, base rounded, entire, obtuse or retuse, densely pilose. Inflorescence axillary and terminal stout spikes, to 10-20 cm; flowers in clusters, distant woolly-pubescent.

Occasional in Southern Eastern Ghats. Fl. & Fr.: Throughout the year.

Kavalur, Javadi hills (NA), *MBV* 859 (CAL); Vandalur RF (CGP), *ANH* 47151 (CAL). Kadiri (NA), *MBV* 952 (MH), Sathanur forest (NA), *EV* 52076 (MH).

INDIA: Peninsular India, Rajasthan.

WORLD: Sri Lanka.

Pupalia lappacea (L.) Juss. var. **velutina** (Moq.) Hook. f., Fl. Brit. India 4: 724. 1885; Gamble 2: 1173. 1925. *P. velutina* Moq. in DC., Prodr. 13(2): 332. 1849.

Erect herb, densely or villously tomentose. Leaves broadly elliptic, 3-7 x 2.5-4 cm, lower side tomentose when young, hairy when mature. Flowers in branched spikes, densely white woolly.

Common. Fl. & Fr.: August– January.

Pennahobilam (ATP), *TP* & *NY* 334; Thummalapalle (CDP), *TP* & *SSR* 33608; Macherla RF (GNT), *VRK* 3253; Balapalli (CDP), *JLE* 14968 (MH); Nagarjunakonda (GNT), *KT* 9619 (CAL); Bhakarapeta (CTR), *DRC* 1747 (MH); Cuddapah forests (CDP), *CBDM s.n.* (MH); Santhamaredupalli (SM), *South India Flora* 9735 (MH).

INDIA: Peninsular India, Odisha.

WORLD: Widespread in the Old World Tropics.

TRICHURIELLA Bennet

Trichuriella monsoniae (L. f.) Bennet, Indian J. Forest. 8: 86. 1985. *Trichurus monsoniae* (L.f.) C. Towns., Kew Bull. 29: 466. 1974 & in Dassan & Fosb., Rev. Handb. Fl. Ceylon 1: 36. 1980; Venkata Raju & Padmavathi in Pullaiah & Moulali, Fl. Andhra Pradesh 2: 808. 1997. *Illecebrum monsoniae* L.f., Suppl. Pl. 161. 1781. *Celosia monsoniae* Retz., Observ. Bot. 2: 13. 1781. *Aerva monsoniae* (Retz.) Mart., Nova. Act. Phys.-Med. Acad. Caes. Leop.-Carol. Nat. Cur. 13: 291. 1826; FBI 4: 728. 1885; Gamble 2: 1178. 1925.

Annual, spreading herb, up to 90 cm long, stem wiry, villous-hairy. Leaves whorled, filiform, up to 2.5 cm long, sessile, covered with villous-hairs. Flowers in axillary and terminal globose heads, pinkish white, spike to 1 cm, flowers bisexual

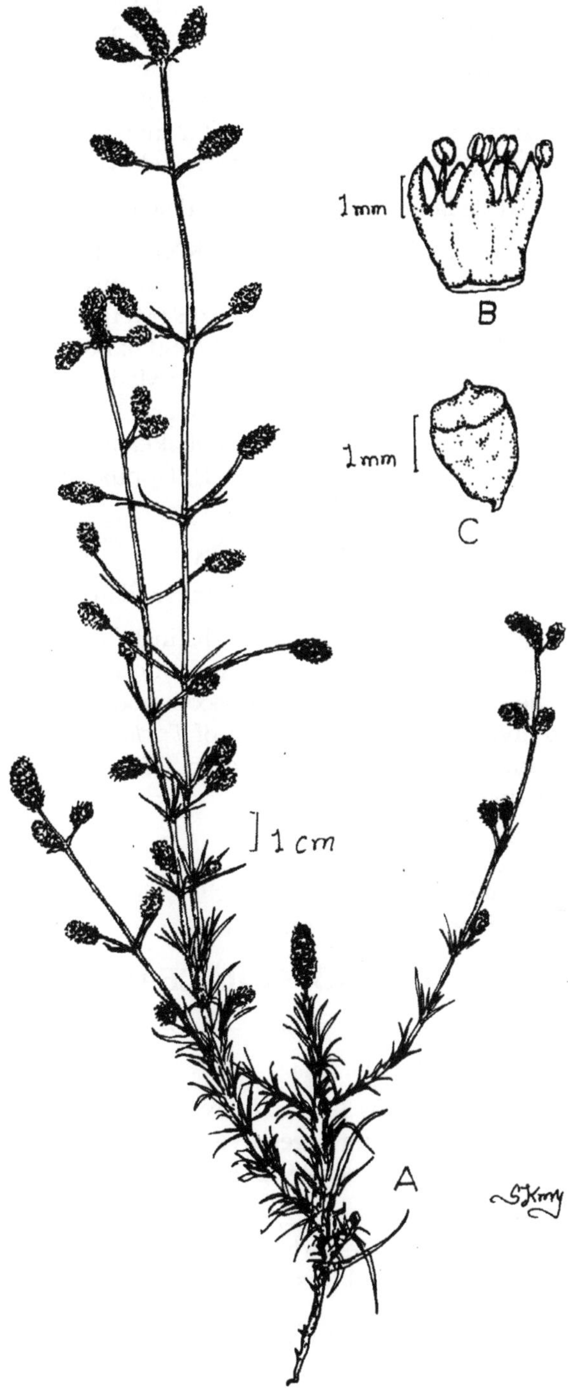

Figure 4. **Trichuriella monsoniae** (L. f.) Bennet
A. Habit; B. Staminal column; C. Pistil

or polygamous; bracts and bracteoles lanceolate, 1-1.5 × 0.5 mm, hyaline, with a green midrib; perianth lobes 4, lanceolate, subclavate, 2-2.5 × 0.7 mm, hyaline with a green midrib, dorsally lanate, curving towards the apex of the spike, persistent; stamens 4, filaments 1-1.5 mm long, alternating with triangular to subquadraate pseudostaminodes; ovary depressed globose, 0.2 mm across, style simple, stigma capitate. Utricle circumscissile, ovoid, to 1 mm; seed ovoid, inverse.

Occasional in waste lands, fields, along road sides. Fl. & Fr.: June – January.

Thalaricheruvu (ATP), *TP* 919; Gani RF (KNL), *RVR* 1778; Chintala thanda (GNT), *VRK* 3238; Srisailam (KNL), *JLE* 16903 (MH) Diguvametta (KNL), *JLE* 32451 (MH); on way to Amudalakona (CTR), *GVS* 46759 (MH); Chamala RF (CTR), *GVM* 112986 (BSID); after N.Peta (NLR), *PV* 112174 (BSID); Polavaram (EG), *CAB* 4797 (MH); Burmakonda RF (VSKP), *KCJ* 17179 (MH); Madanancheri (NA), *MBV* 775, 780 (CAL); Vazhiyur (NA), *EV* 54595 (CAL); Takarai RF (SA), *KRM* 52816 (CAL); Malakangiri (Odisha), *DCSR* 1406 (CAL); Kambakkam hills (CGP), *S. India Flora* 10239 (MH); Takarai RF (SA), *KRM* 52816 (MH), Tippukadu RF (NA), *KRM* 17618 (MH), Vazhiyur (NA), *EV* 54595 (MH).

INDIA: C, E. and Peninsular India.

WORLD: Sri Lanka, Bangladesh, Myanmar, Thailand.

Note: *Aerva monsoniae* (Retz.) Mart. is quite distinct from other species of *Aerva* Forssk. in several characters of external morphology. Townsend (1974) erected a new genus, *Thichurus* to accommodate this species. *Trichurus* Townsend (1974) is a later homonym of *Trichurus* Clements (1896) which belongs to fungi. Bennet (1985) proposed *Trichuriella* as a substitute name for the later homonym.

Telanthera ficoidea (Lam.) Moq. is cultivated in gardens for its crimson coloured leaves. **Iresine herbstii** Hook. ex Lindley, is also an ornamental plant cultivated for its blood red coloured leaves.

CHENOPODIACEAE

1. Fruiting perianth winged .... **Kochia**
1. Fruiting perianth not winged:
    2. Strongly aromatic; stigmas 3-5; top of ovary and fruit studded with glands; young vegetative parts and outside of perianth without powdery white or pink vesicles .... **Dysphania**
    2. Not or faintly smelling; stigmas 2; top of ovary and fruit glandless, young vegetative parts and outside of perianth with powdery white or pink vesicles .... **Chenopodium**

CHENOPODIUM Linnaeus

1. Branchlets green or with purple streaks; tepals ovate-rhomboid; cultivated .... **C. album**
1. Branchlets not streaked; tepals ovate or oblong; wild .... **C. murale**

Chenopodium album L., Sp. Pl. 219. 1753; FBI 5: 3. 1886; Gamble 2: 1181. 1925.

Erect herbs, up to 50 cm high, branchlets with green or purple streaks, glabrous or thinly clothed with white vesicles. Leaves ovate-lanceolate, 2.5-6 × 1-2 cm; margin entire or usually more or less toothed or lobulate, apex acute or subacute. Flowers in axillary and terminal panicles, 2 mm across, bisexual; tepals 5, ovate-rhomboid, 1.3 mm; stamens 5, filaments 0.25 mm, anthers 0.3 mm; stigmas 2, spreading. Utricle 1.5 mm across; seed horizontal, obscurely keeled, 1.2 mm across.

Cultivated as leafy vegetable and found along the streams, often as an escape. Fl. & Fr.: November – April. Vern.: Tel.: *Pappukura, Chakravarthi kura*; Ori.: *Bathua;* Tam.: *Chakaravarthikkeerai.*

Pulagampalli (ATP), *BR* & *ANS* 38424; Saniapay (CDP), *BR* & *DV* 41479; Ganjam, *S. Coll. s.n.* (MH).

INDIA: Throughout India

WORLD: Europe, Continental Asia, Africa, Australia, America.

Chenopodium murale L., Sp. Pl. 219. 1753; FBI 5: 4. 1886; Gamble 2: 1181. 1925.

Erect herbs, to 1 m high. Leaves alternate, rhomboid to deltoid-ovate, 2-7.5 × 1-4.5 cm, petiole 3 cm, base attenuate, margin sharply toothed, apex acute. Flowers bisexual, greenish, minute, in axillary spikes in spreading cymes, polygamous; bract and bracteoles 0; tepals 5, hooded, green, bluntly keeled, enclosing the fruit. Utricle enclosed in the perianth; seed horizontal, black, testa coriaceous.

Occasional weed. Fl. & Fr.: September – November.

Thalaricheruvu (ATP), *TP* & *NY* 1125 (MH); Kalasamudram RF (ATP), *EC* 2924; Penukonda hills (ATP), *BR* & *ANS* 38107; Araku (VSKP), *NPBK* 10732 (MH).

INDIA: Karnataka, Andhra Pradesh.

WORLD: From C and S Europe, N Africa to Caucasia, Iran, Afghanistan, Turkmenistan, Ceylon and to the Arabian peninsula; introduced in subtropical and tropical C, E and S Africa, America, Australia and Oceanic islands.

DYSPHANIA R.Brown

Dysphania ambrosioides (L.) Mosyakin & Clements, Ukrayisk. Bot. Zhurn. 59: 382. 2002. *Chenopodium ambrosioides* L., Sp. Pl. 219. 1753; FBI 5: 4. 1886; Gamble 2: 1181. 1925.

Erect, strongly aromatic herbs, up to 60 cm high, stem striate, glandular-pubescent. Leaves alternate, oblong-lanceolate, 0.5-3.5 × 0.2-0.5 cm, base cuneate, margin sinuate-dentate, upper half entire, apex acute; petiole 2 mm. Flowers green, in axillary and terminal, simple or panicled spikes; bracts up to 3 × 0.6 cm, foliaceous, lower surface glandular, entire, base cuneate, tip acute; 4-5 or rarely 6-merous; perianth 1.5 mm diam., tepals elliptic, acute, connate at base; stamens 4-6, 1.5 mm long; ovary globose, minutely papillose above, stigmas 3-5. Utricle enclosed in the perianth, seed horizontal, smooth, testa coriacious.

Rare on hill slopes, along river banks. Fl. & Fr.: Throughout the year. Vern.: Tam.: *Mannennaipoondu*.

Sunkarimetta RF (VSKP), *NVP* 6628; Mulyaguda to Borra caves (VSKP), *EC* 7415; Araku (VSKP), *GVS* 28266 (MH); Kooppukottai, Araku valley (VSKP), *DDSR* 95897 (MH); Bodalanka (EG), *DN* 85566 (BSID); Kooppukottai, Kottur RF (NA), *MBV* 661 (CAL); Chittarkonda dam site (KPT), *DCSR* 757 (CAL); Kottur RF, Yelagiri hills (NA), *MBV* 661 (MH); Hogainakkal (SM), *EV* 23520 (MH), 20597 (MH); Shevaroys (SM), *S. India Flora* 14800 (MH); Kollimalai (TCP), *S. India Fl.* 12943 (MH); Sirumalai hills (DGL), *SKS* 81, 314 (SGH); Damanjodi (KPT), *SX* & *MB* 6577 (RRL-B); Thuamul, Rampur (KHD), *MB* & *Dhal* 7693 (RRL-B).

INDIA: Throughout India.

WORLD: Native of Tropical America, widely introduced in Europe, Asia, Africa, Australia.

KOCHIA Roth

Kochia indica Wight, Icon. Pl. Ind. Orient. t. 1791. 1852; FBI 5: 11. 1886; Gamble 2: 1182. 1925.

Annual or biennial herb, up to 2 m high, softly villous or pubescent, stem pale, brown or yellowish, with many spreading branches from the base. Leaves alternate, sessile, 5-15 x 1-5 mm, lanceolate, oblong or linear, apex acute or sub-acuminate, margin entire, base cuneate, soft hairy to subglabrous. Flowers solitary or in clusters, arranged in loose, leafy spikes, bisexual, bracts linear, leaf-like, longer than clusters; tepals woolly, connivent, up to 3 mm in fruit. Utricle membranous; seed horizontal, albumen scanty.

Ornamental. Fl. & Fr.: Throughout the year.

INDIA: North-Western and Peninsular India.

WORLD: N.Africa, Sudan, S.W.Asia to Pakistan

Beta vulgaris L. (Beet root) is rarely cultivated for fleshy roots. **Spinacea oleracea** L. is cultivated as a leafy vegetable.

BASELLACEAE

1. Inflorescence unbranched; style simple .... **Basella**
1. Inflorescence branched; style trifid .... **Anredera**

ANREDERA Jussiaeu

Anredera cordifolia (Ten.) Steenis, Fl. Males. 5: 303. 1957; R. Erikss., Kew Bull. 62: 311. 2007; Rasingam & Lakshminarasimhan, Rheedea 22 (1): 16-17. 2012. *Boussingaultia cordifolia* Ten., Ann. Sci. Nat. Bot. Ser. 3, 19: 355. 1853.

Evergreen vine; stem semi-succulent, producing small axillary tubers (bulbils). Leaves simple, ovate to subcordate, 2- 8 x 1- 7 cm, rounded or cordate at base, acute at apex, thin-fleshy; petioles very short. Racemes axillary, pendent, simple or branched, many-flowered; rachis 7 - 25 cm long; each flower subtended by a persistent bract;

lower bracteoles broadly triangular, acute at apex, hyaline, persistent; upper bracteoles orbicular to broadly elliptic, shorter than perianth, flattened, greenish white. Flowers ca 6 mm across, fragrant; pedicels 2- 3 mm long; perianth inflexed, patent in anthesis, white; segments 5, ovate or oblong to elliptic, ca 3 - 2 mm, blunt at apex; stamens 5, opposite perianth segment, white; filaments apically reflexed in bud, spreading in anthesis; styles white, split into 3 stigmatic arms, each with 1 club-shaped or broadly ellipsoid stigma. Fruits sub-tended at base by persistent, patent perianth.

Common invasive plant in Sirumalai hills of Tamil Nadu. Fl. & Fr.: June – October.

Sirumalai hills, Pudur (DGL), *SKS* 811 (SGH).

INDIA: Tamil Nadu, Meghalaya.

WORLD: Australia, America, China, Malaysia, Pacific Islands and South Africa.

BASELLA Linnaeus

Basella alba L., Sp. Pl. 272. 1753; FBI 5: 20. 1886; Gamble 2: 1185. 1925.

Twining, succulent herb, stem and petioles red (var. *rubra*) or green (var. *alba*). Leaves broadly ovate, ovate or orbicular, 3.5 – 12 × 2.2-9 cm, base rounded to shortly decurrent, apex gradually acute; petiole to 1.5 cm. Flowers pinkish or purple in spikes; perianth urceolate, 3 mm across, tube 2.5 mm, lobes 5; stamens 5, inserted below the lobes, filaments 1.5 mm. Pseudoberry globose or lobed, 4-7 mm high, 5-10 mm wide, black with deep purple juice.

Commonly cultivated. Also found in wild state on rocky grassy hillsides. Fl. & Fr.: October – February. Vern.: Tel.: *Bachali;* Ori.: *Poi, Pui;* Tam.: *Paruppukeerai, Pasalaikeerai, Kodipasalai.*

Kulluru (NLR), *SSY* 16320; Nallapadu (GNT), *VRK* 6767; Nagapatla RF (CTR), *KS* 6864 (MH); Rapur hills (NLR), *PV* 110022 (BSID); Karakakonda, Narsipatnam (VSKP), *GVS* 29655 (MH); Rambha (GJM), *SX* & *MB* 7307 (RRL-B); Chinnakoktai, Yelagiri hills (NA), *MBV* 680 (CAL); Theerthamalai RF (DMP), *EV* 51814 (CAL); Pakamalai RF, Gingee (SA), *KRM* 52891 (CAL); Pennagaram RF (SM), *EV* 22408 (MH); Javadi hills (NA), *MBV* 902 (MH); Yelagiri hills, Mangalam RF (NA), *EV* 52035 (MH); Parigam RF (SA), *SR Raju* & *Naganathan* 17901 (MH); Gungavanar RF (SA), *KRM* 60352 (MH).

INDIA: Throughout India, wild and cultivated.

WORLD: Africa, Tropical SE Asia.

POLYGONACEAE

1. Climber; inflorescence tip tendrillar .... **Antigonon**
1. Other than climbers; inflorescence tip not tendrillar:
    2. Branchlets modified into phylloclades; leaves deciduous ... **Muehlenbeckia**
    2. Branchlets otherwise; leaves persistent:

3. Flowers polygamous; perianth 6-cleft; fruit a triquetrous nutlet..**Rumex**
3. Flowers bisexual; perianth 3-5-cleft; fruit a compressed or trigonous nutlet:
    4. Flowers in axillary fascicles...**Polygonum**
    4. Flowers in racemes or heads...**Persicaria**

ANTIGONON Endlicher

Antigonon leptopus Hook. & Arn., Bot. Beechey Voy. 308. 1838; Gamble 2: 1192. 1925.

Climbing shrub, with tendrils, root tuberous, stem angular, glabrescent. Leaves alternate, simple, elliptic-ovate, 3-8 × 1.5-5 cm, base hastate, margin entire, apex acute, stipules conspicuous, to 2 mm; petiole 1.5 cm. Flowers in axillary and terminal panicles, bisexual, pink or white, inflorescence ending into a branched tendril; tepals 5, 6-15 × 3-7 mm; stamens 8; carpels 3, ovary ovoid, trigonous, styles 3, stigma capitate. Nut ca 5 mm long.

Cultivated in gardens, often as an escape, runs wild in Visakhapatnam hills. Fl. & Fr.: Throughout the year. Vern.: Tel.: *Teegarosa;* English: Railway creeper.

Pulagampalli (ATP), *BR* & *ANS* 38423; V.P.South (GNT), *VRK* 3934; Chandragiri (CTR), *S. India Flora* 10032 (MH).

INDIA: Widely cultivated and naturalized.

WORLD: Indigenous to S. America, widely grown in gardens.

MUEHLENBECKIA C.F. Meissner *nom. cons.*

Muehlenbeckia platyclados (F.V. Muell.) Meissn., Bot. Zeitung (Berlin) 22: 313. 1865; Gamble 2: 1192. 1925. *Polygonum platycladum* F.Muell., Trans. & Proc. Philos. Inst. Victoria 2: 73. 1858.

Erect herb or sub shrub, to 55 cm high, branchlets green, flattened, jointed, modified into phylloclades, nodes flattened. Leaves lanceolate, 2-5 × 0.5-1.5 cm, deciduous. Flowers greenish yellow, bisexual, in axillary, short lateral spikes; tepals greenish-yellow, subequal; stamens 6-9, ovary 3-gonous, style arms 3, stigmas simple. Nutlets trigonous

Cultivated in gardens. Fl. & Fr.: Throughout the year. Vern.: Tel.: *Kurathutty.*

Simhachalam temple garden (VSKP), *NVP* 6618; Yelagiri hills-Bus stand (NA), *MBV* 1373 (MH).

INDIA: Grown as ornamental.

WORLD: Native of Solomon Islands, grown as ornamental elsewhere.

PERSICARIA Linnaeus

1. Leaf margin retrorsely prickled; peduncles scabrid .... **P. strigosa**
1. Leaf margin entire or hairy; peduncles not strigose:
    2. Plants erect:
        3. Bracts and ochreae glabrous .... **P. glabra**
        3. Bracts and ochreae ciliate or hispid:
            4. Flowers in heads .... **P. dichotoma**
            4. Flowers in racemes or panicles:
                5. Nutlets biconvex or biconcave:
                    6. Leaves silky-tomentose; flowers pinkish .... **P. pulchra**
                    6. Leaves other than silky tomentose; flowers white .... **P. orientalis**
                5. Nutlets trigonous:
                    7. Bracts closely imbricate:
                        8. Leaves linear-oblong; flowers in racemes; cilia of the ochreal tube much shorter than the tube:
                            9. Stamens 5; ochrea shorter than the petiole .... **P. stagnina**
                            9. Stamens 6; ochrea longer than the petiole .... **P. lapathifolia**
                        8. Leaves lanceolate; flowers in spikes; cilia of the ochreal tube either almost equaling or longer than the tube; leaves glabrous or if pubescent then only on the margins and midrib .... **P. barbata**
                    7. Bracts distant:
                        10. Flowers pink; leaves glabrous beneath, without pink blotches .... **P. hydropiper**
                        10. Flowers white; leaves minutely pubescent beneath, with pink coloured blotches.................. **P. decipiens**
    2. Plants other than erect:
        11. Straggling shrub; ochreae 4 cm long; stamens 8:
            12. Fruiting perianth fleshy and accrescent:
                13. Flowers in terminal, corymbose capitate heads............ **P. chinensis**
                13. Flowers in bracteate clusters on racemes............................ **P. minor**
            12. Fruiting perianth neither fleshy nor accrescent................ **P. nepalensis**

11. Prostrate herb; ochreae less than 1 cm long; stamens 4-5:
    14. Leaves hastate at base; petiole minutely prickled............ **P. perfoliata**
    14. Leaves cordate to obtuse at base; petiole glabrous.............. **P. capitata**

Persicaria barbata (L.) H. Hara, Fl. E. Himalaya 70. 1966. *Polygonum barbatum* L., Sp. Pl. 1: 362. 1753; FBI 5: 37. 1886; Gamble 2: 1189. 1925.

Key to Varieties

1. Spikes 1-3 in each raceme, narrower; stem weak and slender.........var. **barbata**
1. Spikes 5-6 in each raceme, broader; stem robust .... var. **gracilis**

Persicaria barbata (L.) H. Hara var. **barbata**

Erect herbs, to 1.5 m high, stem glabrescent, rooting at nodes, nodes thickened; ochreae tubular, truncate, 1.5 cm, strigose, coarsely ciliate. Leaves lanceolate, 3.5-15 × 1-2.5 cm, glabrous, base acute to cuneate, margin entire, apex acuminate; petiole 1 cm. Flowers pink, in terminal and axillary, to 13 cm long spikes, bisexual; bracts truncate, ciliate along margins; perianth hyaline, lobes 5, oblong, to 2.5 mm long; stamens 5-8; ovary trigonous; styles 5, stigmas capitate. Nutlet trigonous, 3 × 1.5 mm, brown.

Common along the streams, growing in association with *P. glabra*. Fl. & Fr.: Throughout the year. Vern.: Tel.: *Niru ganneru*.

Gorantla (ATP), *BR* & *ANS* 36624; Gundlabrahmeswaram (KNL), *RVR* 2385; Nadyal (KNL), *RVR* 14547; Palakonda hills (CDP), *CS* 8264; Seetharampuram (NLR), *PMR* 23081; way to Mamandur (CTR), *KS* 6919 (MH); Rampa (EG), *VNS* 621 (CAL); Maredumilli (EG), *MM* 100816 (MH); Near Savaraguda stream (SKLM), *GVS* 62446 (MH).

INDIA: Throughout the hotter parts of India.

WORLD: Tropical Africa, W. Asia, China, Japan, Malesia, Australia.

Persicaria barbata (L.) H. Hara var. **gracilis** (Danser) H. Hara, Enum. Fl. Pl. Nepal 175. 1982. *Polygonum barbatum* L. var. *gracile* (Danser) Steward, Contr. Gray Herb. 88: 55. 1930. *P. barbatum* L. subsp. *gracile* Danser, Bull. Jard. Bot. Buitenzorg, ser, 3, 8: 146 f. 2. 1927. *P. serrulatum sensu* Hook. f., Fl. Brit. India 5: 38. 1886. *p.p.* non Lagasca 1816.

Herb with woody base, rooting at nodes below. Leaves alternate, sub-sessile, drying green, linear to linear-lanceolate, 7-8 × 1-2.5 cm, acute to acuminate, glabrous or sparsely adpressed-hairy on midrib and margins. Flowers in paniculate dense racemes. Nutlet trigonous, 1.5 mm, smooth and shiny.

Occasional along streams. Fl. & Fr.: Throughout the year.

Gundlabrahmeshwaram (KNL), *BSS* & *SKB* 34132 (BSID); Perayyakunta (PKM), *BSS* & *SKB* 32671; Bairani falls, Diguvametta (KNL), *JLE* 32464 (MH); way to Diguvametta (KNL), *JLE* 22159 (MH); Peddayedlagadi (KSN), *P.Venkanna*

116527 (BSID); Maredumilli (EG), *MM* 100816 (BSID); Araku valley (VSKP), *NPBK* 10727 (MH), *GVS* 19708 (MH); S. Kota (VSKP), *GVS* 32788 (MH); Donkarayi, near Mangapadu (VSKP), *GVS* 29661 (MH); Between Gummada & Seethampeta (SKLM), *GVS* 62446 (MH).

INDIA: Andhra Pradesh, Tamil Nadu, Maharashtra,

WORLD: Pakistan, Taiwan

Persicaria capitata (Buch.-Ham. ex D.Don) H. Gross, Bot. Jahrb. Syst. 49: 277. 1913. *Polygonum capitatum* Buch.-Ham. ex D.Don, Prod. Fl. Nep. 73. 1825; FBI 5: 44. 1886; Haines, Bot. Bihar & Odisha 2: 780. 1924.

Perennial herb; stems 15-20 cm long, many from a short woody rootstock. Leaves ovate or broadly elliptic, 1.8-4.5 cm long, acute or obtuse, margin closely ciliate, base suddenly contracted to a very short petiole, with 2 auricles at its base with ciliate margins; stipules 5-7.5 mm, cupular. Inflorescence capitate or subcapitate; heads pink, 1-3 terminal, 3.7-5 mm diam. or sometimes ovoid, 7.5 mm long, without involucral leaf.

Pottangi, Koraput (Dubey and Panigrahi, Proc. 7th Ann. Conf. Odisha Bot. Soc. 18. 1982).

INDIA: Assam, Manipur, Nagaland, Bihar, Odisha.

WORLD: Bhutan, China, Nepal.

Persicaria chinensis (L.) H. Gross, Bot. Jahrb. Syst. 49(2): 269, in obs 269. 1913. *Polygonum chinense* L., Sp. Pl. 363. 1753; FBI 5: 44. 1886, incl. var. *ovalifolia* Meissn; Gamble 2: 1190. 1925.

Straggling shrub, up to 1.5 m high; stems glandular hairy, striate. Ochreae 4 cm, membranous, obliquely cleft, abruptly acuminate, extending to the base of the higher node. Leaves broadly elliptic-oblong, 5-15 × 3-7 cm, glabrous above, scabrous along nerves beneath, base rounded, margin entire, apex shortly acuminate; petiole 2 cm. Flowers in terminal corymbose panicle; peduncles up to 0.15 cm long, slender, glandular-hairy; pedicels short, articulated below the perianth; bracts up to 0.5 cm long, glabrous, bracteoles up to 0.35 cm long; perianth up to 0.3 cm long, segments 4, obovate, rounded; stamens 8, adnate to the base of perianth-segments, filaments slender; ovary glabrous, styles united below, stigma capitate. Nutlet trigonous, 1.5 × 1 mm, smooth, testa shiny.

Common in dry rocky beds near streams. Fl. & Fr.: September – January. Vern.: Tam.: *Pulichankodi.*

Anantagiri (VSKP), *NVP* 6619; Minumuluru (VSKP), *GVS* 28273 (MH); forest near Sunkarimetta (VSKP), *NPBK* 10897 (MH); Adativalasa (VSKP), *GVS* 21620 (MH); Gudem (VSKP), *SKW* 2061 (BLAT); Rampa (EG), *GNS* 475 (CAL); Mahendragiri (GJM), *SX* & *MB* 2217, 3566, 3995 (RRL-B); Similipahr (MBJ), *SX* & *MB* 4119, 5582 (RRL-B); Sirumalai hills (DGL), *SKS* 783 (SGH).

INDIA: Throughout India.

WORLD: China, Japan, Malesia.

Persicaria decipiens (R.Br.) K.L.Wilson, Telopea 3: 178. 1988. *Polygonum decipiens* Prodr. Fl. Nov. Holl. 420.1810. *P. serrulatum* Lag., Gen. Sp. Nov. 14. 1816; FBI 5: 38. 1886.

Decumbent herb, 20-50 cm high. Leaves narrowly linear to linear-oblong, 5-12.5 × 0.25-1.5 cm, glabrous or slightly strigillose on margins and nerves; stipules somewhat strigose. Flowers white, in slender erect racemes; bracts bristle ciliate; stamens 5-8; styles 3. Nut deep purple, triquetrous, polished.

In wet places in hilly areas in Eastern Ghats in Odisha. Fl. & Fr.: September – January.

Lanjigarh (KHD), *HFM* 1742 (DD).

INDIA: Odisha.

WORLD: Europe, temporate and tropical Asia south world to Australia. Induced in N. America.

Persicaria dichotoma (Blume) Masam., Sci. Rep. Kanazawa Univ. 2(2): 82. 1954. *Polygonum dichotomum* Blume, Bijdr. 529. 1826. *P. pedunculare* Wall. ex Meisn. in Wall. Pl. Asiat. Rar. 3: 58. 1832; FBI 5: 48. 1886; Gamble 2: 1191. 1925. *Persicaria peduncularis* (Wall. ex Meissn.) Nemoto, Fl. Jap. Suppl. 174. 1936.

Erect herbs, stems striate, minutely prickly. Ochrea glabrous, up to 2.5 cm long, mouth shortly ciliate, cilia up to 0.1 cm long. Leaves alternate, up to 11 × 0.9 cm, narrow lanceolate, base narrowed, truncate or hastate, margin entire, apex acuminate; petiole up to 0.6 cm long, minutely prickly. Flowers in terminal heads; peduncle up to 3.1 cm long, slender, glandular hairy; bracts up to 0.35 cm long, sparsely hairy on the lower surface, ciliate apex; pedicels up to 0.15 cm long, articulate below the perianth; perianth white or pink, up to 0.3 cm long, segments 4, divided halfway down, unequal, ovate-oblong, apex rounded; stamens 5, filaments slender, anthers oblong with a large gland at back; ovary ovoid, biconvex, styles 2, united in the lower half, stigmas capitate. Nuts up to 0.2 cm long, yellowish-brown, ovoid.

Common in sandy wet places. Fl. & Fr.: May - December.

Upper Barakamada, Similipahar (MBJ), *SX* & *MB* 4189 (RRL-B).

INDIA: Assam, Karnataka, Madhya Pradesh, Maharashtra, Meghalaya, Tamil Nadu. Odisha.

WORLD: Sri Lanka, east to China, Malesia, Australia.

Persicaria glabra (Willd.) M.Gomez, Anales Inst. Segunda Ensen 2: 278. 1896. *Polygonum glabrum* Willd., Sp. Pl. 2: 447. 1799. var. *glabrum*: FBI 5: 34. 1886; Gamble 2: 1189. 1925.

Sub-shrub, up to 1 m high, stem terete, ribbed, dark chocolate brown on drying. Ochreae tubular, 2 cm, glabrous. Leaves alternate, narrowly lanceolate, 3.5-18 × 1-3 cm, base cuneate, margin entire, apex acuminate, minutely gland-dotted, glabrous; petiole up to 1.7 cm long, canaliculate. Flowers bisexual, in terminal lax spikes or from upper axils; bract 1, hyaline, subtending 2 flowers; bracteole hyaline, 2-lobed, subtending sessile floret; perianth lobes 5, hyaline, boat-shaped; anthers 7-8, dithecous; ovary compressed; style 2-fid, 2 mm, stigma capitate. Nutlet compressed, 1.5 × 1 mm, faces flattended, black, shining.

Common near the streams and tanks or in water logged places in most districts. Fl. & Fr.: Throughout the year. Vern.: Tel.: *Buradagogu.*

Penakacherla (ATP), *BR & DV* 43435; Rangapuram RF (KNL), *RVR* 1428; Japali – Tirumala (CTR), *KH* 13940; Ethipothala falls (GNT), *VRK* 3581; Mulyaguda to Borra caves (VSKP), *TP & EC* 7417; Narsapally (ATP), *JSG* 15267 (MH); Gurram konda (CDP), *JSG* 21339 (MH); Tirumala hills (CTR), *SKW* 8323 (BLAT), *GVS* 31943 (MH), *DRC* 969 (MH); Somasila dam (NLR), *PV* 110202 (BSID); Upputeru – Mandapadu (KSN), *P. Venkanna* 116553 (BSID); Rampa (EG), *VNS* 101 (CAL); Rampachodavaram (EG), *GVS* 24460 (MH); Araku (VSKP), *GVS* 19729 (MH); Nattavaram (VSKP), *CAB* 1900 (MH); Harishankar (BGR), *SX & MB* 6191 (RRL-B); Bhawanipatna (KHD), *MB & Dhal* s.n. (RRL-B); Sirumalai hills (DGL), *SKS* 317 (SGH).

INDIA: Assam, West Bengal westwards to Indus, Himalayas, Odisha, Andhra Pradesh.

WORLD: Pantropical.

Persicaria hydropiper (L.) Delabre, Fl. Auvergne ed. 2: 518. 1800. *Polygonum hydropiper* L., Sp. Pl. 361. 1753 var. *flaccidum* Steward, Contr. Gray. Herb. 88: 59 1930. *P. flaccidum* auct. non Roxb. 1832; FBI 5: 39. 1886 incl. var. *hispida;* Gamble 2: 1190. 1925.

Erect herb, up to 70 cm high, glabrous, often glandular, nodes often swollen. Ochreae tubular, strigosely hispid, tube up to 1.3 cm long, cilia of the mouth up to 0.8 cm long. Leaves oblong-lanceolate, 9-10.5 × 2-5 cm, subsessile, base cuneate, margin entire, apex acuminate, glabrous above, scabrid beneath; petiole 1.1 cm long, canaliculate. Flowers bisexual, in terminal, slender racemes, interrupted at base; bracts up to 0.3 cm long, sparsely hairy without, gland-dotted, cilia of the mouth up to 0.15 cm long; pedicels up to 0.3 cm long, slender, articulated a little below the perianth; perianth up to 0.3 cm long, gland-dotted, white or white with pink tinge, segments 4, ovate; stamens 8, filaments slender, anther oblong; ovary trigonous, glabrous, styles 3, connate below, stigmas capitate. Nutlet trigonous, brown, granulate.

Common in moist and water logged places in north coastal districts. Fl. & Fr.: August – March.

Mulyaguda to Borra caves (VSKP), *EC* 7418; Rampa (EG), *VNS* 355 (CAL); Maredumilli (EG), *MM* 100798 (BSID); way to Peddalova, Tadepalli (EG), *NRR & DN* 84339 (BSID); Paderu (VSKP), *SKW* 4275 (BLAT); Sunkarametta (VSKP), *NPBK* 10948 (MH); Anantagiri (VSKP), *GVS* 19480, 21775 (MH); Gudem camp (VSKP), *VNS* 528 (CAL); Madugula (VSKP), *GVS* 47274 (MH); near Addatigala (VSKP), *GVS* 68596 (MH); Dohu ghat (GJM), *JSG* 13661; Tickkapalli to Linepada (GJM), *CAB* 1186 (MH).

INDIA: Plains and hills of India.

WORLD: Tropics, Western Europe.

Persicaria lapathifolia (L.) Delarbre, Fl. Auvergne ed. 2: 519. 1800. *Polygonum lapathifolium* L., Sp. Pl. 360. 1753; Mooney, Suppl. Bot. Bihar & Odisha 130. 1950.

Annual, erect herb. Leaves elliptic-ovate to lanceolate, glandular beneath, margins ciliate, papillose, strigose or hispidulous on the nerves beneath and margins, stipules more or less ciliate at mouth. Flowers in lax, panicled racemes; peduncle papillose; pedicels usually glandular; bracts ciliate at mouth; perianth greenish, glandular, strongly nerved; stamens usually 6; styles mostly 2. Nut biconvex.

In Sal forests near Toda, Bonai, by the side of a perennial stream under shade (Mooney, *loc. cit.*).

INDIA: From Bengal to Sutlej, Himalaya, Concan.

WORLD: South America, Europe, Asia, Africa, Pacific Islands.

Persicaria minor (Huds.) Opiz, Seznam 72. 1852. *Polygonum minus* Huds., Fl. Angl. 148. 1762; FBI 5: 36. 1886; Gamble 2: 1191. 1925.

Slender, ascending herb, 15-25 cm long. Leaves linear, 1.8-3.7 cm long, tapering each end, stipules sparsely or densely strigosely bristly. Flowers white or pinkish, in slender, erect racemes; bracts 1.5-2.5 mm; perianth 1.5-1.8 mm in flower, slightly larger in fruit; stamens 5, included; style branches 2, stigma papillose, capitellate. Nut broadly ellipsoid, biconvex, 1.5 mm, polished brown.

Occasional near water-courses in Northern Eastern Ghats of Odisha (Panigrahi *et al.*, 1964). Fl.: July.

INDIA; Throughout hotter parts of India.

WORLD: Europe, Asia.

Persicaria nepalensis (Meisn.) Miyabe, J. Fac. Agric. Hokkaido Imp. Univ. 26: 514. 1934. *Polygonum nepalense* Meisn., Monogr. Polygon. 84. t. 7 f. 2. 1827. *Persicaria nepalensis* (Meisn.) H.Gross, Bot. Jahrb. Syst. 49(2): 277. 1913. *Polygonum alatum* Buch. - Ham. ex Spreng., Syst. Cur.Post. 4(2): 154. 1827; FBI 5:41. 1886 incl. var. *metziamum. P. punctatum* Buch.-Ham. ex D. Don, Prodr. Fl. Nepal 72. 1825, non Rafin. 1820; Gamble 2: 1190. 1925.

Spreading herb, stems glabrous. Ochreae tubular, membranous, truncate, obliquely cleft, hairy at base. Leaves ovate-rhomboid, 2-3.5 × 1-2.5 cm, chartaceous, base truncate or acute, margin entire, apex acute, lateral nerves 4-6 pairs, faint, glabrous; petiole up to 1.5 cm long, winged with 2 amplexicul basal auricles. Flowers in capitate clusters; peduncles up to 3 cm long, slender, gland-tipped hairy, subtended by a sessile, ovate-lanceolate, amplexicaul leaf; bracts up to 0.4 cm long, broadly ovate, glabrous; pedicels up to 0.5 mm long, slender; perianth pink, divided halfway down, segments 4, ovate, rounded; stamens 4-5, filaments slender; ovary trigonous, glabrous, styles 3, free only at apex, stgmas capitate. Nutlet compressed, circular, 2.5 mm across, faces convex to flat, minutely pitted in lines.

Weed in cultivated fields and along drainages in northern Eastern Ghats. Fl. & Fr.: November – December.

Chintapalli (VSKP), *GVS* 28248 (MH); Way to Mamandur (CTR), *KS* 6919 (MH); Mahendragiri (GJM), *SX* & *MB* 3661 (RRL-B), *VNS* 5578 (MH). *Fischer* & *Gage* Acc. No. 374220 (CAL); Melpat (SA), *CAB s.n.* (MH).

INDIA: Hilly districts throughout India.

WORLD: Tropical Africa, Afghanistan, China, Japan, Malesia.

Persicaria orientalis (L.) Spach, Hist. Nat. Veg. 10: 537. 1841. *Polygonum orientale* L., Sp.Pl. 362. 1753.

Subshrubby, to 75 cm high, branchlets densely sericeous pubescent. Leaves elliptic-ovate, 9-15 x 4-7 cm, ochreae tubular, 1.5-2 cm. Flowers white, in terminal and axillary, 10 cm long racemes; perianth 4 mm across; style-arms 3, stigma capitellate. Nutlets compressed.

Kolli hills, 1300 m, river bed, introduced (Matthew, 1991). Fl. & Fr.: Febraury.

INDIA: Assam, Upper Gangetic Plain, West Bengal, Himayala.

WORLD: Myanmar, East Asia, Malesia.

Persicaria perfoliata (L.) H. Gross, Beih. Bot. Centralbl. 37(2): 113. 1919; Prakasa Rao & Satish, Tropical Plant Res. 3(2): 249. 2016. *Polygonum perfoliatum* L., Syst. Nat. ed. 102: 10005. 1759.

Climbing annual vine, usually grows up to 6–15 m; stems scandent, slightly ribbed, glabrous, and glaucous; prickles cover both stem and petioles. Prickles 0.4–1–2 mm long, slightly recurved. Ochrea, round, glaucous, perfoliate, present at both base of the each petiole and inflorescence. Leaves alternate, petiole 3.5–7 cm, glaucous, ochreae green, foliaceous, peltate, not prickled, margins entire, eciliate, surface glabrous; leaf shape triangular, base truncate, apex acute. Inflorescence spike, 5–12 × 5–10 mm; peduncle 10–50 mm, retrorsely prickly; ochrea overlapping, margins eciliate. Pedicels ascending, 1–3 mm. Flowers 1–4 per ochreate fascicle end of the spike has 4 to 11 flowers; perianth creamy greenish white, glabrous, accrescent, becoming fleshy and pale blue in fruit; tepals 5, broadly elliptic, 2–3.5 mm long, apex acute to obtuse; stamens (6–)8, filaments distinct, free; anthers ovate; styles 3, connate proximally. Seeds shiny black, spherical, peanut sized.

Along road sides on moist soils in hills of Visakhapatnam district. Fl. & Fr.: Mid-April to late July.

Kantaram village, Paderu hills (VSKP), *J.Prakasa Rao* 20490 (AUH).

INDIA: Andhra Pradesh, Odisha, Sikkim, Assam.

WORLD: North Amercia.

Persicaria pulchra (Blume) Sojak, Preslia 46: 154. 1974. *Polygonum pulchrum* Blume, Bijdr. Fl. 530. 1826; R.A. Graham in Fl. Trop. East Africa (Polygonaceae) 20. 1958. *P. tomentosum* Willd., Sp. Pl. 2: 477. 1799. non Scharm 1789; FBI 5: 30. 1886; Gamble 2: 1189. 1925.

Erect herb, up to 1.5 m high. Ochreae tubular, 2-3 cm, margin truncate, chartaceous, softly ciliate. Leaves lanceolate or linearly obovate, 10-15 × 2-4 cm, silky tomentose, base cuneate, margin entire, apex gradually acuminate; petiole 2 cm. Flowers pinkish, in terminal panicles; perianth 4 mm across, lobes 5, equal, ovate, 3.5 mm, shortly connate at base; stamens 5, filaments 2 mm; style 1 mm, stigma capitellate. Nutlet biconvex, 3 mm across.

Occasional in wet places and ditches. Fl.: December-February. Fr.: Most part of the year.

Mamundur (CTR), *KS* 6919 (MH); Pedayedlagadi (KSN), *PV* 116527 (BSID); Gajam (GJM), *SX* & *MB* 4062 (RRL-B); Parlakhemudi (GJM), *MB* 2954 (RRL-B); Mathpad (KPT), *SX* & *MB* 6702.

INDIA: Assam, West Bengal, Odisha, South India.

WORLD: Tropical and South Africa, Sri Lanka, Malesia.

Persicaria stagnina (Buch.-Ham. ex Meisn.) Qaiser, Fl. Pakistan 205: 44. 2001. *Polygonum barbatum* L. var. *stagninum* (Buch.-Ham. ex Meisn.) Steward, Contr. Gray Herb. 88: 54. 1930; Saxena & Brahmam, Fl. Orissa 3: 1533. 1995. *P. stagninum* Buch-Ham. ex Meisn. in Wall., Pl. Asiat. Rar. 3: 56. 1832; FBI 5: 37. 1886.

Erect or spreading herb, 60-90 cm high, branches strigose in the upper part. Leaves linear-oblong, 7-18 × 0.8-3 cm, strigose tomentose more on the lower surface, less on the upper surface, base acute-acuminate. Flowers white or pinkish in racemes; pedicels 2-4 mm long, ochreae 2-5 mm; tepals 5, white, oblanceolate-obovate; stamens 8, disc glands prominent, styles 3, connate at base, stigmas 3, capitate. Nut trigonous, black, shining.

Rare along streams and rivers in north coastal districts (Saxena and Brahmam, loc cit.). Fl. & Fr.: Most part of the year.

INDIA: Throughout the hotter parts of India.

WORLD: Myanmar.

Persicaria strigosa (R.Br.) Nakai, Rigakkwai 24: 299. 1926. *Polygonum strigosum* R.Br., Prodr. 420. 1810; FBI 5: 47. 1886; Gamble 2: 1190. 1925.

Sub erect herb, up to 50 cm high. Leaves linear or linear-oblong, 4-6.5 × 1.5-3.2 cm, hairy on veins and margins; petiole 8-25 mm long, base truncate or slightly cordate, margin retrorsely prickled, apex acuminate or acute. Ochrea with scattered antrorse hairs. Flowers in terminal, subglobose clusters; tepals 3-3.7 mm long, pink or white. Achene slightly trigonous to globose, included in the perianth, black.

Rare in wet places in Northern Eastern Ghats. Fl. & Fr.: November – December.

Donkarayi near Magapadu (VSKP), *GVS* 29670 (MH); Araku (VSKP). *GVS* 19736 (MH); Jorandam, Similipahar (MBJ), *SX* & *MB* 5228 (RRL-B).

INDIA: Tropical Himalayas, Khasia hills, Bihar, Tamil Nadu.

WORLD: Sri Lanka, Bangladesh, Malay Archipelago, China.

POLYGONUM Linnaeus *nom. cons.*

Polygonum plebeium R. Br., Prodr. Fl. Nov. Holl. 420. 1810 var. **plebeium;** FBI 5: 27. 1886. incl. var. *indica;* Gamble 2: 1188. 1925. *P. indicum* Heyne ex Roth, Nov. Pl. Sp. 208. 1821.

Annual, prostrate or decumbent, much branched herbs, 5-30 cm long. Ochreae tubular, 2 mm, hyaline, lacerate. Leaves alternate, oblong, obovate or linear, 2-5 × 1-2 mm, subsessile, base attenuate, margin entire, apex acute to obtuse. Flowers bisexual, pink, in axillary fascicles, pedicel to 1 mm; perianth lobes 5, 1-2 mm, lobes hooded at apex, stamens 8; ovary glabrous, styles 3. Nutlet strongly trigonous, 2 × 1 mm with persistent style.

Common weed in harvested rice fields and dried stream beds. Fl. & Fr.: Throughout the year. Vern.: Tel.: *Chanti saga, Muthi saga;* Ori.: *Chanti saga, Muthi saga.*

M.V.Palli (ATP), *BR* & *ANS* 38161; Kammavaripalli (KNL), *TP & GO* 2800; Horsley hills (CTR), *RVR* 14963; Bodhi (PKM), *BSS* & *SKB* 34382; Vinukonda RF (GNT), *VRK* 6873; Chelama (KNL), *JLE* 23735 (MH), *JLE* 16807 (MH); Dongalacheruvu (CDP), *KS* 6321 (MH); Nerabylu (CTR), *GVS* 46027 (MH); Nandavaram (NLR), *MSR* 1169 (MH); Kondapalle RF (KSN), *PV* 5583 (MH); Koruturu (WG), *DN* 84300 (BSID); Polavaram (WG), *DCSR* 47 (CAL); Devipatnam (EG), *GVS* 24433 (MH); Madugula (VSKP), *GVS* 47282 (MH); Way to Punyagiri hill (VSKP), *GVS* 19394 (MH); Nattavaram (VSKP), *CAB* 1916 (MH); Shevaroy hills (SM), *JSG* 381 (CAL); Mettappanur hill (NA), *MBV* 651 (CAL); Podagriha, Similipal (MBJ), *Bahali & Agrawala* 537 (CAL); Tickapalli to Linepada (GJM), *CAB* 1220 (MH); Koijam to Kusadandi (GJM), *CAB* 1272 (MH); Aska (GJM), *JSG* 14174 (MH); Nilavur-Yelagiri hills (NA), *MBV* 1374 (MH); Mangalam RF – Yelagiri hills (NA), *MBV* 651 (MH).

INDIA: Throughout India.

WORLD: Africa, West Asia east to China and Ryukyu, Malesia, Australia.

RUMEX Linnaeus

1. Annual herbs:
    2. Leaves linear-oblong; inflorescence branch rigid .... **R. nigricans**
    2. Leaves elliptic; inflorescence branch not rigid .... **R. vesicarius**
1. Perennial herbs:
    3. Stem branched above; leaf margin long dentate .... **R. dentatus**
    3. Stem branched from base; leaf margin denticulate .... **R. nepalensis**

Rumex dentatus L., Mant. Pl. 2: 226. 1771; FBI 5: 59. 1886; Haines, Bot. Bihar & Odisha 2: 782. 1924.

Erect herb, 60-90 cm high, stems angled and ridged. Leaves oblong, 5-10 cm long, upper ones oblanceolate or linear, base rounded or cordate, apex obtuse. Flowers bisexual, in racemes; fruiting tepals triangular, 4-5 mm long, with broad, closely reticulate wings and strong irregular spreading teeth; style terminal.

Banks of Mahanadi, Angul (Haines *loc. cit.*). Fl. & Fr.: April - June.

INDIA: Assam, West Bengal, Odisha, Western Peninsula, Himalayas.

WORLD: Afghanistan, India, Kazakhstan, Kyrgyzstan, Nepal, Russia; N Africa, SE Europe.

Rumex nepalensis Spreng., Syst. Veg. 2: 159. 1825; Matthew, Flora of Central Tamil Nadu 427. 1991.

Stout, erect herb, to 1 m high. Basal leaves: petiole 4-10 cm; leaves broadly ovate, 10-15 × 4-8 cm, both surfaces glabrous or abaxially minutely papillate along veins, base cordate, margin entire, apex acute; cauline leaves shortly petiolate, ovate-lanceolate; ochrea fugacious, membranous. Flowers bisexual, in panicles; outer tepals elliptic, ca. 1.5 mm; inner tepals enlarged in fruit; valves broadly ovate, base truncate, each margin with 7 or 8 teeth, apex acute; apex hooked or straight. Achenes brown, shiny, ovoid, sharply trigonous, ca. 3 mm, base truncate, apex acute.

Servarayns (Yercaud, Kaka shola) in marshy places. Fl. & Fr.: February –May.

Yercaud (SM), *SKK* 26878 (MH).

INDIA: Himalayas.

WORLD: S.W. Europe, W. Asia, Himalaya, India, east to W. and C. China

Rumex nigricans Hook. f., Fl. Brit. India 5: 59. 1886; Gamble 2: 1192. 1925.

Erect, slender, annual herb, 20-60 cm high; stems angled and grooved. Leaves linear-oblong, 5-20 × 2-5 cm, contracted above cordate base, sometimes lower ones linear-oblong, thin, apex obtuse. Flowers in clusters along terminal leafy panicles; pedicels short, 0.12-0.15 cm long, recurved, jointed below the middle; perianth green; tepals 0.3-0.45 cm long, lanceolate, venose, turning black when dry, with very large oblong, dark tubercles and lanceolate tips; margins very narrow, entire or with 1-2 short, straight spines. Nuts acutely trigonous, smooth, enclosed in accrescent perianth.

Rare in Northern Eastern Ghats. Fl. & Fr.: July - December.

Polavaram (WG), *DCSR* 41 (CAL); Devipatnam (WG), *JSG* 15883 (MH).

INDIA: Meghalaya, Uttarakand, Bihar, Odisha, Andhra Pradesh.

WORLD: Bangladesh, China, Myanmar

Rumex vesicarius L., Sp. Pl. 1: 336. 1753; FBI 5:61. 1886; Gamble 2: 1192. 1925.

Annual, erect, succulent herb, up to 50 cm high, stem fleshy, stipules membranous. Leaves alternate, elliptic, 2-7 × 0.5-2 cm, base cordate to hastate, margin crisped, apex obtuse to acute, succulent, glabrous, dotted, 3-5-nerved from the base; petiole 4 cm. Flowers pink, in axillary and terminal panicles, arranged in clusters, polygamous; pedicels slender; inner fruiting tepals large, not tubercled, orbicular, very membranous and reticulate, white or pink, 2-lobed at each end, without teeth or marginal nerve; styles on the angles of the ovary. Seed ovoid, 1 × 0.75 mm, black.

Culivated in gardens as leafy vegetable. Fl. & Fr.: July – December. Vern.: Tel.: *Chukkakura*.

Penakacherla (ATP), *NY* 595; Kalyanadurgam (ATP), *BR* & *DV* 45328; Nellore (NLR), *SS* 16333.

INDIA: Cultivated and escape in many parts of India.

WORLD: Pakistan, Afghanistan, Iran, North Africa.

PODOSTEMACEAE

1. Tepals 3, free or united; spathe absent .... **Indostricha**
1. Tepals 0, the young buds covered or encircled by a spathe;
    2. Capsule smooth:
        3. Flowers with tepals .... **Hydrobryopsis**
        3. Flowers without tepals .... **Cladopus**
    2. Capsule ribbed:
        4. Plants not thalloid; leaves ribbon like; flowers solitary............... **Dalzellia**
        4. Plants thalloid, ribbon like; flowers numerous, pin headed, in spathe:
            5. Capsule oblique, anisolobus .... **Zeylanidium**
            5. Capsule not oblique, isolobus .... **Polypleurum**

CLADOPUS H. Möller

Cladopus hookeriana (Tul.) C.Cusset in Fl. Camb., Laos and Vietn., 14: 71. 1973. *Mniopsis hookeriana* Tul., Arch. Mus. Hist. Nat. 6: 147. 1852. *Griffithella hookeriana* (Tul.) Warm., Kongel. Danske Vidensk. Selsk. Skr., Naturvidensk. Math. Afd. VI 11: 13. 1901; Gamble 2: 1197. 1925. *Podostemon hookerianus* Wedd. in DC., Prodr. 17: 74. 1873; FBI 5: 65. 1886.

A minute moss like plant, 2.5-4 cm high, thallus variable, attached and closely creeping, secondary shoots marginal or on central parts, small with included axis on vegetative condition, ultimately slightly exserted. Leaves 0.5 cm long. Floral buds usually crowded, bracts usually 4-6, spathe tubular or funnel-shaped. Flowers zygomorphic, perianth absent; stamens 2, monadelphous, staminodes 2; ovary very oblique, stigmas 2, simple with small papillae. Capsule smooth, globose, variable in size.

Rare in running stream in Pamuleru near Maredumilli in East Godavari district (Krishna Prasad, 1999). Fl. & Fr.: October – February.

Maredumilli (EG), *KNReddy* & *CSR* 2431 (KU).

INDIA: South India. Endemic.

DALZELLIA Wight

Dalzellia ceylanica (Gardner) Wight, Icon. Pl. Ind. Orient. 5(2): 34. 1852; Nagendran & Arekal, Bull. Bot. Surv. India 23: 229. 1981. *Tristicha zeylanica* Gardner, Calc.

J. Nat. Hist. 7: 177. 1846. *Terniola zeylanica* (Gardner) Tul., Arch. Mus. Hist. Nat. Paris 6: 190-192. t. 13. f. 3. 1852; FBI 5: 62. 1886.

Perennial, aquatic herbs, roots attached to rocks. Leaves linear, 2-5 mm long, acute, entire. Flowers solitary; pedicels emerging from terminal, axial leaf cupules, 2-5 mm long; perianth 3-lobed, ca 1 mm long; anthers exserted, filaments very slender, flattened; stigma subulate. Capsules obovoid, ca 1 mm long, 9-ribbed; seeds small, ca 0.5 mm long.

Rare, occurs on smooth rocky substratum of running water in streams and rivers in Northern Eastern Ghats. Fl. & Fr.: August - February.

Near Laxmipuram, Paderu agency (VSKP), *NRR* & *DN* 84203 (MH).

INDIA: Maharashtra, Karnataka, Andhra Pradesh, Odisha.

WORLD: Sri Lanka.

HYDROBRYOPSIS Engler

Hydrobryopsis sessilis (Willis) Engl. in Engl. & Prantl, Nat. Pflanzenfam. 18a: 60. 1930; Nagendran & Arekal, Bull. Bot. Surv. India 23: 232. 1981. *Hydrobryum sessile* Willis, Ann. Roy. Bot. Gard. (Peradenia) 1: 239. 1902.

Thalloid, aquatic herbs; thallus crustaceous, thin, appressed to the rock, less than 1 cm wide and profusely, alternately branched; secondary shoots marginal. Leaves 4-6, linear-subulate. Flowers sessile, subtended by 2 pairs of thick, keeled, subulate bracts, 1-3 mm long. Spathe split along dorsal side; stamens 2, monadelphous, staminodes linear; ovary included in the bracts, styles 2. Capsule smooth, globose, ca 1 mm across, seeds many.

Rare, occurs in moist rocky places in higher elevations of Visakhapatnam district. Fl. & Fr.: October – January.

It is included on the authority of Subramanyam and Sreemadhavan (1969).

INDIA: Andhra Pradesh, Goa, Karnataka, Kerala, Maharashtra. Endemic.

INDOTRISTICHA P. Royen

Indotristicha ramosissima (Wight) P. Royen, Acta Bot. Neerl. 8: 475. 1959. *Dalzellia ramosissima* Wight, Icon. Pl. Ind. Orient. 5: 34. 1852. *Tristicha ramosissima* (Wight) Willis, Ann. Roy. Bot. Gard. (Peradenia) 1(3): 208. 1902; Gamble 2: 1194. 1921. *Terniola ramosissima* (Wight) Wedd. in DC., Prodr. 17: 47. 1873; FBI 5: 63. 1886.

Submerged aquatic herb with long floating, thread like filamentous thallus, creeping stems and roots attached by feet, secondary shoots very many, often more or less paired on opposite sides. Sometimes side branches quite long, many-flowered and frequently branched, very often vegetative only. Branches two kinds, long ones with the structure of main axis and short ones consisting of a delicate axis with very small leaves often tristichous. Flowers terminal, on pedicels subtended by 2-3 large or several ordinary leaves, the upper ones often more or less united, pedicel and ovary deciduous, which shrivels after flowering, the small flowers usually emerging through the water at the beginning of the dry season.

Perennial water bodies of the hills and water stagnant areas in outhern Eastern Ghats. Endemic to Peninsular India. Fl. & Fr.: Usually premonsoon (June-July).

Palakonda hills (CDP), *SKS* 27942; Sirumalai hills (DGL) *SKS* 794 (SGH).

INDIA: Goa, Karnataka, Kerala, Maharashtra, Tamil Nadu, Andhra Pradesh. Endemic.

POLYPLEURUM Warming

Polyplerum filifolium (Ramam. & J. Joseph) A.S.Rao & Hajra, Bull. Bot. Surv. India 21: 195. 1981. *Dicraea filifolia* Ramam. & J. Joseph, Bull. Bot. Surv. India 6: 333. 1964; Sastry & Joseph, Indian Forest. 93: 800. 1967.

Thallus thin, beset with numerous, c. 2 mm long, erect, filiferous secondary shoots, 10-12 mm long. Leaves subulate-filamentd, with sheathing bases. Flowers solitary, subsessile, irregular; stamens 2, united, staminodes 2. Fruit ribbed, with 3-4 mm long stalk.

Occasional in Dabata Parvat near Jeypore in Odisha (Sastry and Joseph, *loc. cit.*).

IUCN Status: Vulnerable B1ab(iii,v)+2ab(iii,v) ver 3.1 (Rehel, 2011).

INDIA: Odisha, Kerala (Endemic).

ZEYLANIDIUM (Tulasne) Engler

1. Roots crustaceous, shoots scattered on upper surface of root .... **Z. olivaceum**

1. Roots ribbon-like, sympodial, shoots at sinuses of distinctly unequal root branches .... **Z. lichenoides**

Zeylanidium lichenoides (Kurz) Engl. in Engl. & Prantl, Nat. Pfamilien 18a: 62. 1930; Rao & Hajra, Bull. Bot. Surv. India 17: 207. 1975. *Hydrobryum lichenoides* Kurz, J. Asiat. Soc. Beng. 42: 103. 1873; Gamble 2: 1199. 1925. *Podostemon microcarpus* Wedd. in DC., Prodr. 17: 76. 1873; FBI 5: 66. 1886.

Small, green, thalloid, ribbon-like, closely attached to rocks in rapid water, green to pinkish green, branching dichotomously, secondary shoots marginal with 3-8 bracts, usually prostrate. Leaves simple, subulate. Flowers numerous, pinhead like, flowering shoots with 2-3 pairs of imbricating bracts, spathe boat-shaped, splitting along the upper side; stamens 2, unequal, ovary subglobose, 2-celled, stigmas 2. Capsule 3-4 mm long, ellipsoid to subglobose, brown. Seeds small, many, ellipsoid, brown.

Rare in hill streams attached to rocks at Pamuleru in Maredumilli in East Godavari district (Krishna Prasad, 1999) and Jeypore in Odisha (Subramanyam, 1962).

Fl. & Fr.: January-February.

IUCN Category: Least concern (Anitha, 2011)

INDIA: Karnataka, Kerala, Maharashtra, Meghalaya, Odisha, Tamil Nadu.

WORLD: Sri Lanka, Myanmar, Thailand.

Zeylanidium olivaceum (Gardner) Engl., Nat. Pflanzenfam. (ed. 2) 18a: 62. 1930. *Podostemon olivaceum* Gardner, Calcutta J. Nat. Hist. 7: 181. 1847. *Hydrobryum olivaceum* (Gardner) Tul. Ann. Sci. Nat. Bot. Ser. 3. 11: 104. 1849; Gamble, 2: 1198. 1925.

Thallus thin, disc-like, crustaceous, green, up to 12 cm in diameter, growing by marginal increase. Leaves many, subulate, all over dorsal surface, clustered, 2-16, sometimes the cluster with a stalk 4-6 cm long. Each leaf 1-6 cm long, caducous at flowering. Flowering shoots many, arising at the region of leaf clusters, 4-5 mm, each with 4-8, boat-shaped, imbricate, ascending large bracts. Flower(s) one per shoot, zygomorphic, bisexual, naked; perianth thin, translucent, oblong, opening on upper side, stamens 2, linear, 3 mm, one on either side of common filament, ovary globose, smooth, 2-4 mm long, shortly stalked, bicarpellary syncarpous, bilocular; placenatation axile; stigmas 2, sessile, either lobed or toothed. Capsules 3-5 mm long, 8-ribbed, anisolobous; pedicels 2-3 mm long; seeds small, dark brown, ovoid, non-endospermous; embryo dicotyledonous completely occupying the thin reticulate testa.

Hills, above 900 m, on the floor of streams on rocks in running water in Central Tamil Nadu (Matthew, 1991). Locally abundant. Fl. & Fr.: November – February.

Kolli hills, Periyasamikoil (NMK), *SKS* 1018 (SGH).

INDIA: Tamil Nadu.

WORLD: Sri Lanka.

ARISTOLOCHIACEAE

ARISTOLOCHIA Linnaeus

1. Cultigens; flowers large; perianth limb up to 7 cm across .... **A. littoralis**
1. Wild plants; flowers small; perianth limb up to 1 cm across:
    2. Prostrate herbs .... **A. bracteolata**
    2. Twining herbs or shrubs:
        3. Leaves linear to oblong, base rounded to truncate .... **A. indica**
        3. Leaves ovate or cordiform, base with deep narrow sinus .... **A. tagala**

Aristolochia bracteolata Lam., Encycl. 1: 258. 1783. *A. bracteata* Retz., Obs. Bot. 5: 29. 1788; FBI 5: 75. 1886; Gamble 2: 1201. 1925.

Perennial, slender, prostrate herb, to 50 cm high; branches angular with ridges, white glaucous. Leaves reniform-cordiform, 3–5 × 4–6 cm, glaucous below, glabrous above, base cordate to broadly sagittate, margin entire to undulate, apex obtuse to sub acute; lateral veins 5 at base, palmately divergent; petiole to 3 cm long. Flowers axillary, solitary, 5 cm long, purple; bract cordate-orbicular, 1 × 1 cm; pedicel to 0.5 cm long, tube 2 cm long, curved, glabrous without, hirsute within, base inflated, 0.5 cm across; lip dark purple, linear, 2 cm long, with revolute margin, apex obtuse; mouth 0.4 cm across; stamens 6; anthers oblong, 0.6 mm long; ovary 8 × 0.3 mm; style 2 mm long; stigmatic lobes 6, connivent, acute. Capsule about 2.5 cm long,

1.5 cm across, glabrous, ellipsoid-oblong, 12-ribbed, septicidally opened; seeds cordate, 4 mm across, partly covered by the remains of the placenta, gland-dotted; testa thin, albumen fleshy.

Common weed of cultivated fields and waste places. Fl. & Fr.: June – December. Vern.: Tel.: *Gadidagadapa;* Ori: *Panari, Panairi, Gopa kanna, Ishwar mula, Oyidi, Bal-bolena;* Tam.: *Aaduthinnapaalai.*

Rangapuram RF, Erramalai (KNL), *RVR* 1502; Palakonda hills (CDP), *CS* 8297; Velpula (CDP), *KRKS* & *KRS* 37673; Lankamalai RF (CDP), *SRSR* & *KL* 13091; Panapakam (CTR), *GVS* 31880 (MH); Raikal, Nagarjuna Nagar (GNT), *MRRP* 512 (MH); Upputeru – Mandapadu (KSN), *PV* 116550 (BSID); Ragupalem (VSKP), *CAB* 1588 (MH);); Kandavanur RF (SA), *KRM* 50628 (MH).

INDIA: West Bangal, Bihar, Odisha, Peninsular India.

WORLD: Sri Lanka, Nepal, Bangladesh.

Aristolochia indica L., Sp. Pl. 960. 1753; FBI 5: 75. 1886; Gamble 2: 1202. 1925.

Twining shrubs, stems long, slender, woody at the base, grooved, glabrous. Leaves variable, 4-7 × 2-4 cm, oblong to obovate – oblong, base rounded, truncate or sub-cordate, margin entire, apex obtusely acuminate, glabrous. Flowers greenish white, tube with globose inflated base, then bent at right angle and suddenly narrowed into a cylindric tube, to 2 cm, with oblique trumpet-shaped mouth, to 4 cm across, gradually passing into a long, narrow, linear-oblong, obtuse brownish lip, lip to 1.5 cm; stamens 6; anthers oblong, 1 mm; ovary 1 × 2.5 cm, glabrous; style 3 mm; stigmatic lobes 6, erect, 1.5 mm, acute. Capsule 4 × 2 cm, oblong or globose – oblong, opening from below upwards, 6-valved; seeds many, flattened, laterally winged.

Occasional in forests. Fl. & Fr.: September – March. Tam.: *Urikaikodi, Thalaichuruli, Kozhikunduchedi.*

Kalasamudram (ATP), *EC* 2920; Gundam, Erramalai Hills (KNL), *TP* & *RVR* 3015; Rachakuntapalle (CDP), *KRKS* & *KRS* 37908; Kailasakona (CTR), *VSRao* & *MVS* 30112; Udayagiri durgam (NLR), *SSY* 16217; Peddapenta (PKM), *RVK* 15872; Vinukonda (GNT), *VRK* 6933; Bhupatipalem -Rampa forest (EG), *TP & KJ* 12509; Balapalli (CDP), *JLE* 15019 (MH); way to upper Ahobilam, Nallamalais (KNL), *JLE* 32574 (MH); Sydapuram – Podalakur (NLR), *PV* 112187 (BSID); Sirumalai hills (DGL), *SKS* 611 (SGH).

INDIA: Throughout India.

WORLD: Sri Lanka, Nepal, Bangladesh.

Aristolochia littoralis Parodi, Anal. Soc. Cient. Argent. 5: 155.1878; Loigier, Rhodora 67: 328. 1965; Raizada, Suppl. Fl. Upper Gang. Pl. 241.1976. *A. elegans* Mast., Gard. Chron. 2: 301.1885; Bailey, Man. Cult. Pl. 346.1949.

Slender, perennial, twiner with numerous striate branches. Leaves broadly ovate to sub-reniform, 5-10 cm across, cordate at base with broad sinus, with undulate margins, apex obtuse, green above, somewhat glaucous beneath; petioles

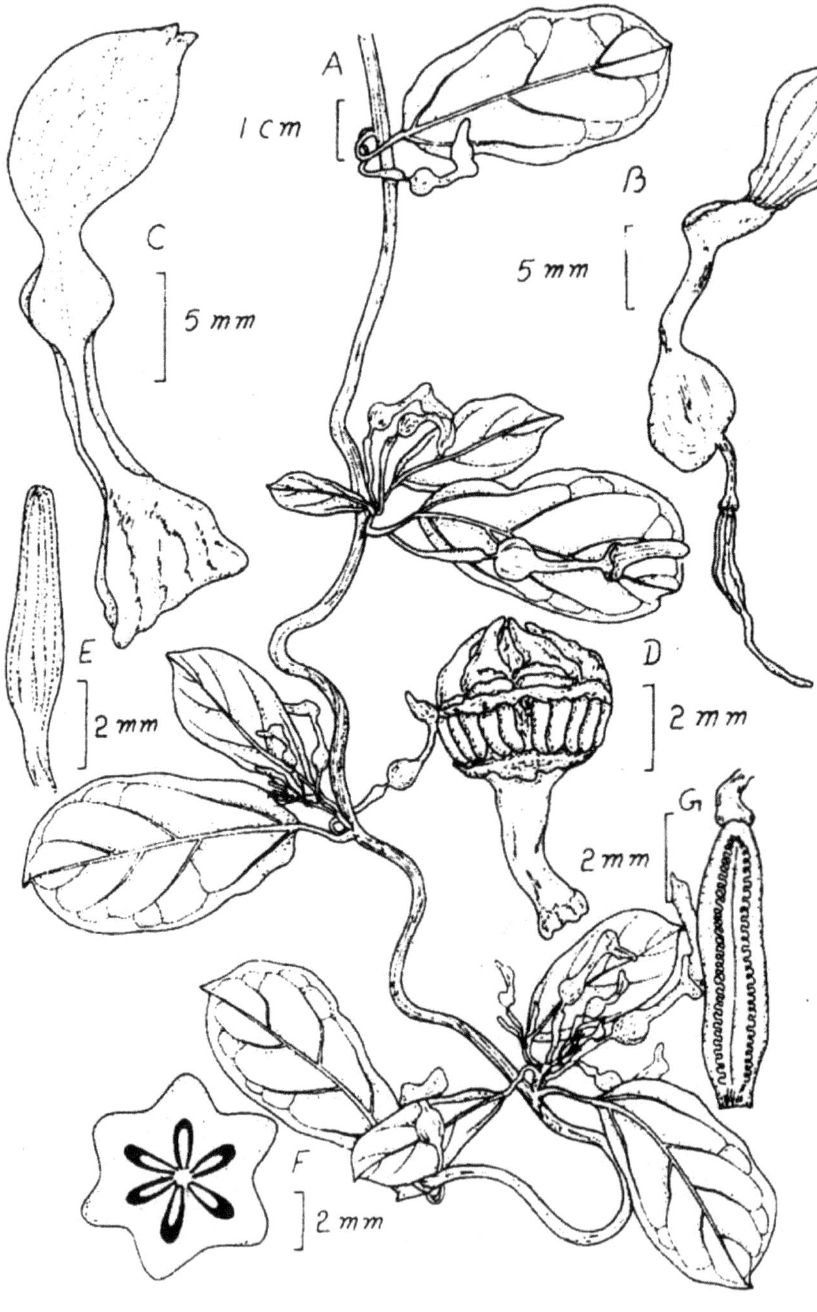

Figure 5. **Aristolochia indica** L.
A. Twig, B. Flower, C. Perianth, D. Stylar column with stamens, E. Anther, F. T.s. of ovary, G. L.S. of ovary.

2-4 cm long; stipules foliaceous. Flowers solitary, axillary; perianth ash-grey or white with inflated base and yellowish-green cylindric tube; limb 6-7 cm across, blotched with violet-brown, orbicular. Capsules ovoid-oblong, 3-4 × 2-2.5 cm, 6-grooved, greenish-brown; seeds ovoid, 2-3 mm long, flat, black.

Cultivated in gardens for its showy flowers. Fl. & Fr.: September– March.

INDIA: Cultivated.

WORLD: Native to Brazil, cultivated in many tropical countries.

Aristolochia tagala Cham., Linnaea 7: 207, t. 5. f. 3. 1832; Gamble 2: 1202. 1925. *A. acuminata* Lam., Encycl. 1: 254. 1783; Bosser, Adansonia ser. 3, 19(1): 170. 1997. *A. roxburghiana* Klotzsch., Monatsfer. Konigl. Preuss. Akad. Wiss. Berlin 596. 1859; FBI 5: 75. 1886.

Climbing shrubs; stem grooved, stout, glabrous. Leaves 7- 19 × 3.5-12 cm, ovate or cordiform, alternate, base deeply cordate, with a deep and narrow sinus, margin entire, apex acute to acuminate, glaucous beneath, glabrous above; petioles up to 7 cm long, striate. Flowers dark purple with greenish-yellow tube, 7 cm long, axillary, peduncles and pedicels slender, long, finely brown-pubescent; bracts up to 0.4 × 0.2 cmn, foliaceous, stalked, deltoid, acuminate, brown-pubescent on both sides, entire, ciliate; perianth up to 6.8 cm long, ventricose portion of the perianth-tube up to 0.8 cm long, cylindrical neck 2 cm long, mouth oblique, limb 4 cm long, margins of mouth and limb recurved; stamens 6, extrorse; stigmatic-lobes 6, forming a cone. Capsule 6 × 3 cm, globose-pyriform or oblong ellipsoid, glabrous; seeds deltoid, flattened

Occasional in hills, above 750 m, by ravines, on thickets, *etc.* in northern Eastern Ghats. Fl. & Fr.: September – December. Vern.: Tam.: *Perumarunthukodi, Eswaramooli.*

Towards Padalammagudi (VSKP), *GVS* 42572 (MH, BSID); Minimuluru (VSKP), *GVS* 30018 (MH); Kolli hills, Kannimarkoil odai (NMK), *SKS* 1038 (SGH).

INDIA: NE, E and Peninsular India, Andaman and Nicobar Islands.

WORLD: Sri Lanka, East Asia, Malesia.

PIPERACEAE

1. Leaves alternate, not punctate, nerves prominent:
    2. Spikes on an axillary peduncle, several together in an umbel-like cluster, rarely solitary .... **Lepianthes**
    2. Spike leaf-opposed, rarely terminal, never axillary, solitary .... **Piper**
1. Leaves opposite, pellucid punctate, nerves obscure .... **Peperomia**

LEPIANTHES Rafinesque

Lepianthes umbellatum (L.) Raf., Sylva. Tellur. 85. 1838; Matthew, Fl. Palni hills 3: 1053. 1999; Kottaimuthu *et al.*, Ethnobotanical Leaflets 12: 301. 2008.

Subshrubs, up to 1 m tall; branchlets subsucculent, lepidote. Leaves alternate, cordate-orbicular, 17-36 x 12-30 cm, palmately 10-15-nerved from base; apex rounded to subacute; petiole 10-20 cm long, sheathing at base. Spikes 1-6, umbellate,

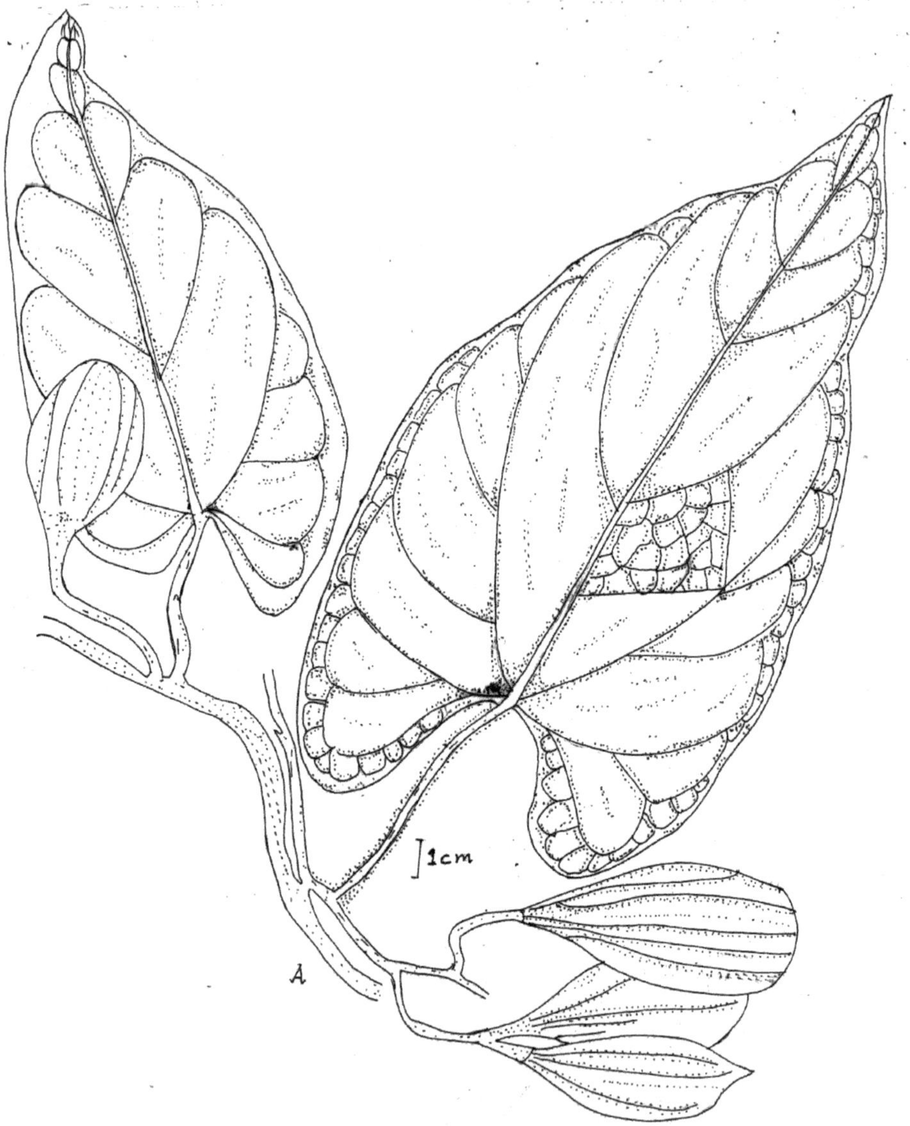

Figure 6. **Aristolochia tagala** Cham. (Syn *Aristolochia acuminata* Lam.)
A. Twig.

on stout axillary peduncle, 5-12 x 0.3 cm; bracts peltate, densely white ciliate at margins; stamens 2; anther 4-lobed; ovary ellipsoid or obovoid, punctuate; stamens 3, caducous.

Fl.: September.

On way to Kannimaruthu odai, Karandamalai (DGL), *R.Kottaimuthu* and *K.Natarajan* 2010 (ATREE).

INDIA:Tamil Nadu, Kerala, Karnataka.

WORLD: Sri Lanka, Africa, South-East Asia.

PEPEROMIA Ruiz & Pavon

1. Leaves coriaceous, exclusively in whorls of 4; spikes pubescent, ribs not distinct.. **P. tetraphylla**
1. Leaves fleshy, alternate or opposite, or upper most in whorls of 3, ribs distinct:
    2. Spikes rather stout, peduncle up to 2.5 cm long.................. **P. portulacoides**
    2. Spikes slender, peduncle to 1.8 cm long:
        3. Leaves and stems crisped-pubescent, leaves obscurely nerved and not very thin and membranous when dry, base rounded to cuneate .... **P. blanda**
        3. Leaves and stems glabrous; leaves clearly nerved and very thin and membranous when dry, base often slightly cordate:
            4. Pellucid glands prominent; fruit glabrous with minutely striate.. **P. pellucida**
            4. Pellucid glands absent; fruit rugose with peaked stigma .... **P. wightiana**

Peperomia blanda (Jacq.) Kunth, Nov. Gen. Sp. 1: 67. 1816. *P. blanda* (Jacq.) Kunth, var. *floribunda* (Miq.) Hüber in Dassan. and Fosberg, Rev. Handb. Fl. Ceylon 6: 294. 1987. *P. dindygulensis* Miq., Syst. Piperac. 122. 1843; FBI 5: 98. 1886 ('*dindigulensis*'); Gamble 2: 1210. 1925; Pullaiah, Fl. Andhra Pradesh 2: 820. 1997.

Erect, rather large, succulent herbs, conspicuous by its pubescence, stems rooting at lower nodes. Leaves 2.5-5 × 1.3 – 1.5 cm, elliptic-ovate, or obovate, decussate, upper ones in whorls, base cuneate, margin entire, apex rounded or sub-acute, chartaceous; petiole up to 1.2 cm long, crispate-hairy, canaliculated. Flowers in spikes 1 or 2, from the upper leaf axils or terminating the branchlets; bracts small, orbicular, attached in the centre by a short stalk; perianth wanting; stamens 2, filaments thick, anther oblong; ovary set in deep socket in the axis, acute. Berry globose, 1 mm across, permanently punctuate.

Occasional in hills, above 900 m in shade; on rotting wood, crevices of wet rocks, or on tree branches, especially along streams. Fl. & Fr.: July – December. Vern.: Tam.: *Musthakkuchedi.*

Tirumala, Papanasanam hills (CTR), *KS* 6959 (MH), *DRC* 2218 (MH); Talakona (CTR), *JHFB* 112940 (BSID); Forest near Sunkarimetta (VSKP), *NPBK* 10905 (MH); Sunkarimetta (VSKP), *GVS* 20667 (MH); Mahendragiri (GJM), *SX* & *MB* 3960 (RRL-B). Sirumalai hills (DGL), *SKS* 626 (SGH).

INDIA: Deccan Peninsula.

WORLD: Widely distributed throughout the Old World Tropics.

Peperomia pellucida (L.) Kunth, Nov. Gen. Sp. 1: 64. 1816; Gamble 2: 1210. 1925. *Piper pellucidum* L., Sp. Pl. 30. 1753

Annual succulent, delicate, glabrous herb, membranous when dry; stems weak, erect, 6-15 cm high. Leaves spiral, rarely opposite, greenish above, whitish beneath, pellucid, broadly ovate deltoid, distinctly 5-nerved, 1-2.5 cm wide and long, base slightly cordate, entire, apex subacute; petiole 0.5-1.5 cm long. Inflorescence leaf-opposed, terminal spike, 2-4.5 cm long, glabrous, peduncle *ca* 5 mm long; stamens 2, filaments very short, anther cells confluent; ovary superior, 1-carpellary with 1 basal orthotropous ovule; stigma penicillate, apical. Fruits 0.8 mm long, indehiscent nutlet, striate, with 10 vertical ridges and minute transverse raised lines across the pericarp; seed minute, testa membranous, albumen floury.

A shade and moist loving weed of gardens, pots and damp walls, particularly during monsoon period, till December (C.S. Reddy and V.S. Raju. 2002). Fl. & Fr.: September – December.

INDIA: Naturalized in many parts of India.

WORLD: S. America, Pantropical Weed.

Peperomia portulacoides (Lamk.) A. Dietr., Sp. Pl. (ed. 6) 1: 172. 1831; FBI 5: 98. 1886; Gamble 2: 1210. 1925. *Piper portulacoiedes* Lam., Ill. 1: 882.1791.

Glabrous, erect, succulent, terrestrial herbs, 15-20 cm high. Leaves oppposite, membranous, upper opposite, lower alternate, smaller, to 3.5 × 2.5 cm, elliptic-ovate or obovate, apex acute, base attenuate, glabrous, 3-nerved from base. Flowers greenish in longer, rather stout, terminal spikes, often in terminal panicles; peduncle 2.5 cm long; bracts orbicular, sessile, margins white. Nutlets globose, embedded in the rachis, punctuate.

Frequent herb on rocky crevices of slow streams in East Godavari (Rolla Rao *et al.*, 1999) and Vizianagaram districts. Fl.: June – September.

INDIA: South India.

WORLD: Madagascar to SW India.

Peperomia tetraphylla (G. Forst.) Hook. & Arn., Bot. Beechey Voy. 97. 1832. *Piper tetraphylla* G. Forst., Fl. Ins. Austr. 5. 1786. *Peperomia reflexa* (L.f.) A. Dietr. in L., Sp. Pl. (ed. 6.) 180. 1831, non Kunth, 1815; FBI 5: 99. 1886; Gamble 2: 1209. 1925.

Much branched, succulent herbs, stems rooting at lower nodes, erect or decumbent, quadrangular, minutely pubescent, grooved. Leaves coriaceous, 0.8-1.5 × 0.6-1 cm, orbicular or ovate to rhomboid, whorled at a node, base cuneate, margin entire, apex acute, nerves obscure; petiole very short, to 0.3 cm. Flowers in terminal stout spikes, bracts orbicular, peltate, 3 mm; perianth absent; stamens 2, filaments 0.15 mm; ovary set in a deep socket in the axis, ovate, acute, style wanting stigma simple. Berry obovoid, glabrous, brown.

Common in hills above 1200 m, in shade. Epiphytic on tree trunks or on rotting wood. Fl. & Fr.: November – April.

Minumuluru to Kappa konda (VSKP), *GVS* 29630 (MH); Chintapalli (VSKP), *JLE* 37196 (MH); Mahendragiri (GJM), *CECF* 93 (CAL); Kolli hills, Kulivalavu (NMK), *SKS* 1017 (SGH).

INDIA: Assam, Bihar, Odisha, Andhra Pradesh, Central India, Kerala

WORLD: Pantropical.

Peperomia wightiana Miq., London J. Bot. 5: 548. 1846; FBI 5: 98. 1886; Gamble 2: 1210. 1925.

Epiphytic, succulent herb, up to 10 cm high, Leaves alternate, upper 1 or 2 pairs sometimes opposite, elliptic-ovate or orbicular, 1.5-2.5 × 0.7-1.5 cm, 3-nerved. Flowers in 3-5 cm long spikes; bract ovate, decurrent to the peduncle. Berry minutely rugose and scurfy, beaked with stigma.

Servarayans at 1600 m, Yercaud, in shade (Matthew, 1991). Fl.: December

INDIA: Peninsula.

WORLD: Sri Lanka.

PIPER Linnaeus

1. Erect or subscadent under shrubs .... **P. longum**
1. Climbing herbs or shrubs:
    2. Flowers bisexual .... **P. nigrum**
    2. Flowers unisexual:
        3. Leaves crispately pubescent beneath; ovary ellipsoid .... **P. hymenophyllum**
        3. Leaves glabrous; ovary ovoid .... **P. attenuatum**

Piper attenuatum Buch.-Ham. ex Miq., Syst. Piperac. 306. 1843; FBI 5: 92. 1886; Gamble 2: 1207. 1925. *P. sylvestrea* Wight, Icon. Pl. Ind. Orient. t. 1937. 1853; FBI 5: 93. 1886. *P. trioicum* Roxb., Fl. Ind. 1: 151. 1820.

Slender climbers with often somewhat zig-zag branchlets, stems terete, striate. Leaves about 12 × 9 cm, ovate or broadly elliptic-ovate, alternate, base rounded to broadly acute, rarely sub-cordate, margin entire, apex acuminate, glabrous, 5-7-nerved; petiole up to 1 cm long, canaliculate. Flowers unisexual, green or white,

0.5 mm long, in leaf-opposed spikes, about 11 cm long; bracts up to 0.2 cm long, more or less linear-oblong, margins scarious, free, overlapping, medianly and longitudinally attached to the rachis; stamens 2-4, filaments short, thick, anthers 2-celled, confluent; ovary ovoid, style wanting, stigma simple. Berry smooth, globose, 3.4 mm across, granulate with persistent style.

Common in forests in damp valleys. Fl. & Fr.: July – September.

Gundlabrahmeswaram (KNL), *BR, TSS* & *MO* 43248, *JLE* 16912 (MH); Talakona water falls (CTR), *KJ & SSR* 12569; Narayanagiri, Tirumala (CTR), *DRC* 1645 (MH); Durgam (NLR), *PV* 111137 (BSID); Peddarutla stream (PKM), *VBH* 84942 (BSID); Madaliveedu, Tadepally (EG), *NRR & DN* 84375 (MH); Bison hills (EG), *CAB* 5051 (MH); Marripakala (EG), *GVS* 24528 (MH); forest near Araku (VSKP), *NPBK* 10881 (MH); Galikonda (VSKP), *GVS* 19647 (MH).

INDIA: Eastern Tropical Himalaya, Sikkim, Assam, Peninsular India.

WORLD: Bhutan, China, Java.

Piper hymenophyllum (Miq.) Wight, Icon. Pl. Ind. Orient. 6: t. 1942. 1853; FBI 5: 93. 1886; Gamble 2: 1207. 1925.

Climbing herbs, stems terete, grooved, glabrous, nodes swollen, Leaves 7-13 × 4-6 cm, elliptic – lanceolate, alternate, base rounded, often unequal sided, margin entire, apex shortly acuminate, coriaceous, thinly pubescent below, 7-ribbed; petiole 1 cm long. Flowers green, in slender, 10 cm long, drooping spikes, male spike shorter; bracts orbicular to ovate, ca 2 mm, adnate to the rachis with free margis, hairy; stamens in groups of 3; ovary ellipsoid, stigma 3-lobed. Berry globose, 4 mm across, when dry with persistent styles.

Occasional in forests. Fl. & Fr.: February– December. Vern.: Tam.: *Kaatumilagu*.

Talakona (CTR), *KJ* & *SSR* 12569, *GVS* 31997 (MH); Gundlabrahmeshwaram (KNL), *BR* & *BSS* 34159, *JLE* 16957 (MH); Sirumalai puthur (DGL), *SKS* 649 (SGH).

INDIA: South India.

WORLD: Sri Lanka.

Piper longum L., Sp. Pl. 29. 1753; FBI 5: 83. 1886; Gamble 2: 1205. 1925.

Slender, erect or subscandent under shrubs, creeping and rooting below; the branches erect. Leaves membranous, the lower with longer petiole, the upper sessile, subscandant; upper leaves ovate, base deeply and widely equally cordate, often unequal, 7-nerved, glabrous; petiole up to 3 cm long, puberulous, striate, canaliculate. Flowers in spikes; female spikes up to 2.5 cm long including the peduncles, leaf-opposed, short, cylindric, thick; bracts up to 0.1 cm across, peltate, attached to the rachis in the centre, margins free; ovary sunk in the rachis, style 0, stigma irregularly 3-5-lobed. Fruits small, blackish green, sunk in fleshy spikes, red when ripe.

Cultivated in damp and shady places in the valleys in Northern Eastern Ghats. Fl. & Fr.: June – September. Vern.: Tel.: *Pippallu*; Tam.: *Thippili*.

Near Kakur Teak plantation (VSKP), *GVS* 68503 (MH); near Surhiguda, Lakshmipuram (VSKP), *NRR* & *DN* 84399 (BSID); Sirumalai hills (DGL), *SKS* 527 (SGH).

INDIA: Native to North East India. Cultivted throughout India.

WORLD: Malay Peninsula, Philippines, Sri Lanka, Timor.

Piper nigrum L., Sp. Pl. 28. 1753; Hook.f., Fl. Brit. India 5: 90. 1886; Gamble 2: 1206. 1925.

A stout glabrous climber; stems terete, much thickened at the nodes, rooting at nodes. Leaves 12-20 × 9-14 cm, broadly ovate, thick, leathery, glabrous, base rounded, more or less oblique, margin entire, apex shortly acuminate, veins 5-7; petiole 1-2 cm. Flowers bisexual, in leaf-opposed spikes; bracts spathulate-oblong, adaxially adnate to the rachis; stamens 2, on either side of pistil, filaments flat, 0.5 mm; ovary globose, stigmas 3 or 4, rarely 5. Berry ovoid-globose, 3-4 mm in diam., sessile, at first yellow, afterward becoming red when fully ripe.

Cultivated for its spicy fruits. Fl. & Fr.: June – April. Vern.: Tel.: *Miriyalu*; Tam.: *Kurumilagu, Milagu.*

Maredumilli (EG), *TP* & *KJ* 12516; Gundlabramheswaram (KNL), *JLE* 22210 (MH); near Lakshmipuraam (VSKP), *NRR* & *RAN* 79438 (BSID); Sirumalai hills (DGL), *SKS* 530 (SGH).

INDIA: Native to South India. Cultivated in South India.

WORLD: Cultivated in Sri Lanka, Malay Peninsula, Philippines, Borneo, South America.

Piper betle L. (Betel leaf) is cultivated for its leaves. Vern.:Tel.: *Thamalapaku*; Tam.: *Vetrilai.*

MYRISTICACEAE

1. Inflorescence unbranched (or bifurcate); filaments connate into a peltate disk, aril entire or lacerate at apex .... **Knema**
1. Inflorescence dichotomously branched; fiaments connate into a column; aril deeply laciniate almost to base .... **Myristica**

KNEMA Loureiro

Knema attenuata (Hook.f. & Thoms.) Warb., Nova Acta Acad. Caes. Leop.-Carol. German. Nat. Cur.68: 590. 1897; Gamble 2: 1215. 1925. *Myristica attenuata* Wall. ex Hook. f. & Thoms., Fl. India 157. 1855; FBI 5: 110. 1886.

Evergreen tree, up to 2 m tall, bark brownish, usually smooth, irregularly flaky. Leaves coriaceous, oblong – lanceolate, acuminate, 12-23 × 4-8 cm, apex gradually acute to acuminate, margin entire, base rounded or acute, pubescent beneath, chartaceous to subcoriaceous; petiole 0.6-1.3 cm, canaliculate. Flowers small, unisexual, dioecious, female many, in fascicles from axils of the leaves, pedicel up to 0.8 cm long, male few, larger, on short axillary peduncle, fulvous stellate tomentose,

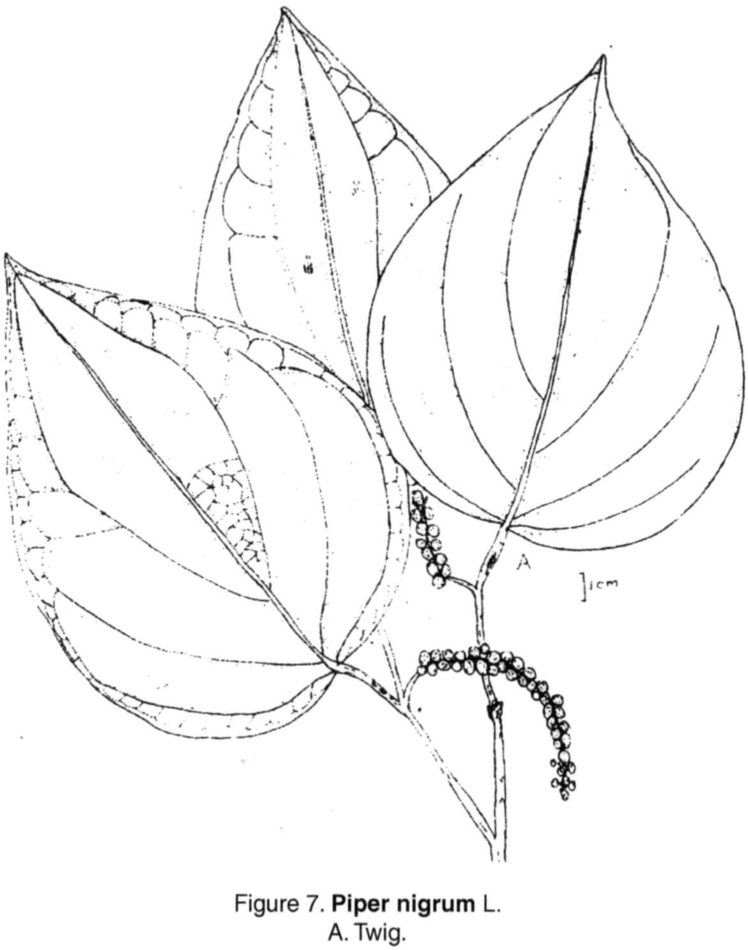

Figure 7. **Piper nigrum** L.
A. Twig.

anthers 10-20. Capsule, elliptic, to 4 cm long, 2-valved, apiculate, fulvous tomentose; seed one with crimson red lacinated aril.

Rare in moist valleys in slopes of Dummakonda in East Godavari District (Rolla S. Rao *et al.*, 1999). Fl.: February. Vern.: Tam.: *Chora pathiri.*

IUCN Status: Lower Risk/least concern ver 2.3 (2017).

INDIA: Maharashtra, Karnataka, Kerala, Tamil Nadu. Endemic.

MYRISTICA Gronov

Myristica dactyloides Gaertn., Fruct. Sem. Pl. 1: 195. 1788. *M. beddomei* King, Ann. Roy. Bot. Gard. (Calcutta) 3: 291. t. 118. ff. 1-8. 1891; Gamble 2: 1214. 1925. *M. laurifolia* Hook.f. & Thoms. Fl. India. 163. 1855. var. *lanceolata* Hook.f.. Fl. Brit. India 5: 103. 1886.

Densely foliaceous tree, to 20 m high, girth to 1.6 m. Leavess elliptic to broadly lanceolate, 15-25 × 6-10 cm, Male flowers 8-14 in umbels, on short axillary shoots; perianth column produced beyond anthers, anther bearing part dome-shaped. Female flowers fewer, ovary ovoid-globose, appressed-pubescent; style arms 2. Drupe broadly ovoid to elliptic, 4-6 × 3-4 cm; seed ovoid, arillate, aril yellowish.

Occasional in Shola forest in Kolli hills (Matthew, 1983; Ramya *et al.*, 2015). Fl.: December – February; Fr.: Dehiscing by August. Vern.: Tam.: *Kaatuchathikkai.*

Kolli hills, Kulivalavu shola (NMK), *SKS* 1021 (SGH).

IUCN Status: Vulnerable A2cd ver 3.1 (Haridasan *et al.*, 2015).

INDIA: Maharashtra, Goa, Karnataka, Tamil Nadu.

WORLD: Sri Lanka.

Myristica fragrans Houtt. is cultivated for its fragrant seeds and arils in TAMPCOL garden of Kolli hills (*SKS*, 1061 (SGH) and planted in estates of Sirumalai hills of Eastern Ghats.

LAURACEAE

1. Leafless twining parasite .... **Cassytha**
1. Leafy shrubs or trees:
    2. Inflorescence dense, clustered or umbelliform; all anthers introrse:
        3. Fertile stamens 6 .... **Neolitsea**
        3. Fertile stamens 9 or 12:
            4. Bracts persistent until anthesis; fruits subtended by unchanged calyx tube; leaves penninered, non-subverticillate .... **Litsea**
            4. Bracts caducous before anthesis; fruits subtended by cyathoform calyx tube; appendaged by residue of lobes; leaves 3-plinerved, usually verticillate .... **Actinodaphne**

2. Inflorescence laxly cymose or panicle; inner most whorl of perfect stamens with extrorse anthers:
    5. Berry ellipsoid; stigma peltate; staminal gland subsessile .... **Alseodaphne**
    5. Berry globose; stigma simple or capitate; staminal glands stipitate:
        6. Anthers 2-celled; perianth deciduous .... **Beilschmiedia**
        6. Anthers 4-celled; perianth persistent:
            7. Leaves 3-plinerved; fruiting perianth enlarged into a cupule embracing the fruit:
                8. Leaf apex abruptly acuminate; petiole long .... **Cinnamomum**
                8. Leaf apex caudate-acuminate; petiole short .... **Neocinnamomum**
            7. Leaves penninerved; fruiting perianth hardly enlarged:
                9. Perianth lobes spreading or reflexed, not hardened .... **Persea**
                9. Perianth lobes erect, hardened .... **Phoebe**

ACTINODAPHNE Nees

1. Leaves with lowest pair of nerves opposite; fruiting perianth much enlarged, thick .... **A.maderaspatana**
1. Leaves penninerved; fruiting perianth slightly enlarged .... **A. angustifolia**

Actinodaphne angustifolia Nees, Pl. Asiat. Rar. 3: 34. 1832; FBI 5: 152. 1886; Saxena & Brahmam, Fl. Orissa 3: 1559.1995.

Evergreen tree. Leaves subverticillate, mostly 6 in a whorl, oblong, lanceolate to narrowly oblanceolate, rarely obovate-oblong, 10-30 × 2.5-8 cm, glaucous glabrous beneath, dark shining above, apex acute to acuminate, base narrowed; petiole 0.7-2.5 cm long. Flower in the axils of fallen bud-scales, inflorescence silky. Male flowers clustered or shortly racemose; perianth rotate or subcampanulate; tepals ovate to oblong; stamens far exserted. Female flowers in clusters or in subsessile, simple or slightly branched umbels; perianth-tube campanulate, tepals spreading; ovary and style hirsute. Fruit globose or ellipsoid, seated on the cupular, truncate, swollen perianth tube.

In shady valleys of Eastern Ghats of Odisha. Fl.: August; Fr.: November – May.

Saptasajya (DKL), *SX* 2112 (RRL-B); Similipahar (MBJ), *SX* & *MB* 6335 (RRL-B).

Actinodaphne madraspatana Bedd. ex Hook. f., Fl. Brit. India 5: 149. 1886. A. *hookeri* Beddome, Fl. Sylv. S. India t. 296. 1872; Gamble 2: 1231.1925.

Evergreen tree, up to 12 m high; branchlets pubescent. Leaves chartaceous, elliptic or oblong-lanceolate, 13-21 × 4.7-6.7 cm, glabrous, shining above, galucous below, base acute, margin entire, apex acute, lowest pair of nerves just above the base; secondary nerves 7-10 pairs; petiole to 2 cm. Flowers yellow, dioecious; male flowers ca. 8 in clusters; female flowers in umbellate cymes; stamens 9 in male flowers, arranged in 3 rows; staminodes 9 in female flowers, also arranged in 3 rows. Fruit ellipsoid.

Common in Seshachalam hills at 700-900 m. Endemic to Eastern Ghats. Fl.: August – September. Fr.: February. Vern.: Tel.: *Kovangutti.*

Mogalipenta (CDP), *BR* & *MVS* 38610, *JSG* 21331 (MH); Tirumala hills (CTR), *SSR* 13783, *KJ* & *SSR* 12561, 12562, *SKW* 5635 (BLAT); Talakona (CTR), *BR* & *AMR* 26019; *DRC* 421 (MH); Dharmagiri, Tirumala (CTR), *DRC* 1236 (MH); Tirumala (CTR), *GVS* 31904 (MH), Near Eethakayalamandapam (CTR), *GVS* 46910 (MH), Kambakkam hills (CTR), *s. coll. s.n* (MH); N.Arcot hills (NA), *CBD s.n* (MH).

INDIA: Peninsular India. Endemic.

ALSEODAPHNE Nees

Alsedaphne semecarpifolia Nees in Wallich, Pl. Asiat. Rar. 2: 72. 1831; FBI 5: 144. 1886; Gamble 2: 1226. 1925.

Tree, 6-18 m tall. Leaves alternate or clustered at apices, obovate to oblanceolate, 6-11 × 2-6 cm, base acute to cuneate, apex (sub) acute; petiole to 1.5 cm. Panicles much branched, terminal, to 18 cm; pedicel to 0.5 cm. Flowers bisexual, 2.5 mm across, tepals 6, equal, 3 + 3, ovate, concave, 2 mm, puberulous without, pubescent within, ciliate, acute, fertile stamens 9; filments 0.7 mm; glands ovate, sessile; anthers 4-celled, 1 mm, in 3rd series the upper cells latrorse; staminodes 3, to 1.5 mm, ovary 0.8 × 0.5 mm; style 1 mm; stigma peltate. Berry ellipsoid, 2.5 × 1.3 cm; seed oblong-globular.

Hills above 1000 m in Tamil Nadu Eastern Ghats. Common and locally abundant; even one of the dominants in places. Fl. March-April and June-July. Fr.: Through the year. Vern.: Tam.: *Thumbaraa, Karinjikkadaa, Manjaya mulaamaram, Vandukadipattai.*

Way to Sirumalai Top (MDR), *KS* 5766 (MH); Singarapettai hills (NA), *MBV* 1406 (MH); Sirumalai puthur (DGL), *SKS* 799 (SGH).

INDIA: Peninsular India.

WORLD: Sri Lanka.

BEILSCHMIEDIA Nees

1. Bark greenish brown; lateral nerves of leaf 6-8 pairs .... **B. bourdillonii**
1. Bark dark brown; lateral nerves of leaf 4-5 paris:
   2. Leaf apex acuminate; fruit ellipsoid .... **B. sikkimensis**
   2. Leaf apex obtuse or bluntly acuminate; fruit oblong-obovoid .... **B. roxburghiana**

Beilschmiedia bourdillonii Brandis, Indian trees 528. 1906; Gamble 2: 1221. 1925.
B. fagifolia Beddome, Fl. Sylv. S. India t. 263. 1872, non Nees, 1830.

Semievergreen tree to 25 m. Leaves alternate to subopposite, elliptic to obovate, 12 × 4 cm, penninerved, lateral nerves 10-12 pairs, glabrous, pale below, acute at both ends; petiole to 2 cm. Panicles (sub) terminal, villous pubescent. Flowers bisexual, 4 mm across; tepals 6, (sub) equal, 3 + 3, pubescent without, acute; fertile stamens 9; filaments bearded, 1 mm; glands stipitate; anthers 2-celled. 1 mm; staminodes 3, cordate or arrow-shaped, 2 mm, stalked, ovary 0.7 mm, pubescent; style 1.5 mm; stigma (sub) capitate.

Kollimalais (Mathikettan shola) 1300 m (Matthew, 1983) and Sirumalai hills. Fl.: March.

Kollimalais (NMK), *KMM* 26750 (RHT); Sirumalai hills, Upper slopes (DGL), *J.M. Pallithanam* 1107 (RHT).

INDIA: Peninsular India. Endemic.

Beilschmiedia roxburghiana Nees in Wall., Pl. Asiat. Rar. 2: 69. 1831; FBI 5: 121. 1886; Gamble 2: 1221. 1925.

Evergreen, moderate-sized tree; wood white, even-grained; branchlets blackish-brown. Leaves chartaceous, elliptic, narrowly elliptic or elliptic-lanceolate, 10-12 × 4-8.7 cm, apex obtuse or bluntly acuminate, margin entire, base rounded; triplinerved, tertiaries parallel; petiole to 2 cm. Flowers whitish in long-pedunculate racemes; bracts narrow, deciduous; tepals variable in size, linear-oblong to oblong-oblanceolate, pubescent on both sides; perfect stamens in 3-4 series, staminodes white, fleshy; ovary glabrous. Berry oblong-obovoid, 3.3 cm long, purple black.

In shady valleys of Eastern Ghats. Fl. & Fr.: April – May.

Tirumala hills (CTR), *KJ* & *SSR* 12557; Mahendragiri (GJM), *JSG* 14024 (CAL);

INDIA: Andhra Pradesh, Assam, Bihar, Odisha

WORLD: Bhutan, Nepal, China, Myanmar, Thailand.

Beilschmiedia sikkimensis King ex Hook. f., Fl. Brit. India 5: 122. 1886.

Tree, branchlets whitish. Leaves coriaceous, lanceolate or elliptic-oblong, 7.5 × 2.5-6 cm, glabrous, base obtuse, margin entire, apex obtuse or bluntly acuminate, secondary nerves 10-12 pairs, irregular; petiole to 8 mm. Flowers in axillary, tomentose racemes. Berry ellipsoid, 2.7 × 1.4 cm.

Occasional, in forests of East Godavari district. Fl. & Fr.: April – May.

From Ramavaram to Nulakamaddi (EG), *GVS* 27449 (MH).

INDIA: Sikkim, Himalayas, Andhra Pradesh.

WORLD: Bhutan.

CASSYTHA Linnaeus

Cassytha filiformis L., Sp. Pl. 35. 1753; FBI 5: 188. 1886; Gamble 2: 1241. 1925.

Perennial, parasitic, green, glabrous, leafless, very long, wiry, twining herb; branches slender, green, attached to hosts by suckers or haustoria, younger ones

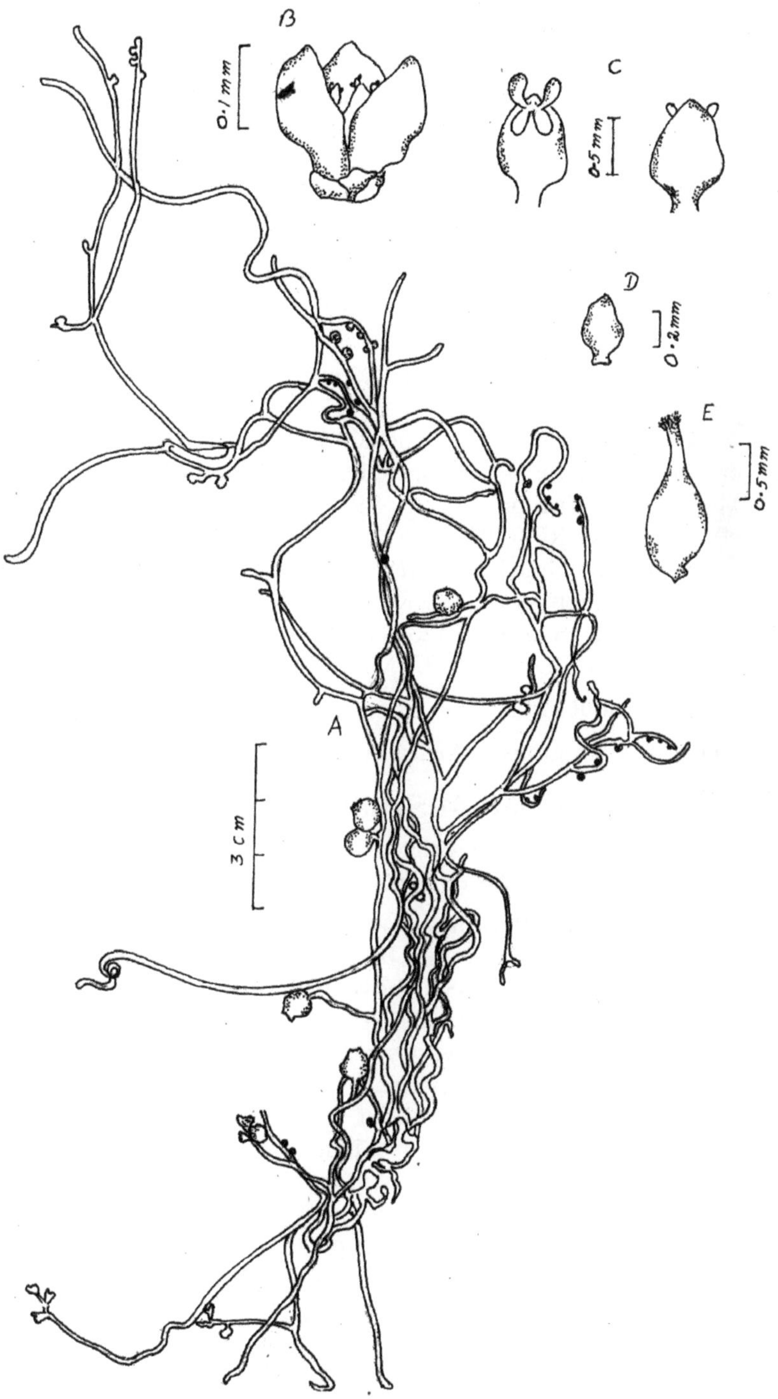

Figure 8. **Cassytha filiformis** L.
A. Twig, B. Flower, C. Stamens, D. Staminodes, E. Pistil.

yellow. Flowers in spikes, white, 3-merous, bisexual, 2 mm across; tepals 6, free, unequal, 3 + 3, truncate, obtuse at apex; outer lobes ovate-orbicular, 1 mm long, ciliate; inner lobes obovate, 1.5 mm long; fertile stamens 9; filaments 1.5 mm; glands sessile; anthers 2-celled, 1 mm long; staminodes 3, to 1 mm long; ovary 1.5 mm; style 0.5 mm long; stigma capitate. Drupe globose, enclosed within inflated perianth, crowned by lobes, white when mature, 0.5 across; seed 1, glabrous, globose; testa thin, hard, black, 3 mm across.

A common parasite on *Carissa spinarum, Vitex negundo, Acacia latronum, Ziziphus mauritiana, Azadirachta indica, etc.* Common in scrub jungles. Fl. & Fr.: July-December. Vern.: Tel.: *Sitummavari jadalu, Antaravalli, Pachi teega, Nullu teega*; Ori.: *Nirmuli*; Tam.: *Andhrakodi, Vellarikothu, Kotthan.*

Muchukota RF (ATP), *TP* & *NY* 230; Gani RF (KNL), *TP* & *RVR* 1783; Owk RF (KNL), *RVR* 1461; Rachakuntapalle (CDP), *KRKS* & *KRS* 37908; Narasimhakonda (NLR), *SSY* 18696; Macherla RF (GNT), *VRK* 3288; Nemalakunta (KNL), *VBH* 86654 (BSID); Balapalle (CDP), *JLE* 14307; Way to Duggamakonda, Lankamalai RF (CDP), *KS* 6338 (MH); Way to Tirumala (CTR), *KS* 6922 (MH) Kambakkam (CTR), *MCB* 45147 (MH); Palamaner RF (CTR), *KCJ* 439 (MH); Rajupalem (NLR), *PV* 109887 (BSID); Brooks hill (Sambalpur), *SP* & *APD* 183 (CAL); Ganagakur (GJM), *JSG* 13629 (CAL); Forest near Kallakurichi (SA), *KRM* 77303 (CAL); Tippukadu RF (NA), *KRM* 16564 (MH); Gingee RF (SA), *KMS* 12376 (MH); Kadu vanur RF (SA), *KRM* 50619 (MH); Vandalur RF (CPT), *ANH* 45532 (MH); Alagar hills (MDU), *SKS* 84 (SGH).

INDIA: Throughout India.

WORLD: Tropical regions of Africa, Asia, Australia.

CINNAMOMUM Schaeffer *nom. cons.*

Cinnamomum camphora (L.) J.S. Priroz, Rostlin. 2: 47.t.8. 1825; FBI 5: 134. 1886; Gamble 2: 1225. 1925. *Laurus camphora* L., Sp. Pl. 1: 369. 1753.

Evergreen tree, up to 10 m high; much branched with a strong, persistent, camphoraceous odour, wood yellowish brown. Leaves ovate or elliptic lanceolate, 3-10 × 1.5-4 cm, glabrous, dark green above, glaucous-waxy beneath, base acute, margin entire, apex abruptly acuminate, 3-5-nerved, bullate at nerve-axils; petiole to 2.5 cm, reddish. Flowers whitish, very sweet scented in pseudo-terminal panicles, to 7 cm; pedicels up to 0.2 cm long; perianth up to 0.25 cm long, lobes 6, subequal, up to 0.2 cm long, ovate, glabrous without, white woolly within; stamens 9, in 3 whorls, all fertile, outer 2 whorls introrse, inner extrorse, filaments up to 0.8 mm long, hairy with stipitate glands, anthers up to 0.5 mm long, 4-celled, staminodes up to 0.5 mm long; ovary ovoid, up to 1 mm long, style thick, stigma peltate. Berry black, subglobose, seated on the perianth-cup.

Occasionally planted in Agency areas. Fl.: November. Vern.: Tel.: *Gamburamu, Karpuramu, Kappuramu, Himavaluka;* Tam.: *Lavangampattai.*

Araku (VSKP), *GVS* 62360 (MH), *GVS* 19709 (MH); Araku valley (VSKP), *GVS* 22665 (MH); Raktachandanam thota (SKLM), *GVS* 62630 (MH).

INDIA: Widely cultivated.

WORLD: Japan, Korea, Vietnam; introduced or cultivated in many countries around the world.

Cinnamomum verum Presl. (*C. zeylanicum* Garc. ex Blume) (Tel.: *Dalchini chekka;* Tam.: *Briyani-ilai*) is cultivated for spicy bark and leaves. **Cinnamomum tamala** Nees is planted occasionally.

CRYPTOCARYA R.Brown

Cryptocarya beddomei Gamble, Kew Bull. 1925: 127. 1925 & Fl. Madras 2: 1222.1925; Murugesan *et al.*, J. Econ Taxon. Bot. 33: 553. 2009.

Small evergreen trees, up to 12 m tall, branchlets glabrous or greyish-pubescent, terete. Leaves alternate, ovate or elliptic-oblong, rarely obovate, 4-10 × 2-5 cm, base rounded, oblique or unevenly truncate, margin entire, apex obtusely acute, rounded, retuse or emarginate, thin-coriaceous, shining above, reddish glaucous beneath; lateral nerves distant, impressed above; petioles 2-4.5 cm long. Flowers in spreading axillary racemose panicles, fulvous-pubescent; peduncle ca 7.5 cm long; bracts and bracteoles ovate-lanceolate, apex acute, early caducous. Perianth tube turbinate, grey-pubescent, ca 2 mm long; lobes 6 in two rows, subequal, ovate-lanceolate, ca 2.2 × 1 mm, apex acute, glabrous within, grey-pubescent, without; stamens 9 in 3 rows, perfect; staminal glands 6, stipitate, ca 0.9 mm long; staminodes 3, stipitate; ovary sessile, at the bottom of the perianth-tube, style shortly exserted. Drupe globose-oblong, enclosed in the accrescent perianth-tube; seed solitary, cotyledons fleshy.

Rare in evergreen and semievergreen forests at about 1300 m above msl in Minchigully valley. A rare tree whose occurrence is restricted to stream banks.

Fl.: December – February; Fr.: March – May.

Minchigully valley (Erode District), *P.Balasuramanian* & *M.Murugesan* 1660 (SACON).

INDIA: Karnataka, Kerala, Tamil Nadu. Endemic.

LITSEA Lamarck *nom. cons.*

1. Perianth incomplete or aborted:
    2. Leaves thin, glabrous beneath or nearly so; umbels corymbose or racemose, few-flowered, puberulous:
        3. Umbel simple; leaves with 7-12 pairs of lateral nerves .... **L. floribunda**
        3. Umbel compound; leaves with 5-7 pairs of lateral nerves ..... **L. glutinosa**
    2. Leaves thick, fulvous tomentose beneath; umbels solitary:
        4. Berry depressd-globose; perianth persistent and enlarged in fruits .... **L. oleoides**
        4. Berry globose; perianth deciduous .... **L. deccanensis**

1. Perianth lobes complete, segments 6:
   5. Leaves subcoriaceous, glabrous and glaucous beneath, base narrowed .... **L. laeta**
   5. Leaves chartaceous, tomentose, not glaucous beneath, base acute or rounded .... **L. monopetala**

Litsea deccanensis Gamble, Fl. Madras 2: 1235. 1925. *Tetranthera tomentosa* Nees in Wall., Pl. Asiat. Rar. 2: 65. 1830. *p.p. Litsea tomentosa* Heyne ex Wall., Pl. Asiat. Rar. 2: 65. 1861; non Blume, 1826; FBI 5: 157. 1886.

Evergreen tree, up to 12 m high; bark thin, smooth; wood yellowish, wavy-grained, branchlets stringy, fulvous tomentose. Leaves coriaeous, crowded towards the end of branches, elliptic-oblong, 10-22 × 5-7 cm, glabrous above, softly and densely tomentose, base obtuse-acute, margin entire, apex acute, secondary nerves to 14 pairs, conspicuous beneath; petiole to 2 cm, tomentose. Flowers in globose, pedunculate, many-flowered umbels, tepals 2-4 or more, linear, deciduous; stamens 20 or more, filaments 5 mm, anthers 1.3 mm. Berry globose, 1 cm across, black.

Occasional, in moist deciduous forests at 700-800 m. Fl.: October – December; Fr.: March. Ven.: Tel.: *Nara Mamidi*; Tam.: *Kaattuppila, Aenaava maram.*

Horsley hills (CTR), *SSR* 13796, *JSG* 20984 (DD), *DRC* 1119 (MH); Tirumala hills (CTR), *KJ* & *SSR* 12591; Dharmagiri-Tirumala (CTR), *DRC* 2412 (MH); Kambukudi (NA), *KS* 7439 (CAL), *EV* 51980 (CAL). Chengam Range, Javadi hills (NA), *MBV* 1019; Guttirayanbetta, Kodagarai (DMP), *T. Ravi Shankar* 95504 (MH), 84134 (MH); Chinna Kalrayan (SM), *R. Ansari* 49962 (MH); Kollimalai (TCP), *S. India Flora* 11235 (MH); Sirumalai (DGL), *S.India Flora* 9152 (MH); Way to Mukkangera from Betta mugalalam (DMP), *T. Ravi Shankar* 84151 (MH); Melpat (SA), *CAB* 1081 (MH); Way to Kambakudi (NA), *KS* 7439 (MH); Sirumalai hills (DGL), *SKS* 813 (SGH).

INDIA: Western India, Deccan Peninsula.

WORLD: Pakistan, Sri Lanka.

Litsea floribunda (Blume) Gamble, Fl. Madras 2: 1238. 1925; Murugesan *et al.*, J. Econ Taxon. Bot. 33: 553. 2009.

Large, densely foliaceous, evergreen trees, up to 25 m tall; branchlets densely minutely fulvous – tomentose, terete. Leaves alternate, elliptic-lanceolate or obovate-oblong, rarely oblanceolate, 7-17 × 2.5-7 cm, glabrous or rarely sparsely tomentose above, densely fulvous-tomentose beneath, base truncate, cuneate or acute, margin entire or slightly revolute, apex acute or abruptly acuminate, lateral nerves 10-15 pairs, joining at margins forming an intramarginal loop, impressed above, raised beneath, prominent on both surfaces; lateral nerves prominent on lower surface, obscure on upper surface; petioles ca 1.5 cm long, densely rusty-tomentose. Flowers in axillary racemes, densely rusty-tomentose; peduncle 1.5-5.5 cm long, stout, rusty-tomentose; perianth densely rusty-tomentose; tube cupular, ca 5 mm long; lobes 3+3, ovate, apex acute. Male flowers: stamens 9, filaments bearded, staminodes 3, glands 6; pistillode globose, beaked. Female flowers: ovary ca 2 mm across, ovule

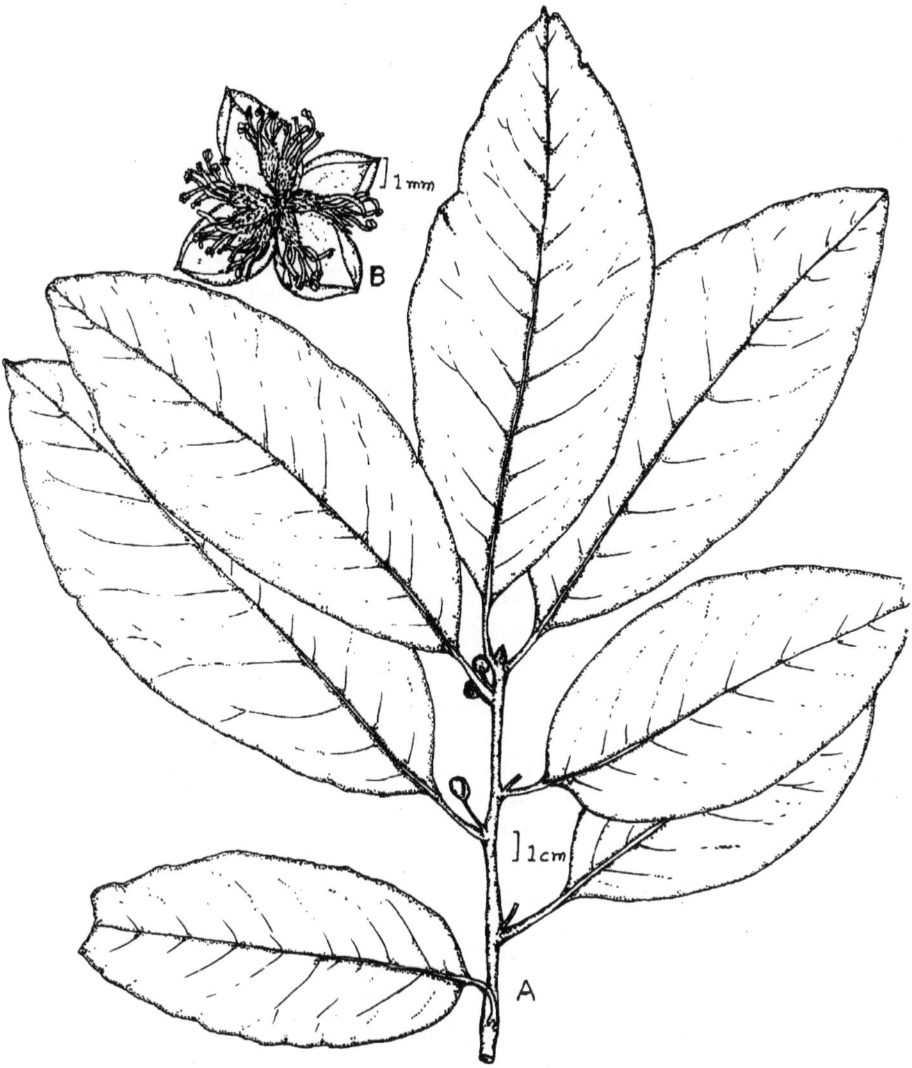

Figure 9. **Litsea deccanensis** Gamble
A. Twig; B. Umbelule.

1, stigma small, discoid; staminodes 9-12. Berries ovoid-oblongoid, ca 2.5 cm across, with a basal rim of perianth, seated on the turbinate perianth-tube with thick pedicel, densely rusty-tomentose; seed solitary with fleshy cotyledons.

Occasional in evergreen forests at about 1400 m above msl. Fl.: April – Sept.; Fr.: Through out the year.

Minchigully valley (Erode district), *P.Balasubramanian* & *M.Murugesan* 1665 (SACON).

INDIA: South Western Ghats, South Eastern Ghats. Endemic.

Litsea glutinosa (Lour.) C.B. Rob., Philipp. J. Sci. 6: 321. 1911. *Sebifera glutinosa* Lour., Fl. Cochinch. 2: 783. 1790. *Litsea chinensis* Lam., Encycl. 3.574. 1789; FBI 5: 157. 1886; Gamble 2: 1235. 1925. *L. sebifera* Pers., Syn. 2:4 1807; FBI 5. 157. 1886.

Evergreen trees, up to 12 m high; bark brown, somewhat corky; wood greyish-brown, fairly straight grained, medium and even textured; branchlets pubescent. Leaves coriaceous, very variable in shape, elliptic, obovate, or elliptic-oblong, glabrous, base acute, margin entire, apex obtuse or acute, secondary nerves 7-12 pairs; petiole 1-3.5 cm. Flowers white or yellowish, small in compound, few-flowered umbels; peduncles pubescent or glabrescent, 0.7-2 cm long, axillary or from the axils of caducous bracts on the new shoots below the leaves; bracts concave, orbicular; perianth-tube minute in flower, thickened and disc-like in continuation with the pedicel, 3.7-4.4 mm diam. in fruit, lobes 0 or incomplete; stamens 9-20, with long villous filaments. Fruit globose, black, 1.2 × 0.8 cm.

Occasional, in the hills, in moist deciduous forests at 650-1100 cm. Fl.: June. Fr.: September. Vern.: Tel.: *Meda nara-naliki, Narva- alagi, Kanuju, Burra manu, Kanuja nalike;* Ori.: *Ledha chhali;* Tam.: *Pisinpatty.*

Velgode to Gundlabrahmeswaram (KNL), *BR* & *SKB* 34085; Talakona top (CDP), *BR* & *SKB* 32736; Tirumala hills (CTR), *SSR* 13736, 13715, *KJ & SSR* 12597, *DRC* 2097 (MH); Diguvametta (PKM), *RVK* 15740; Borra (VSKP), *SSR* 15937; Mogalipenta (CDP), *JSG* 21276 (DD); Talakona (CTR), *GVS* 91464 (MH); 46955 (MH); Japali, Tirumala hills (CTR), *DRC* 1487 (MH); Tirumala (CTR), *GVS* 31936 (MH); Tadepalli (EG), *NRR & TRS* 86113 (MH); Rampachodavaram (EG), *GVS* 27249 (MH); Cheedipalem (EG), *GVS* 24524 (MH); Gannavaram (EG), *GVS* 24529 (MH); Addatigala (EG), *GVS* 24543 (MH); Krisnapuram (VSKP), *CAB* 44673 (MH); near Torikivalasa, Kurimi (SKLM), *NRR* 86145 (MH); Alagar hills (MDU), *KS* 3419 (MH); Near Kambukudi, Inner Javadi RF (NA), *KS* 6082 (MH).

INDIA: Maharashtra, Kerala, Tamil Nadu, Andhra Pradesh, Odisha.

WORLD: Sri Lanka, Bhutan, Nepal, China, Myanmar, Philippines, Vietnam.

Litsea laeta (Nees) Hook. f., Fl. Brit. Ind. 5: 169. 1886; Gamble 2: 1236. 1925. *Tetranthera laeta* Nees in Wall., Pl. Asiat. Rar. 2: 67. 1831.

Small tree; branchlets slender, pubescent. Leaves sub-coriaceous, elliptic-oblong or lanceolate, 15-25 cm long, green above, glaucous beneath, base narrowed, margin entire, apex acute, secondary nerves 12 pairs, reticulations faint. Flowers small, sessile in short peduncled, puberulous umbels. Berry globose, seated on the perianth tube.

Mahendragiri hills, Rampa hills and Tirumala hills.

Papanasanam (CTR), *SKB* & *BR* 30212.

INDIA: Arunachal Pradesh, Assam, West Bengal, Sikkim, Odisha, Andhra Pradesh,

WORLD: Bhutan, Bangladesh.

Litsea monopetala (Roxb.) Pers., Syn. Pl. 2: 4. 1806. *Tetranthera monopetala* Roxb., Pl. Coromandel 2: 26. t. 148. 1798. *Litsea polyantha* Juss., Ann. Mus. Hist. Paris B: 211. 1805; FBI 5: 162. 1886; Gamble 2: 1237. 1925.

Evergreen tree, up to 15 m high, bark grey, exfoliating in corky scales, wood olive-grey, smooth, fairly straight grained, medium- fine and quite even-textured; branchlets pubescent. Leaves chartaceous, elliptic-oblong, rarely brown when dry, base acute or rounded, margin entire, apex obtuse or slightly acute, rarely emarginated, secondary nerves 7-12 pairs, transverse nervules parallel, prominent beneath; petiole to 1.2 cm. Flowers small, sessile, 5-6 in each umbellate head; fruiting pedicel 6.2-18 mm; perianth lobes usually 5, linear-oblong, nearly free; stamens 9-13, filaments hairy, reduced to 2-glandular staminodes in female flower. Berry black or deep purple, ovoid, 0.7 cm long, seated on 1 mm long, saucer-shaped perianth.

Occasional, in valleys, chiefly near streams and shady places of moist forests of Srikakulam to East Godavari districts at 1000-1200 m. Fl.: February–March. Fr.: April – May. Vern.: Tel.: *Chiru mamidi, Bhunjada*; Ori.: *Bastura, Baghoari, Kulya, Burlo, Mosonea.*

Anantagiri (VSKP), *SSR* 15940, 18605; Kusimi forest (SKLM), *SSR* 15597; Maredumilli (EG), *GVS* 24409, 27262, 27288 (MH); Valamuru (EG), *MM* 101289 (MH); on the way to Mallavaram (EG), *PVS & NRR* 76869 (MH); Minumuluru (VSKP), *GVS* 92231; Lambasingi (VSKP), *HS* 20748 (BLAT); Sunkarimetta (VSKP), *HSP* 20564 (BLAT); Korai Road (VSKP), *GVS* 22632 (MH); Minumuluru towards Ontally (VSKP), *GVS* 30023 (MH); Khotnighada area-Anantagiri (VSKP), *GVS* 19509 (MH); Near Laxmipuram stream (VSKP), *NRR* 83696 (MH); Duggeru forest (SKLM), *GVS* 62514 (MH); Ushakothi RF (SBP), *SP* & *APD* 541 (CAL).

INDIA: Arunachal Pradesh, Assam, Tripura, Meghalaya, West Bengal, Sikkim, Bihar, Jharkhand, Odisha, Andhra Pradesh, Madhya Pradesh, Maharashtra, Uttarakhand, Andaman and Nicobar Islands.

WORLD: Nepal, Bhutan, China, Myanmar, Thailand, Malaysia, Cambodia, Vietnam.

Litsea nitida (Roxb.) Hook.f., Fl. Brit. India 5: 174. 1886; Saxena & Brahmam, Fl. Orissa 3: 1567. 1995. *Tetranthera nitida* Roxb., Fl. Ind. 3: 818. 1832.

Moderate-sized tree, branched low down, branches 5-angled or subulate. Leaves oblong or lanceolate, 30-35 × 7.5-10 cm, shining, glabrous beneath; petiole stout,

1.8-3.7 cm long. Flowers in umbels, arranged in axillary racemes, male racemes 2.5-10 cm long, female 1.2-5 cm long; perianth-tube tawny pubescent, tepals 4, unequal, glandular. Fruit c. 1.2 cm long, red when ripe, invested by the enlarged, fleshy cupular perianth.

In damp valleys in Eastern Hills of Bonai in Odisha. Fl. & Fr.: May – September.

INDIA: Eastern Himalayas, Assam, Bihar, Odisha.

WORLD: Bhutan, Bangladesh, Myanmar.

Litsea oleoides (Meissn.) Hook.f., Fl. Brit. India 5: 175. 1886; Gamble 2: 1236. 1925. *Tetranthera oleodes* Meissn. in DC. Prodr. 15(1): 195. 1864.

Lofty evergreen tree up to 20 m tall. Leaves oblong to lanceolate, midrib thick and prominent when compared to lateral nerves, green above, drying brown below, base acute to attenuate, apex obtuse to acute. Leaf galls common. Berry depressed, globose, 2.5 cm across: fruiting perianth enlarged, rim undulate.

Kollimalais, above 1200 m, in moist sholas. Occasional. Vern.: Tam.: *Arambaamaram.*Fr.: March-April.

INDIA:Kerala, Tamil Nadu. Endemic.

NEOCINNAMOMUM H. Liou

Neocinnamomum caudatum (Nees) Merr., Contrib. Arnold. Arbor. 8: 64. 1934. *Cinnamomum caudatum* Nees in Wall., Asiat. Rar. 2. 76. 1831; FBI 5: 134. 1886; Gamble 2: 1225. 1925.

Evergreen tree, 3-6 m high, twigs terete, wiry, glabrous. Leaves ovate or orbicular, 6-10 × 3-5 cm, glabrous, base acute or rounded, margin entire, apex caudate-acuminate, 3-5-ribbed; petiole up to 1.3 cm long, canaliculated, tawny tomentose. Flowers pale yellow, small, in short, axillary, few to 2-flowered, pubescent cymes; perianth up to 0.25 cm long, tawny or ferrugineous-tomentose on both sides, lobes 6, unequal; stamens 9, all fertile, in 3 whorls, outer 2 whorls introrse, innermost whorl extrorse, 2-glandular, staminodes up to 0.05 cm long, hairy; ovary ovoid, syle thick, short, stigma capitate. Berry globose or ellipsoid, fleshy; persistent perianth lobes reflexed.

Rare, in dense moist forests of Northern Eastern Ghats above 1000 m. Fl. & Fr.: August – December.

Paderu (VSKP), *GVS* 42594 (MH); Anatagiri (VSKP), *NPBK* 10998 (MH), *GVS* 32839 (MH); Gudem (VSKP), *SKW* 2055 (BLAT); Minumuluru to Kappakonda (VSKP) *GVS* 29635 (MH); Sigumari (VSKP), *GVS* 44358 (MH); Minumuluru (VSKP), *GVS* 28269 (MH); Similipahar (MBJ), *SX* 551 (RRL-B), *SX* & *MB* 4266 (RRL-B); Mahendragiri (GJM), *SX* & *MB* 2242, 3893 (RRL-B).

INDIA: E. Himlaya, Sikkim, Assam, Karnataka, Andhra Pradesh

WORLD: Bhutan, Nepal, Myanmar, Thailand, Vietnam.

NEOLITSEA (Bentham) Merril *nom. cons.*

1. Leaves elliptic-oblong, apex acutely acuminate; fruiting perianth very small compared with the berry:
    2. Flowers in terminal and axillary panicles .... **N. cassia**
    2. Flowers in axillary, solitary umbels .... **N. gamblei**
1. Leaves ovate-lanceolate, apex obtusely acute; fruiting perianth rather large compared with the berry:
    3. Umbels with 1-3 flowers .... **N. scrobiculata**
    3. Umbels with 3-6 flowers .... **N. umbrosa**

Neolitsea cassia (L.) Kosterm., J. Sci. Res. (Jakarta) 1: 85. 1952. *Laurus cassia* L., Sp. Pl. 369. 1753. *Neolitsea zeylanica* (Nees) Merr., Philipp. J. Sci. (Bot.) 1 (Suppl.) 57. 1906; Gamble 2: 1239. 1925. *Litsea zeylanica* Nees, Cinnam. Disp. 1: 58. 1823 (*ceylanica*).

Evergreen tree, up to 20 m high; bark pale brown with darker spots, rough; wood reddish-white, darkening on exposure, straight grained, medium and even textured; branchless glabrous. Leaves coriaceous, elliptic or oblong-lanceolate, 8-20 cm, glabrous and shining above, glaucos beneath, base slightly oblique and rounded, margin entire, apex acute or rounded, secondary nerves 4-8 pairs; petiole to 3.8 cm. Flowers greenish-white, 1 cm across, pedicellate in terminal or axillary, 17 cm long panicles. Berry dark purple or black, globose, 1.3 cm across, glabrous with a basal persistent lobed calyx; seed 1, globose, smooth, 1.5-2 × 0.83 cm, black.

Rare, in semi-evergreen and moist deciduous forests. It is a shade bearer. Fl.: November- January. Fr.: April- June. Vern.: Tel.: *Nara.*

Talakona (CTR), *SSR* 15554 & 19602; Way to Thumburatheertham (CTR), *SSR* 16177; Galikonda (VSKP), *SSR* 15921; Valamuru – Maredumilli (EG), *MM* 100868 (MH, BSID); Sunkarimetta (VSKP), *SSR* 15913, *GVS* 21681 (MH); Anantagiri (VSKP), *GVS* 22689 (MH); Forest near Anantagiri (VSKP), *NPBK* 10960 (MH); Chintapalli (VSKP), *GVS* 28239 (MH); Minumuluru to Kappakonda (VSKP), *GVS* 29625 (MH); On way to Ontally (VSKP), *GVS* 42646 (MH); Vanadurga temple, Minumuluru (VSKP), *GVS* 47354 (MH); Mahendragiri (GJM), *SX* 1258 (RRL-B), *SX & MB* 3544, 3614, 3630, 3924 (RRL-B), *Fischer & Gage* 154 (CAL); Guttirayanbetta -Kodagarai (DMP), *T. Ravi Shankar* 95502 (MH); Kambukudi (NA), *KS* 6505 (MH); Melpat (SA), *S. coll.* 939 (MH); Mahendragiri (GJM), *VNS* 5617, 5641 (MH).

INDIA: Maharashtra, Karnataka, Tamil Nadu, Kerala, Andhra Pradesh, Odisha.

WORLD: Sri Lanka, Bangladesh, Myanmar, Java,Sumatra.

Neolitsea gamblei Chakrab. & P. G. Diwakar, J. Econ. Taxon. Bot. 33: 345. 2009. *N. foliosa* (Nees) Gamble var. *caesia sensu* Gamble, Fl. Madras 2: 1240. 1925.

Tree, all parts except young shoots (which are yellow sericeous) glabrous, terminal buds perulate, branchelts brown, terete, 2-4 mm thick, glabrous. Leaves narrowly oblong-elliptic or occasionally widely so, 7-20 × 2-7 cm, base cuneate or acute, slightly decurrent into petioles, margin entire, apex caudate (cauda 8-20 mm long, acute), chartaceous, pale greenish brown above when dry, brown to pale brown and glabrous to glaucescent beneath, 3-plinerved, midrib slightly raised beneath, lateral primary veins slender, ascending 50-60 per cent way along the lamina, lateral nerves 3-4 per side above the basal, faint above, prominent beneath, brochidodromus, tertiary nerves obscure (or) faint above, inconspicuous beneath, scalariform, petioles 10-20 × 1-1.5 (2) mm, channeled above. Involucral bracts orbiculer, 4-5 mm across, appressed ochraceous-puberulous; male umbels in the axils of fallen leaves, arising singly, each 5-flowered; pedicels 3-4 × ca.0.6 mm, ochraceous-sericeous; sepals 4, lanceolate; oblong, ca 2 × 0.7 mm, puberulous at base outside; stamens 6, filament 1.5-2 mm long. Infructescences: Umbels arisinig singly, each ca 5-flowered. Fruits subglobose, ca 7×6 mm, glabrous, yellow-brown, wrinkled when dry, fruiting pedicels 8-10 mm long, cupure ca 2×5 mm, entire.

Rare in Rampa hills in East Godavari district. Endemic to Andhra Pradesh. Fl. & Fr.: October-February.

Without precise locality (EG), *VNS* 123 (CAL); Rampa hills, *MSR* 1536 (CAL); Ethakonda (EG), *VNS* 593 (CAL).

INDIA: Andhra Pradesh, endemic.

Neolitsea scrobiculata Gamble, Fl. Madras 2: 1240. 1925; Matthew, Mat. Fl. Tamilnadu Carnatic 323, 1981 & Ill. Fl. Tamilnadu Carnatic t.602. 1982. *Litsea zeylanica* sensu Hook. f., Fl. Brit. India 5: 178. 1886, p.p., non Nees 1823. *L. scrobiculata* Meissn. in DC. Prodr. 15(1): 223. 1864.

Densely foliaceous, dioecious, evergreen tree with a dense crown. Leaves obovate to lanceolate, 8-13 × 2-6 cm, dark green above, glaucous below, alternate to subverticillate, 3-nerved from base. Flowers 3-8 in umbellules; involucral bracts 4. Male flowers: tepals 4; stamens 3+3, all fertile, opposite tepals, pistillode linear. Female flower: staminodes 3 + 3, ovary globose, 7 x 5 mm; style 2 mm; stigma discoid. Drupe globose, 0.8 cm across, with basal rim of perianth tube, brownish black when dry.

Occasional in hills in Sholas above 1000 m in Eastern Ghats of Odisha and Tamil Nadu. Fl.: April – July; Fr.: ripening by March.

On the way to Puliyar – Javadi hills (NA), *MBV* 1249 (MH); Kottur RF- Yelagiri hills (NA), *MBV* 914 (MH); Kaveri peak – Yercaud (SM), *D.B.Deb* 31269 (MH); Kiliyur water falls – Yercaud (SM), *AVN Rao* 26956 (MH); Mahendragiri (GJM), *JSG* 14109 (CAL); Sirumalai hills (DGL), *SKS* 113 (SGH).

INDIA: Peninsula. Endemic

Neolitsea umbrosa (Nees) Gamble, Pl. Wilson 2: 79. 1914. *N. foliosa* (Nees) Gamble, Fl. Madras 2: 1240. 1925. *Litsaea foliosa* Nees, Syst. Lour. 622. 1836.

Evergreen, medium-sized tree. Leaves subcoriaceous or chartaceous, ovate-lanceolate, 5-11 × 3-5 cm, glabrous, glaucous beneath, base acute, margin entire, apex acute, triplinerved; petiole 1-2 cm. Flowers yellow, small in axillary and lateral, sessile clusters of 4-5-flowered umbels. Berry globose, 8 mm in dia.

Occasional in forests at 600-800 m. Fl.: September-November. Fr.: January-March. Vern.: Tel.: *Bodeda, Aku patrikamu, Buroyida.*

Mamandur to Thumburatheertham (CTR), *SSR* 16177; Tirumala (CTR), *KJ* 12554, *DRC* 1147 (MH); Papavinasanam (CTR), *LR* & *M.S.Rao* 2527 (BSID); Seshachlam hills (CDP), *CECF* 4881 (CAL); Rampa hills (EG), *MSR* 1536 (CAL); Talakona (CTR), *BR* & *SKB* 32710; Japalitheertham (CTR), *JHFB* 117496 (BSID); Mahendragiri (GJM), *SX* & *MB* 2169 (RRL-B).

INDIA: Deccan Peninsula, Madhya Pradesh.

WORLD: Malesia.

PERSEA Miller *nom. cons.*

1. Leaves tawny villous beneath; petiole 1.2-1.8 cm .... **P. villosa**
1. Leaves glabrous, glaucous beneath; petiole 18-3.2 cm .... **P. macrantha**

Persea macrantha (Nees) Kosterm., Reinwardtia 6: 193. 1962 & 7: 537. 1969. *Machilus macrnatha* Nees in Wall., Pl. Asiat. Rar. 2: 70. 1831 & 3: 31. 1831; FBI 5: 140. 1886; Gamble 2: 1227. 1925.

Evergreen tree, up to 20 m high. Leaves alternate, elliptic to oblong, 8-18 × 3-8 cm, base acute to obliquely truncate, margin entire, apex obtuse to acute; nerves up to 12 pairs, sometimes with incomplete ones in between, veins arched forming intramarginal nerve, midrib impressed above, raised below; petiole up to 3.6 cm long, canaliculated, glabrous. Flowers in terminal panicles, bisexual; tepals 6, equal or the outer whorl smaller, 3 + 3, obovate, 4-5 mm, puberulous; fertile stamens 9, filaments pubescent, 2.5 mm, staminodes 3. Berry globose, 1.5 cm across, with a basal, persistent rim of perianth, lobes reflexed in young fruits, deciduous later, epicarp red when ripe; seed 1, globose.

Occasional in forests above 1100 m. Fl.: February- March, Fr.: May-June. Vern.: Tam.; *Injinaari, Kolamavu.*

Sunkarimetta (VSKP), *SSR* 15913; Vanadurga temple, Minumuluru (VSKP), *GVS* 47354 (MH); Yercaud (SM), *KCJ* 018063 (MH), on the way to Nagaluru-Yercaud (SM), *AVN Rao* 27466 (MH); Way to Manja Kuttai, Yercaud (SM), *DB Deb* 31390 (MH); Yercaud to Salem (SM), *DB Deb* 31432 (MH); Sirumalai hills, near Vellimalai (DGL), *SKS* 201 (SGH).

INDIA: Peninsular India.

WORLD: Sri Lanka.

Persea villosa (Roxb.) Kosterm., Reinwardtia 6: 194. 1962. *Laurus villosa* Roxb., Fl. Ind. 2: 310. 1832. *Machilus villosa* (Roxb.) Hook.f., Fl. Brit. India 5: 140. 1886.

Large, usually much branched tree; bark dark grey, nearly smooth; shoots tawny tomentose. Leaves crowded towards the ends of branches, lanceolate, elliptic or elliptic-obovate, 10-17.5 cm long, glossy and glabrous above, tawny villous beneath; apex acuminate, base acute; petiole 1.2-1.8 cm long. Flowers in panicles at the base of new shoots, tawny-tomentose; perianth campanulate, tepals elliptic-obovate. Fruit black, globose, 1 cm in diam.

Occasional in Eastern Ghats of Odisha. Fl. & Fr.: April – May.

Bhanjabasa (MBJ), *Bal* 114 (DD); Dudurchampa (MBJ), *Haines* Acc. No. 89151 (DD); Kalandi (KHD), *HFM* 2447 (DD).

INDIA: Sikkim, Himalaya; Upper Assam,

WORLD: Bangladesh.

Persea americana Miller, is cultivated in estates for its fruits (Avocado) in Yercaud, Kollimalais and Sirumalai hills.

PHOEBE Nees

1. Inflorescence and leaves rusty-tomentose beneath .... **P. wightii**
1. Inflorescence and leaves glabrous... **P. lanceolata**

Phoebe lanceolata Nees, Syst. Laur. 109. 1836; FBI 5: 1411. 1886; Gamble 2: 1228. 1925.

Small tree, bark grey, nearly smooth, twigs brown with persistent leaf-scars and lenticels. Leaves lanceolate or oblong-lanceolate, 11-23 × 2-5 cm, glabrous, apex acuminate or caudate, base tapering; petiole 1-2 cm long. Flowers in axillary panicles; peduncle slender, 9 cm long; perianth 3-4 mm long, glabrous, erect and hard, clasping the base of the fruit. Berry narrowly ellipsoid, black.

In shady valleys at Similipahar in Odisha. Fl.: Mar-May; Fr.: May – July.

Similipahar (MBJ), *SX* 361 (RRL-B), *SX* & *MB* 4319, 4402, 4474, 4541, 4581, 5584, 5658 (RRL-B).

INDIA: Meghalaya, Odisha.

WORLD: Myanmar.

Phoebe wightii Meissn. in DC., Prodr. 15(1): 38. 1864; Gamble 2: 1228. 1925. *P. paniculata auct. non* Nees.; FBI 5: 142. 1886 *p.p.*

Densely foliaceous, evergreen tree, up to 10 m high; bark thin, brown; wood brown, smooth with brown medullary patches. Leaves crowded at the ends of branchlets, coriaceous, lanceolate-elliptic, 9-17 × 4-5.8 cm, glabrous above, glaucous and minutely puberulous beneath, apex acuminate, margin entire, base cuneate to acute; secondary nerves 6 pairs, the nerves and transverse nervules distant; petiole to 4 mm. Flowers yellowish, 1cm across in terminal or axillary cymose panicles; peduncle slender, 7-10 cm long, usually lateral from the branches below the upper most axils; perianth densely rusty-pubescent, 3-3.5 mm long; fertile stamens 9, in 3 whorls, outer 2 whorls introrse, eglandular, inner 3rd whorl extrorse with stipitate glands, 4th inner shorl is of 3 staminodes; ovary globose, style up to 1.5 mm long stigma discoid. Berry red, globose, 1 cm across, persistent perianth lobes erect, cupular; seed 0.7 cm across.

Common, in semi-evergreen forests. Fl.: April-May. Fr.: Throughout the year. Vern.: Tam.: *Karunthagiri*.

Galikonda (VSKP), *GVS* 19631 (MH); Kuthadya hill, *VNS* 5918 (MH); Bhanjanagar, Dudurchampa-Badomukkabadi, *SX* & *MB* 4997 (RRL-B); Kolli hills (NMK), *SKS* 109 (SGH).

INDIA: Peninsular India. Endemic.

HERNANDIACEAE

GYROCARPUS Jacquin

Gyrocarpus americanus Jacq., Select. Stirp. Amer. Hist. 282. t. 178. f. 80. 1763; Gamble 2: 1242. 1925. *G. jacquini* Roxb., Pl. Coromandel t. 1. 1795; FBI 2: 461. 1878.

Deciduous tree, to 20 m high; trunk cylindric, bark whitish; wood light, soft, white; young shoots pubescent. Leaves broadly ovate, 3-lobed, 5–10 × 8–15 cm, chartaceous, glabrous above, thinly pubescent below with appressed hairs, base truncate to sub-cordate, margin entire, apex acute to acuminate; lateral veins forking, extending to $^{2}/_{3}$ of the leaf; veins flattened to impressed above, slightly raised below; petiole 8–12 cm long. Flowers small, white or pale yellow, unisexual in large, branched cymes; males numerous; females few. Male flowers 2–3 mm across; tepals 4–7, unequal, ovate, concave, 2 mm, thinly pubescent without, acute; stamens 4–7; filaments 3 mm, sparsely hairy; anthers 0.5 mm long, valves opening from base upwards; staminodes flat, obovate-spathulate, 1 mm across, obtuse. Female flowers 3 mm across; tepals 2, basally adnate to ovary, 2-winged above, appressed-pubescent, obtuse. A few of bisexual flowers scattered among a mass of male flowers, yellow or creamish. Gland white, disc-shaped, 1 mm across. Drupe obovoid-globose, longitudinally ribbed, crowned by 2, spathulate wings; wings 7 × 0.7 cm, apex obtuse, divaricate; seed 1; cotyledons large, convolute.

Common, in dry deciduous forests and along the roadsides, to 500 m. Fl: January-March, Fr.: May-September. Vern.: Tel.: *Tella poniki, Kummara poliki, Thanaku, Thanaki, Poliki*; Ori.: *Pitela, Sutorono*; Tam.: *Kaathadimaram, Tanakku, Vellathannakku*.

Kalasamudram (ATP), *NY* 400; Thimmammamarrimanu (ATP), *KRKS* & *KRS* 37932; Gundam RF (KNL), *TP* & *RVR* 3013 and 3017; Guvvalacheruvu RF (CDP), *RVR* 8157; Chandragiri hills (CTR), *PSPB* 9125; Chintala (PKM), *RVK* 15772; Ethipothala (GNT), *VRK* 3903; Kovur (WG), *SSR* 15257; Chelama RF (KNL), *JLE* 23732 (CAL); Bakarapeta (CDP), *JSG* 21245 (MH); Palakonda range (CDP), *KS* 7805 (MH); Chandragiri fort (CTR), *DRC* 2276 (MH); Rapur hills (NLR), *PV* 110013 (BSID); Kondapalli Fort (KSN), *PV* 6028 (MH); DM Bhavi, Kondapalli (KSN), *PV* 5648 (MH); Devipatnam (EG), *GVS* 27216 (MH); Sathanur dam (NA), *EV* 55667 (MH); Hogainakkal (SM), *SI Flora* 11622 (MH), *KCJ* 018025 (MH); Hogainakkal forest (SM), *EV* 20583 (MH); Gomukhi-Vellimalai Ghat Section (SA), *KRM* 77359 (MH); Parsuramkund (GJM), *SX, MB* & *Panigrahi* (RRL-B); Sirumalai hills, Kudladampatti (DGL), *SKS* 859 (SGH).

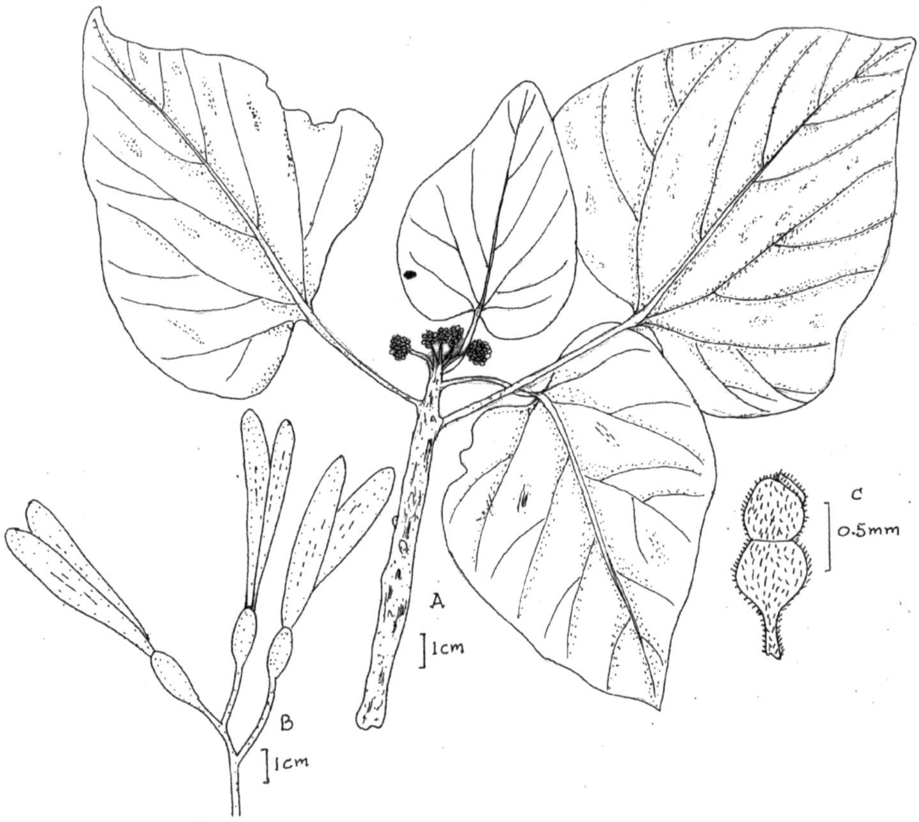

Figure 10. **Gyrocarpus americanus** Jacq.
A. Twig, B. Fruit, C. Female flower.

INDIA: Deccan Peninsula, Odisha, West Bengal.

WORLD: Malaysia, Philippines, Lesser Sunda Island. Tropical W. and S. Africa, S. America.

PROTEACEAE

GREVILLEA R.Brown ex R.A.Salisbury

Grevillea robusta A. Cunn. ex R. Br., Suppl. Prodr. Fl. Nov. Holl. 24. 1830; Gamble 2: 1243.1925.

Evergreen tree, 10-25 m high; sparingly branched at the top; young branches hoary tomentose. Leaves 2-pinnatifid, fern like, the secondary lobes entire or again lobed, lanceolate, margins recurved, silvery beneath. Flowers orange with slender pedicelled in unilateral racemes; pedicel 1-1.5 cm long; sepals 1.5-2 cm long, hooded, at first all fused together except on one side, later fused in twos basally and apically, free for greater length in the middle; stamens sessile; disc annular; gynophores about 2-3 mm long; ovary glabrous, styles 1-2.5 cm long; stigma conical. Fruit broad, very oblique, 2 to 2.5 cm long, dehiscing on one side; seeds 1-2, often winged, ca. 1 cm in diam.

Planted, originally introduced in the hills as a shade tree for coffee plantations. It is now well established. It is also a fast growing tree and reproduces naturally. Fl.: May-June; Fr.: July - August. Vern.: Tel.: *Parana, Perane*; Tam.: *Adhichavukku, Kathichavukku*. Eng.: Silver Oak.

Tirumala (CTR), *KJ* & *SSR* 12564; Anantagiri (VSKP), *GVS* 19461 (MH).

INDIA: Grown as an ornamental, road-side tree and as a shade or wind break tree in tea and coffee gardens.

WORLD: Endemic to E. Australia; extensively cultivated in tropics and sub-tropics as an ornamental, road-side tree and as a shade or wind break tree in tea and coffee gardens.

THYMELAEACEAE

GNIDIA Linnaeus

Gnidia glauca (Fresn.) Gilg., Bot. Jahrb. Syst. 19: 265. 1894. *Lasiosiphon eriocephalus* (Wall. ex J. Graham) Decne., Voy. Inde 4: 148. 1844. *L. eriocephalus* var. *zeylanicus* Meisn., Prodr. 14(2): 598. 1857. *L. hugelii* Meisn., Prodr. 14(2): 598. 1857. *L. metzianus* Miq., Anal. Bot. Ind. 2: 15-16, pl. 1 15. 1850. *L. rivae* (Gilg) H. Pearson, Fl. Trop. Afr. 6(1): 232. 1910. *Gnidia monticola* Miq., Flora 32: 557. 1849.

Small trees up to 7 m tall; young branchlets angular to terete, pubescent when young, later glabrous. Leaves simple, alternate, spiral, clustered at tips; petiole 0.3 cm long, slender, pubescent when young, narrow elliptic to elliptic-ovate, apex acute with mucronate tip, base obtuse, margin entire, pubescent beneath. Flowers bisexual, bright yellow, in terminal white silky compact heads surrounded by an involucres of silky bracts; pedicels 2 mm, densely pilose; calyx tube to 1.5 cm, densely long pilose without; lobes 5, 3 mm, oblong, pilose without; petaloid appendages 8, yellow, terete, fleshy; stamens 8-10, in 2 rows; ovary superior, sessile, oblong, 1-celled; style filiform, stigma capitate. Fruit a berry, 4 mm long, ellipsoid, densely pilose, enclosed in the calyx. Fruit indehiscent, dry, small, 1-seeded; seed brown, ridged laterally.

Occasionally found in Sirumalai hills. Fl. & Fr.: February - July. Vern.: Tam.: *Nachinar, Paatainaar.*

Sirumalai hills (DGL), *SKS* 422 (SGH).

INDIA: Karnataka, Kerala and Tamil Nadu.

WORLD: Tropical Africa, Madagascar, Arabia and Sri Lanka.

ELAEAGNACEAE

ELAEAGNUS Linnaeus

1. Fascicle more than 5 flowers; nut winged.. **E. kologa**
1. Fascicle less than 5 flowers; nut not winged:
    2. Leaf margin entire; scales shining, silvery only................................ **E. indica**
    2. Leaf margin revolute; scales shining or dull coppery and silvery.. **E. conferta**

Elaeagnus conferta Roxb., Fl. Ind. 1: 460. 1820; Gamble 2: 1246. 1925. *E. latifolia* auct non L., FBI 5: 202. 1890 *p.p.*

Evergreen small trees or scandent shrubs; bark smooth; branchlets, petioles and underside of leaves densely clothed with silvery circular, dentate and lobed scales. Leaves variable, 4-10 × 1.5-5 cm, broadly elliptic to elliptic-lanceolate, alternate, base acute or rounded, margin revolute, apex acute or acuminate, coriaceous, the upper surface pale green clothed with small coppery scales, the lower surface glistening, silvery white. Flowers small, straw coloured, many in a cluster, pedicels clothed with silvery scales. Nut ellipsoid, 2.5 cm, succulent, red or yellow when ripe, edible.

Occasional in forests in Araku valley. Fl.: March – April; Fr.: Through the year. Vern.: Tel.: *Konda jilledu*; Tam.: *Kulangai, Kulari.*

Sunkarimetta (VSKP), *GVS* 22675 (MH); Sapparlagedda (VSKP), *GVS* 42779 (MH); Melpat (SA), *CAB* 932 (MH), *CAB* 1079 (MH).

INDIA: Maharashtra, Karnataka, Kerala, Tamil Nadu, Andhra Pradesh.

WORLD: Bhutan, Nepal, China, Bangladesh, Myanmar, Malaysia, Indonesia, Laos, Vietnam.

Elaeagnus indica Servett., Bull. Herb. Boissier (Ser.2) 8: 393. 1908; Gamble 2: 1246. 1925.

Slender straggling shrubs, silvery stellate scales on the branches of inflorescence and under-surface of the leaves especially. Leaves 4-7 × 1.5-4 cm, oblanceolate to elliptic, alternate, base acute to truncate, margin entire, apex acute, chartaceous, scales silvery white, stellate, closely appressed below. Flowers small, creamish, in clusters of 3-4 flowers, 3.5 mm across. Nut ellipsoid-oblong.

Occasional in forests. Fl. & Fr.: March – June.

Towards Minumuluru (VSKP), *GVS* 47336 (MH); Inner Javadi hills RF (NA), *KS* 6206 (MH); Sirumalai (MDU), *KS* 5792 (CAL).

INDIA: Peninsula.

WORLD: Indo-Malesia.

Elaegnus kologa Schlecht. in DC. Prodr. 14: 611. 1857; Gamble 2: 1246. 1925. *E. latifolia* L. Sp. Pl. 121. 1753, p.p.; FBI 5: 202. 1886 p.p.

Large, sometimes thory, straggling or climbing shrub; thorns strong, straight or reflexed; tender terminal shoots rusty-silvery. Leaves elliptic or elliptic-oblong, 3.8-10 × 2-5 cm, silvery beneath, apex obtuse or rarely acute, margin entire or subentire, base obtuse or rounded; petiole 2.5-6.2 mm. Flowers yellowish-green or greenish-white, clustered; perianth-tube broadly urceolate, 5-7.5 mm long. Nut oblong-ellipsoid, 1.2-1.8 cm long, ribbed, orange-red.

Occasional in forests in Koraput, Araku valley, Kollimalais and Servarayans, above 1200 m. Fl.: March – April. Fr.: Through the year. Vern.: Tam.: *Kulangi, Perunkuliri;* Ori.: *Dibaguda.*

Deomali (KPT), *SX* & *MB* 6604 (RRL-B).

LORANTHACEAE

1. Corolla choripetalous; petals free completely or almost to the base, but sometimes pseudosympetalous or sometimes coherent .... **Helixanthera**

1. Corolla sympetalous; petals fused up to the middle or higher, sometimes with corolla-tube deeply split on one side:

   2. Corolla 5-merous, slightly zygomorphic, not deeply split on one side:

      3. Floral tube constricted in the middle; corolla lobes not reflexed .... **Macrosolen**

      3. Floral tube constricted above the middle; corolla lobes reflexed .... **Dendrophthoe**

   2. Corolla mostly 4-merous, zygomorphic with a deep split on the inner side of the curved tube:

      4. Inflorescence usually 3-10-flowered simple raceme, rarely reduced, 2-flowered umbel; fruit pyriform, obovoid or clavate, distinctly stipitate, not woody .... **Scurrula**

      4. Inflorescence a few-flowered simple 2-5-rayed umbel; fruit urceolate or ellipsoid, not stipitate, warty .... **Taxillus**

DENDROPHTHOE Martius

1. Branches trigonous; corolla less than 2 cm long, cylindric throughout; calyx flask-shaped .... **D. trigona**
1. Branchlets terete; corolla more than 3.5 cm long, inflated above the middle; calyx tubular .... **D. falcata**

Dendrophthoe falcata (L.f.) Ettingsch., Denskschr. Kaiserl. Akad. Wiss. Wien Math.-Naturwiss. Kl. 32: 52. 1872. var. **falcata**; Rajasekaran in N.P.Balakr. *et al.*, Fl. India 23: 3. 2012. *Loranthus falcatus* L.f., Sp. Pl. Suppl. 211. 1781. *L. longiflorus* Desr. in Lam., Encycl. 3: 598. 1792; FBI 5: 214 1886. *L. longiflorus* Desr. var. *falcata* (L.f.) Kurz, Forest Fl. Burma 2: 321. 1877; FBI 5: 215. 1886; Gamble 2: 253. 1925.

Key to Varieties

1. Leaves linear, up to 2.5 cm wide. .... var. **falcata**
1. Leaves broader, 2.5 cm wide:
   2. Leaves sessile or more than subsessile, amplexicaule at base; flowers glabrous throughout .... var. **amplexifolia**
   2. Leaves petiolate, not amplexicaule at base; flowers covered with malpighian hair when young .... var. **pubescens**

Dendrophthoe falcata (L.f.) Ettingsch. var. **falcata**

Large bushy, usually glabrous branch parasites; bark grey, smooth. Leaves very variable in shape and venation, 5-12 × 2-3.5, lanceolate or elliptic-ovate, alternate and opposite, base acute to cuneate, margin undulate, apex obtuse, thick, coriaceous. Flowers in racemes, racemes to 6 cm; pedicel short, to 0.6 cm; bracts broadly ovate, 1.5-2 × 0.8-1 mm; calyculus-tube subentire to denticulate, ca 1.2 cm long; petals 5, united, 2.5-5 cm long, yellow-scarlet at base, green at apex, tube terete, lobes 5, linear-elliptic, shorter than tube; stamens 5; ovary oblong, style exserted, stigma capitate. Fruit ellipsoid-ovoid to oblong, with persistent apical calyculus-tube, glabrous, bright pink.

Occasional on trees all over Eastern Ghats. A partial stem parasite on *Albizia lebbeck, Chloroxylon swietenia, Hardwickia binata etc.* Fl. & Fr.: November– March. Vern.: Tel.: *Badanika chettu, Bajinika, Jiddu, Yelinga;* Ori.: *Madang, Malang, Bridhango;* Tam.: *Pulluruvi, Ottu.*

Gundumala RF (ATP), *TP & NY* 743; Thummalapalle (CDP), *TP & SSR* 33647; Palakonda hills (CDP), *CS* 8208; Bollapalli (GNT), *VRK* 5749; Rampachodavaram (EG), *TP & KJ* 12504; Diguvametta (KNL), *JLE* 16768 (MH); Balapalli (CDP), *JLE* 15761, 14986 (MH); Kambakam hills (CTR), *MCB* 45125 (MH); North of Mamadur FRH (CTR), *GVS* 46915 (MH); Kakalakonda RF (NLR), *KCJ* 386 (MH); Enamadala (KSN), *PV* 5509 (MH); Maredumilli (EG), *MM* 105062 (MH); Devipatnam (EG), *GVS* 24420 (MH); Sirivaka (EG), *CAB* 5134 (MH); Chintapalli (VSKP), *GVS* 42786 (MH); Lakshmipuram (VSKP), *CAB* 1772 (MH); Simhachalam (VSKP), *GVS* 19347 (MH);); Near Singapuram (SKLM), *GVS* 62351 (MH); Kholbilung (SBP), *SP & APD* 314 (CAL); Javadi hills (NA), *MBV* 965 (MH); Sathanur dam (NA), *EV* 55669 (MH);

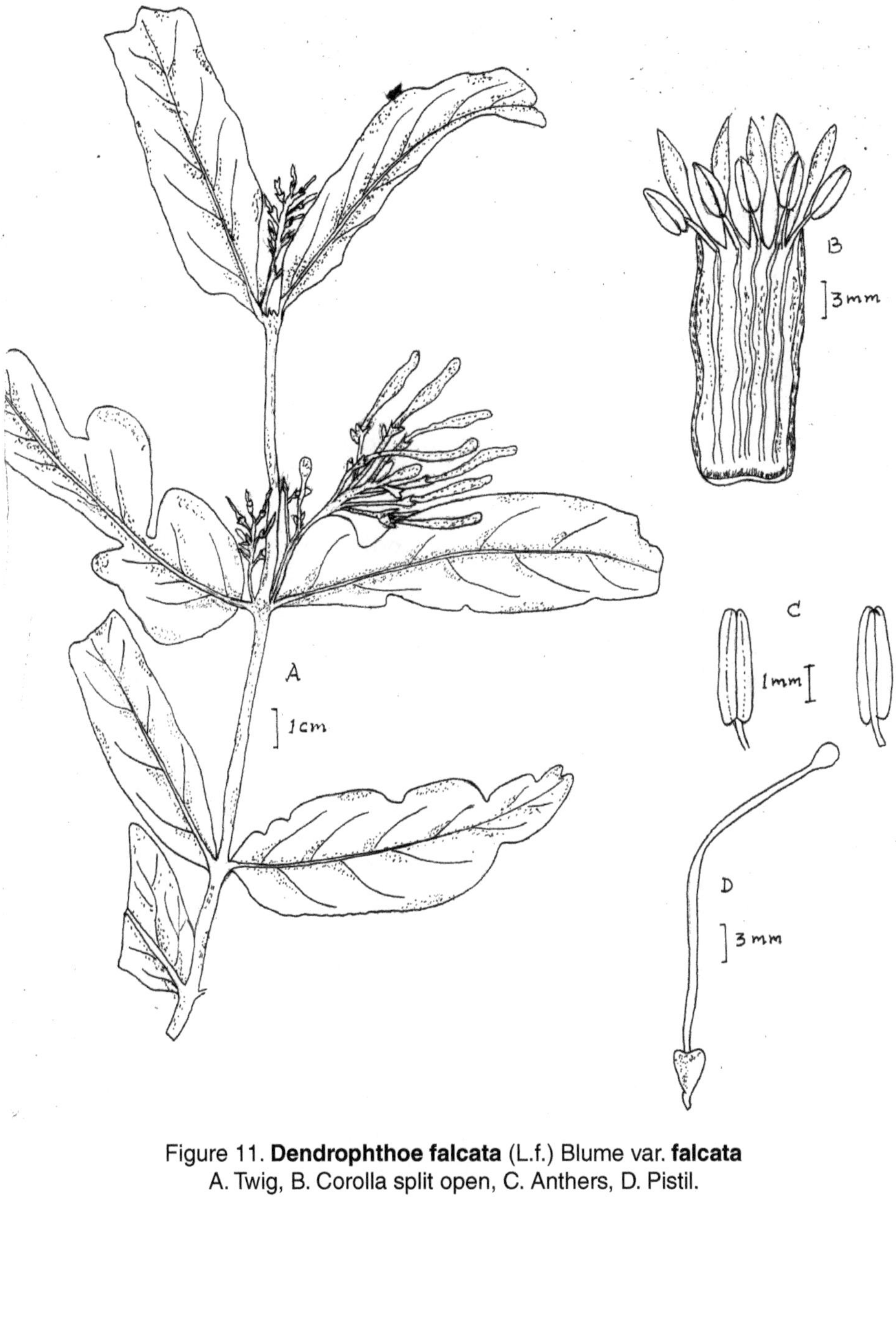

Figure 11. **Dendrophthoe falcata** (L.f.) Blume var. **falcata**
A. Twig, B. Corolla split open, C. Anthers, D. Pistil.

Hogainakkal forest (SM), *EV* 2924 (MH); Tickapalli to Linepada (GJM), *CAB* 1198 (MH); Sirumalai hills (DGL), *SKS 204* (SGH).

INDIA: Through out India.

WORLD: Sri Lanka, Bangladesh, Myanmar, Indo-China, Thailand, throughout Malesia to N.Australia and Solomon Islands.

Dendrophthoe falcata (L.f.) Ettingsch. var. **amplexifolia** (DC.) Rajasekaran, J. Econ. Taxon. Bot. 31: 182. 2007 & in N.P. Balakr. *et al.*, Fl. India 23: 4. 2012. *Loranthus amplexifolius* DC., Prodr. 4: 305. 1830. *L. longiflorus* var. *amplexifolia* (DC.) Thwaites, Enum. Pl. Zeyl. 134. 1864; FBI 5: 215. 1886.

This variety is more robust than var. *falcata*. The leaves are larger, sessile, orbicular to oblong, 10-18 cm across and cordate-amplexicaul at base.

Occasional in Eastern Ghats. Fl. & Fr. November - April.

Bhupatipalem, Rampa hills (EG), *TP* & *KJ* 12510; Maredumilli (EG), *TP* & *KJ* 12525.

INDIA: Maharashtra, Andhra Pradesh, Karnataka, Kerala and Tamil Nadu.

WORLD: Sri Lanka.

Dendrophthoe falcata (L.f.) Ettingsch. var. **pubescens** (Hook.f.) V.Chandras. in A.N.Henry *et al.*, Fl. Tamil Nadu, Ser. I Analysis 2: 215. 1987; Rajasekaran in N.P.Balakr. *et al.*, Fl. India 23: 5. 2012. *Loranthus longiflorus* var. *pubecens* Hook.f., Fl. Brit. Brit. India 5: 215. 1886.

This variety is characterised by the presence of short hairs on young flowers.

In deciduous forests. Fl. & Fr.: January - December.

Madhavaram (KNL), *TP* & *RVR* 1324; Lankamala RF (CDP), *SRSR* 13084; Talakona (CTR), *BR, MSR* & *BSS* 29559; Diguvametta (PKM), *AJR* & *PANR* 20416; Ethipothala (GNT), *VRK* 3892.

INDIA: Andhra Pradesh, Maharashtra, Karnataka, Kerala and Tamil Nadu.

WORLD: Sri Lanka.

Dendrophthoe trigona (Wight & Arn.) Danser ex Santapau, Rec. Bot. Surv. India 16(1): 263. 1953. *Loranthus trigonus* Wight & Arn., Prodr. Fl. Ind. Orient 1: 386. 1834; FBI 5: 219.1886. *Macrosolen trigonus* (Wight & Arn.) Tiegh., Bull. Soc. Bot. France 42: 442. 1895 in obs.; Rajasekaran in N.P.Balakr. *et al.*, Fl. India 23: 21. 2012.

Shrubs, branchlets trigonous, branching lateral. Leaves opposite, subopposite or whorled, broadly ovate-elliptic or suborbicular, 10-14 × 8-11 cm, base cuneate or rounded, margin entire or undulate, apex obtuse or rounded. Flowers in pseudoracemes; calyx flask –shaped; stamens 5, filaments coherent to petals at base, free above; ovary ellipsoid, stigma capitate. Fruits usually ellipsoid or obovoid with persistent calyculous tube.

Massive clumps in Balaghat in Tamil Nadu (Matthew - Central Tamil Nadu). Fl.: September; Fr.: October-November.

INDIA: Peninsula. Endemic.

HELICANTHES Danser

Helicanthes elasticus (Desv.) Danser, Verh. Kon. Ned. Akad. Wetensch., Afd. Natuurk. Tweede Sect. 2 29(6): 55. 1933; Rajasekaran in N.P.Balakr. *et al.*, Fl. India 23: 10. 2012. *Loranthus elasticus* Desv., in Lam. Encycl. 3: 599. 1792; FBI 5: 216. 1886; Gamble 2: 1254. 1925.

Branchlets woody, dichotomous or trichotomous, swollen at nodes. Leaves decussate, obovate to elliptic, 5-12 x 2.5-6.5 cm, sessile or subsessile, coriaceous, base acute, apex rounded to acute. Flowers in short, axillary dichasial fascicles, sunken in nodal cavities, bract ovate, concave, calyx flask-shaped, 3.5 mm, narrowed above the middle, corolla 1 cm across, tube 1.5 cm, split lengthwise towards one side, lobes 5, linear-lanceolate stamens 5, red, anthers narrowly oblong; ovary subspherical with depressed sides, 2-2.5 mm long, style curved; stigma papillate. Berry red when ripe.

Hills, abundant on shola trees (Matthew, 1991). Fl.: December – April and July – September; Fr.: Throughout the year.

Sirumalai top (MDU), *KS* 5773 (CAL) (MH); Near Doddikere, Kodagarai (DMP), *T Ravi sankar* 95501 (MH); Melpat (SA), CAB 1009 (MH); Kolli hills, road sides of Kulivalavu shola (NMK), *SKS* 1206 (SGH).

INDIA: Peninsula. Endemic.

HELIXANTHERA Loureiro

Helixanthera wallichiana (Schult.f.) Danser, Bull. Jard. Bot. Buitnzorg. Ser. 3, 10: 319. 1929; Rajasekaran in N.P.Balakr. *et al.*, Fl. India 23: 18. 2012. *Loranthus wallichianus* Scult.f., Syst. Veg. 7: 100. 1829; FBI 5: 204. 1886.

Shrubs, branches and leaves hairy when young, glabrous when old. Leaves opposite or alternate, elliptic to elliptic-ovate, 4-13 × 2.5 - 6 cm, rarely orbicular, base attenuate, margin entire or denticulate, apex subacute or obtuse. Flowers scarlet red, in axillary pseudoracemes; calyculous truncate, petals 4, almost free, linear, 4-5 mm long, subglabrous; stamens 4, filaments adnate to petals; ovary urceolate, style 4-gonous. Fruit urceolatee, later turning ovoid or ellipsoid, overtopped by persistent base of calyculous.

In Andhra Pradesh and Tamil Nadu (Rajasekaran *loc. cit.*). Fl. & Fr.: April - November.

INDIA: Andhra Pradesh, Maharashtra, Karnataka, Kerala and Tamil Nadu. Endemic.

MACROSOLEN (Blume) Reichenbach

Macrosolen cochinchinensis (Lour.) Tiegh., Bull. Soc. Bot. France 41: 122. 1894; Rajasekaran in N.P.Balakr. *et al.*, Fl. India 23: 20. 2012. *Loranthus cochinchinensis* Lour., Fl. Cochinch. 1: 195. 1780. *L. globosus* Roxb., Fl. Ind. 2: 206. 1824; FBI 5: 220. 1886. *L. ampullaceus* Roxb., Fl. Ind., eds. Carey & Wall. 1: 189. 1820.

Elytranthe ampullacea (Roxb.) G.Don, Gen. Hist. 3: 425. 1834. *E. cochinchinensis* (Lour.) G.Don, l.c. 426. 1834

Glabrous parasitic shrub. Leaves opposite, elliptic-lanceolate, ovate or elliptic, 5-10 × 2-5 cm, acute to acuminate, glabrous, coriaceous, obscurely nerved when green, base acute or obtuse; petiole 0.5-1.5 cm, thick, rugose. Flowers greenish-orange, ca. 1.2 cm long before the reflexion of the tepals, in 3-7-flowered short racemes, 1.2-1.8 cm long, axilary and from the leaf-scars; bract 1, 1.5 mm long; bracteoles 2, similar, rather smaller, connate with one another about half-way up. Hypanthium with calyculus 3.7 mm long, truncate. Perianth-tube oblong, 5-6-angled and inflated; lobes 5-6, linear-spathulate, about as long as the tube; ridges sharp, extending whole length of the perianth in bud and with triangular projections about middle of the tube, anthers short, oblong, truncate much shorter than free part of filaments. Fruits yellow, globose.

Occasional parasite on *Diospyros melanoxylon* and *Flacourtia indica* in Eastern Ghats of Odisha.

Similipahar (MBJ), *SX* 322, 504 (RRL-B).

INDIA: Bihar, West Bengal, Sikkim, Assam, Tripura, Meghalaya, Odisha and Andaman and Nicobar Islands.

WORLD: Nepal, Bhutan, Bangladesh, Myanmar, China, Thailand, Cambodia, Vietnam, Malaysia to New Guinea.

SCURRULA Linnaeus

1. Leaves white tomentose on both surfaces:
    2. Fruit pyriform; corolla reddish brown:
    2. Fruit baccate; corolla orange yellow .... **S. atropurpurea**
        3. Leaf base cordate .... **S. cordifolia**
        3. Leaf base narrowed or rounded .... **S. pulverulenta**
1. Leaves glabrous above, rusty tomentose beneath .... **S. parasitica**

Scurrula atropurpurea (Blume) Danser, Bull. Jard. Bot. Buitenzorg 10: 349 349 1929; Rajasekaran in N.P.Balakr. *et al.*, Fl. India 23: 23. 2012. *Loranthus atropurpureus* Blume, Verh. Batav. Genootsch. Kunsten 9: 186. 1823. *Cichlanthus philippensis* (Cham. & Schltdl.) Tiegh., Bull. Soc. Bot. France 42: 253, 1895. *Loranthus philippensis* Cham. & Schltdl., Linnaea 3: 204 204 1828. *Scurrula philippensis* (Cham. & Schltdl.) G.Don, Gen. Hist. 3: 442 442 1834.

Semi parasitic shrub; branchlets rusty brown to the elongated hairs mixed with stellate hairs. Leaves opposite, broadly ovate, rounded or slightly narrowed at base, acute or subacute at apex, 8-12 x 3-6 cm, coriaceous, glabrous above, rusty stellate tomentose beneath; lateral nerves 5-7 pairs; petiole to 7 mm long. Inflorescence axillary, fascicled racemes; peduncle 1 or 2 in each axil, 8-10-flowered, up to 5 cm long. Flowers erect, stalked, ventricose at base, 1.7-1.9 cm long; pedicels 3 mm long; bracts ovate, 1.5-2 x 0.5-0.7 mm; perianth united, tube ventricose, medium split extending up to two-third the length, 10-12 mm long; lobes narrowly elliptic,

4-5 mm long, stamens erect, filaments 2 mm long; anthers linear, 3 mm long, ovary obcuneate, 4 mm long; style tetragonous; stigma subglobose. Fruits pyriform, brown, densely stellate hairy with reticulate tubercles on surface, 9-10 x 4 mm.

Usually parasite on *Ficus* species and *Artocarpus heterophyllus* in Sirumalai hills. Fl. & Fr.: November – February.

Sirumalai hills, Maanamuna bend (DGL), *SKS* 883 (SGH).

INDIA: Tamil Nadu, Kerala, Karnataka.

WORLD: Sri Lanka, South East Asia.

Scurrula cordifolia (Wall.) G. Don, Gen. Hist. 3: 421. 1834; Rajasekaran in N.P.Balakr. *et al.*, Fl. India 23: 23. 2012. *Loranthus cordifolius* Wall. in Roxb., Fl. Ind. (ed.) Carey & Wall. 2: 222. 1824; FBI 5: 209. 1886; Gamble 2: 1252. 1925.

Large parasitic shrubs, branchlets white tomentose. Leaves opposite, 9-11 × 5-7 cm, ovate, alternate, base cordate, margin entire, apex obtuse, coriaceous, white tomentose on either side. Flowers orange in fascicles, zygomorphic, variable in length; calyculus with annular hairy rim; petals united; tube with median split extending up to half the length, 1.5-1.8 cm long, lobes 4, narrowly elliptic-ovate; stamens recurved; ovary clavate, style 4-gonous, stigma conical. Fruit pyriform, tomentose with rusty brown hairs.

Occasional in forests. Fl. & Fr.: March – May.

Kalasamudram RF (ATP), *KRKS* 41886; Nallamalai forest (KNL), *KJ* 12593; Tirumala hills (CTR), *KJ* & *SSR* 12559; *DRC* 1213 (MH); way to Tirumala (CTR), *KS* 6929; Nellore (NLR), *PV* 111160 (BSID); Valamuru, Maredumilli (EG), *MM* 100895 (BSID), Rampachodavaram (EG), *MM* 101344 (BSID); Anantagiri (VSKP), *GVS* 19570 (MH); Araku (VSKP), *NPBK* 10753 (MH).

INDIA: Jammu & Kashmir, Uttar Pradesh, West Bengal, Arunachal Pradesh, Odisha, Andhra Pradesh, Karanataka, Tamil Nadu and Kerala

WORLD: Nepal and Sri Lanka.

Scurrula parasitica L., Sp. Pl. 110. 1753; Rajasekaran in N.P.Balakr. *et al.*, Fl. India 23: 27. 2012. *Loranthus scurrula* L., Sp. Pl. (ed.2) 472. 1762; FBI 5: 208. 1886. *L. buddleioides* Desv. in Lam., Encycl. 3: 600. 1792; Gamble 2: 1251. 1925.

A perennial much-branched, stem parasite, branches terete, branchelets rusty tomentose. Leaves 4-6.5 × 5-8 cm, elliptic – ovate, opposite, base rounded or sub-acute, margin entire, apex obtuse, coriaceous, glabrous above, rusty tomentose below. Flowers scarlet, in short axillary racemes, zygomorphic; variable in length, 1-2.4 cm long (including calyx and corolla), buds acute; bracts ovate or elliptic-ovate; pedicels 1-2 mm long; calyculous obsolete or up to 4 mm long; petals 4, united, tube gibbous, median split up to two-third the length, dark reddish brown outside, purplish inside, lobes sagittate; stamens 4, filaments 1.8-2.5 mm long; ovary clavate, style up to 1 mm long, stigma broadly conical. Fruit pyriform, pale green, rusty brown due to candelabra hairs.

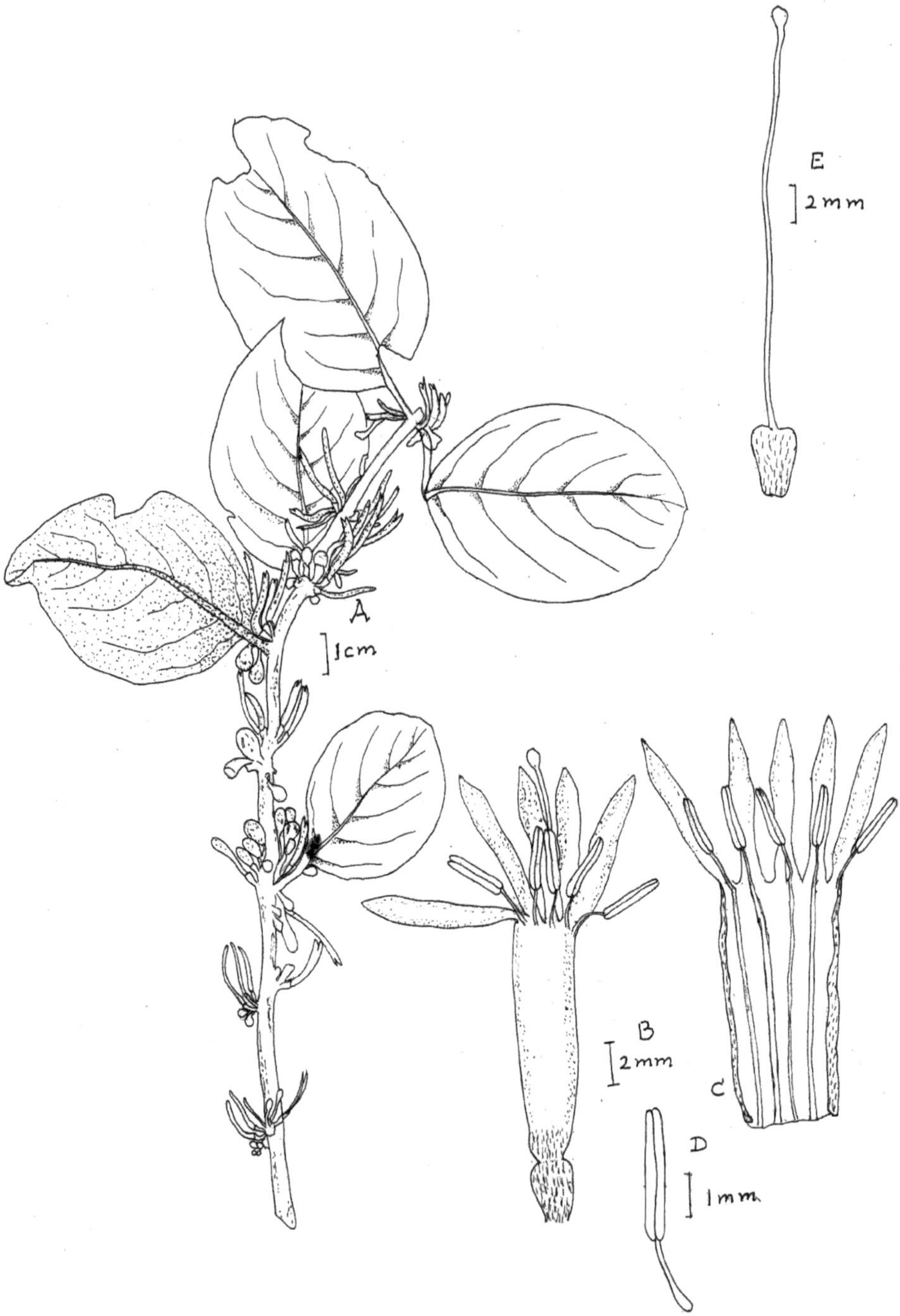

Figure 12. **Scurrula parasitica** L.
A. Twig, B. Flower, C. Corolla split open, D. Anther, E. Pistil.

Occasional in forests. Fl. & Fr.: May – December.

Kalasamudram (ATP), *EC* 2928, *KRKS* 41886; Peddacheruvu (KNL), *BR* & *VSRao* 38889; Tirumala (CTR), *KJ* & *SSR* 12596; Maredumilli (EG), *TP* & *KJ* 12515; Araku valley (VSKP), *NPBK* 10753 (MH).

INDIA: Through out India.

WORLD: Sri Lanka, Nepal, Bhutan, Bangladesh, Myanmar, China, Thailand, Taiwan, Vietnam, Malaysia, Indonesia, Philippines, Moluccaas and Timor.

Scurrula pulverulenta (Wall.) G.Don, Hist. 3: 421. 1834; Rajasekaran in N.P.Balakr. *et al.*, Fl. India 23: 27. 2012. *Loranthus pulverulentus* Wall. in Roxb., Fl. Ind. ed. Carey, 2: 221. 1824; FBI 5: 211. 1886; Gamble 2: 1252. 1925.

Stout, parasitic shrubs, found on deciduous trees. Leaves ovate, to 9 × 6 cm, base narrowed or rounded, margin entire, apex bluntly acute or obtuse, sub-coriaceous; petiole 1.2-2.5 cm long. Flowers in fascicles, corolla slender, curved, green when fresh, cream-coloured when dry, tube terete, split above, lobes 4, connate; stamens 4, erect, filmentss 5-6 mm long; ovary obconical, style filiform, stigma subcapitate. Fruit clavate, pyriform, white stellate hairy.

Rare on forest trees. Gamble reported in forests of Northern Eastern Ghats. Fl. & Fr.: September - March.

Mahendragiri (GJM), *JSG* 13737 (CAL); Kalugia (GJM), *JSG* 13737 (MH).

INDIA: Uttarakhand, Sikkim, Arunachal Pradesh, Manipur, Mizoram, Meghalaya, Madhya Pradesh, Maharashtra, Odisha and Andhra Pradesh.

WORLD: Pakitan, Nepal, Bhutan, Bangladesh, Myanmar, China and Thailand.

TAXILLUS Tieghem

1. Branchlets tawny tomentose; leaves pilose pubescent; flowers in fascicles:
    2. Bracts large; corolla tube inflated at and above the middle .... **T. heyneanus**
    2. Bracts minute; corolla tube inflated below the middle .... **T. tomentosus**
1. Branchlets not so; leaves glabrous; flowers in cymes .... **T. cuneatus**

Taxillus cuneatus (Roth) Danser, Bull. Jard. Bot. Buitenzorg. Ser. 3. 10: 354. 1929; Rajasekaran in N.P.Balakr. *et al.*, Fl. India 23: 32. 2012. *Loranthus cuneatus* Heyne ex Roth, Nov. Pl. Spec. 193. 1821; FBI 5: 214. 1886; Gamble 2: 1253. 1925.

Bushy shrubs, bark grey, branches long, slender, numerous, with prominent nodes, parasitic on great variety of trees and shrubs. Leaves 1.5-3.5 × 1-2 cm, obovate-spathulate, alternate, base cuneate, decurrent into a short petiole, margin entire, apex rounded or obtuse, coriaceous, glabrous. Flowers bisexual, 2-5 in axillary sessile or pedunculate cymes; bracts elliptic to ovate-elliptic, clasping ovary; calyculous tube subcylindrical, irregularly 5-toothed; petals 5, united; stamens 5; ovary oblong-cylindric, style tetragonous, stigma subcapitate. Fruit obconic, truncate, crowned with persistent calyculus, red.

Occasional in forests as stem parasite on various trees and shrubs. Fl. & Fr.: July – October.

Velgode to GBM (KNL), *BR* & *SKB* 34031; Jyothi (CDP), *BSS* & *SKB* 30857; Harsapalli, *JSG* 15249 (MH); Ragupalem (VSKP), *CAB* 1584 (MH); Sirumalai hills, Vellimalai (DGL), *SKS* 637 (SGH).

INDIA: Andhra Pradesh, Maharashtra, Karnataka and Kerala.

WORLD: Sri Lanka.

Taxillus heyneanus (Schult.f.) Danser, Bull. Jard. Bot. Buitenzorg, ser. 3, 10: 355. 1929; Rajasekaran in N.P.Balakr. *et al.*, Fl. India 23: 33. 2012. *Loranthus heyneanus* Schult. f., Syst. Veg. 7: 106. 1829 (non Schult.f. 1830). *L. bracteatus* Heyne ex Wall. in Roxb., Fl. Ind. 2: 220. 1824; FBI 5: 213. 1886; Gamble 2: 1252. 1925. *Taxillus bracteatus* (Wall.) Tiegh., Bull. Soc. Bot. France 42: 256. 1895; Jayamma in Pullaiah & Moulali, Fl. Andhra Pradesh, 2: 831. 1997.

Large, semiparasitic shrubs with haustoriferous runners, to 1 m high on trees; stem brown tomentose, corky; branches whitish, tawny-tomentose. Leaves alternate, elliptic, sub-orbicular, 1.5–2.5 × 1–2 cm, pubescent, base truncate-obtuse, acute, margin entire, apex obtuse; lateral veins 3–5 pairs, mostly raised from base; petiole to 0.5 cm long, tomentose. Inflorescence axillary umbellate raceme or cyme. Flowers 2.5 cm long, subtended by a lateral bract, bract oblanceolate, 7 × 3 mm; calyx 4 mm long, margin obscurely 2–4-toothed, white-pilose without; corolla greenish yellow, red inside, 0.4 cm across; tube 2.4 cm long, inflated above middle, pilose without; lobes 5, linear-lanceolate, reflexed, 7 × 1.5 mm, subacute, red within; stamens 5; filaments 3 mm long; anthers 2 mm long, basifixed; ovary oblong-cylindric, 2 × 1.5 mm; style 2.2 cm long; stigma capitate, light yellow. Fruit oblong or ellipsoid, up to 2 cm long, red, with persistent bract, annulate above by persistent calyx; seed 1; yellowish, oblong, to 1.7 cm long, sticky nature.

Occasional in forests as stem parasite on various trees. Fl. & Fr.: April – October. Vern.: Tel.: *Jana-bhajanika.*

Batrepalli (ATP), *KRKS* & *BSS* 37977; Owk RF (KNL), *RVR* 1464; Rollapenta (KNL), *PSPB* 9134; Yerramalai (KNL), *JSG* 17706 (MH); Chelama (KNL), *JLE* 16791 (MH); Kapilatheertham (CTR), *LR* & *M.S. Rao* 2517 (BSID); Lakshmipuram (VSKP), *CAB* 1734 (MH).

INDIA: Andhra Pradesh, Karnataka, Kerala and Tamil Nadu. Endemic.

Taxillus tomentosus (Heyne ex Roth) Tiegh., Bull. Soc. Bot. France 42: 256. 1895; Rajasekaran in N.P.Balakr. *et al.*, Fl. India 23: 33. 2012. *Loranthus tomentosus* Heyne ex Roth, Nov. Pl. Sp. 191. 1821; FBI 5: 212. 1886. *L. gibbosus* Talbot, Syst. List Trees, Shrubs Bombay ed. 2: 289. 1902.

Semi-parasitic shrub; branchlets rusty hairy, intermixed with white stellate hairs, glabrous when old. Leaves alternate, obovate-oblanceolate or broadly elliptic-ovate, cuneate or sometimes rounded at base, 2.5-7 x 1.5-3.5, thickened along margin, obtuse at apex, coriaceous; petiole 5-10 mm long. Inflorescence axillary, flowers in dichasial, sessile or shortly peduncled with single or in paris. Usually flowers 5-merous, up to 2.2 cm long; bracts broadly ovate; petals 5, united, tube ventricose at abse, with median split open up to two-third the length; lobes narrowly elliptic to lanceolate, acute, 3-4 mm long, stamens 5, filament up to 2 mm long; anthers oblong, 1-2 mm long; ovary subturbinate, 2-3 mm long; style up to 1.5 mm long, stigma bilobed, papillate. Fruits urceolate, crowned with persistent unevenly toothed calyculus, 10 x 5 mm, smooth, browny hairy.

Occasionlly found on the *Wendlandia thyrsoidea* and *Careya arborea* in Sirumalai hills of Tamil Nadu. Fl. & Fr.: January – August.

Sirumalai hills, Oordai (DGL), *SKS* 793 (SGH).

INDIA: Himachal Pradesh, Uttarakhand, Sikkim, Arunachal Pradesh, Assam, Meghalaya, Kerala, Karnataka and Tamil Nadu.

WORLD: Nepal, Myanmar and China.

VISCACEAE

VISCUM Linnaeus

1. Plants leafless:
    2. Internodes flattened .... **V. articulatum**
    2. Internodes angular .... **V. angulatum**
1. Plants leafy, internodes terete:
    3. Leaves present only on lower branches .... **V. ramosissimum**
    3. Total plant leafy:
        4. Leaf apex subacute .... **V. monoicum**
        4. Leaf apex obtuse:
            5. Central flower male; leaves cuneate, subtruncate ovate, obovaate or lanceolate;
                6. Young fruits warty; internodes longitudinally grooved; leaves ovate, obovate or oblong .... **V. heyneanum**
                6. Young fruits smooth; internodes not grooved; leaves lnceolate, cuneate or subtruncate .... **V. capitellatum**
            5. Central flower female; leaves lanceolate, oblong or obovate, acute .... **V. orientale**

Viscum angulatum B. Heyne ex DC., Prodr. 4: 283. 1830; FBI 5: 225. 1886; Sanjai & N.P. Balakr. in N.P.Balakr. *et al.*, Fl. India 23: 49. 2012.

Monoeciouss, green or yellowish, herbaceous or shrubby, semi-stem-parasites, usually hanging, stems slender, angular; branches numerous, more than 2 or rarely decussate at nodes. Leaves scaly, mostly not visible, up to 0.5 mm long. Flowers unisexual. Male flowers: globose, laterally compressed and wedge-shaped towards the base, perianth lobes 4, broadly ovate; stamens 4, sessile. Female flowers: perianth lobes 4, broadly ovate, ovary inferior, style short, conical and stigma nipple-shaped. Fruit baccate, subglobose to globose, up to 4 mm in diameter, with persistent bracts at base.

Southern Eastern Ghats. Fl. & Fr. February - August.

Melpatti (NA), *KS* 6126 (CAL); Sirumalai hills (DGL) *SKS* 216 (SGH).

INDIA: Gujarat, Maharashtra, Goa, Karnataka, Kerala, Tamil Nadu and Andhra Pradesh.

WORLD: Asia and Australia.

Viscum articulatum Burm. f., Fl. Indica 311. 1768; FBI 5: 226. 1886; Gamble 5: 1257. 1925; Sanjai and N.P.Balakr. in N.P.Balakr. *et al.*, Fl. India 23: 50. 2012. *V. nepalense* Spreng., Syst. Curi-post. 47. 1827.

Subshrub, moderate sized, semiparasite, evergreen, up to 1 m long, drooping; branches usually dichotomously branched, sometimes more than 2 branches at each node, yellowish green, internodes flattened, at right angle to the preceding one, gradually broadening from the base to tip, twisted near the base, longitudinally prominently grooved. Leaves absent. Flowers sessile triads in axillary fascicles, unisexual, actinomorphic, central flower of the triad female and lateral male. Male flowers sessile, usually 2 in a triad on either side, up to 3 mm across, actinomophic, epigynous; perianth lobes 4, green, triangular, 1.5 mm long, valvate; stamens 4, 1 × 0.7 mm, opposite and adnate to the perianth lobes; filaments absent; pollen sacs many, dehisce introrsely. Female flower sessile, centre in the triad, up to 4 mm across, actinomorphic; perianth lobes 3 or 4, green, triangular, valvate, up to 2 mm across; ovary inferior; style absent; stigma nipple-shaped, truncate. Fruit berry, globose, green when young, yellowish when mature, up to 4 mm across, apex truncate; seed 1, globose, 2.5 mm across.

A partial stem parasite on spiny bushes such as *Ziziphus mauritiana, Z. oenoplia, Mangifera indica etc.* Occasionally found in the deciduous forests. Fl. & Fr.: August – December. Vern.: Ori.: *Malang.*

Kekati RF (ATP), *TP* & *NY* 751, 752; Kottakota RF (ATP), *TP* and *NY* 2114; Madhavaram RF (KNL), *TP* & *RVR* 1323; Chintala (PKM), *RVK* 15709; Ganagavaram (EG), *TP* & *KJ* 12537; Sikharam, near Srisailam (KNL), *KJ* 12581; Srisailam (KNL), *JLE* 16882 (MH); Guvvalacheruvu RF (CDP), *KS* 6847 (MH); Nerabylu (CTR), *GVS* 46971 (MH); Mamandur (CTR), *GVS* 31947 (MH); Cheedipalem (EG), *GVS* 27436 (MH); Rampachodavaram (EG), *GVS* 24477 (MH), *MM* 101340 (MH, BSID); Barmakonda RF (VSKP), *KCJ* 17158 (MH); Karaka (VSKP), *CAB* 1607 (MH); Anchetty forest, Kunthukottai (DMP), *EV* & *R.Rajan* 57986 (CAL); Mantharakuttai (NA), *EV* 53479 (CAL); Kambukudi (NA), *EV* 51986 (CAL); Yercaud (SM), *DB Deb* 31468 (MH); Gundur Road, Yercaud (SM), *SKK* 26820 (MH); Tickepalli to Linepada (GJM), *CAB* 1204 (MH); Bhaminigam to Damudia (GJM), *CAB* 1456 (MH).

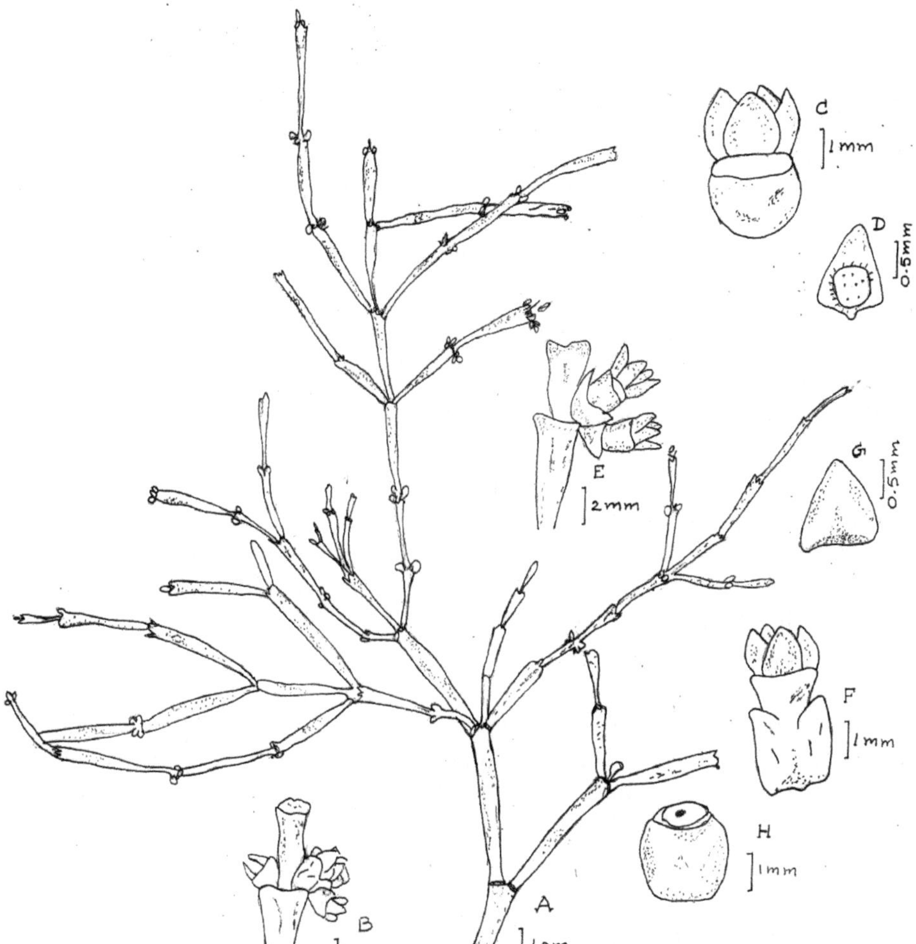

Figure 13. **Viscum articulatum** Burm.f.
A. Twig, B. Node with male flowers, F. Flower, D. Tepal,
E. Node with female flowers, F. Flower, G. Tepal, H. Fruit.

INDIA: Through out except Punjab and southern areas of Himachal Pradesh.

WORLD: Sri Lanka, Nepal, Bhutan, China, Myanmar, Thailand, Vietnam, Malesia and Australia.

Viscum capitellatum Sm. in Rees Cyclop. 37: Viscum no. 18. 1818; FBI 5: 225. 1886 p.p.; Sanjai & N.P.Balakr. in N.P.Balakr. *et al.*, Fl. India 23: 51. 2012.

Evergreen parasitic herb, up to 15 cm long, densely branched; internodes terete, smooth, swollen at nodes, ca 6x 0.5 cm. Leaves partly normally developed, sessile, opposite, sometimes lacking or reduced to scales, normal leaves 1-2 x 0.5 cm, orbicular-ovate or spathulate, thick, margin entire. Flowers in axillary cymes, the central flower male, the laterals female. Male flowers: sessile, obovate, perianth lobes 2, stamens 2. Female flowers: bracteolate, oblong, perianth lobes 3, free, subulate, ovary inferior, style very short, stigma conical. Drupe ellipsoid or ovoid, smooth.

Occasional in Eastern Ghats of Tamil Nadu. Fl. & Fr.: February – July.

Javadi hills Komattiyeri (Vellore), *CECF* 4000 (CAL); Hogainakkal (SM), *EV* 23564 (MH).

INDIA: Karnataka, Kerala and Tamil Nadu.

WORLD: Sri Lanka.

Viscum heyneanum DC., Prodr. 4: 278. 1830; Sanjai & N.P.Balakr. in N.P.Balakr. *et al.*, Fl. India 23: 51. 2012. *V. verruculosum* Wight & Arn., Prodr. 379. 1834; FBI 5: 224. 1886; Gamble 2: 1258. 1925. *V. orbiculatum* Wight, Icon. Pl. Ind. Orient. t. 1016. 1845; FBI 5: 224. 1886; Gamble 2: 1258. 1925.

Semiparasitic herbs, up to 40 cm long, much branched, branchlets terete. Leaves obovate-oblong, coriaceous, 1.5-3 x 0.6 - 1.5 cm, base attenuate, apex obtuse. Flowers axillary in fascicles. Male flowers: obovoid, perianth lobes 4, deltoid; stamens 4. Female flowers: slightly compressed, perianth lobes 4, ovary inferior, style short, stigma nipple shaped. Berries oblong, warty, with a crown of persistent or subpersistent perianth lobes.

Rare in forests. Fl. & Fr.: November – May. Vern.: Tel.: *Badanika*.

Komaticheruvu (CTR), *GVS* 458415 (MH); Melpat (SA), *CAB* 1051 (MH); Sirumalai (MDR), *S.India Flora* 16715 (MH); Way to Sirumalai (MDR), *KS* 5774 (MH).

INDIA: Andhra Pradesh, Karnataka, Kerala, Tamil Nadu and Andaman and Nicobar Islands

WORLD: Sri Lanka.

Viscum monoicum Roxb. ex DC., Prodr. 4: 278. 1830; FBI 5: 13. 1886; Gamble 2:1258. 1925; Sanjai & N.P.Balakr. in N.P.Balakr. *et al.*, Fl. India 23: 52. 2012.

Semi parasitic shrubs on other plants; stems dichotomously branched, branches terete, others angled, grooved, nodes swollen. Leaves 1.5-2.5 × 0.5-1 cm, obovate-spathulate, flattened, base cuneate, margin entire, apex sub-acute, deciduous;

chartaceous. Flowers minute, green, in axillary or lateral clusters of cymes. Male flowers: sessile, oblong, perianth lobes 4, bearing an anther on its inner surface; stamens 4. Female flowers: clavate or obovate; perianth with 4 triangular lobes; ovary inferior, style short, stigma capitate. Berry 5 mm across, apex truncate.

Frequent parasite on trees and shrubs in the hills. Fl. & Fr.: February – August. Vern.: Ori.: *Malang, Monjodama.*

Thimmammamarrimanu (ATP), *KRKS* 37966; Gundlabrahmeswaram (KNL), *JLE* 16914 (MH); Peddaupparapalli (CTR), *DRC* 1956; (MH), Talakona (CTR), *DRC* 2149 (MH); Mahendragiri (GJM), *SX* & *MB* 3560 (RRL-B), 390 (RRL-B); Simlipahar (MBJ), *SX* 506 (RRL-B); Gonasika forest, Keojhar, *MB* & *Dhal* 8473 (RRL-B); Daringam badi (GJM), *CAB* 1361 (MH).

INDIA: Through out India.

WORLD: Sri Lanka, Bhutan, Bangladesh, Myanmar, China, Vietnam and Thailand.

Viscum orientale Willd., Sp. Pl. 4: 737. 1806; FBI 5: 224. 1886; Gamble 2: 1258. 1925; Sanjai & N.P.Balakr. in N.P.Balakr. *et al.*, Fl. India 23: 54. 2012.

Large shrubs, parasitic on many kinds of trees, *Albizzia, Pongamia etc.*, stems much branched, internodes terete, faintly striate, thickened at the nodes; twigs slender, glabrous. Leaves 2.5-7 × 1-3 cm; obovate-elliptic or spathulate, base tapering into a short, often obscure petiole, opposite, margin entire or undulate, apex obtuse or acute, coriaceous, glabrous. Flowers in short axillary dichasia, usually triads, green in colour. Male flowers: raised, perianth lobes 4 with anther attached on the inner side; stamens 4. Female flowers: sub sessile, perianth lobes 4; ovary inferior, style short, stigma capitate. Berry obovoid.

Common parasite on forest trees. Fl. & Fr.: February – March, and August – October. Vern.: Tel.: *Bajinika*; Ori.: *Malang.*

Chendrayunikona (ATP), *BR* & *ANS* 35265; Gundlabrahmeswaram (KNL), *PSPB* 9131, *JLE* 16914 (MH); Gangavaram (EG), *TP* & *KJ* 12538; Simhachalam (VSKP), *KJ* & *SSR* 12590; Pecheruvu (KNL), *JLE* 42273 (MH); Balapalli (CDP), *JLE* 14983 (MH); Venkatagiri (NLR), *ASR* 6511 (BSID); Ramapatnam (NLR), *JSG* 12380 (DD); Polavaram (EG), *JSG* 15866 (MH); Jidikuppa (EG), *CAB* 5337 (MH); Ragupalem (VSKP), *CAB* 1570 (MH); Karaka (VSKP), *CAB* 1609 (MH); Forest near Anantagiri (VSKP), *NPBK* 10973 (MH); Kaliyaguoa (GJM), *s.coll. s.n.* (CAL); Tickapalli to Linepada (GJM), *CAB* 1205, 1206 (MH); Chamudya (GJM), *VNS* 5861 (MH); Melpat (SA), *CAB s.n.* (MH); Melpatti to Puliyar (NA), *KS* 6124 (MH); Hogainakkal (SM) *EV* 21966 (MH), *EV* 20619 (MH); Pennagaram RF (SM), *EV* 20723 (MH).

INDIA: Bihar, Odisha, Madhya Pradesh, Maharashtra, Andhra Pradesh, Karnataka, Kerala, Tamil Nadu and Andaman and Nicobar Islands.

WORLD: Sri Lanka, China and Australia.

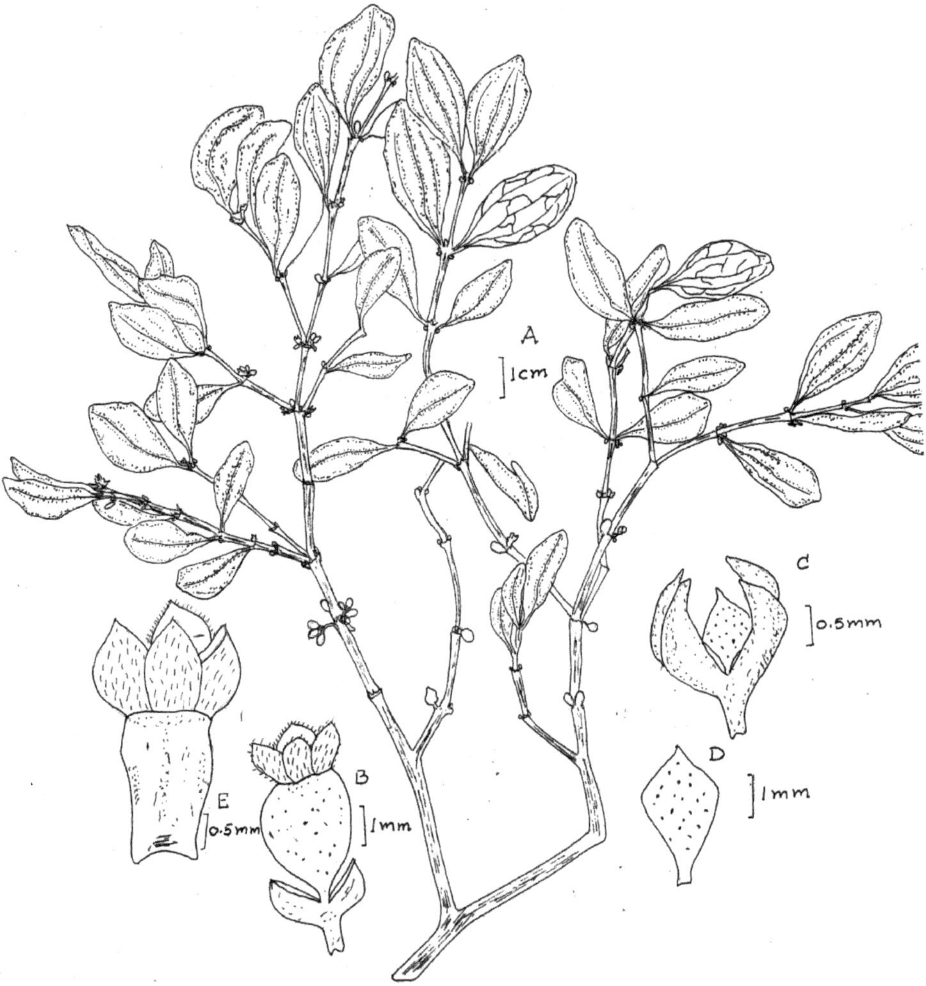

Figure 14. **Viscum orientale** Willd.
A. Twig, B. Male flower, C. Corolla lobes enclosing stamen, D. Stamen.

Viscum ramosissimum Wall. ex Wight and Arn., Prodr. Fl. Ind. Orient.380. 1834; FBI 5: 225. 1886; Gamble 2: 1258. 1925; Sanjai & N.P.Balakr. in N.P.Balakr. *et al.*, Fl. India 23: 55. 2012.

Large, semi-parasitic, many slender branching, sub-shrub, up to 1.5 m tall, hanging broomlike. Branches slender, cylindrical, numerous; nodes swollen, internodes terete, basally larger, up to 5 cm long, apically smaller, up to 2 cm long. Leaves usually absent, when present only on lower branches at young stage, obovate, spathulate, flattened, 1.5–2.5 × 0.4–0.8 cm, faintly 3-nerved from base. Flower triads in axillary fascicles, small, sessile, unisexual, central female and laterals male. Male flowers compressed, up to 3 mm across, actinomorphic; perianth lobes 4, free, ovate, valvate, up to 2 mm long; stamens 4, opposite and adnate to the perianth lobes; staminal filaments absent; pollen sacs many, dehisce introrsely, pistillode absent. Female flowers oblong, larger than male, present in the center of the triad, 0.4 cm long, actinomorphic, epigynous; perianth lobes 4, free, conical, staminodes absent; ovary inferior; style reduced; stigma capitate, minute, nipple-shaped. Fruit berry, small, globose, up to 5 mm across, green, apex truncate; seed 1, 0.4 cm across, globose, whitish.

Rare on forest trees in the southern part of Andhra Pradesh. Fl. & Fr.: February-July.

Thimmammamarrimanu (ATP), *KRKS* 37966; Nerabyulu way to Talakona (CTR), *PSPB* 9122; Peddaupparappalli (CTR), *DRC* 1956; Talakona (CTR), *DRC* 2149 (SVU); Melpatti (NA), KS 6126 (MH); Vazhakadu (NA), *EV* 53428 (MH); Pennagaram RF (SM), *EV* 22406 (MH); Hogainakkal (SM), *EV* 23527 (MH); Woddapatti RF (SM), *EV* 20704 (MH); Koinpur to Kusadandi (GJM), *CAB* 1257 (MH).

INDIA: Maharaashtra, Karnatak, Kerala, Tamil Nadu and Andhra Pradesh.

WORLD: Sri Lanka.

SANTALACEAE

1. Leaves opposite; perianth lobes 4 .... **Santalum**
1. Leaves alternate; perianth lobes 3 .... **Osyris**

OSYRIS Linnaeus

Osyris quadripartita Salzm. ex Decne., Ann. Sci. Nat. Bot. 2. 6: 65. 1836; Gandhi in Sald. & Nicolson, Fl. Hassan district 302. 1976; Sanjai & N.P.Balakr. in N.P.Balakr. *et al.*, Fl. India 23: 61. 2012. *O. wightiana* Wall. ex Wight, Icon. Pl. Ind. Orient. t. 1853. 1852. *O. arborea* Wall. ex DC., Prodr. 14: 633. 1857; FBI 5: 232. 1886; Gamble 2: 1262. 1925.

Evergreen, much branched, glabrous shrubs; young branches sharply triangular. Leaves 2-5 × 1.5-3 cm, elliptic-ovate to lanceolate, alternate, base acute to cuneate, sessile, margin entire, apex mucronate, coriaceous, glabrous to glaucescent. Flowers minute, polygamous, bi or unisexual, solitary or in short, axillary cymes; Male flowers: perianth lobes 3 or 4, triangular, concave, enclosing prominent disc in the midde; stamens stalked. Female and bisexual flowers: mostly solitary, axillary, perianth lobes 3, urceolate, adnate to the ovary, style very short, thick, stigma peltate. Drupe ovoid-globose or broadly ellipsoid, with prominnt disc at top, yellow to reddish when ripe.

Rare in forests. Fl. & Fr.: February – May.

Tirumala on the way to Kumaradhara (CTR), *DRC* 2155 (MH); Nallamalais (KNL), *JSG* 46070 (MH), Anantagiri (VSKP), *GVS* 21739 (MH); Cherukonda (VSKP), *GVS* 28180 (MH); Mahendragiri (GJM), *SX* & *MB* 2161, 3545, 3570, 3726 (RRL-B), *VNS* 5571 (MH), *VVS* 5571 (MH); on the way to Kambukudi (NA), *KS* 6488 (MH); Yelagiri hills (NA), *MBV* 549 (MH); Balmadies estate,Yercaud (SM), *KS* 7541(MH); Karadiyar via Nagalur-Yercaud (SM), *AVN* Rao 6789 (MH); Kaveri peak-Yercaud (SM), *DB Deb* 31315 (MH) Kollimalai, (TRP), *SI Flora* 112569 (MH); Sirumalai hills, Karuppukoil (DGL), *SKS* 645 (SGH).

INDIA: Himachal Pradesh, Uttarakhand, Sikkim, Arunachal Pradesh, Nagaland, Manipur, Odisha, Madhya Pradesh, Andhra Pradesh, Goa, Kerala and Tamil Nadu.

WORLD: Sri Lanka, Nepal, Bhutan, Myanmar, Thailand, Indo-China, China, Africa and S. Europe.

Note: Hondrych in Tutin *et al.*, Fl. Europaea 1: 70. 1964 and Gandhi in Saldhana and Nicolson, Fl. Hasan 302. 1976 treat *Osyris quadripartita* and *O. wightiana* as conspecific and name this species as *Osyris quadripartita* Salzm. ex Decne, Ann. Sci. Nat. Bot. (ser. 2). 6: 65. 1836, but Tebbs (in Hara, Enum. Fl. Pl. Nepal 3: 192. 1982) treats them as distinct species.

SANTALUM Linnaeus

Santalum album L., Sp. Pl. 349. 1753; FBI 5: 231. 1886; Gamble 2: 1261. 1925; Sanjai & N.P.Balakr. in N.P.Balakr. *et al.*, Fl. India 23: 64. 2012.

Small, evergreen, semiparasitic, glabrous tree with slender drooping branches, up to 12 m high, bark grey, cracked. The sap wood white and odorless, the heart wood yellowish brown, strongly scented. Leaves chartaceous, 4-6.5 × 2-3 cm, elliptic ovate to lanceolate, opposite, base acute, margin entire, apex sub-acute, sub-coriaceous. Flowers maroon or brownish-purple within, green without in 6 cm long, pedunculate cymes, shorter than leaves, trichotomous, terminal and axillary; perianth lobes 4 or 5, broadly triangular; stamens 4 or 5, filaments short, inserted at the mouth of perianth-tube; ovary at first free, later semi-inferior, 1-loculed, stigma 3-lobed. Drupe subglobose, 1 cm across, glabrous, purple black when ripe, endocarp hard.

Occasional in open dry deciduous and scrub forests, frequently planted in gardens. Fl. & Fr.: January – October. Vern.: Tel.: *Chandanam, Srigandham, Harichandanam*; Ori.: *Chandan*, Tam.: *Santhanam.*

Kalasamudram RF (ATP), *NY* 682; near Sikharam, Srisailam (KNL), *KJ* & *SSR* 12580; North Dhone RF (KNL), *RVR* 2511, *JSG* 10777 (DD); Palakonda hills (CDP), *CS* 8244; Tirumala hills (CTR), *KJ* 12560, *SSR* 13738, *DRC* 220 (MH), *GVS* 31909

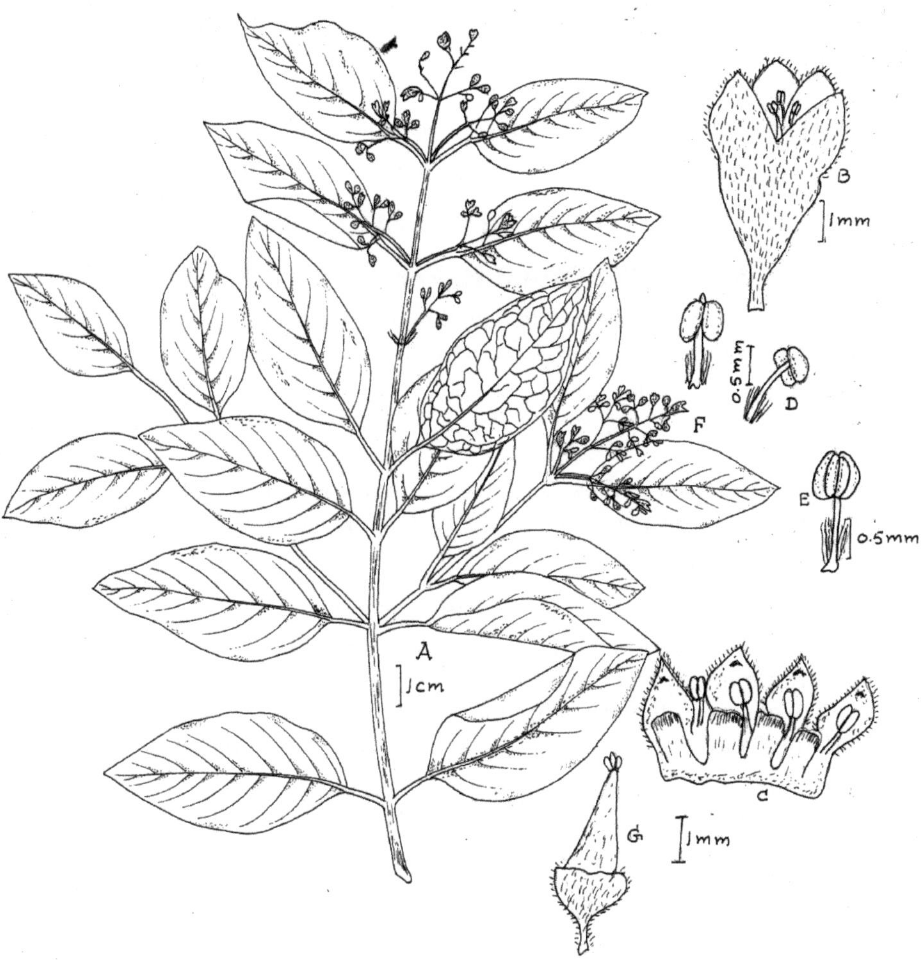

Figure 15. **Santalum album** L.
A. Twig, B. Flower, C. Corolla split open, D-F. Stamens, G. Pistil.

(MH); Chinarutla (PKM), *SKB* & *BSS* 30616; Diguvametta (KNL), *JLE* 32494 (MH); Balapalli (CDP), *JLE* 14251 (MH); Cuddapah forests (CDP), *CBDM* s.n. (MH); Udayagiri (NLR), *BSN* 4649 (BSID); Kondapalli RF (KSN), *PV* 5298 (MH); Chintapalli (VSKP), *JLE* 37107 (MH), Anantagiri (VSKP), *NPBK* 1098 (MH); Javadi RF (NA), *KS* 6133 (MH); Gomukhi – Vellimalai (SA), *KRM* 77385 (MH); Melpat (SA), *CAB* 697 (MH); Kotheru RF, Yelagiri hills (NA), *MBV* 547 (MH); Kottaiyur RF (NA), *MBV* 547 CAL (MH); Takarai RF (SA), *KRM* 52842 (CAL); Nrusimhanath (SBP), *SP* 1032 (CAL); Hogainakkal (SM), *EV* 21947 (MH); Woddapatti RF (SM), *EV* 20694 (MH); Yercaud, Kaveri peak (SM), *DB Deb* 31318 (MH); Yercaud, orchidarium (SM), *AVN* Rao 18227 Anchetti (SM), *KCJ* 18039 (MH).

INDIA: Deccan Peninsula, cultivated elsewhere.

WORLD: Sri Lanka.

BALANOPHORACEAE

BALANOPHORA J.R. Forster & G. Forster

1. Plants monoecious, male and female flowers in the same inflorescence .... **B. fungosa**
1. Plants dioecious, male and female flowers in different inflorescences .... **B. polyandra**

Balanophora fungosa J.R. Forst. & G. Forst., Char. Gen. Pl. 50. 1775; Prakasa Rao *et al.*, J. Threatened Taxa 7: 6943. 2015; N.P.Balakr. in N.P.Balakr. *et al.*, Fl. India 23: 74. 2012.

Key to Subspecies

1. Bracts subtending flowers free or few are not developed, bracts rudimentary or not developed; inflorescences ovoid to ellipsoid; flowers actinomorphic with isomorphic perianth lobes .... **B. fungosa** subsp. **fungosa**
1. Bracts developed and stout, flowers long pedicellate; anther cells usually U-shaped; perianth lobes ovate, less than 2 mm; synandria hemispheric .... **B. fungosa** subsp. **indica**

Balanophora fungosa J.R. Forst. & G. Forst. subsp. **fungosa**

The plants are dioecious. Rhizome yellowish-brown, surface with granular warts and scattered yellow stellate lenticels; branches subglobose, depressed, up to 3 cm. Scapes yellowish, up to 30. Leaves alternate, broadly ovate, 1.5×3.2 cm, apex obtuse. Male inflorescences ellipsoid, conically ovoid, up to 6 cm; male flowers actinomorphic; pedicel up to 1.2 cm long; perianth lobes 4 or 5, ovate-elliptic, apex acute; synandria subglobose, anthers 4 or 5, U-shaped, pollen white, powdery.

Rare, near perennial streams of moderately dense semi evergreen forest, at 1100 m altitude in Visakhapatnam district. Fl. & Fr.: December – April.

Raktakonda, Araku Valley (VSKP), *J. Prakasa Rao* & *K.V. Satish* 21264, 21286 (AU, BDH).

INDIA: Andhra Pradesh, Manipur.

WORLD: China, Japan, Indonesia, Philippines, New Guinea, NE Austraalia and Pacific Islands.

Balanophora fungosa Forst. & Forst.f., Char. Gen. Pl. 100. t. 50. 1776, subsp. **indica** (Arn.) B. Hansen, Dansk Bot. Ark. 28: 100. 1972. *Langsdorffia indica* Arn., Ann. Nat. Hist. 2: 37. 1838. *Balanophora indica* (Arn.) Griff., Trans. Linn. Soc. London 20: 95. 1846; FBI 5: 237. 1886; Gamble 2: 1263. 1925. var. *indica;* N.P.Balakr. in N.P.Balakr. *et al.*, Fl. India 23: 76. 2012. *B. dioica* auct. non R.Br. Royle, 1839; sensu Gamble 1263. 1925; Pullaiah & Moulali, Fl. Andhra Pradesh 2: 835. 1997.

Dioecious herb, 15-25 cm high, stem stout, cylindric, irregularly lobed below. Leaves 7-15, closely spiral, orbicular, concave or slightly cucullate, 1.5-2.5 × 1.5-2 cm, thick, subacute or obtuse. Male flowers cream to yellow, in ovoid to ellipsoid, cylindric heads; tepals 4, oblong to lanceolate, 4-6 × 3 mm, thin-fleshy, shortly connate; synandrium cylindric, 4-6 × 3 mm, anthers 4, horse-shoe-shaped. Female flowers rusty-brown, in globose heads, ovary globose, style filiform; spadicles club-shaped, 0.8-1.5 mm.

Occasional in hills above 1000 m, as root parasite. Fl.: January- March; Fr.: February Onwards. Vern.: Tam.: *Vaerkaalan*.

Peddavalasa (VSKP), *JSG* 21783 (DD); On the way to Swamimalai-Yelagiri hills (NA); Shavaroy (SM), *MBV* 1357 (MH); Bankite hills-Yercaud (SM), *AVN Rao* 27481 (MH); Kolli hills, Kulivalavu (NMK), *SKS* 1029 (SGH).

INDIA: Assam, Manipur, Andhra Pradesh, Maharashtra, Karnataka, Kerala and Tamil Nadu.

WORLD: Sri Lanka, Bangladesh, Myanmar, SW China, Vietnam, Laos, Thailand, Malaysia, Indonesia, Philippines, Australia and Pacific Islands.

Balanophora polyandra Griff. Trans. Linn. Soc. London 20: 94. t. 7. 1851; FBI 5: 238. 1886; N.P.Balakr. in N.P.Balakr. *et al.*, Fl. India 23: 77. 2012.

Glabrous fleshy herb, rootstock tuberous, lobes with peduncle breaking through a rounded closely lenticellate lobe which becomes cup-shaped and embraces the base of the peduncle with a subregularly lobed sheath. Peduncles 7.5-15 cm, several, clothed with few large, oblong-lanceolate, imbricating and sheathing scales, peduncles monoecious, swollen at the end into a densely crowded above with small stipitate knob-shaped protuberances each surrounded by most minute female flowers. Male perianth 3.7-5 mm diam, with 4, occasionally 6 tepals, posterior and anterior tepals broadly oblong, lateral smaller; androecium a fleshy mass with numerous confluent anther-cells. Female flower a solitary stipitate fusiform ovary with a hair-like style.

Mahendragiri, 1200 m. (Mukherjee, J. Indian Bot. Soc. 14: 306. 1935). Fl.: September-October.

Bhatodih (Keonjhar), *HFM* 2786 (DD).

INDIA: Sikkim, West Bengal, Assam, Arunachal Pradesh, Nagaland, Manipur, Maharashtra, Odisha.

WORLD: Nepal, Bhutan, Myanmar, SW. China.

BUXACEAE

SARCOCOCCA Lindley

Sarcococca pruniformis Lindl., Bot. Reg. 12: t. 1012. 1826; FBI 5: 266. 1887 *p.p.* *S. trinervia* Wight, Icon. Pl. Ind. Orient. t. 1871. 1852; Gamble 2: 1265. 1925; Pullaiah & Moulali, Fl. Andhra Pradesh 2: 835. 1997. *S. coriacea* (Hook.) Sweet, Hort. Brit. ed. 1, 491. 1826; Chakrab. & N.P.Balakr. in N.P.Balakr. *et al.*, Fl. India 23: 83. 2012. *Pachysandra coriacea* Hook., Exot. Fl. 2: t. 148. 1825. *Sarcococca saligna* Müll.Arg. in DC., Prodr. 16(1): 11. 1869.

Large shrub, 1.5-2 m high, branches green, terete, glabrous. Leaves ovate-lanceolate, lanceolate or elliptic, 6-15 × 2-6.2 cm, base acute or obtuse, rounded to cuneate, apex acuminate or long-acuminate, coriaceous, glabrous, petiole 5-15 mm long. Flowers yellow or white in racemes; uisexual; bracts ovate, triangular or deltoid. Male flowers: sessile or pedicels up to 1.5 mm long, perianth lobes 4, oblong; stamens 4, twice as long; female flowers: bracteoles 4-8, in subopposite pairs, triangular, ovate-triangular or deltoid; perianth lobes 4-6, similar to bracts, ovary ovoid, styles recurved above. Fruit variable, ovoid or ellipsoid, ca. 7.5 cm long, fleshy.

In damp, usually shady ravines in Northern Eastern Ghats. Fl. & Fr.: November – May.

Vankachinata (VSKP), *GVS* 29580 (MH); Near Gannela (VSKP), *NRR* & *RAN* 79453 (BSID); Mahendragiri (GJM), *SX* & *MB* 2239, 3575, 3722, 3955 (RRL-B); near Pottangi (KPT), *HFM* 4090 (DD).

INDIA: Himachal Pradesh, Arunachal Pradesh, Nagaland, Manipur, Mizoram, Tripura, Meghalaya, Odisha, Andhra Pradesh, Karnataka, Tamil Nadu and Kerala.

WORLD: Sri Lanka, Nepal, Bhutan and Myanmar.

EUPHORBIACEAE

1. Inflorescence a cyathium; tepals 0 or minute; stamen 1, jointed:
    2. Branchelets zig-zag; cyathia zygomorphic, scarlet red................**Pedilanthus**
    2. Branchelets not zig-zag; cyathia actinomorphic, green or reddish:
        3. Cyathia with distinct glands; involucre without a basal rim .... **Euphorbia**
        3. Cyathia without distinct glands; involucre with a basal rim .... **Synadenium**

1. Inflorescence other than a cyathium; tepals conspicuous; stamens 2-10 or many, filaments not jointed:
   4. Cells of ovary 1-ovuled:
      5. Filaments inflexed in bud.. **Croton**
      5. Filaments erect, rarely inflexed:
         6. Outer tepals (sepals) valvate in male flowers:
            7. Inner tepals (petals) present in male flowers:
               8. Herbs; stamens 5-15.. **Chrozophora**
               8. Trees; stamens 15-32.. **Aleurites**
            7. Inner tepals (petals) absent in male flowers:
               9. Climbers:
                  10. Fruiting perianth without stinging hairs:
                     11. Stamens 12-30 .... **Plukenetia**
                     11. Stamens 2 (or 3) .... **Pachystylidium**
                  10. Fruiting perianth with stinging hairs:
                     12. Flowers in spikes; stamens 3:
                        13. Sepals of male flower adaxially inflexed to form a pseudo-disc; stamens 3, anthers introrse, connective thick; style massive.. **Cnesmone**
                        13. Sepals of male flowers not adaxially inflexed; stamens many, anthers extrorse, connective not thickened; style slender.. **Tragia**
                     12. Flowers in involucrate clusters; stamens 20-25... **Dalechampia**
               9. Erect herbs, shrubs or trees:
                  14. Filaments much branched, staminal bundles indefinite:
                     15. Leaves simple;
                        16. Plants dioecious; female flowers in spike .... **Homonoia**
                        16. Plant monoecious; female flower solitary .... **Lasiococca**
                     15. Leaves palmately lobed .... **Ricinus**
                  14. Filaments not branched:
                     17. Anthers vermiculiform or linear; styles laciniate or unblobed:

18. Female bract foliaceous; anther theca twisted .... **Acalypha**

18. Female bract not foliaceous; anther theca normal .... **Claoxylon**

17. Anthers not vermiculiform or linear:

19. Anthers 4-celled:

20. Indumentum simple; anther connective enlarged, pollen grains reticulate .... **Agrostistachys**

20. Indumentum malpighiaceous; anther connective not enlarged, pollen grains other than reticulate:

21. Styles bifid; anthers globose, 2 or 4-locular, apiculate; pollen grains vermiculate rugulose .... **Cleidion**

21. Styles simple; anthers cross-shaped, 3 or 4-locular, not apiculate; pollen grains spinulose-rugulose .... **Macaranga**

19. Anthers 2-celled:

22. Stamens numerous:

23. Styles linear; fruit indehiscent .... **Trewia**

23. Styles plumose; fruit a capsule:

24. Sap watery; fruits with glandular hairs .... **Mallotus**

24. Sap milky; fruit without glandular haris .... **Manihot**

22. Stamens 2-9:

25. Filaments slender .... **Alchornea**

25. Filaments inflexed .... **Epiprinus**

6. Outer tepals (sepals) imbricate in male flowers:

26. Inner tepals (petals) present in male flowers:

27. Inner tepals free:

28. Flowers glomerate spicate or racemose........ **Dimorphocalyx**

28. Flowers paniculate, panicles branching dichasially .... **Jatropha**

27. Inner tepals (petals) connate .... **Givotia**

26. Inner tepals (petals) in male flowers absent:

29. Herbs or undershrubs:

30. Leaves elliptic-ovate, 2-glandular at base ...... **Baliospermum**

30. Leaves linear-lanceolate, eglandular .... **Microstachys**

29. Shrubs or small trees:

31. Inflorescence terminal:

32. Leaves not crowded at the ends of the branchlets, elliptic-lanceolate or acute-rounded at base, entire, subacuminate; pedicels of the flowers long; style very short, pointed on top of the ovary .... **Triadica**

32. Leaves crowded at the ends of the branchlets, attenuate at base, crenate, acuminate; pedicels of the flowers very short; styles short, recurved .... **Falconeria**

31. Flowers in axillary racemes:

33. Stamens 3, filaments 2 mm .... **Excoecaria**

33. Stamens 25-32, filaments 0.5 mm......... **Suregada**

4. Cells of ovary 2-ovuled:

34. Inner tepals (petals) present, small:

35. Ovary 2-celled; fruit a drupe with 2 pyrenes............................. **Bridelia**

35.Ovary 3-celled; fruit a capsule.. **Cleistanthus**

34. Inner tepals (petals) 0; outer tepals (sepals) imbricate in bud:

36. Disc wanting:

37. Flowers monoecious; Leaves trifoliate................................ **Bischofia**

37. Flowers dioecious; Leaves simple:

38. Male and female flowers in clusters, stamens connate by their connectives in a column, style also connate in a column .... **Glochidion**

38. Male flowers in clusters, female sub-solitary, stamens free or rarely connate by their filaments:

39. Branchlets pendulous; disc wanting; stamens 2-3; drupe yellowish, pendulous .... **Putranjiva**

39. Branchlets spreading; disc present; stamens 6-12; drupe red or scarlet, not pendulous .... **Drypetes**

36. Disc conspicuous:

40. Flowers in spikes or recemes .... **Antidesma**

40. Flowers in clusters:

41. Tepal lobes 4 in both male and female flowers ...... **Margaritaria**

41. Tepals 5 or 6:

42. Stamens 5:

43. Filaments connate .... **Meineckia**

43. Filaments free .... **Flueggea**

42. Stamens 3, rarely 5, filaments connate or sometimes not:

44. Tepals gland-dotted, stamens born on a gynophore .... **Sauropus**

44. Tepals not gland-dotted, stamens sessile or born on a filament:

45. Tepals of male flower turbinate, ovary truncate .... **Breynia**

45. Tepals spreading, ovary globose .... **Phyllanthus**

ACALYPHA Linnaeus

1. Male and female flowers on separate spikes:

2. Female flowered spikes unbranched, less than 3 cm long; petiole shorter than the leaf blade.. **A. alnifolia**

2. Female flowered spikes branched, more than 5 cm long; petiole longer than the leaf blade.. **A. paniculata**

1. Male and female flowers on same spike:

3. Petiole shorter than the leaf blade; leaf apex gradually acuminate .... **A. fruticosa**

3. Petiole as long as or longer than the leaf blade; leaf apex acute or abruptly acuminate:

4. Bracts 3 to 5-fid, lanceolate, margin entire, glandular hairy .... **A. brachystachya**

4 Bracts entire, campanulate, margin dentate or aciculate:

5. Spikes less than 4 cm long; bracts aciculate:

6. Female bracts almost tubular, fimbriate along the rim with long subulate hispid teeth, not gland tipped ..... **A. ciliata**

6. Female bracts campanulate, not tubular, shortly toothed along margins, bordered with long hispid gland-tipped hairs .... **A. malabarica**

5. Spikes more than 4 cm long; bracts dentate:

7. Herbs; leaf base shortly attenuate .... **A. indica**

7. Undershrubs; leaf base obtuse....................................... **A. lanceolata**

Acalypha alnifolia Klein ex Willd., Sp. Pl. 4: 525. 1805; FBI 5: 415. 1887; Gamble 2: 1330. 1925. N.P.Balakr. & Chakrab., Fam. Euphorb. India 124. 2007; Susila Rani & N.P.Balakr. in N.P.Balakr. *et al.*, Fl. India 23: 91. 2012.

Undershrub, up to 70 cm high, monoecious; branchlets terete, striate tomentose. Leaves 1-2.5 × 0.6-2 cm, ovate, base rounded or cordate, margin crenate-serrate, apex acute; petiole 1.5 cm; stipules linear. Male flowers many, minute, in slender, elongate, cylindric, lower axillary spikes; flowers 0.7 mm across; pedicel 0.6 mm, hirsute; tepals 4, 0.6 × 0.4 mm, free, broadly ovate, concave, acute, hirsute without; stamens many, filaments short, anthers 0.35 mm, vermiculiform. Female flowers few, in short terminal spikes; flowers 1.3 × 5 mm; pedicel 1 mm; bracts 6.5 × 5 mm, cupular, dentate, ciliate; tepals 3, 1.5 × 0.6 mm, free, minute; pistil 1, ovary 2 mm across, hirsute, styles 3, to 1 cm, each with many filiform hairs, some further divided. Capsule 5 mm across, globose, hirsute; seeds globose, smooth.

Common in scrub jungles, in bushes and rocky hill areas. Fl. & Fr.: April – September, Fr.: Throughout the year. Vern.: Tel.: *Chiru kuppinta;* Tam.: *Chinni.*

Muchukota RF (ATP), *TP* & *NY* 229; Pennahobilam (ATP), *TP* & *NY* 345; Dhone RF (KNL), *RVR* 1394; Palakonda (CDP), *CS* 7612; Saddicudumadugu (CTR), *PSPB* 9171; Penchalakona (NLR), *PSPB* 9105; Macherla RF (GNT), *VRK* 3518; Nagapatla RF (CTR), *KMS* 6853 (CAL); Guvvalacheruvu (CDP), *EKK* 79298 (DD); Veligonda (NLR), *PV* 109866; Nagarjunakonda (GNT), *KT* 81674 (CAL); Kondapalli (KSN), *SKW* 3307, 3308 (BLAT); Addatigala (EG), *SKW* 2999 (BLAT); Alagar hills (MDU), *SKS* 604 (SGH).

INDIA: Andhra Pradesh, Karnataka, Kerala and Tamil Nadu. Endemic.

Acalypha brachystachya Hornem., Cat. Hort. Hafn. Willd. Enum. Hort. Berol. 992. 1809; FBI 5: 416. 1887; Gamble 2: 1330. 1925. *A. superba* Forssk., Fl. Aegypt.-Arab. 162. 1775; Govaerts *et al.,* World Checkl. & Bibl. Euphorbiaceae 1: 40. 2000; Susila Rani & N.P.Balakr. in N.P.Balakr. *et al.*, Fl. India 23: 103. 2012.

Herb, up to 60 cm high; monoecious; branchlets striate. Leaves broadly ovate, 1.5-5 × 1-3 cm, base rounded-subcordate, margin serrate, apex subacuminate. Flowers in axillary spikes, terminating in an allomorphic flower (female). Male flowers many, from above the middle, tepals 4, ovate-lanceolate, concave, tomentose-glandular; stamens many, subsessile, anthers vermiculiform. Female flowers: few, at the lower half of the spike; bracts lanceolate, 3 to 5-fid, glandular-hairy; sepals lanceolate, hyaline; ovary 1-1.5 mm across, densely pubescent, styles 3, each with 2-3 filiform hairs. Capsules 3-lobed, hispid, totally concealed inside large geen bracts.

Rare in moist and shady places in hills in Prakasam and Visakhapatnam districts. Fl. & Fr.: October-February.

Thummalabailu (PKM), *RKM* 0412 (MH); Ranajalledu (VSKP), *GVS* 44397 (MH); Kilagada (VSKP), *GVS* 21536 (MH); Kothuru (VSKP), *NPBK* 871 (MH).

INDIA: Jammu & Kashmir, Himachal Pradesh, Uttarakhand, Uttar Pradesh, Meghalaya, Odisha, Andhra Pradesh, Maharashtra, Karnataka, Kerala and Tamil Nadu.

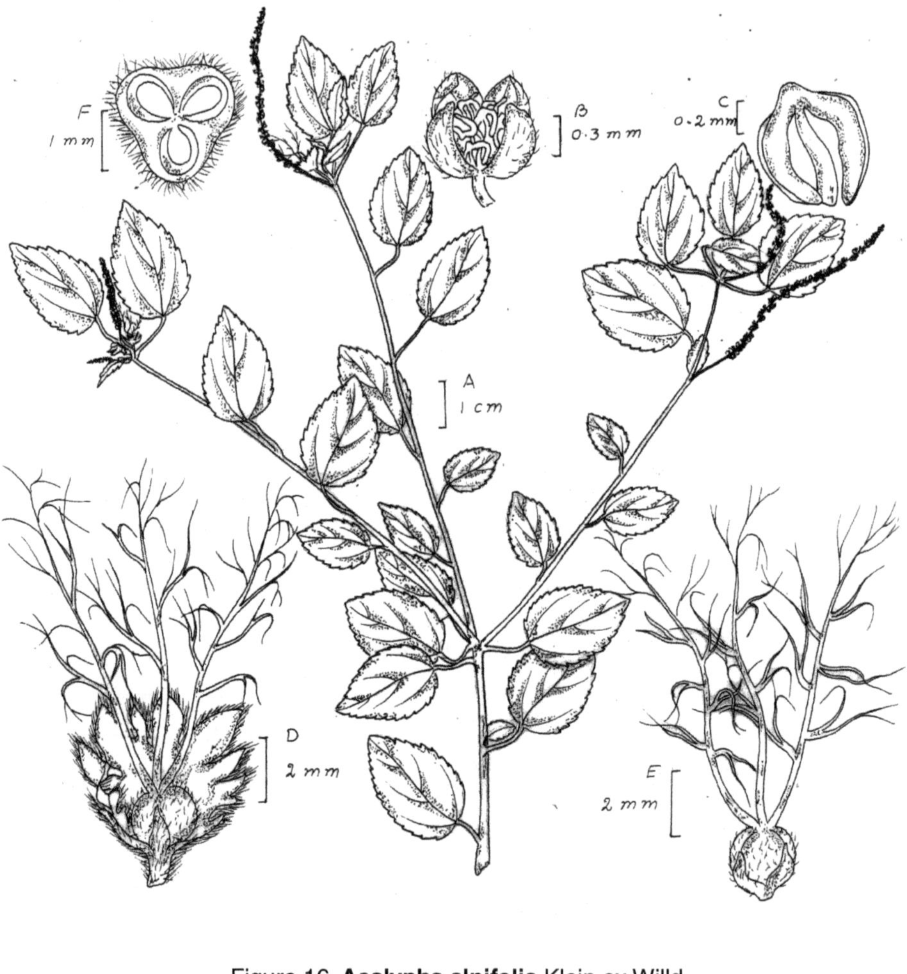

Figure 16. **Acalypha alnifolia** Klein ex Willd.
A. Twig, ♂ B. Flower, C. Stamen, ♀ D. Flower with bract, E. Pistil, F. T.s. of ovary

WORLD: Tropical C and E Africaa, Pakistan, Sri Lanka, Nepal, Bhutan, Myanmar, Indo-China, Thailand, Sumatra, Java, Lesser Sunda Islands, Timor and Celebes.

Acalypha ciliata Forssk., Fl. Aegypt. – Arab, 162. 1775; FBI 5: 417. 1887; Gamble 2: 1330. 1925; N.P.Balakr. & Chakrab., Fam. Euphorb. India 125. 2007; Susila Rani & N.P.Balakr. in N.P.Balakr. *et al.*, Fl. India 23: 94. 2012.

Undershrub, up to 80 cm high, monoecious; branchlets terete, striate. Leaves 3-9 × 2.5 cm, ovate, base rounded, margin serrate, apex abruptly acuminate; petiole to 5 cm. Flowers in axillary short spikes. Male flowers many, minute, from above the middle, terminating in an allomorphic flower; flowers 0.5 mm across; pedicel 0.1 mm; tepals 4, 0.5 × 0.4 mm, free, broadly ovate, acute, muricate without; stamens 8, filaments short, anthers 0.3 mm, vermiculiform. Female flowers few, clustered at the lower half of the spikes; flowers 4 × 2.5 mm; pedicels 0.6 mm; bracts 4 × 2.6 mm, folded, campanulate, aciculate; tepals 4-5, 0.7 × 0.3 mm, lanceolate, acute, hairy without; pistils 2, one sterile, ovary 1.2 mm across, hispid, styles 3, to 1.6 mm, each with 3-4 filiform hairs. Capsule 2 mm across, globose, glabrous; seeds globose, smooth.

Occasional in waste lands and as forest undergrowth in scrub jungles. Fl. & Fr.: August-November. Vern.: Tel: *Tellamurupindi.*

Kalasamudram RF (ATP), *KRKS* 40074 (BSID); Ahobilam (KNL), *TP* & *RVR* 2594; Rajampeta (CDP), *BR* & *DV* 43133; Nallagutta (PKM), *RKM* 625 (CAL); Kondapalli (KSN), *SKW* 3155, 3156 (BLAT); Papikonda (WG), *DCSR* 510 (CAL); Rampachodavaram (EG), *MM* 102665 (BSID); Mahendragiri (GJM), *SX* & *MB* 3327, 3509 (RRL-B).

INDIA: Throughout India.

WORLD: Tropical Africa, Southern Arabia, Pakistan, Nepal, Sri Lanka and Java.

Acalypha fruticosa Forssk., Fl. Aegypt.-Arab. 161. 1775; FBI 5: 415. 1887; Gamble 2: 1331. 1925; N.P.Balakr. & Chakrab., Fam. Euphorb. India 125. 2007; Susila Rani & N.P.Balakr. in N.P.Balakr. *et al.*, Fl. India 23: 96. 2012.

Shrub, up to 90 cm high, monoecious; branchlets terete, striate. Leaves 1-4 × 0.5-2.5 cm, ovate, base truncate-rounded, margin crenate-dentate, apex sub-acuminate, thinly pubescent, glandular below; petiole to 1.5 cm. Flowers in axillary short spikes. Male flowers many, minute, from above the middle; flowers 0.4 mm across; pedicel 0.1 mm; tepals 4, 0.6 × 0.4 mm, free, broadly ovate, acute, concave, muricate without; stames 8-10, filaments short, anthers 0.2 mm, vermiculiform. Female flowers few, at the lower end of the spike; flowers 3 × 2 mm; pedicel 1 mm; bracts folded, campanulate, crenate; tepals 3, 1.1 × 0.45 mm, lanceolate, acute, hairy without; pistils 2, one sterile, ovary 0.7 mm across, hispid, styles 3, to 1.5 mm across, globose, glabrous. Capsule tomentose with large yellow glands in the sinuses; seeds globose, smooth.

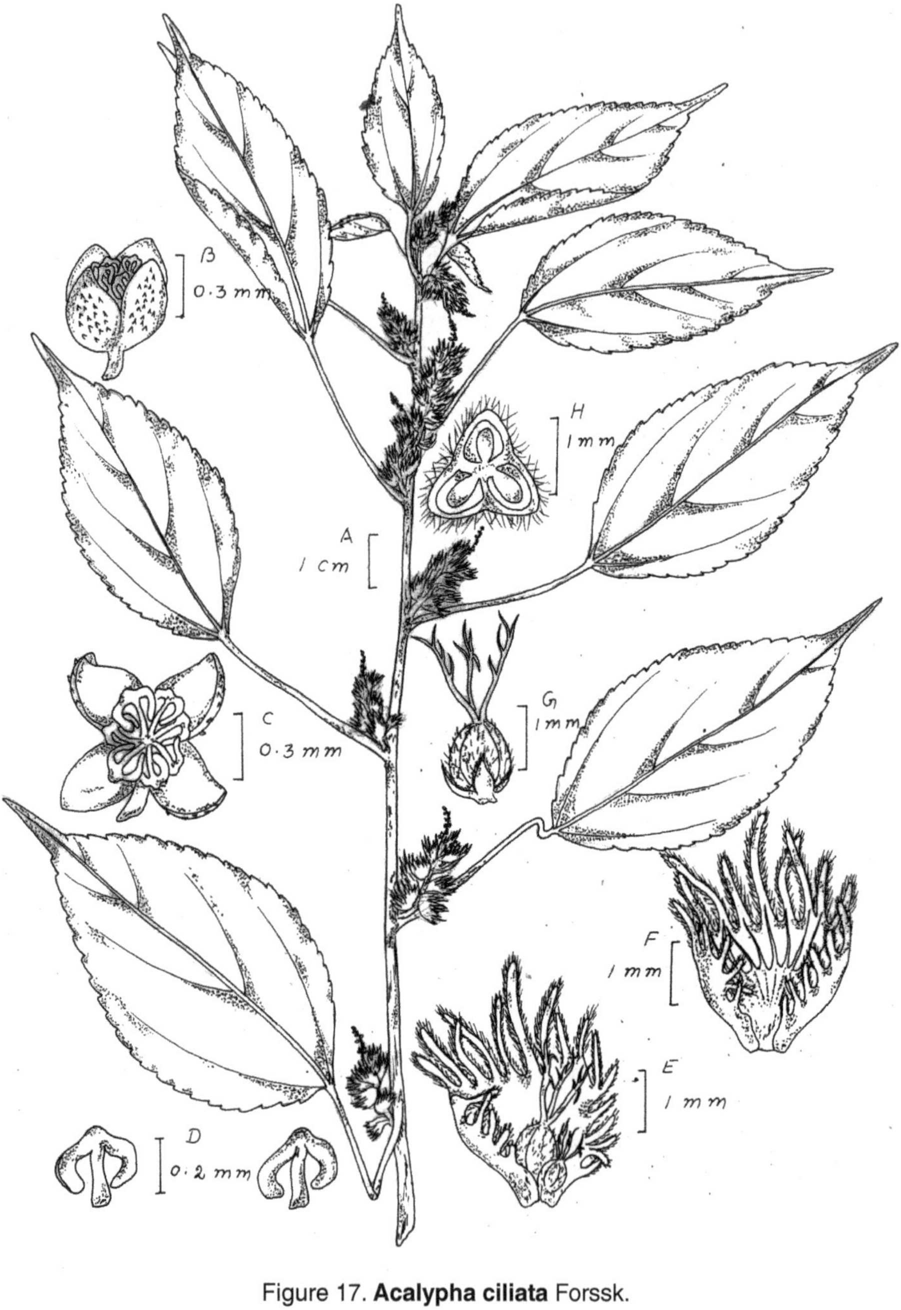

Figure 17. **Acalypha ciliata** Forssk.
A. Twig, ♂ B and C. Flowers, D. Stamens. ♀ E. Flower with bract, F. Bract, G. Pistil, H. Ovary T.s.

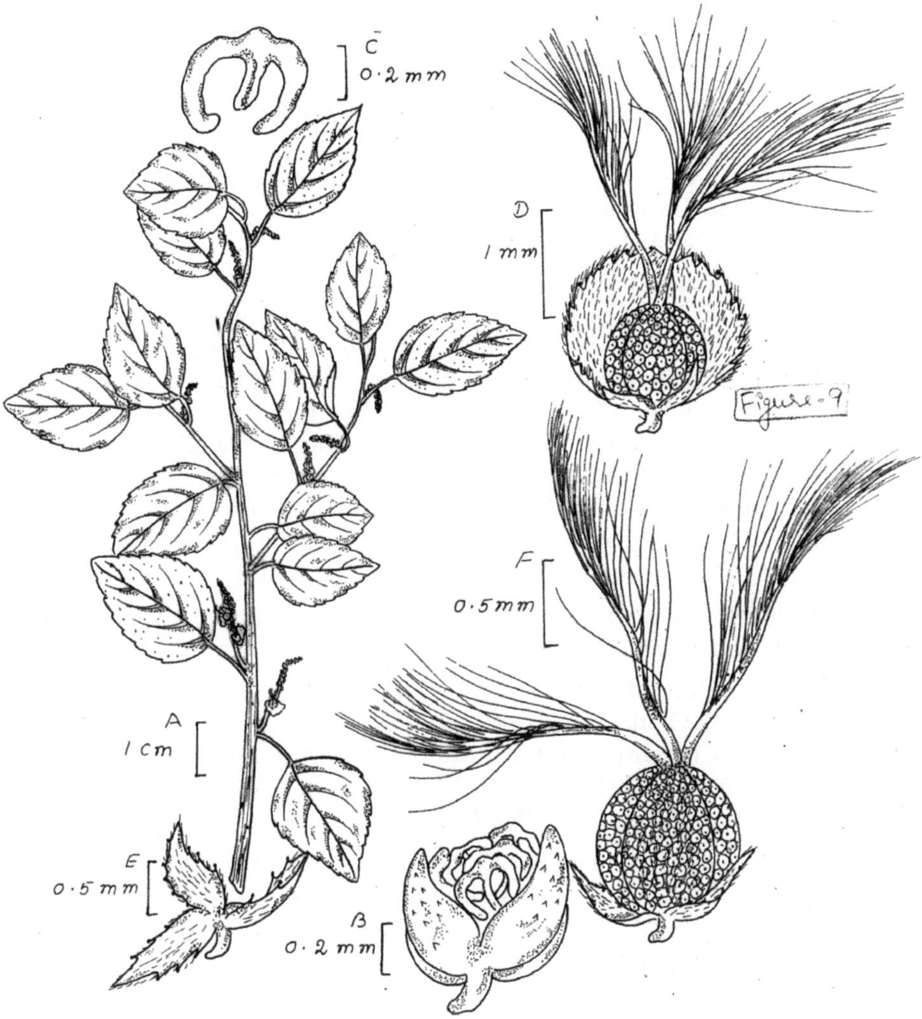

Figure 18. **Acalypha fruiticosa** Forssk.
A. Twig, ♂ B. Flower, C. Stamen. ♀ D. Flower with bract, E. Tepals, F. Flower.

Along the roadsides near the foothills. Fl. & Fr.: July - December. Vern.: Tel.: *Mabbaku*; Tam.: *Aathathalai*.

Chinnapalli (ATP), *BR* & *SSCM* 45770; Thummalapalle (CDP), *TP, KRKS, KRS* & *SSR* 33654; Foot hills of Guvvalacheruvu (CDP), *PSPB* 9195; Gangammasirassu falls (CTR), *DRC* 1933 (MH); Somasila (NLR), *BSN* 6074 (BSID); Durgam (NLR), *PV* 111156 (BSID); Kolli hills (NMK), *SKS* 936 (SGH).

INDIA: Andhra Pradesh, Karnataka, Tamil Nadu and Kerala.

WORLD: Tropical Africa, Arabia, Sri Lanka, Myanmar and Java.

Acalypha indica L., Sp. Pl. 1003. 1753; FBI 5: 416. 1887; Gamble 2: 1330. 1925; N.P.Balakr. & Chakrab., Fam. Euphorb. India 125. 2007; Susila Rani & N.P.Balakr. in N.P.Balakr. *et al.*, Fl. India 23: 96. 2012.

Herb, up to 60 cm high, monoecious; branchlets terete, striate, thinly pubescent. Leaves 2-5 × 1-4 cm, broadly ovate, base shortly attenuate, margin crenate-serrate, apex acute; petiole to 6 cm; stipules small, caducous. Flowers in elongate, axillary spikes. Male flowers many, minute, from above the middle, terminating in an allomorphic flower; flowers 0.7 mm across; pedicel 0.1 mm; tepals 4, 0.4 × 0.3 mm, broadly ovate, concave, thinly pubescent without; stamens 8, filaments short, anthers 0.3 mm, vermiculiform. Female flowers 12-15, at the lower half of the spike; flowers 6 mm across; pedicel 0.1 mm; bracts 6.3 × 5 mm, folded, shortly dentate, ciliate; tepals 3, 0.7 × 0.35 mm, broadly ovate, acute, ciliate; pistils 2, one sterile, ovary 1.2 mm across, hispid, styles 3, to 3.2 mm, each with 2-3 filiform hairs.

A common weed in wastelands. Fl. & Fr.: Throughout the year. Vern.: Tel.: *Muripinda, Kuppinta*; Tam.: *Kuppaimaeni*.

Rayadurg (ATP), *PSPB* 9194; Palakonda (CDP), *CS* 9372; Thummalapalle (CDP), *KRKS* & *KRS* 33680; Obulapuram thanda (PKM), *BSS* & *SKB* 32796; Macherla RF (GNT), *VRK* 3275; Chandragiri (CTR), *KMS* 9861 (CAL); Rapur to Rajampet (NLR), *PV* 110077 (BSID); Kothapeta (PKM), *RKM* 1.37 (CAL); Maredumilli (EG), *MM* 102689 (BSID); Simhachalam (VSKP), *SKW* 2010 (BLAT).

INDIA: Gujarat, Rajasthan, Uttar Pradesh, Bihar, West Bengal, Assam, Odisha, Madhya Prdesh, Maharashtra, Karnataka, Kerala, Tmil Nadu and Andaman and Nicobar Islands.

WORLD: Old World Tropics.

Acalypha lanceolata Willd., Sp. Pl. 4: 524. 1805; Gamble 2: 1331. 1925; N.P.Balakr. & Chakrab., Fam. Euphorb. India 126. 2007; Susila Rani & N.P.Balakr. in N.P.Balakr. *et al.*, Fl. India 23: 94. 2012. *A. fallax* Müll.Arg., Linnaea 34: 43. 1865; FBI 5: 416. 1887.

Undershrub, up to 70 cm high, monoecious; branchlets terete, striate, glabrous, pubescent when young. Leaves 4.5 – 6.5 × 1.5-3 cm, ovate-lanceolate, base obtuse, margin crenate-serrate, apex acute; petiole to 3 cm; stipules small, lanceolate, ciliate. Flowers in axillary spikes. Male flowers few, minute, at top; flowers 0.6 mm across; pedicel 0.1 mm; tepals 4, 0.6 × 4 mm, broadly ovate, concave, acute, muricate without;

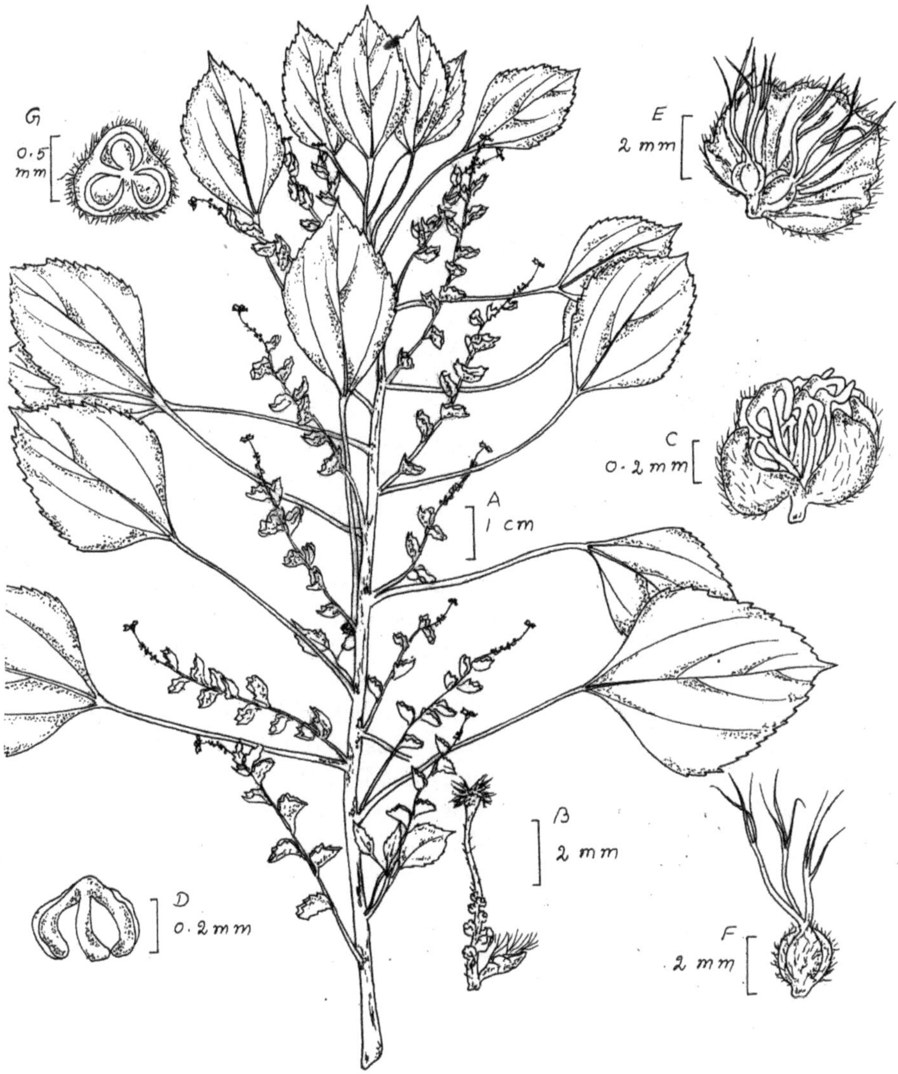

Figure 19. **Acalypha indica** L.

A. Twig, B. Spike ending with an allomorphic flower. ♂ C. Flower, D. Stamen. ♀, E. Flower, F. Pistil, G. Ovary T.s.

stamens 8, filaments short, anthers 0.25 mm, vermiculiform. Female flowers many; flowers 2.5 mm across; pedicel 0.05 mm; bracts 2 × 2.8 mm, dentate, teeth 10-12, ciliate, short, glandular tipped hairs without; tepals 3, 0.5 × 0.25 mm, free; pistil 1, ovary 1.5 mm across, hispid, styles 3, to 1.4 mm, each with 1-3 filiform hairs. Capsule 2 mm across, globose, hispid; seeds globose, smooth.

Occasional in wastelands, near moist places, shady moist places along road sides and fields. Fl. & Fr.: April-August.

Pennahobilam (ATP), *TP* & *NY* 350; Dhone RF (KNL), *RVR* 1617; Palakonda (CDP), *CS* 7604; Rachakuntapalli (CDP), *KRKS* 40177; Ethamukala (PKM), *KR* & *RG* 88850; Rangapuram (WG), *SKW* 6839, 6940; *HS* 21335 (BLAT); Gollarayi Konda (EG), *GVS* 87547 (CAL).

INDIA: Gujart, Rajasthan, Madhya Pradesh, Uttar Pradesh, West Bengal, Odisha, Andhra Pradesh, Maharashtra, Karnataka, Kerala, Tamil Nadu and Andaman Islands.

WORLD: Sri Lanka, Bangladesh, Myanmar, China, Thailand, Indo-China, Malesia to Polynesia.

Acalypha malabarica Müll.Arg., Linnaea 34: 42. 1865; FBI 5: 416. 1887; Gamble 2: 1330. 1925; N.P.Balakr. & Chakrab., Fam. Euphorb. India 126. 2007; Susila Rani & N.P.Balakr. in N.P.Balakr. *et al.*, Fl. India 23: 100. 2012; V.S.Raju, J.Econ. Taxon. Bot. 4: 296. 1983.

Herb, up to 40 cm high, monoecious; branchlets striate, thinly pubescent. Leaves 2.5-4.5 × 1.5-3 cm, ovate or elliptic lanceolate, base cuneate-obtuse, margin crenate-serrate, apex acute; petiole to 4.5 cm. Flowers in axillary spikes. Male flowers few at top; flowers 0.6 mm across; pedicel 0.4 mm; tepals 4, 0.6 × 0.4 mm, broadly ovate, acute, muricate without at middle; stamens 8, filaments short, anthers 0.2 mm, vermiculiform. Female flowers many, clustered at the base; flowers 3.5 mm across; pedicel 0.2 mm; bracts 3 × 4.2 mm, folded, dentate, teeth 12-13, ciliate, short gland tipped hairs within near the margins; tepals 3, 0.8 × 0.45 mm, broadly ovate, acute, hairy without; pistils 2, one sterile, ovary 1.5 mm across, thinly hispid at top, styles 3, to 2 mm, each with 4-5 filiform hairs. Capsule globose, glabrous or thinly hispid at top.

Occasional in moist shady places as forest undergrowth near the foothills in Southern Eastern Ghats. Fl. & Fr.: August-November.

Pennahobilam (ATP), *CJP* & *VS* 14133; Kalasamudram (ATP), *KRKS* 39529; Bhumayyagaripalle (CDP), *KRKS* & *KRS* 37615; Rachakuntapalle (CDP), *KRKS* & *BSS* 40190; Penchalkona (NLR), *BSN* 5099 (VV).

INDIA: Gujarat, Rajasthan, Uttar Pradesh, Madhya Pradesh, Odisha, Andhra Pradesh, Maharashtra, Karnataka, Kerala and Tamil Nadu. Endemic.

Acalypha paniculata Miq., Fl. Ned. Ind. 1(2): 406. 1859; FBI 5: 415. 1887; Gamble 2: 1330. 1925; N.P.Balakr. & Chakrab., Fam. Euphorb. India 126. 2007; Susila Rani & N.P.Balakr. in N.P.Balakr. *et al.*, Fl. India 23: 100. 2012. *A. racemosa* Wall. ex Baill., Etude Euphorb. 443. 1858, *nom. illeg.*, vide Müll.Arg. in DC. Prodr. 15(2): 802. 1886; Subba Rao & Kumari, Fl. Visakhapatnam 2: 137. 2008.

Herb, up to 90 cm high, monoecious; branchlets striate, thinly pubescent. Leaves 3-7 × 1.5-4 cm, broadly ovate, base truncate- rounded, margin crenate-serrate, apex acuminate; petiole to 6 cm. Male flowers many, minute, in slender, elongate, cylindric, lower axillary spikes; flowers 0.6 mm across; pedicel 0.3 mm; tepals 4, 0.6 × 0.4 mm, ovate, acute, muricate without at middle; stamens 7-8, filaments short, anthers 0.3 mm, vermiculiform. Female flowers many, in terminal panicles; flowers 2 mm across; pedicel 0.4 mm; bracts minute, acute, each subtending 3 flowers; tepals 3 or 6, 0.6 × 0.3 mm, free, lanceolate, acute, hairy without; pistil 1, ovary 0.6 mm across, glandular ciliate, styles 3, to 0.3 mm, each with many filiform hairs. Capsule globose, glandular ciliate.

Near moist shady places of hill areas. Fl. & Fr.: Throughout the year. Vern.: Tel.: *Adavi kuppinta.*

Borra (VSKP), *PSPB* 11824, *GVS* 44275 (MH); Chamala RF (CDP), *GVM* & *JHF* 115110 (MH); Bakarapeta (CTR), *JHFB* 115110; Durgam (NLR), *PV* 111129 (BSID); Anantagiri (VSKP), *HS* 21478 (BLAT); Bispuram stream (VSKP), *GVS* 44344 (MH); Kolli hills, Kulivalavu (NMK), *SKS* 119 (SGH).

INDIA: Madhya Pradesh, Andhra Pradesh, Karnataka, Kerala and Tamil Nadu.

WORLD: Tropical Africa, Sri Lanka.

Acalypha amentacea Roxb. subsp. **wilkesiana** (Müll.Arg.) Fosberg is cultivated in gardens as ornamental.

ACTEPHILA Blume

Actephila excelsa (Dalzell) Müll.Arg., Linnaea 32: 78. 1863 & DC., Prodr. 15(2): 222. 1866; FBI 5: 283. 1887; Gamble 2: 1283. 1925. var. **brevifolia** N. Balach. *et al.*, J. Econ. Taxon. Bot. 33: 717. F.1. 2009; Chakrab. & N.P. Balak. in Balakrishnan *et al.*, Fl. India 23: 500. 2012. *Anomospermum excelsum* Dalzell, Hooker's J. Bot. Kew. Gard. Misc. 3: 228. 1851. *Actephila neilgherrensis* Wight, Icon. Pl. Ind. Orient. t.1910. 1852.

Shrubs to 1.5 m tall; branchlets glabrous. Leaves elliptic to cuneate-obovate, 4-9 x 1.5-3 cm, acute, obtuse at base, entire along margins, obtuse to acuminate at apex, coriaceous to chartaceous, glossy above; lateral veins 5-12 pairs; petiole short, 0.2-1 cm long; stipules minute, triangular to ovate. Inflorescence axillary. Male flowers in fascicles; pedicels 2-5 mm long, slender; sepals 5, ovate or oblong, 1.5 x 0.8 mm; petals 5, spathulate, 1 x 0.3 mm; disc 5-lobed; stamens 2.5-3 mm long; filaments connate below into a column; anthers ca 0.7 mm long; pistillode ovoid. Female flowers solitary; pedicels 1-6 cm long, slender; sepals 5, ovate, elliptic or

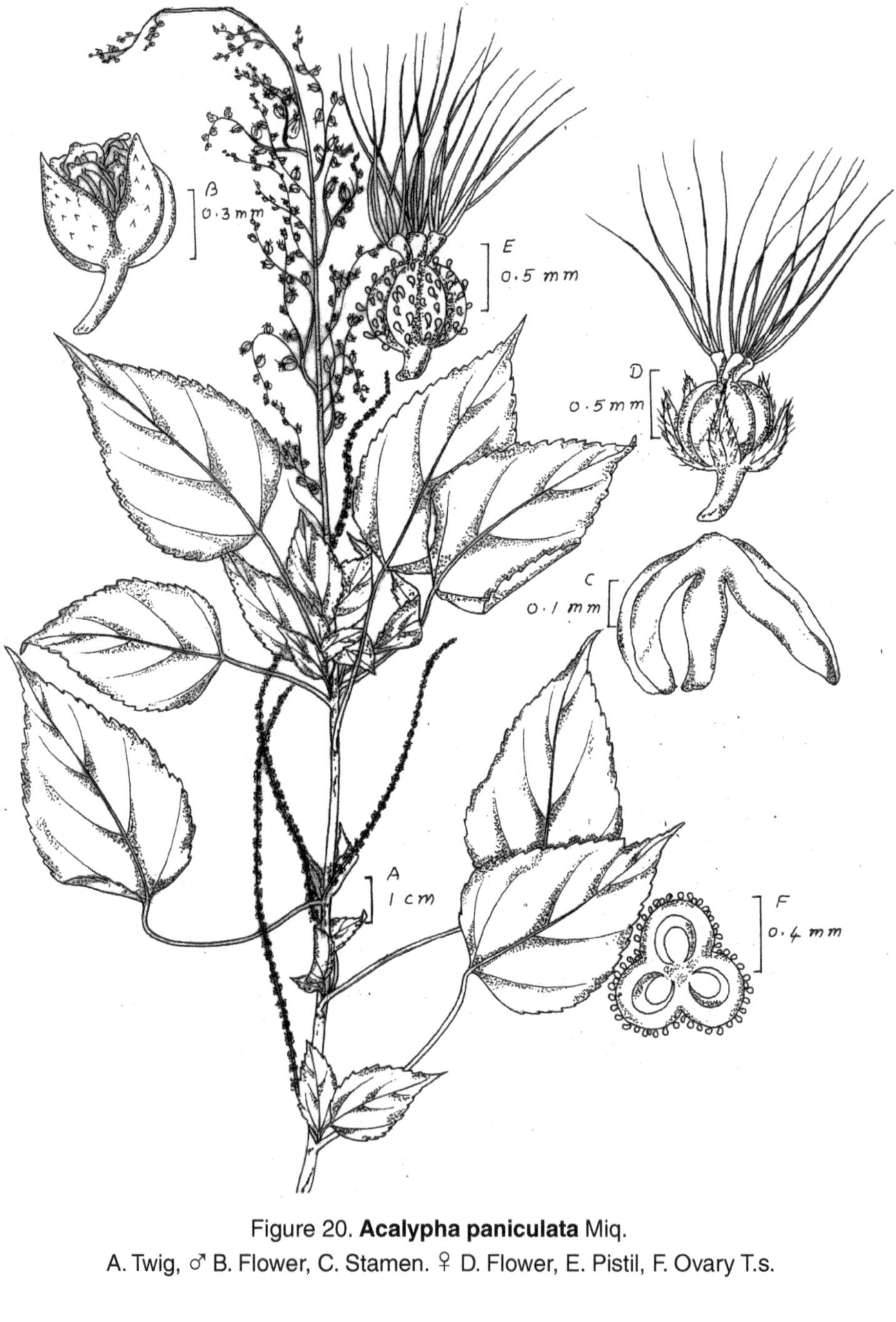

Figure 20. **Acalypha paniculata** Miq.
A. Twig, ♂ B. Flower, C. Stamen. ♀ D. Flower, E. Pistil, F. Ovary T.s.

oblong, 2-3 x 1-2 mm; petals 5, spathulate, 1.5-0.5 mm; disc flat; ovary subglobose, ca 2 mm across; styles 2 mm long, bifid above. Fruit subglobose, depressed, 1.5-3 cm across, 3 to 6-lobed, usually verrucose; pedicels to 6 cm long.

Occasional on water course of Sirumalai hills in Tamil Nadu. Fl. & Fr.: December – June.

Sirumalai hills, above Kudladampatty falls (DGL), *SKS* 426 (SGH).

INDIA: Kerala, Tamil Nadu and Andaman Islands.

WORLD: Sri Lanka.

AGROSTISTACHYS Dalzell

Agrostistachys meeboldi Pax & K.Hoffm., Pflanzenr. IV. 147. VI: 100.1912; Gamble 2: 1318. 1925; N.P.Balakr. in N.P.Balakr. *et al.*, Fl. India 23: 172. 2012. *Sarcoclinium longifolium* Wight, Icon. Pl. Ind. Orient. 5: 24. t. 1887-88. 1852 non *A.longifolia* Kurz, 1875. *Agrostistachys longifolia* (Wight) Benth. ex Hook.f., Fl. Brit. India 5: 407. 1887 excl. var. *malayana* & var. *latifolia* (non Kurz, 1875). *A. borneensis* sensu K.M.Matthew & Rani in Matthew, Fl. Tamil Nadu Carnatic 3: 1409. 1983.

Dioecious evergreen tree, to 12 m high. Leaves quasispiral, obovate, spathulate, 8-20 x 4-9 cm, thick-coriaceous, dark green below, apex shortly acuminate; stipules scaly; petioles 5-15 mm long. Flowers brownish yellow in axillary spikes; male flowers few, female flower solitary; tepals 5+5, stamens 10, exserted; disc glands 5, free, ovary 3-locular, styles 3, each bifid; disc glands 6-8, partly connate. Capsule bivalved, depressed-globose, red; seeds globose.

Occasional in Kolli hills, Periya Shola, above 1400m. Fl. & Fr.: January – April.

Kolli hills, on the way to Kulivalavu (NMK), *SKS* 1024 (SGH).

INDIA: Karnataka, Kerala, Tamil Nadu. Endemic.

Note: Several recent regional floras have treated this species as *A. borneensis* Becc. which is different from Wight's *Sarcoclinium longifolium* and distributed only from Myanmar onwards to Malesia and extending to peninsular India. *A. longifolia* var. *malayana* and var. *latifolia* given by Hook.f. (loc. cit.) ar their synonyms, but not *A. longifolia* var. *longifolia* (N.P.Balakr. *loc.cit.*).

ALCHORNEA Swartz

Alchornea mollis (Benth.) Müll.Arg., Linnaea 34: 168. 1865; FBI 5: 420. 1887; Gamble 2:1324.1925; N.P.Balakr. & Chakrab., Fam. Euphorb. India 159. 2007; N.P.Balakr. in N.P.Balakr. *et al.*, Fl. India 23: 174. 2012. *Stipellaria mollis* Benth., Hooker's J. Bot. Kew Gard. Misc. 6: 3. 1854.

Small tree, up to 5 m high; branches pubescent. Leaves coriaceous, broadly ovate, 6-22 × 3.5-16.5 cm, pubescent, base cuneate or rounded, margin dentate, apex caudate-acuminate, 3-5-nerved from base, 4 glandular, in between the basal nerves; stipels at base of leaf up to 0.2 cm long, subulate, pubescent, persistent; petiole to 15 cm. Flowers small in terminal, long spikes or racemes; males solitary or in clusters,

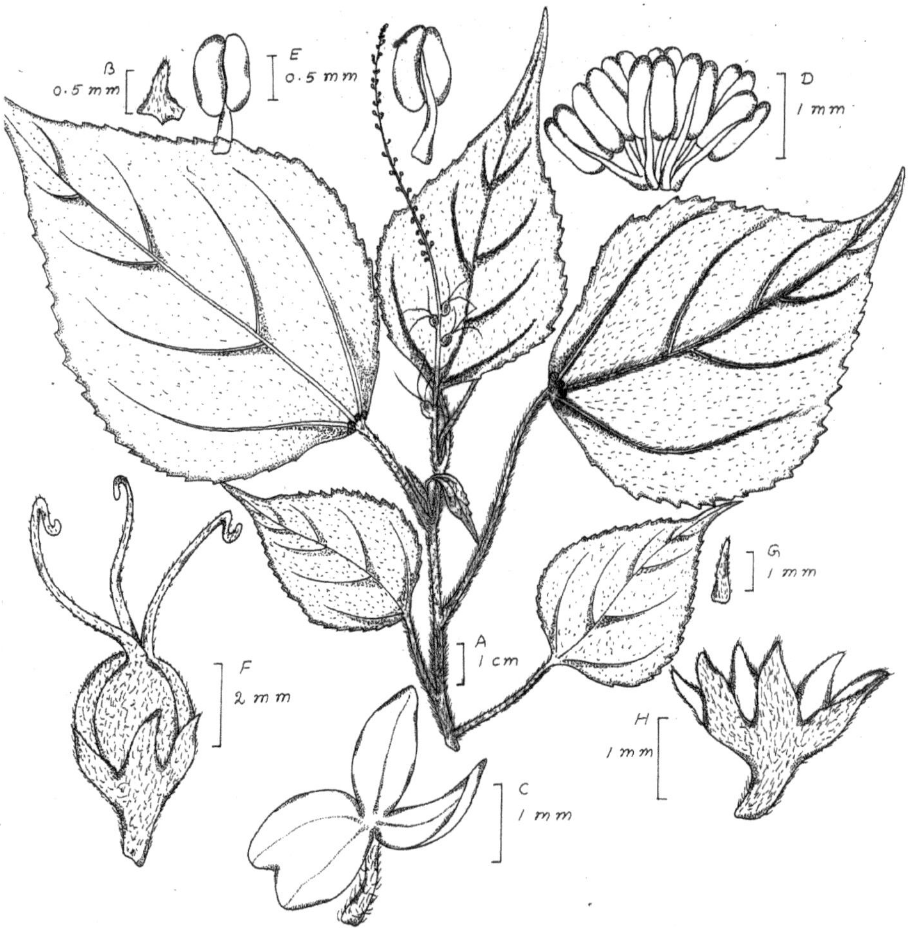

Figure 21. **Alchornea mollis** (Benth.) Müll.Arg.,
A. Twig, ♂ B. Bract, C. Tepals, D. Androecium, E. Stamens. ♀ F. Flower,
G. Bract, H. Tepals.

females solitary. Male flowers: tepals 4, broadly ovate, minutely pubescent without, reflexed in fully grown flowers; stamens 6, free. Female flowers: bracts up to 0.3 cm long, lanceolate, chocolate-brown, pubescent without, eglandular, subtending a single flower, persistent; bracteoles up to 1.5 mm lon, similar to bracts, persistent; tepals 5-6, unequal, up to 0.2 cm long, lanceolate, chocolate-brown, pubescent without, ciliate, persistent, reflexed in fruit; ovary up to 0.2 × 0.2 cm, globose, densely white-velvety, styles 3, up to 1 cm long, entire, chocolate-brown, minutely white-pubescent, erect, bent or curled, connate at base up to 0.1 cm length; stipe up to 1 mm long, white-velvety Capsule gobose, 9 mm, grey-pubescent; seeds subglobose, rugulose.

Occasional, in the hills especially in moist valleys above 700 m in northern Eastern Ghats. Fl. & Fr.: June-August. Vern.: Tel.: *Badiki*; Ori.: *Tabhari.*

Pedavalasa (VSKP), *SSR* 19667; Korapur (VSKP), *GVS* 44401; Minumuluru (VSKP), *GVS* 30042 (MH); Vankachinta (VSKP), *GVS* 47345 (MH); Bhanjanagar, Barakmada, Similipahar (MBJ), *SX* & *MB* 4409 (RRL-B); near Janchua (KHD), *HFM* 2177 (DD).

INDIA: C. and E. Himalayas, West Bengal, Sikkiam, Assam, Odisha and Andhra Pradesh.

WORLD: Nepal, Bhutan and China.

ALEURITES J.R.Forster & J.G.A.Forster

Aleurites moluccana (L.) Willd., Sp. Pl. 4: 590. 1805; FBI 5: 384. 1887; Gamble 2: 1318; N.P.Balakr. & Chakrab., Fam. Euphorb. India 192. 2007; Chakrab. & N.P.Balakr. in N.P.Balakr. *et al.*, Fl. India 23: 207. 2012. *Jatropha moluccana* L., Sp. Pl. 1006. 1753.

Evergreen tree; young shoots densely clothed with brownish or grey stellate tomentum. Leaves subcoriaceous, broadly elliptic or ovate to deltoid, tomentose when young, base truncate, apex shortly acuminate; petiole long, 2-glandular at apex. Flowers white in large, cymose, terminal panicles. Male flowers: calyx subglobose in bud, 6-8 mm long, lanceolate or ovate, disc urceolate, with 5 glands; stamens 15-20, verticillate in 3-5 rows, filaments hairy. Female flowers: sessile at the division of panicle; calyx cylindrical and conical, splitting into 2 or 3 segments; petals 5, narrowly oblong, 9-10 mm long; disc urceolate with 5 glands; ovary densely hispid, styles 2, each bifid with linear-subulate arms. Drupe globose, fleshy, 4 cm across with hard stone; seeds large.

Cultivated in the gardens. Fl.: December – February. Fr.: February-May. Vern.: Tam.: *Neikottan*.

Batrepalli (ATP), *KRKS* 36947; Sirumalai hills (DGL), *SKS* 573 (SGH).

INDIA: Assam, West Bengal, Odisha, Andhra Pradesh, Karnataka, Kerala and Tamil Nadu.

WORLD: Native of SE Asia and Pacific Islands, widely cultivated.

ANTIDESMA Linnaeus

1. Stamens 2; pistillode 0 .... **A. acidum**
1. Stamens 3-5; pistillode present:
    2. Leaves round-elliptic, apex obtuse; tepals 5-7 .... **A. ghaesembilla**
    2. Leaves broadly oblong or lanceolate, apex acute or acuminate; tepals 4 or 5:
        3. Leaves 9 or more than 9 cm long; male flowers pedicellate .... **A. montanum**
        3. Leaves less than 9 cm long; male flowers sessile:
            4. Pistillode truncate .... **A. bunius**
            4. Pistillode acute .... **A. alexiteria**

Antidesma acidum Retz., Obs. 5: 30. 1789; Sharma *et al.*, Biol. Mem. 2: 127. 1977; N.P.Balakr. & Chakrab., Fam. Euphorb. India 303. 2007; Chakrab. & M. Gangop. in N.P.Balakr. *et al.*, Fl. India 23: 356. 2012. *Stilago diandra* Roxb., Pl. Coromandel 2: 35. t. 166. 1802 & Fl. Indica 3: 759. 1832. *Antidesma diandrum* (Roxb.) Roth, Nov. Pl. Sp. 369. 1821; FBI 5: 361. 1887; Gamble 2: 1298. 1925.

Small tree, to 6 m high, dioecious; branchlets terete, glabrous. Leaves 4-7.5 × 1.5-3 cm, lanceolate or elliptic lanceolate or obovate, base obtuse, margin entire, apex acute-acuminate, thinly pubescent below; petiole to 3 mm; stipules linear, 4 mm. Flowers in long slender, glabrous or pubescent terminal spikes, pedicellate; bracts ovate, 0.7 mm. Male flowers 3 mm across; tepals cupular, 0.7 mm, with 4 or 5 short lobes, lobes obtuse, sparsely puberulous; stamens 2, filaments 0.8 mm, anthers 0.3 × 0.25 mm, curved. Female flowers 1.5 mm across; tepals cupular, 1.4 mm, with 4 short lobes, lobes minute or rather deeper than in the male, obtuse; ovary 0.7 mm across, obovoid, glabrous, styles 2. Drupe 3 mm across, globose, prominently ribbed.

Common in exposed slopes of dry deciduous forests usually near the streams. Fl. & Fr.: Auust-October. Vern.: Tel.: *Gummadipella, Pulleru*; Ori.: *Nannunia, Nuniari, Jamarla, Manmudi, Kath-marsuri, Jamula.*

Batrepalli (ATP), *KRKS* & *BSS* 39593; Pangadi (KNL) *RVR* 2393; Guvvalacheruvu (CDP), *PSPB* 9196; Tirumala (CTR), *SSR* 13711; Diguvametta (KNL), *JLE* 22189 (MH); Japali (CTR), *GVS* 45897, *DRC* 1485 (MH); Kota (EG), *MSRS* 501 (CAL); Addatigala (EG), *SKW* 4019 (BLAT); Minumuluru (VSKP), *HS* 20700; Paderu (VSKP), *SKW* (BLAT), *GVS* 28073 (MH); Jyotimonidi (VZN), *GVS* 68599 (MH); Kolli hills (NMK), *SKS* 1041 (SGH).

INDIA: Thruough out India.

WORLD: Pakistan, Nepal, Bhutan, Bangladesh, Myanmar, Thailand, S. China, Indo-China and Java.

Antidesma alexiteria L., Sp. Pl. 1027. 1753 *p.p.*; FBI 5: 359. 1887; N.P.Balakr. & Chakrab., Fam. Euphorb. India 304. 2007; Chakrab. & M. Gangop. in N.P.Balakr. *et al.*, Fl. India 23: 357. 2012. *A. zeylanicum* Lam., Encycl. 1: 207. 1782; Gamble 2: 1297. 1925; Surya Prakash Babu in Moulali & Pullaiah, Fl. Andhra Pradesh 2: 843. 1997.

Shrub or small tree, to 6 m high, dioecious; branchlets terete, glabrous. Leaves 5-8 × 2.5-3.5 cm, elliptic to elliptic-lanceolate or oblong-elliptic, base acute-cuneate, margin entire, apex acutely apiculate, subcoriaceous; petiole to 5 mm; stipules lanceolate. Flowers in axillary and terminal spikes. Male flowers 2.5 mm across; tepals 4 or 5, free, concave, thinly pubescent without; stamens 3, filaments 1.5 mm, anthers 0.5 mm across; pistillode acute. Female flowers 1.2 mm across, sessile; tepals 4, free, concave, thinly pubescent without; ovary 0.8 × 1.2 mm, globose, styles 2, one of them 2-fid. Drupe 4-5 × 3-4 mm, globose.

Occasional on either sides of the streams in the deciduous forests. Fl. & Fr.: September-December.

Dhaniyanicheruvu (ATP), *TP* & *NY* 839; Gulemallebad RF (KNL), *RVR* 1418; Gundlabrahmeswaram (KNL), *BR, SKB* & *BSS* 34088; Galikonda (VSKP), *PSPB* 11813, *SSR* 15919 (♂); Alagar hills, Bison valley (MDU), *SKS* 76 (SGH).

INDIA: Andhra Pradesh, Tamil Nadu and Kerala.

WORLD: Sri Lanka.

Antidesma bunius Spreng., Syst. Veg. 1: 826. 1824; FBI 5: 358. 1887; Gamble 2: 1298. 1925; N.P.Balakr. & Chakrab., Fam. Euphorb. India 304. 2007; Chakrab. & M. Gangop. in N.P.Balakr. *et al.*, Fl. India 23: 358. 2012.

Ver.: *Anepu.*

Small tree, to 6 m high, dioecious; branchlets pubescent. Leaves variable, elongate-elliptic or elliptic-lanceolate or oblanceolate, margin entire, apex acuminate, shining above; petiole 1 mm; stipules lanceolate-subulate, pubescent. Flowers in pubescent axillary spikes, male flowers often compound; female flowers simple; bracteate. Male flowers sessile; tepals cupular, with 4 short lobes; stamens 3, anthers curved; pistillode short, truncate; disc lobed, glabrous. Female flowers short pedicellate; tepals tubular, short, truncate; ovary glabrous, styles 2, short, one of them usually 2-fid; disc cupular, glabrous. Drupe globose-ovoid.

Rare in the hills of northern Eastern Ghats. Fl. & Fr.: June-August. Vern.: Tel.: *Anepu.*

Chintapalli (VSKP), *SKW* 2821 (BLAT); Paderu (VSKP), *HS* 20770 (BLAT); Similipahar (MBJ), *SX* & *MB* 4316 (RRL-B); Mahendragiri (GJM), *SX* & *MB* 2212 (RRL-B).

INDIA: Uttarakhand, Uttar Pradesh, West Bengal, Sikkim, Assam, Meghalaya, Mizoram, Odisha, Andhra Pradesh, Maharashtra, Karanataka, Kerala, Tamil Nadu and Andaman Islands.

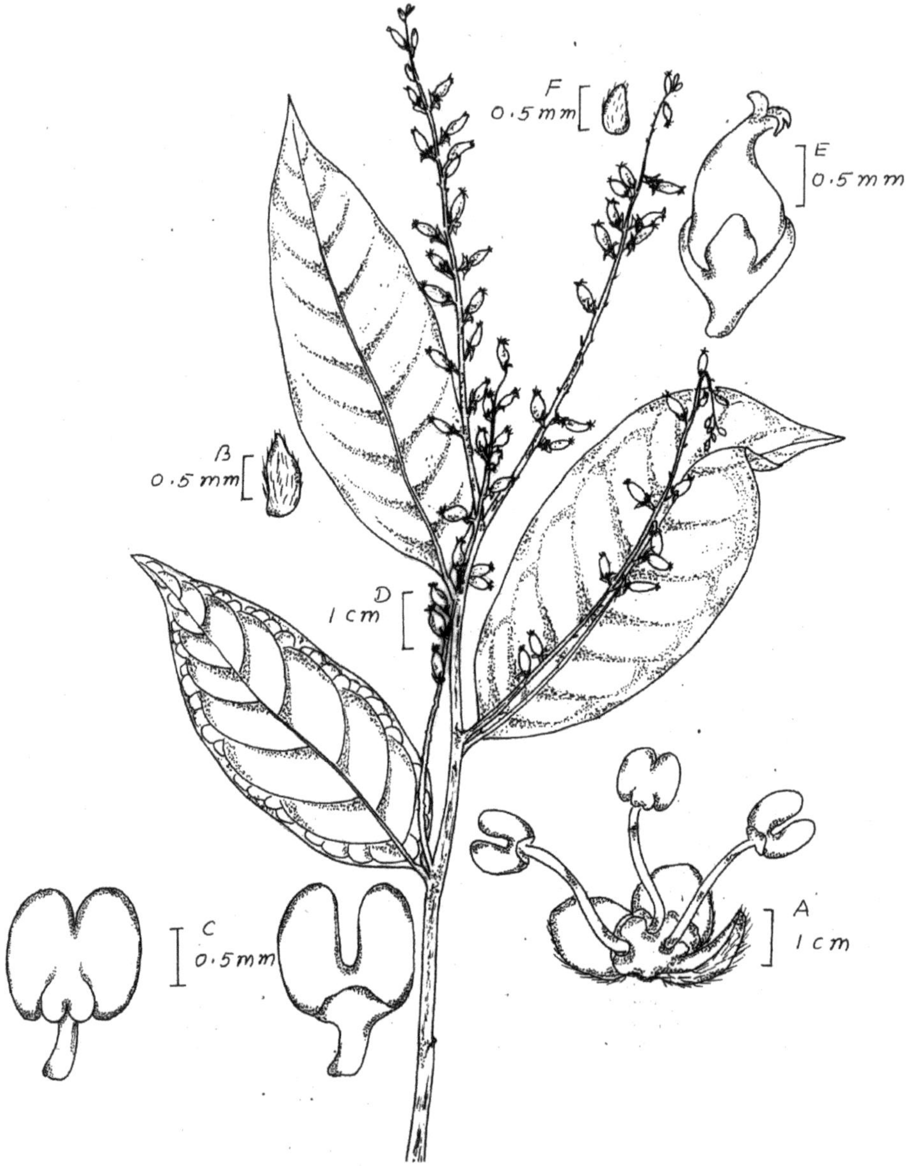

Figure 22. **Antidesma alexiteria** L. (Syn.: *Antidesma zeylanicum* Lam.)
♂ A. Flower, B. Bract, C. Anthers. ♀ D. Twig, E. Flower, F. Bract.

WORLD: Sri Lanka, Nepal, Bhutan, Bangladesh, Myanmar, S. China, Laos, Vietnam, Malesia, to New Guinea, Hawai Islands and N. Australia.

Antidesma ghaesembilla Gaertn., Fruct. Sem. Pl. 1: 189, t. 39. 1788; Müll.Arg. in DC., Prodr. 15: 215. 1866; FBI 5: 357. 1887; Gamble 2: 1298. 1925; N.P.Balakr. & Chakrab., Fam. Euphorb. India 305. 2007; Chakrab. & M. Gangop. in N.P.Balakr. *et al.*, Fl. India 23: 359. 2012.

Small tree, to 10 m high, dioecious; branchlets terete, pubescent or fulvous tomentose. Leaves oblong, broadly round-elliptic, 5-9 × 2.5-5.5 cm, base rounded, margin entire, apex rounded or abruptly acute, often emarginated; petiole 1 cm, pubescent; stipules lanceolate. Flowers in slender paniculate axillary spikes, sessile. Male flowers 1.7 mm across; tepals 5 to 7, densely pubescent; stamens 4-5, filaments 1 mm, anthers 0.4 mm, curved; disc 5-lobed, pubescent. Female flowers 2.2 mm across; tepals as in the male flower; ovary 2 mm across, pubescent, styles 2, short, one of them 2-fid; disc as in the male flower. Drupe 3 mm across, globose, prominently rugose when dry; seeds covered with dried albumen.

Occasional near the streams in the deciduous forests. Fl. & Fr.: June-August. Vern.: Tel.: *Polari, Puttam, Lona, Katha pulitsi, Janupulisaru*; Ori.: *Suankali, Nuniari, Jamula, Kath marmuri.*

Batrepalli RF (ATP), *KRKS* 39593; Gundlabrahmeswaram (KNL), *SSR* 15994, 19555; Talakona (CTR), *SSR* 18592; Tirumala (CTR), SSR 13724; Velagalpaya (PKM), *SKB* & *BSS* 30506; Seethampeta (SKLM), *SSR* 15574; Chettigunta (CDP), *EKK* 77832 (DD); Maredumilli (EG), *GVS* 24404; Chintapalli (VSKP), *GVS* 30083 (MH); Althi (SKLM), *PVS* & *NRR* 76887 (MH); Elwinpet (SKLM), *SKW* 7046, 7047 (BLAT); Itchapur (SKLM), *GVS* 62487 (MH).

INDIA: Through out India except Kashmir and Rajasthan.

WORLD: Sri Lanka, Pakistan, Nepal, Butan, Mayanmar, S. China, Laos, Vietnam, Thailand, Malesia up to N. Australia.

Antidesma montanum Blume, Bijdr. Fl. Ned. Ind. 1124. 1826-27; Mandal & Panigrahi, J. Econ. Tax. Bot. 4: 259. 1983; Chakrab. & N.P.Balakr., J. Econ. Tax. Bot. Add. Ser. 9: 18. 1992; N.P.Balakr. and Chakrab., Fam. Euphorb. India 305. 2007; Chakrab. & M. Gangop. in N.P.Balakr. *et al.*, Fl. India 23: 359. 2012. var. **montanum.** *A. acuminatum* Wight, Icon. Pl. Ind. Orient. 6: 12. t. 1991. 1853; FBI 5: 363. 1887. *A. menasu* (Tul.) Müll.Arg. 257. 1866 *p.p.* tantum quoad Hohenacker 104; Gamble 2: 1297. 1925; Surya Prakash Babu in Pullaiah & Moulali, Fl. Andhra Pradesh 2: 843. 1997. *A. menasu* var. *linearifolia* Hook. f., Fl. Brit. India 364. 1887. *A. wattii* Hook. f., Fl. Brit. India 5: 366. 1887.

Tree, to 13 m high, dioecious; branchlets sparsely fulvous tomentose. Leaves 9-20 × 2-5 cm, oblong, base truncate-broadly acute, margin entire, apex acuminate; petiole to 6 mm; stipules lanceolate. Flowers in axillary or terminal panicles, pedicellate. Male flowers: 3 mm across; tepals 3, free, concave, hairy without, gland tipped margin; stamens 3-5, filaments 2 mm, anthers 0.5 mm across; pistillode

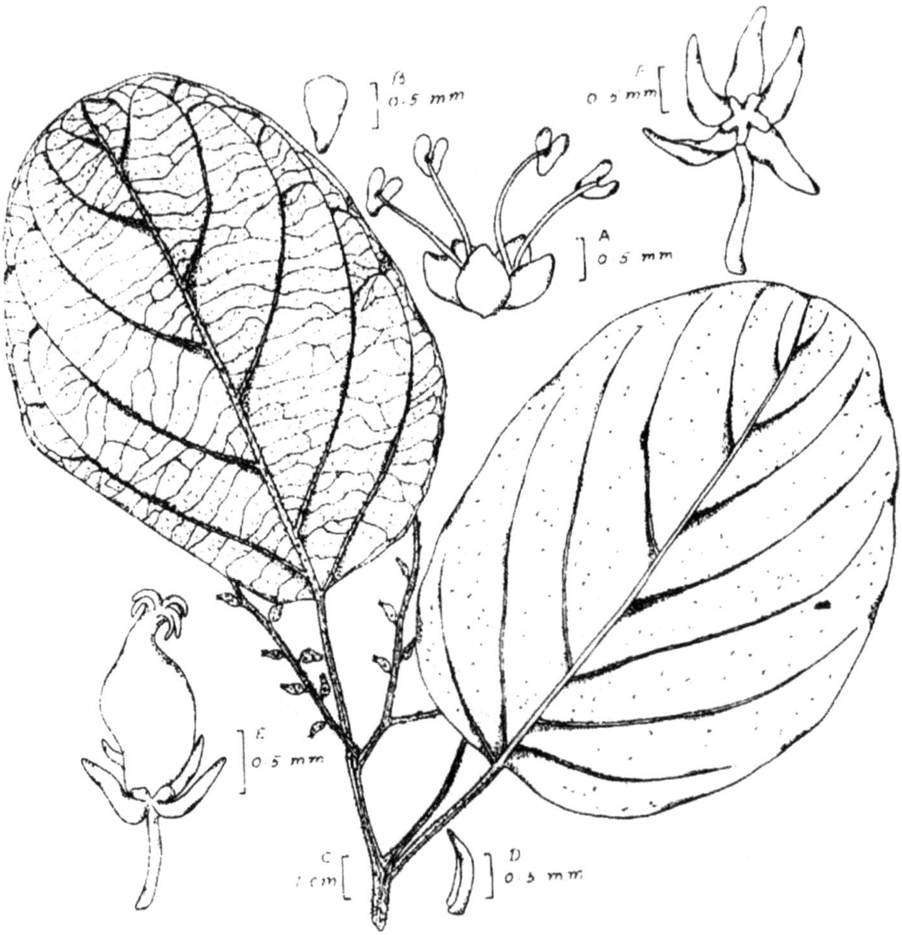

Figure 23. **Antidesma ghaesembilla** Gaertn.
♂ A. Flower, B. Bract. ♀ C. Twig, D. Bract, E. Flower, F. Tepals.

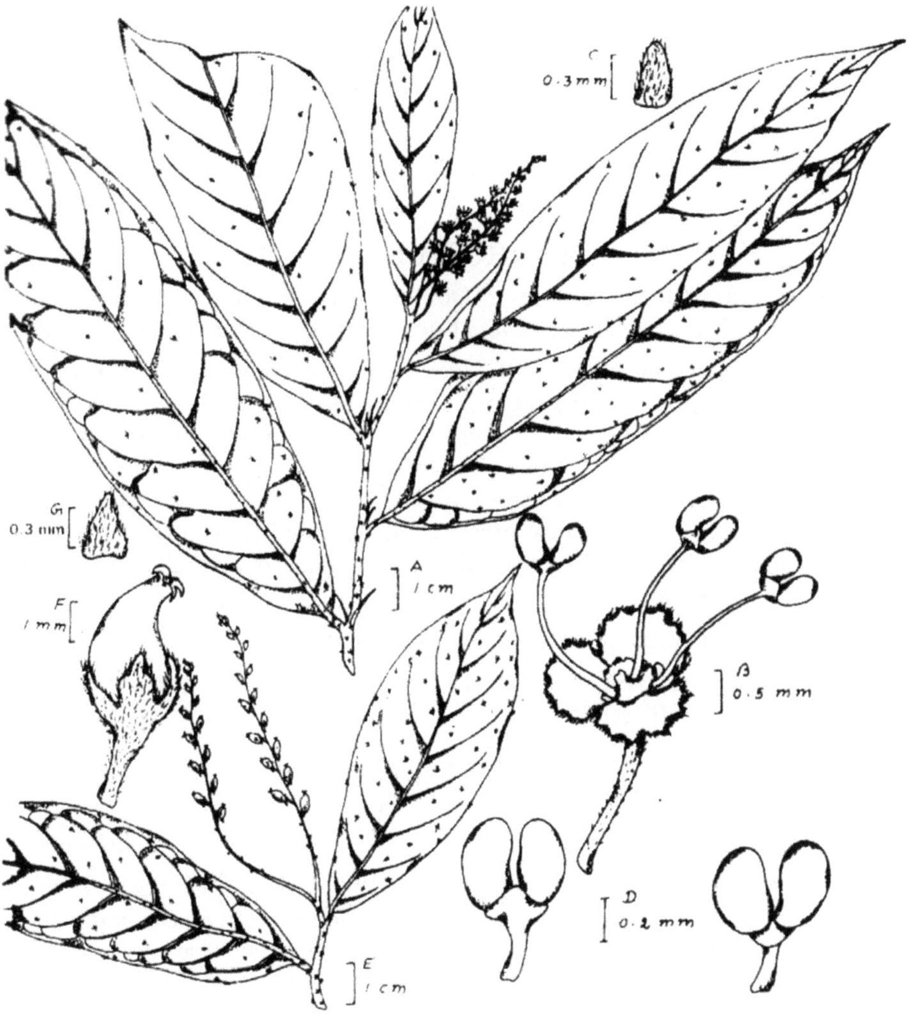

Figure 24. **Antidesma montanum** Blume [Syn.: *Antidesma menasu* (Tul.) Müll.Arg.]
A. Twig, B. Flower, C. Bract, D. Anthers. ♀ E. Twig, F. Flower, G. Bract.

subacute. Female flowers 2 mm across; tepals 3, connate at base, hairy without, subacute; ovary 2.5 × 2 mm, subglobose, styles 2, short, each 2-fid. Drupe globose.

Occasional in the hills of Northern Eastern Ghats. Fl. & Fr.: September-December. Vern.: Ori.: *Kath jamrala;* Tam.: *Aasaripuli.*

Nulakamaddi (EG), *GVS* 24537; Kokkarampalli, Minumuluru RF (EG), *SSR* 15964; Sapparla (VSKP), *SSR* 19647; Chintapalli (VSKP), *GVS* 28200; Galikonda (VSKP), *GVS* 19635; Minumuluru (VSKP), *GVS* 47353; Padalammagudi (VSKP), *GVS* 42586 (MH); Similipahar (MBJ), *SX & MB* 4652, 4755 (RRL-B); Gandhamardan (Balangir), *SX & MB* 5856 (RRL-B).

INDIA: Almost through out India.

WORLD: Sri Lanka, Bhutan,Bangladesh, Myanmar, China, Thailand, Malesia to New Guinea and N. Australia.

APOROSA Blume

Aporosa octandra (Buch.-Ham. ex D.Don) Vickery in Hara *et al.,* Enum. Fl. Pl. Nepal 3: 193. 1982. *Myrica octandra* Buch.-Ham. ex D.Don, Prodr. 56. 1825.

Small tree; bark nearly smooth, grey brown or rough in old trees. Leaves elliptic, oblong, elliptic-oblong or oblong-lanceolate, 6-17.5 × 2.5- 7.5 cm, coriaceous, glabrous, apex obtuse or obtusely acuminate, margin entire or slightly dentate in distal part, base subacute or obtuse; petiole 0.8-1.5 cm long; stipules ovate-oblong 7.5-11 mm. Male flowers minute, in yellowish spikes, from the leaf axils and scars of fallen leaves on old branches; bracts orbicular, slightly brown-hairy. Female flowers sessile or subsessile, in ovoid spikes; bracts ciliate; ovary hairy. Fruit globose, apiculate, 1-1.2 cm diam.

Near streams in Eastern Ghats of Odisha. Fl.: February – March; Fr.: April – May.

Similipahar (MBJ), *SX* 410 (RRL-B), *SX & MB* 4385, 4439, 4539, 5054, 5098 (RRL-B); Nayagarh, *HFM* 1674 (DD).

BACCAUREA Loureiro

Baccaurea ramiflora Lour., Fl. Cochinc. 661. 1790. *Plierardia sapida* Roxb., Fl. Ind. 2: 254. 1832. *Baccaurea sapida* (Roxb.) Mull.Arg. in DC., Prodr. 15: 459. 1866; FBI 5: 371. 1887. *B. wrayi* King ex Hook.f., Fl. Brit. India 5: 374. 1887.

Tree; bark smooth; buds and young twigs densely strigosely hairy. Leaves elliptic, elliptic-obovaate, oblanceolate or obovate, 10-24 × 4.5-9 cm, apex acuminate or obtuse, margin entire or repand, base narrowed; petiole 3-5.5 cm long. Flowers in racemes. Male flowers subsessile, solitary or grouped on very small, lateral branches of the raceme, in the axils or deciduous, lanceolate, stipular bracts; tepals 4, oblong or ovate; pistillode large, 3-lobed. Female flowers yellow; tepals oblong-lanceolate, tomentose; ovary strigosely tomentose; stigma large, peltate, almost fimbriate with large papillae. Fruit globose or globose-oblong, hairy, terminated by 3 stigmatic scars; seed plano-convex, with reed testa and white aril.

Occasional near streams in Eastern Ghats of Odisha. Fl.: April –May; Fr.: May-June. Vern.: Ori.: *Rajkoli.*

Similipahar (MBJ), *Yoganarasimhan* 027 (RRL-B), *SX* & *MB* 4317, 4583, 5037 (RRL-B).

BALIOSPERMUM Blume

Baliospermum solanifolium (J.Burm.) C.R.Suresh in Nicolson *et al.*, Interpr. Van Rheede's Hort. Malab. 106. 1988; N.P. Balakr. & Chakrab., Fam. Euphorb. India 199. 2007; N.P.Balakr. in N.P.Balakr. *et al.*, Fl. India 23: 216. 2012. *Croton solanifolius* J. Burm., Fl. Malab. 6. 1769; Geiseler, Croton Monogr. 74. 1807. *Jatropha montana* Willd., Sp. Pl. 4: 563. 1805. *Baliospermum axillare* Blume, Bijdr. 604. 1825; FBI 5; 461. 1887. *Croton polyandrous* Roxb., Fl. Ind. 3: 682. 1832. *Baliospermum montanum* (Willd.) Müll.Arg. in DC., Prodr. 15: 1125. 1866; Gamble 2: 1342. 1925; Surya Prakash Babu in Pullaiah & Moulali, Fl. Andhra Pradesh 2: 843. 1997.

Herb or undershrub, to 1 m high, monoecious, often dioecious; branchlets terete, striate, tomentose. Leaves 3-21 × 1.5-8 cm, elliptic-ovate, often 3 to 5-lobed, base rounded or cuneate and 2-glandular, margin sinuate-toothed, apex acuminate, sparsely hirtillous; petiole to 6 cm; stipules lanceolate, 2-glandular. Flowers in axillary and terminal racemes; bracts 1.2 mm, widely ovate. Male flowers 2.5 mm across; tepals 5, 2.2 × 1.6 mm, slightly connate at base, suborbicular, concave, rounded; stamens numerous, filaments 1 mm long, anthers 0.5 × 0.8 mm; disc glands 6. Female flowers 3.5 mm across; tepals 6, connate at base, ovate-lanceolate, subacute, hirtillous without; ovary 2.5 mm across. Capsule 0.9 × 0.7 mm, obovate, pubescent; seeds elliptic-obovate, smooth.

Common along the roadsides, in fields and also occasional in shady places in semi moist deciduous forests. Fl. & Fr.: October-February. Vern.: Tel.: *Kondamudam, Nelajidi.* Tam.: *Neeradimuthu.*

Pangadi (KNL), *PSPB* 9132; Velugodu (KNL), *RVR* 2199; Rollapenta (PKM), *RVK* 15900; Rampa (EG), *PSPB* 11840; Maredumilli (EG), *SKW* 1671 (BLAT).

INDIA: Jammu & Kashmir, Rajasthan, Madhya Pradesh, Uttar Pradesh, Uttarakhand, Bihar, West Bengal, Sikkim, Assam, Odisha, Andhra Pradesh, Maharashtra, Goa, Karnataka, Kerala and Tamil Nadu.

WORLD: Pakistan, Sri Lanka, Bhutan, Nepal, Bangladesh, Myanmar, China, Thailand, Cambodia, Laos, Vietnam, Malaysia, Indonesia, Celebes, Moluccas, and Lesser Sunda Islands.

BISCHOFIA Blume

Bischofia javanica Blume, Bijdr. Fl. Ned. India 1168. 1827; FBI 5: 345.2012; N.P.Balakr. & Chakrab. in N.P.Balakr. *et al.*, Fl. India 23: 375. 2012.

Deciduous trees, 5-30 m high; branches spreading forming a large crown. Leaves alternaate, digitately trifoliolate, leaflets obovate, 5-20 × 3-11 cm, coriaceous to chartaceous, margin denticulate-serrulate; petiole up to 11.2 cm long, striate,

Figure 25. **Baliospermum solanifolium** (J.Burm.) C.R.Suresh
[Syn.: *Baliospermum montanum* (Willd.) Müll.Arg.]

A and B. Twigs, ♂ C. Flower, D. Androecium, E. Stamens. ♀ F. Flower, G. Bract, H. Tepals, I. Pistil.

lenticellate, glabrous, estipulate. Male flowers: red in bud, yellowish when open; sepals 5, oblong, ovate or orbicular; stamens 5, free, opposite to and enclosed by sepals. Female flowers: pedicels up to 12 mm long; bracts linear; sepals 5, ovate to orbiculr, caducous; ovary ellipsoid or ovoid, styles 3-6 mm long. Fruit drupaceous, subglobose; seeds ovate, brown, smooth, shining.

In deciduous or semievergreen forestss in Northern Eastern Ghats. Fl. & Fr.: January - December. Vern.: Tam.: *Manchadhai, Cholavaengai.*

Chintapalli (VSKP), *SSR* 16141; Sunkarimetta (VSKP), *SSR* 15914; Similipahar (MBJ), *SX* & *MB* 4423 (RRL-B), *MB* 5449 (RRL-B); Koinpur, below Mahendragiri (GJM), *SX* & *MB* 3908 (RRL-B).

INDIA: Himachal Pradesh, Uttarakhand, Uttar Pradesh, West Bengal, Sikkim, Assam, Meghalaya, Mizoram, Tripura, Odisha, Andhra Pradesh, Karnataka, Kerala, Tamil Nadu and Andaman Islands.

WORLD: Bhutan, Bangladesh, Myanmar, Thailand, Ryu-kyu Islands, China, Malesia.

BREYNIA J.R. Forster & G. Forster *nom. cons*

1. Leaf apex apiculate; perianth greatly enlarged in fruit; capsule globose or depressed globose .... **B. retusa**
1. Leaf apex obtuse-subacute; perianth not or scarcely enlarged in fruit; capsule globose .... **B. vitis-idaea**

Breynia retusa (Dennst.) Alston, Ann. Roy. Bot. Gard. (Peradeniya) 11: 204. 1929; N.P.Balakr. & Chakrab., Fam. Euphorb. India 340. 2007; Chakrab. *et al.*, in N.P.Balakr. *et al.*, Fl. India 23: 410. 2012. *Phyllanthus retusus* Dennst., Schlussel Hortus Malab. 24. 1818. *P. patens* Roxb., Fl. Ind. 3: 667. 1832. *Breynia patens* (Roxb.) Rolfe, J. Bot. 20: 359. 1882; FBI 5: 329. 1887; Gamble 2: 1304. 1925. *B. angustifolia* Hook. f., Fl. Brit. India 5: 330. 1887.

Shrub, to 2 m high, monoecious; branchlets slender, glabrous, angular. Leaves ovate-elliptic or orbicular, up to 2.7 × 1.5 cm, base obtuse-rounded, margin entire, apex apiculate; petiole 2 mm, glabrous; stipules up to 0.2 cm long, deltoid, acute, persistent. Flowers in axillary cymes; monoecious, male in lower and female in upper axils, pedicels filiform. Male flowers: pale yellow, up to 3 in a fascicle, 4 mm across; pedicels filiform, up to 0.5 cm long; tepals 6, connate, lobes minute, triangular, acute; stamens 3, filaments connate, anthers oblong. Female flowers 5 mm across; tepals 6, obovate-orbicular, rounded; ovary cupular, styles 3, each 2-fid. Capsule globose or depressed globose, seated on the accrescent tepals cup; seeds wedge-shaped, smooth.

Common in dry localities near the forests. Fl. & Fr.: February-July. Vern.: Ori.: *Tina, Jajan, Raktatrichuli.*

Rollapenta (KNL), *JLE* 42277 (MH); Horsley hills (CTR), *JSG* 15149 (DD); Maredumilli (EG), *GVS* 24288, 27254(MH); Sunkarimetta (VSKP), *NPBK* 10897 (MH); Elwinpet (SKLM), *GVS* 62467 (MH); Below Deomali (KPT), *SX* & *MB* 6631, 6586

(RRL-B); Karlapat (KHD), *HFM* 3476; Gunpur (KHD), *HFM* 2468 (DD); Pottangi (KPT), *HFM* 3828 (DD).

INDIA: Bihar, West Bengal, Sikkim, Assam, Meghalaya, Manipur, Mizoram, Odisha, Madhya Pradesh, Maharashtra, Goa, Karnataka, Kerala and Tamil Nadu.

WORLD: Sri Lanka, Nepal, Bhutan, Bangladesh, Myanmar, S.W.China, Indo_ China, Thailand and Malaysia.

Breynia vitis-idaea (Burm.f.) C.E.C. Fisch., Bull. Misc. Inf. Kew 1932: 65.1932; N.P.Balakr. & Chakrab., Fam. Euphorb. India 340. 2007; Chakrab. *et al.,* in N.P.Balakr. *et al.*, Fl. India 23: 411. 2012. *Rhamnus vitis-idaea* Burm. f., Fl. Ind. 61. 1768. *p.p. Phyllanthus rhamnoides* Retz., Observ. Bot. 5: 30. 1788. *nom. illeg. P. rhamnoides* Willd., Sp. Pl. 4: 580. 1805. *Breynia rhamnoides* (Willd.) Müll.Arg. in DC., Prodr. 15: 440. 1866 *nom. illeg.*; FBI 5: 330 1887; Gamble 2: 1304. 1925.

Shrub, to 3 m high, monoecious; with many horizontal branches, branchlets angular. Leaves elliptic, 1.3-2.4 × 0.9-1.5 cm, base cuneate, margin entire, apex obtuse-subacute; petiole to 0.5 cm; stipules subulate from a triangular base, persistant. Flowers minute, in axillary cymes. Male flowers: in small fascicles, pedicels filiform, variable in length; flowers 2 mm across; tepals 6, turbinate or subcampanulate, entire or obscurely lobed; stamens 3, filaments connate, staminal column to 6 mm, anthers 1 × 0.6 mm, linear. Female flowers: solitary, 2.5 mm across; pedicels filiform; tepals 6, campanulate, with 6 lobes, acute-obtuse; ovary 2 × 1.2 mm, truncate, style minute, obscure. Capsule 5 mm across, globose; seeds arillate.

Common in open places in deciduous forests. Fl. & Fr.: September- January. Vern.: Tel.: *Errapurugudu, Erraballi*; Ori.: *Jajan, Jhanjikam Pohalkuli.*

Somandepalli (ATP), *BR* & *ANS* 36523; Indireswaram (KNL), *PSPB* 9143; Kambakkam hills (CTR), *PSPB* 9172; Saddikudumadugu (CTR), *PSPB* 9172; Tirumala (CTR), *RVR* & *GO* 2423, *SSR* 13767; Siddulaiah kona (NLR), *SSY* 18611; Pernadu (NLR), *PV* 111225; Maredumilli (EG), *MM* 100813 (BSID); Sunkarimetta (VSKP), *SKW* 2584 (BLAT).

INDIA: Uttar Pradesh, Bihar, West Bengal, Odisha, Andhra Pradesh, Goa, Karnataka, Kerala, Tamil Nadu and Andaman and Nicobar Islands.

WORLD: Pakistan, Sri Lanka, Nepal, Bangladesh, Myanmar, China, Indo-China, Thailand, Malaysia, Ryu-Kyu Islands and Philippines.

BRIDELIA Willdenow

1. Climbers; lateral nerves up to 6 pairs; flowers ca 1 cm in diam; fruit oblong .... **B. stipularis**

1. Trees or shrubs; lateral nerves 6 or more than 6 pairs; flowers less than 8 mm in diam; fruits globose:

   2. Secondary nerves numerous, parallel, straight:

      3. Secondary nerves forked or reticulate within the margin .... **B. glauca**

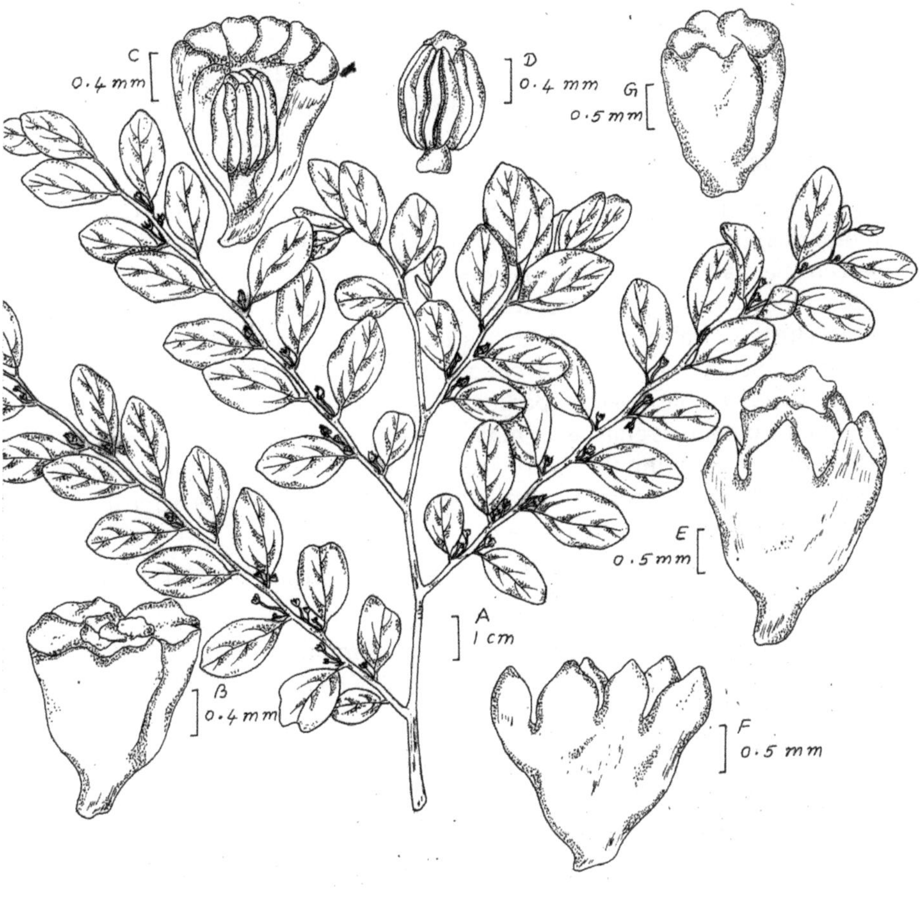

Figure 26. **Breynia vitis-idaea** (Burm.f.) C.Fischer

A. Twig, ♂ B. Flower, C. Flower split open, D. Androecium. ♀ E. Flower, F. Tepals, G. Pistil.

3. Secondary nerves not forked, joining in a marginal thickened one .... **B. retusa**

2. Secondary nerves 6-10 pairs, not markedly straight or parallel:

4. Leaves oblong-lanceolate, secondary nerves unbranched up to the marginal nerve; flowers white, numerous............. **B. tomentosa**

4. Leaves rhombic-obovate or oblanceolate, secondary nerves branched near margin; flowers pale green, few........... **B. montana**

Bridelia glauca Blume, Bijdr. Fl. Ned. Ind. 597. 1826; Dressler, Blumea 41(2): 311.f.4.1996; N.P.Balakr. & Chakrab., Fam. Euphorb. India 318. 2007; Chakrabarty *et al.*, in N.P.Balakr. *et al.*, Fl. India 23: 378. 2012. *B. pubescens* Kurz., J. Asiat. Soc. Beng 42(2): 241. 1873 & For. Fl. Brit. Burma 2:367. 1877; FBI 5: 270. 1887; Surya Prakash Babu in Pullaiah & Moulali, Fl. Andhra Pradesh 2: 846. 1997.

Evergreen tree, up to 15 m high; branchlets tubercled, pubescent. Leaves membranous, elliptic or obovate, 7-15 × 0.5-15 cm, thinly puberulous above, tawny pubescent beneath, base acute-rounded, magin entire, apex shortly acuminate, secondary nerves 6-12 pairs, arching; petiole 6 mm. Flowers pubescent in axillry compact clusters. Male flowers: pedicels 2-3 mm long; sepals tomentellous outside; petals suborbicular, flabellate or obcordate; disc platelliform; staminal column ca 1 mm long. Female flowers: pedicels 1-2 mm long; sepals broadly triangular, tomentellous outside; petals as in male; outer disc ca 2.5 mm in diam., inner disc shortly cupular, covering the ovary; ovary ovoid or conoid, glabrous; styles 1-1.5 mm long, erect, connate below for half way, shortly bifid above. Drupe black, ovoid or oblong with persistent outer tepals.

Occasional, along streams in semi-evergreen forest, 1000-1500 m in northern Eastern Ghats. Fl. & Fr.; August-November.

Minumuluru (VSKP), *GVS* 28262; Madalamma gudi (VSKP), *GVS* 42583; Sunkarimeta (VSKP). *GVS* 22647 (MH); Karlapat (KHD), *HFM* 1704 (DD); Kalahandi (KHD), *HFM* 1227 (DD).

INDIA: Assam, Meghalaya, Mizoram, Tripura, Arunachal Pradesh, Sikkim, West Bengal, Bihar, Odisha and Andhra Pradesh.

WORLD: Nepal, Bhutan, Myanmar, S. China to SE Asia, Malesia to Bismarck Archippelago.

Bridelia montana (Roxb.) Willd., Sp. Pl. 4: 978. 1806; Hook. f., Fl. Brit. India 5: 269. 1887 *p.p.*; Gamble 2: 1281.1925; N.P.Balakr. & Chakrab., Fam. Euphorb. India 319. 2007; Chakrab. *et al.*, in N.P.Balakr. *et al.*, Fl. India 23: 379. 2012. *Clutia montana* Roxb., Pl. Coromandel 2: 36. t. 171. 1802. *Bridelia hamiltoniana* Müll. Arg., Linnaea 34. 77. 1865; FBI 5: 271.1887.

Small tree, up to 8 m high; bark thin, rather rough; branchlets glabrous. Leaves coriaceous, obovate or oblanceolate, 4-10.5 × 2-5.8 cm, glabrous, green above, paler beneath, base cuneate, margin entire, apex acute or obtuse, secondary nerves 7

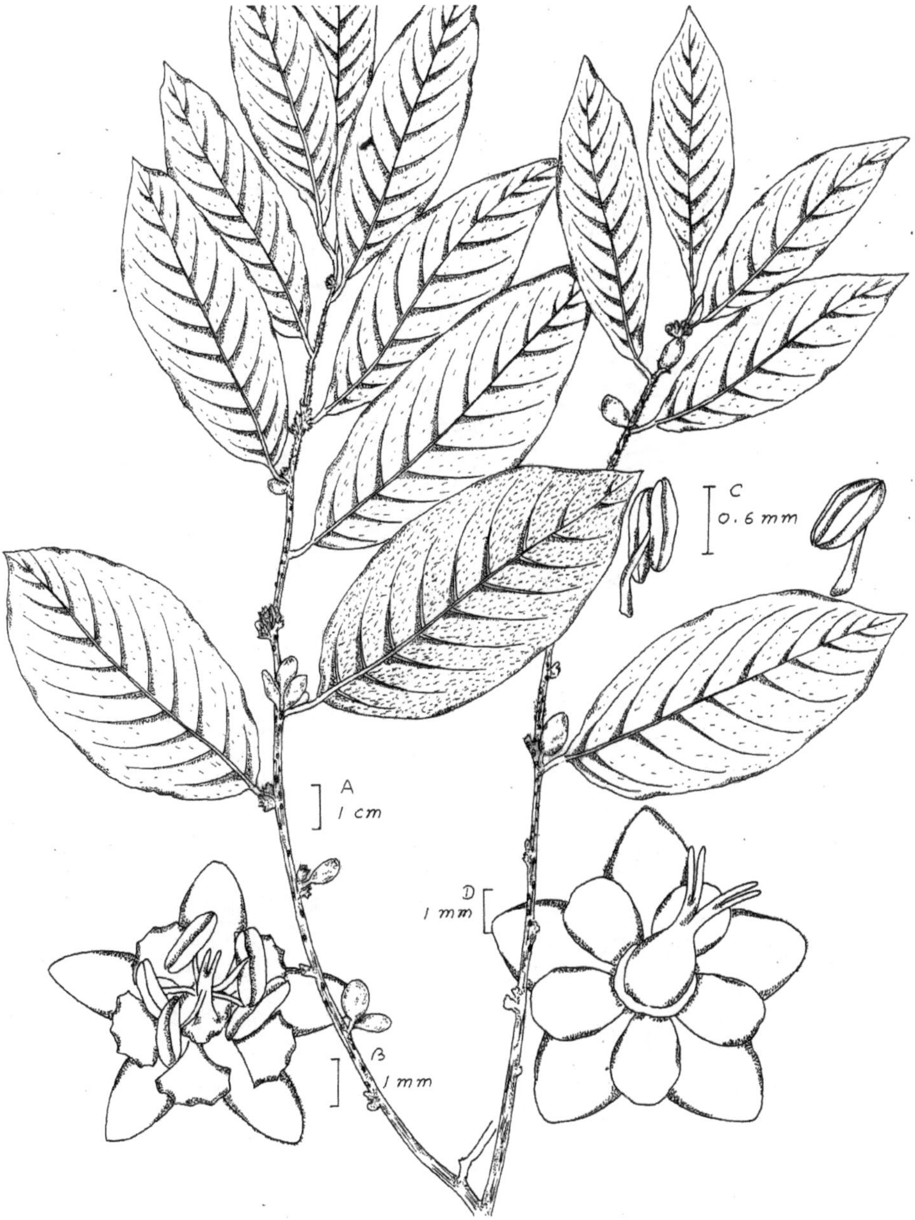

Figure 27. **Bridelia glauca** Blume (Syn.: *Bridelia pubescens* Kurz)
A. Twig, ♂ Flower, C. Anthers, ♀ D. Flower.

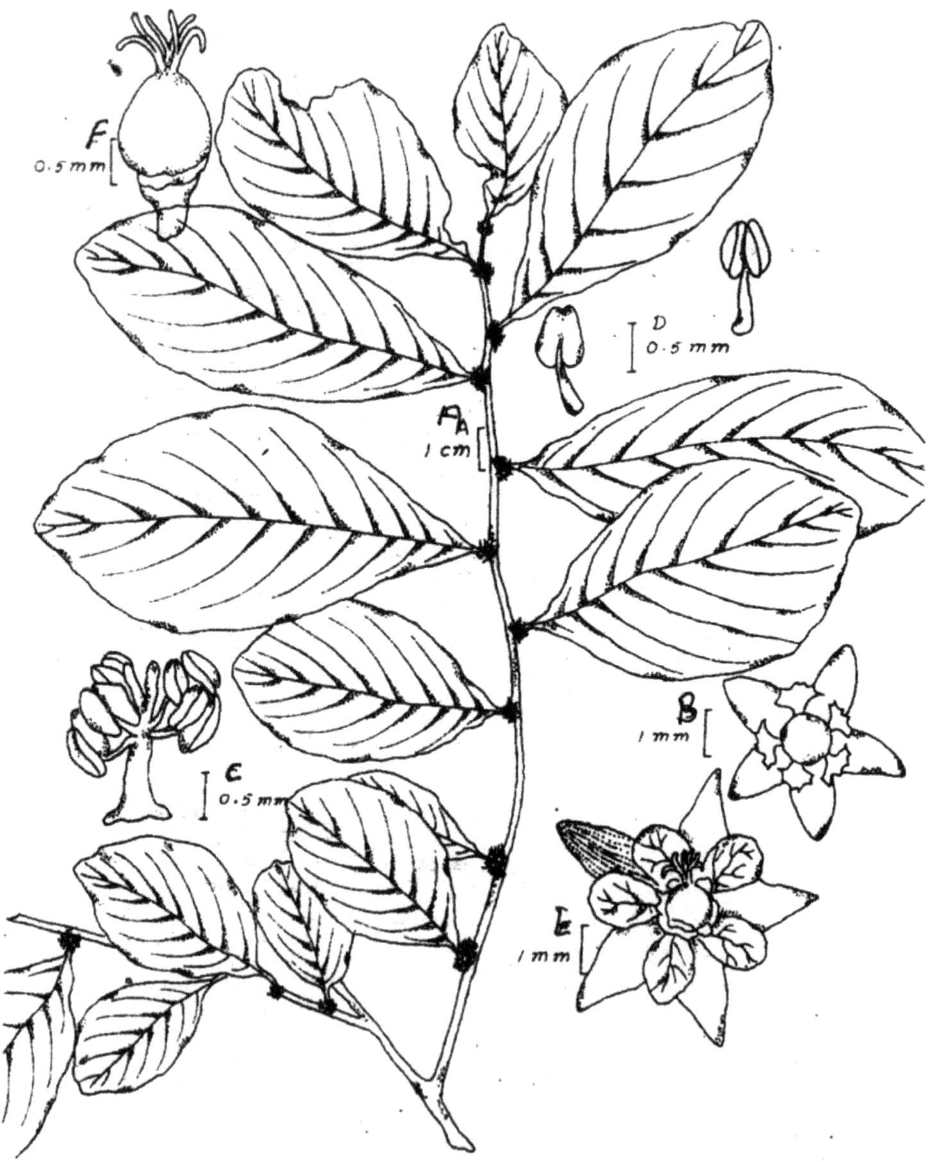

Figure 28. **Bridelia montana** Roxb.

A. Twig, ♂ B. Perianth, C. Androecium with gynophores, D. Anthers.

♀ E. Flower, F. Pistil.

pairs; oblique; petiole to 5 mm. Flowers green, few in leaf axils. Male flowers: sepals triangular; petals flabellate; disc flat, entire; stamens 1.5-1.8 mm long. Female flowers: sepals rarely 6, triangular-oblong; petals 0.5-2 mm long; outer disc platelliform, inner disc urceolate, lobulate, covering the ovary; ovary subglobose or ovoid; styles ca 1 mm long, almost free, deeply bifid. Drupe globose, 0.5 cm in dia, pyrene furrows shallow, the ridge rounded, curving from a central one.

Common, in dry deciduous and degraded forests. Fl.: September- October. Fr.: January - March. Vern.: Tel.: *Panchavoni, Tella yegise, Pantangi, Balli, Kanchep, Sanna koctari, Fundu bigalu*; Ori.: *Marda.*

Alurukona (ATP), *TP* 932; Kalasamudram (ATP), *NY* 1200; Lower Ahobilam (KNL), *TP* & *RVR* 2588; Ramannapenta (KNL), *TP* & *SSR* 15211; Vanipenta (KNL), *SSR* 15299; Palakonda (CDP), *CS* 7628; *KMS* 2386 (DD); Talakona (CTR), *BR* & *SSCM* 45766; Peddamantanala (PKM), *RVK* 15833; Vinukonda (GNT), *VRK* 6898; Guvvalacheruvu (CDP), *KMS* 6412 (CAL); Pedda yerragona (NLR), *PV* 109906; Kondapalli (KSN), *SKW* 3237 (BLAT); Chodavaram (EG), *MSR* 1503 (CAL); Addatigala (EG), *SKW* 3930 (BLAT); Gollarayikonda (EG), *GVS* 62535 (CAL); Kalimela (KPT), *SX* & *MB* 6857 (RRL-B).

INDIA: Uttar Pradesh, Jharkhand, West Bengal, Odisha, Mahdya Pradesh, Chattisgarh, Maharashtra and Karnataka. Endemic.

Bridelia retusa (L.) A. Juss., Euphorb. Gen. 109. t.7, f. 22. 1824; Dressler, Blumea 41(2): 289. 1996; N.P.Balakr. & Chakrab., Fam. Euphorb. India 319. 2007; Chakrbarty *et al.*, in N.P.Balakr. *et al.*, Fl. India 23: 380. 2012. *Clutia retusa* L., Sp. Pl. 1042. 1753 ('*Clutia*'). *Bridelia retusa* (L.) Spreng., Syst. Veg. 3: 48. 1826; FBI 5: 268. 1887. *B. crenulata* Roxb., Fl. Ind. 3: 734. 1832; Surya Prakasa Babu in Pullaiah & Moulali, Fl. Andhra Pradesh 2: 845. 1997. *B. roxburghiana* (Müll. Arg.) Gehrm., Bot. Jahrb. Syst. 41: 30. 1908; Gamble 2: 1280. 1925. *B. retusa* var. *roxburghiana* Müll.Arg. in DC., Prodr. 15. 493. 1866; FBI 5: 268. 1887; Gamble 2: 1280. 1925. *B. retusa* var. *squamosa* (Lam.) Müll.Arg. in DC. Prodr. 493. 1866; FBI 5: 268. 1887. *B. retusa* var. *glauca* Hook.f. Fl. Brit. India 268. 1887. *B.cinerascens* Gehrm., Bot. Jahrb. Syst 41, Beibl 95: 30. 1908; Surya Prakasha Babu in Pullaiah & Moulali, Fl. Andhra Pradesh 2: 845. 1997. *B. airy-shawii* P.T.Li, Acta Phytotax. Sin. 20: 117. 1982, nom. superfl.; Subba Rao & Kumari, Fl. Visakhapatnam 2: 143. 2008.

Deciduous tree, up to 12 m high; spiny when young; bark grey, longitudinally cracked; wood grey to crimson, shallowly or deeply interlocked grained in narrow bands, medium textured; branchlets tomentose. Leaves chartaceous, elliptic-oblong or obovate, 6-19.5 × 3-9 cm, glabrous above, chartaceous and glaucous beneath, base obtuse, margin entire, slightly crenulate, apex acute, or obtuse, secondary nerves 15-25 pairs, conspicuous with finely reticulate veins. Flowers greenish-yellow or purple in 5-10 cm long, terminal spikes or in axillary clusters. Male flowers: pedicels 1.5-2 mm long; sepals triangular; petals suborbicular, toothed or lobed; disc flat, entire; staminal column ca 1.5 mm long, filamentss ca 1 mm long; pistillode conic. Female flowers: pedicels 0.5-2 mm long; sepals narrowly triangular; petals suborbiclar or oblong; outer disc disciform, inner disc conical-cupular; styles 0.8-1.5 mm long,

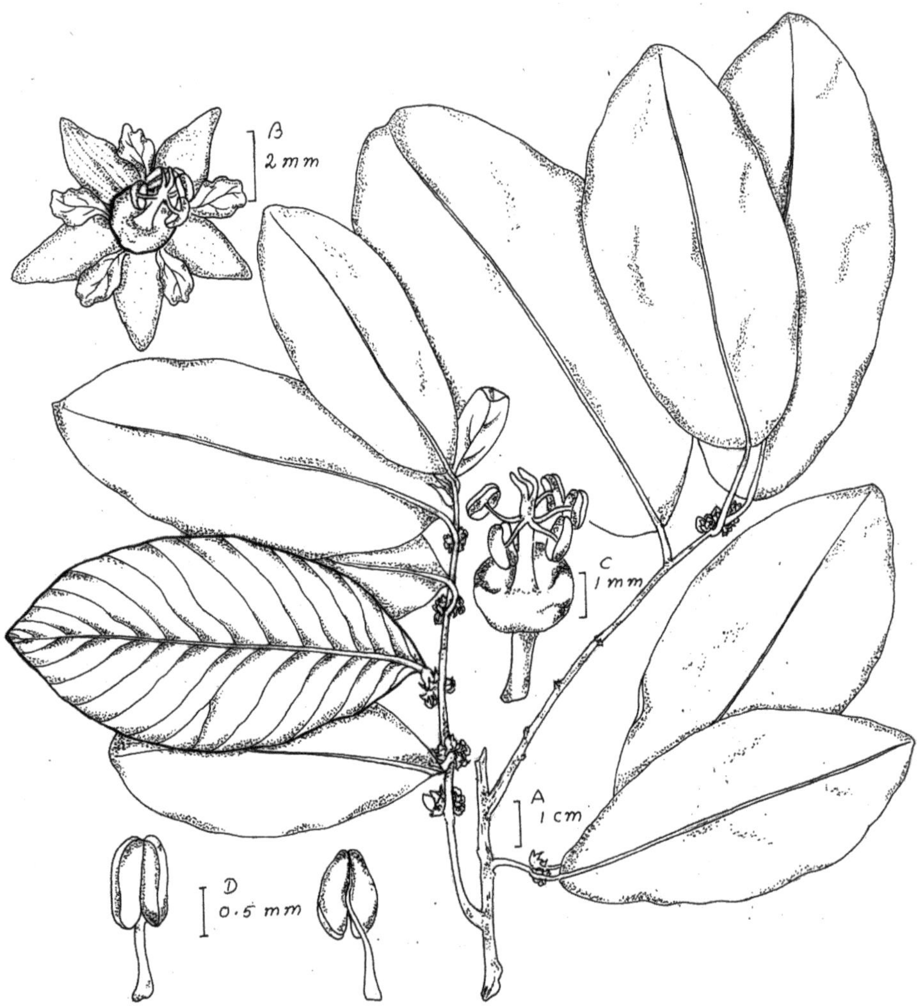

Figure 29. **Bridelia retusa** (L.) A.Juss (Syn.: *Bridelia cinerascens* Gehrm.)
♂ A. Twig, B. Flower, C. Androecium with gynophores, D. Anthers.

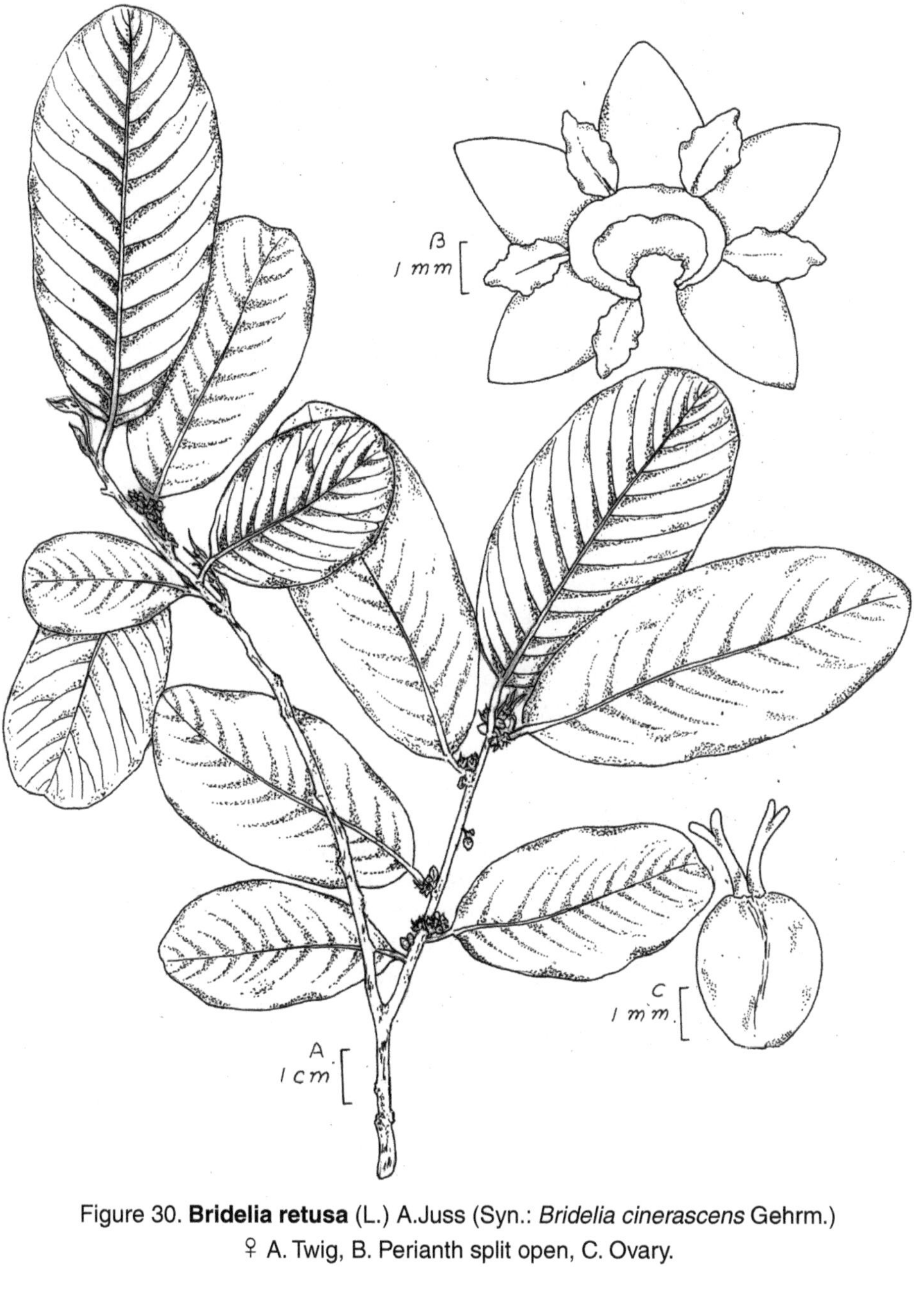

Figure 30. **Bridelia retusa** (L.) A.Juss (Syn.: *Bridelia cinerascens* Gehrm.)
♀ A. Twig, B. Perianth split open, C. Ovary.

shortly connate below and shortly bifid above. Drupe purple or black, seated on enlarged calyx, pyrene furrows curled and deep, 1-2 seeds with fairly thick bony shell, with brownish papery testa.

Common, in mixed and deciduous forests at 600-1200 m. It is drought hardy, produces root-suckers and good coppicer. Fl.: May-July. Fr.: November-January. Vern.: Tel.: *Kora maddi, Mulu maddi, Puta karaka, Anemu, Koramanu, Koderu, Durga maddi, Pia maddi, Bonta yepi, Aanemchettu; Kosengi, Yerra sittuga*; Ori.: *Kasi;* Tam.: *Mulluvaengai.*

Kalasamudram (ATP), *KRKS* 39195, 40023; Upper Ahobilam (KNL), *TP* & *SSR* 15209; Pangadi (KNL), *RVR* 2395; Chelama (KNL), *PSPB* 1181; Guvvalacheruvu (CDP), *PSPB* & *RVR* 8425, *PSPB* 9197; Jyothi (CDP), *BR* & *SKB* 30844; Japali, Tirumala hills (CTR), *SSR* 15274; Penchalakona (NLR), *SSY* 16234; Chintala RF (PKM), *SSR* 16119; Addatheegala (EG), *AJR* & *K. Raja* 20680; Anantagiri (VSKP), *TP* & *EC* 7379; Gandi (CDP), *JSG* 68570 (CAL); Tummala kunta (CTR), *GVS* 4960; Roygudem (WG), *DN* 85507; Arasavelli (SKLM), *NRR* 86699 (BSID); Sirumalai hills (DGL), *SKS* 568 (SGH).

INDIA: Through out the country.

WORLD: Sri Lanka, Pakistan, Nepal, Bhutan, Bangladesh, Myanmar, S.China, Indo-China, Thailand and Sumatra.

Bridelia stipularis (L.) Blume, Bijdr. Fl. Ned. Ind. 2: 597. 1825; FBI 5: 270. 1887; Gamble 2: 1286. 1925; N.P. Balakr. & Chakrab., Fam. Euphorb. India 320. 2007; Chakrab. *et al.*, in N.P. Balakr. *et al.*, Fl. India 23: 378. 2012. *Clutia stipularis* L., Mant. Pl. 1: 127. 1767 (*'Clutia'*). *C. scandens* Roxb., Pl. Coromandel 2(3): 39. t. 173. 1802(*'Clutia'*). *Bridelia scandens* (Roxb.) Willd., Sp. Pl. 4(2): 883.1805; Gamble 2: 1281. 1925.

Large climbing shrub, monoecious; branches fulvous tomentose. Leaves 3-9 × 1.5-4.5 cm, broadly elliptic-ovate or obovate, base rounded, margin entire, apex acute, fulvous tomentose below, thinly below, leaves on flowering shoots small; petiole to 6 mm; stipules deciduous. Flowers in long axillary or terminal spikes, sessile or shortly pedicelled. Male flowers 6 mm across; sepals 3.5 × 1.8 mm, connate at base, ovate, acute, pubescent without; petals 1.8 mm, free, obovate, cuneate, obtuse, fimbriate; stamens 5, on gynophore, anthers 1.2 × 0.5 mm; pistillode bifurcate; disc annular. Female flowers 9 mm across; sepals 3, 8 × 1.8 mm, free, obovate, obtuse; ovary 3 × 1.5 mm, glabrous, styles 2, slightly connate at abase, each 2-fid, from above the middle; disc annular, with a ring of bristles at the base within. Drupe oblong, black.

Occasional in the deciduous forests to mixed evergreen forests. Fl. & Fr.: August-December. Vern.: Ori.: *Darakhia, Gaurkarsi, Nota kasi.*

Dummakonda (EG), *GVS* 68637 (MH); Kadagandi (VSKP), *KCJ* 17172 (MH); Kilagada (VSKP), *GVS* 33310 (MH); Srungavarapukota (VZN), *GVS* 21806 (MH).

INDIA: Through out the country.

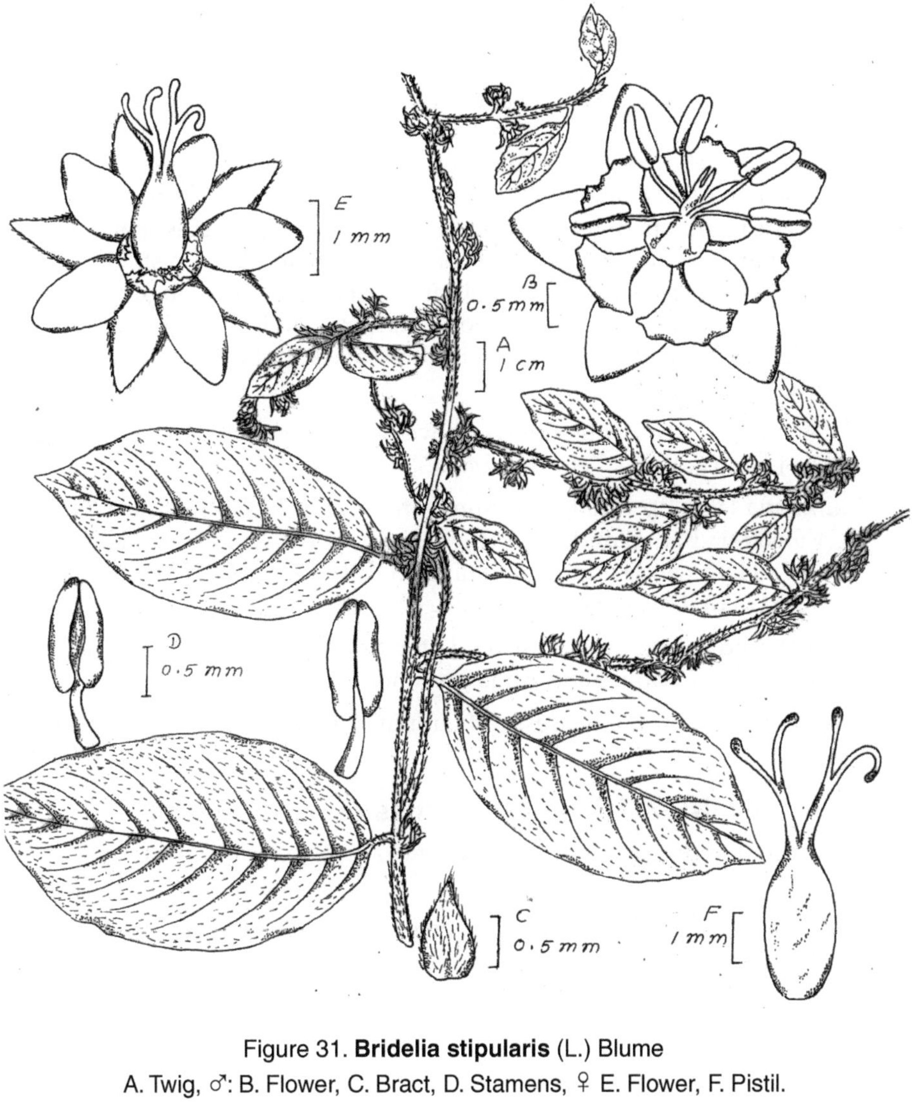

Figure 31. **Bridelia stipularis** (L.) Blume
A. Twig, ♂: B. Flower, C. Bract, D. Stamens, ♀ E. Flower, F. Pistil.

WORLD: Sri Lanka, Nepal, Bhutan, Bangladesh, Myanmar, Thailand, S. China, Taiwan, Malay Peninsula, W. Malesia, Philippines to Lesser Sunda Islands.

Bridelia tomentosa (Bail.) Blume, Bijdr. Fl. Ned. Ind. 597. 1826; FBI 5: 271; Gamble 2; 1281. 1925; N.P. Balakr. & Chakrab., Fam. Euphorb. India 321. 2007. var. **tomentosa;** Chakrab. *et al.*, in N.P.Balakr. *et al.*, Fl. India 23: 384. 2012. *Bridelia monoica* (Lour.) Merr., Philipp. J. Sci. Bot. 13: 142. 1919; Surya Prakash Babu in Pullaiah & Moulali, Fl. Andhra Pradesh 2: 845. 1997. *Clutia monoica* Lour., Fl. Cochinch. 2: 638.1790.

Small tree, up to 8 m high; bark white, smooth; wood red; branches brown tomentose, drooping, slender. Leaves coriaceous, oblong-lanceolate to linear lanceolate, 4.5-13 × 2.8-5.5 cm, base acute, margin slightly wavy, apex acuminate, thinly puberulous above when young, brown tomentose beneath; secondary nerves 8 pairs; petiole to 1 cm. Flowers white, very small in axillary clusters. Male flowers: sessile; calyx ca 3 mm in diam.; sepals triangular, glabrous; petals spathulatee\ to orbicular-ovate; disc discoid; staminal column ca 0.6 mm long; pistillode conical-ovoid. Female flowers: sessile; calyx 3-4 mm in diam.; sepals triangular, glabrous; petals ca 0.6 mm long; ovary subglobose, glabrous, styles ca 0.5 mm long, free, deeply bifid. Berry globose, to 7 mm dia., blue-black, puberulous, pericarp splits into 6 valves; seed black.

Occasional, in deciduous forests above 800 m. Fl.: September-October. Fr.: November-January. Vern.: Tel.: *Talla balli.*

Galikonda (VSKP), *SSR* 15928; Annavaram (EG), *KCJ* 12582; Rajavommangi (EG), *GVS* 68641; Barnakonda (VSKP), *KCJ* 17134 (MH); Paderu (VSKP), *SKW* 4173 (BLAT); Hathiganda and Parsurampur (GJM), *SX* & *MB* 1814, 2847 (RRL-B); Sorisiapada (DKL), *Dhal* & *Rout* 8130 (RRL-B).

INDIA: Andhra Pradesh, Odisha, Uttar Pradesh, Jharkhand, West Bengal, Assam, Meghalaya, Tripura and Andaman and Nicobar Islands.

WORLD: Bhutan, Bangladesh, Myanmar, S.China, Taiwan, Malesia to N. Australia.

CHROZOPHORA A. Jussieu *nom. cons.*

1. Procumbent herbs; flowers in short racemes; capsule 0.5-0.7cm across .... **C. plicata**
1. Erect herbs; flowers on elongated racemes; capsule 0.9 cm across .... **C. rottleri**

Chrozophora plicata (Vahl) A.Juss. ex Spreng., Syst. Veg. 3: 850. 1826. *Croton plicatus* Vahl, Symb. Bot. 1: 78. 1790. *Chrozophora prostrata* Dalzell in Dalzell & A. Gibson, Bombay Fl. 233. 1861; N.P.Balakr. & Chakrab., Fam. Euphorb. India 166. 2007. *C. prostrata* Dalzell var. *parvifolia* (Klotzsch ex Schweinf) N.P. Balakr., Bull. Bot. Surv. Ind. 15: 7: 1973 (1976); N.P.Balakr. in N.P.Balakr. *et al.*, Fl. India 23: 181. 2012. *C. parvifolia* Klotzsch ex Schweinf, Pl. Quaed. Nilot. 11. 1862; Gamble 2: 1316. 1925.

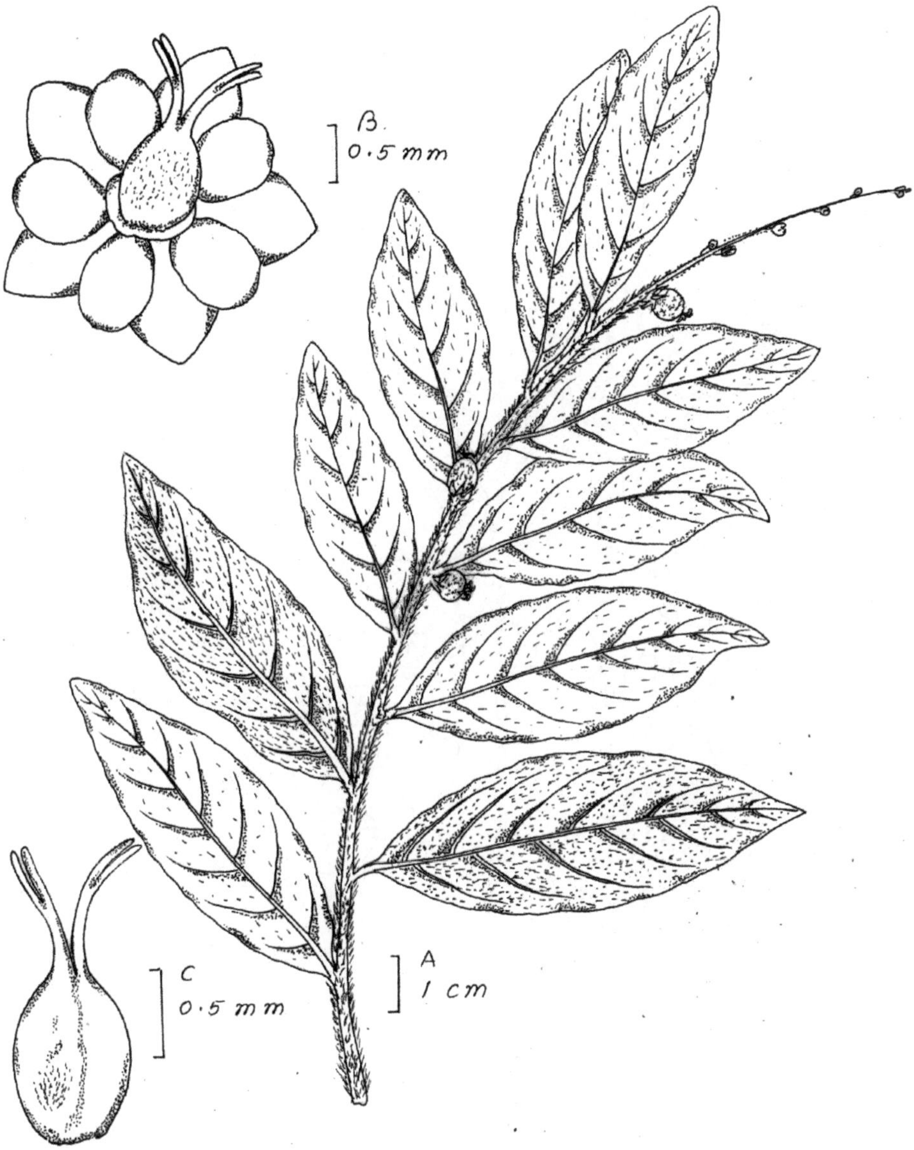

Figure 32. **Bridelia tomentosa** (Bail.) Blume [Syn.: *Bridelia monoica* (Lour.) Merr.]
♀ A. Twig, B. Flower, C. Pistil.

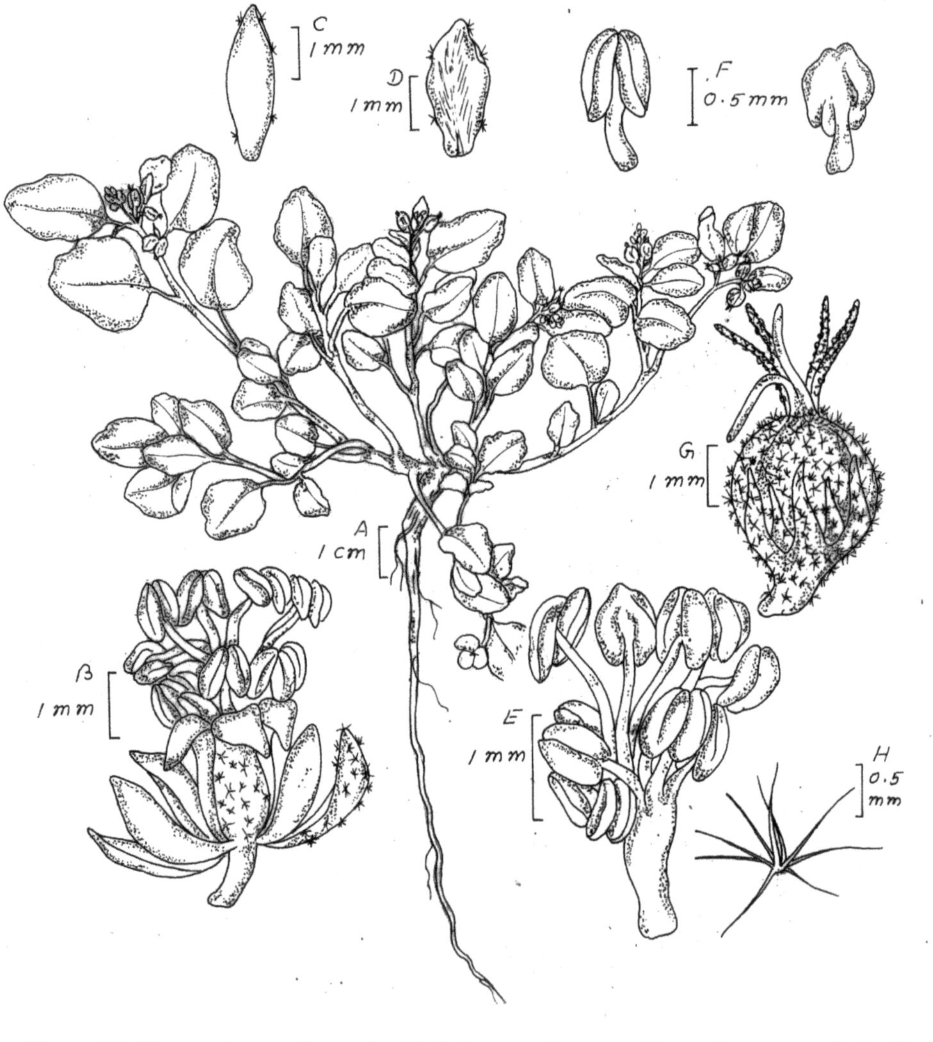

Figure 33. **Chrozophora plicata** (Vahl) A.Juss. ex Spreng., (Syn.: *C. prostrata* Dalzell)
A. Twig, ♂ B. Flower, C. Outer tepal, D. Inner tepal, E. Androecium, F. Anthers.
♀ G. Flower, H. Stellate hair (sessile).

Herb, to 30 cm high, monoecious; stem naked in the lower part, slightly branched above, branchlets stellate-tomentose, stellate hairs stalked. Leaves 1.5-3 × 0.8-2.3 cm, ovate-orbicular, base cuneate-truncate or obtuse, 2 glandular, margin sinuate, apex obtuse, bullate; petiole to 1.5 cm; stipules linear. Flowers in short densely stellate-tomentose racemes; male flowers crowded above; female flowers at base, pedicellate; bracts linear. Male flowers 4 mm across; tepals in two whorls, 5 + 5; outer (sepals) 1.8 × 0.5 mm, free, oblanceolate, rounded, stellate hairy without; inner (petals) 1.6 × 0.6 mm, free, lanceolate, acute-obtuse; stamens 15, biseriate; filaments connate, outer short, inner longer, anthers 0.8 × 0.5 mm; disc glands 5, free. Female flowers 4.5 mm across; tepals in two whorls, 5 + 5; outer 1.5 × 0.5 mm, connate at base, narrowly triangular, acute, stellate hairy without; inner 0.9 × 0.5 mm, connate at base, narrowly triangular, acute, stellate hairy without; ovary 2.5 mm across, globose, stellate hairy, styles 3, slightly connate at base, each 2-fid, near the base, stigma yellow in colour. Capsule 7 mm, densely stellate hairy, without silvery scales.

Occasional in dried up tanks. Fl. & Fr.: June-August. Vern.: Tel.: *Chinna-miriyam, Lingamiriyam.*

Salakamcheruvu (ATP), *PSPB* 9190; Tungabhadra (KNL), *PSPB* 9186.

INDIA: Punjab, Rajasthan, Uttar Prdesh, Bihar, West Bengal, Odisha, Madhya Pradesh, Gujarat, Maharashtra, Andhra Pradesh, Karnataka and Tamil Nadu.

WORLD: Pakistan to North Africa.

Chrozophora rottleri (Geis.) Adr. Juss. ex Spreng., Syst. Veg. 3: 850. 1826; Gamble 2: 1317. 1925; N.P.Balakr. & Chakrab., Fam. Euphorb. India 167. 2007; N.P.Balakr. in N.P.Balakr. *et al.*, Fl. India 23: 181. 2012. *Croton rottleri* Geisel., Croton Monog. 57. 1807. *Chrozophora plicata* (non Adr. Juss.) Voigt, Hort. Sub. Calc. 156. 1846. *C. tinctoria* (non Adr. Juss.) Hook. f., Fl. Brit. India 5: 409. 1887. *p.p.*

Erect herb, to 60 cm high, monoecious; branchlets fulvous-tomentose. Leaves 2-6 × 1.2-4 cm, broadly ovate, base acute-cordate, 2 glandular, margin sinuate, obtuse; petiole to 3 cm; stipules deciduous. Flowers in densely tomentose terminal racemes; male flowers crowded above; female flowers at base, pedicellate; bracts lanceolate. Male flowers 5 mm across; tepals in two whorls, 5 + 5; outer (sepals) 1.2 × 1 mm, connate at base, ovate-lanceolate, concave, acute-obtuse; inner (petals) 2.5 × 1.2 mm, free, lanceolate, obtuse; stamens 12-15, biseriate, filaments connate, outer short, inner long, anthers 1 mm across; disc glands 5, free. Female flowers 5 mm across; tepals in two whorls, 5 + 5; outer (sepals) 2 mm, connate at base, lanceolate-triangular, acute, stellate hairy wihtout; inner (petals) 1.5 mm, connate at base, lanceolate-triangular, acute, stellate hairs without; ovary 3 mm across, globose, stellate hairy, styles 3, slightly connate at base, each 2-fid, more than halfway down, stigma red in colour. Capsule 9 mm across, depressed, 3-lobed, stellate-tomentose; seeds globose, smooth, shiny.

Common weed in cultivated fields especially in black soils. Fl. & Fr.: Throughout the year. Vern.: Tel.: *Erra-meriyam;* Tam.: *Puraporukki.*

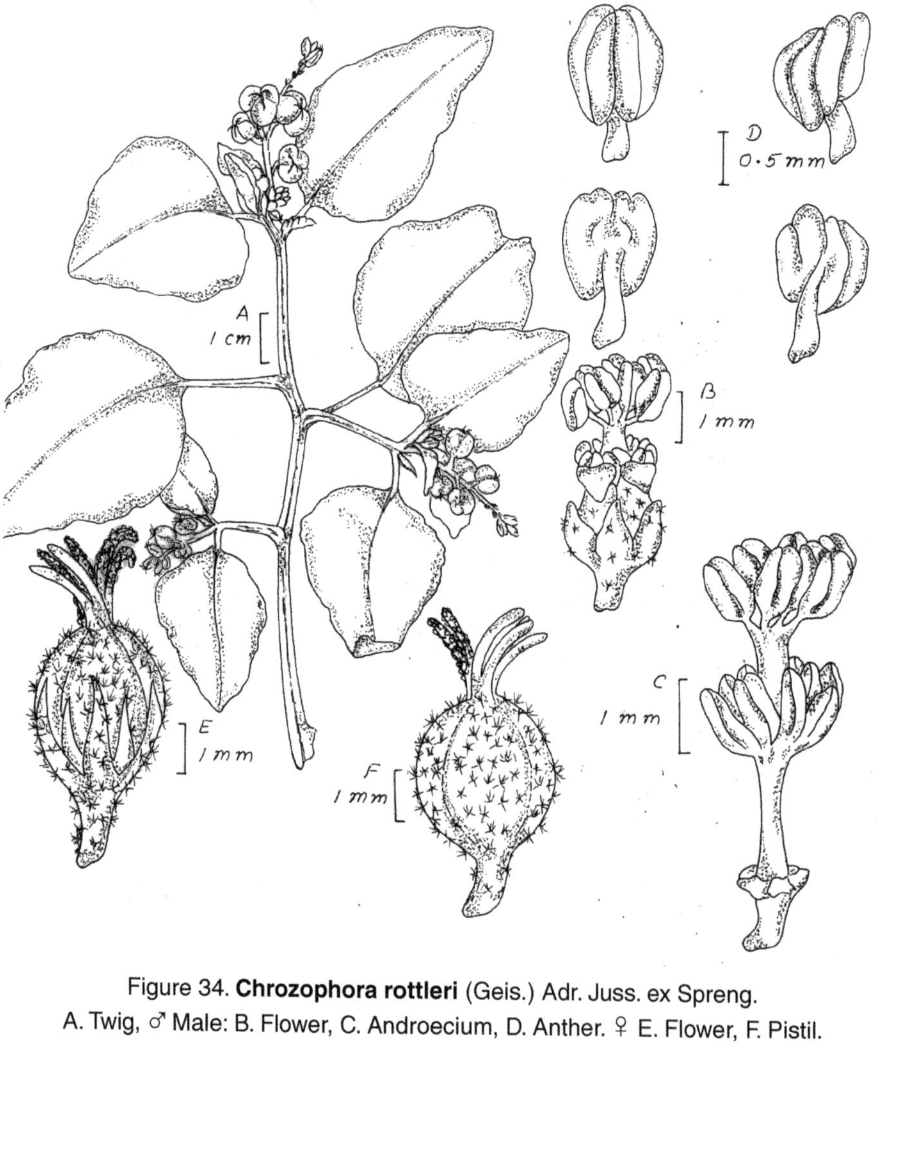

Figure 34. **Chrozophora rottleri** (Geis.) Adr. Juss. ex Spreng.
A. Twig, ♂ Male: B. Flower, C. Androecium, D. Anther. ♀ E. Flower, F. Pistil.

Dhaniyani cheruvu (ATP), *TP* & *NY* 859; Rayadurg (ATP), *PSPB* 9193; Gulemallebad RF (KNL), *RVR* 1418; Indireswaram (KNL), *PSPB* 9138; Puttaparthi (ATP), *SKW* 6221 (BLAT); Vuyyalavada (PKM), *RKM* 537(MH); Alapadu – Kolletikota (KSN), *PV* 116503; Koruturu (EG), *DCSR* 50 (CAL); Salur (VZN), *NPBK* 1160 (CAL); Bankula (DKL), *MB* & *Dhal* 8248 (RRL-B); Nuchciapada (MBJ), *Bal* 816 (DD); Athmallik (Angul), *HFM* 2883 (CAL); Athagarh, bank of Kuradjhor (Cuttack), *HFM* 2884 (DD).

INDIA: Punjab, Rajasthan, Uttar Pradesh, West Bengal, Assam, Odisha, Madhya Pradesh, Gujarat, Maharashtra, Andhra Pradesh, Karnataka, Kerala, Tamil Nadu and Andaman and Nicobar Islands.

WORLD: Afghanistan, Pakistan, Bangladesh, Sri Lanka, Myanmar, Thailand and Java.

CLEIDION Blume

1. Male spike 5-20 cm long; female peduncle 5-7 cm long; fruits mostly didymous .... **C. javanicum**
1. Male spike 10-35 cm long; female peduncle 4-5 cm long; fruits tridymous .... **C. nitidum**

Cleidion javanicum Blume, Bijdr. Fl. Ned. Ind. 613. 1826; FBI 5: 444. 1887; Airy-Shaw in Kew Bull. 36: 279. 1981; N.P. Balakr. & Chakrab., Euphorbicaeae in India 134. 2007; Susila Rani & N.P.Balakr. in N.P.Balakr. *et al.*, Fl. India 23: 120. 2012; Ravikumar *et al.*, J.Threatened Taxa 6: 6112. 2014.

Tree, up to 10 m tall; bark ashy brown, smooth. Leaves alternate, elliptic-oblanceolate, up to16 × 6.5 cm, cuneate at base, abruptly acuminate at apex, distantly crenate, glabrous above, domatia along the nerve axis beneath; petioles ca. 10 cm long. Female flowers solitary, axillary; peduncles ca. 7 cm long, thickened above; styles 2, each divided into 2 filiform arms. Capsules ca. 3 cm across, usually didymous.

Rare in Minumuluru. Fl. & Fr.: September – October.

Minumuluru (VSKP), *KR* & *N. Balachandran* 110877 (FRLHT).

INDIA: Arunachal Pradesh, Assam, Nagaland, Manipur, Mizoram, Tripura, Meghalaya, Sikkim, Andhra Pradesh, Maharashtra, Karnataka, Kerala, Tamil Nadu and Andaman & Nicobar Islands.

WORLD: Sri Lanka, SE Asia to N. Australia.

Note: In India, this species has so far been reported in the northeastern states, throughout the Western Ghats and the Andaman and Nicobar Islands (Balakrishnan and Chakrabarthy, 2007).

Cleidion nitidum (Müll.Arg.) Kurz, J. Asiat. Soc. Bengal Pt.2, Nat. Hist. 42: 245. 1873; FBI 5: 444. 1887; N.P.Balakr. *et al.*, Fl. India 23: 122. 2012; Kottaimuthu, Life Sciences Leaflets 4: 31. 2012. *Mallotus nitidus* Müll.Arg. in DC., Prodr. 15: 979. 1866.

Shrub or small tree, 3- 7 m tall. Leaves ovate, elliptic or oblong-obovate, cuneate, acute or rounded at base, distantly denticulate, 7-16 x 3-9 cm, acute to acuminate at apex, thinly coriaceous, glossy, lateral nerves 5-8 pairs; petiole to 3.5 cm long, glabrous. Male inflorescence axillary, glomerate-spicate, 10-35 cm long, simple or few-branched towards base; flowers sessile; bracts ovate-triangular, acute; calyx subrobicular to ovate, up to 3.5 mm long; stamens 75-80; filaments ca 1 mm long; anthers ca 1 mm across. Female flowers axillary, solitary or sometimes paired or fasciculate-raceme, up to 8 cm long peduncles; pedicels 1-5 cm long; sepals 4, ovate to triangular, 2.5 x 1.2 mm; ovary globose, 2-4 mm across, 2- or 3-loculed, fulvous tomentose; style 3, connate at base to 1-2 mm long column, deeply bifid above. Capsules globose, somewhat depressed, tridymous, up to 1.5 cm across; fruting pedicel up to 8 cm long; seeds subglobose, 6-8 mm across, smooth, white, variegated with brown patches.

Karandamalai of Tamil Nadu. Fl. & Fr.: April – December.

Karandamalai, along the stream banks of Periya-aruvi (DGL), *R. Kottaimuthu* 897, 1108 (Thyiagaraja College, Madurai).

INDIA: Tamil Nadu and Andaman and Nicobar Islands.

WORLD: Sri Lanka.

CLEISTANTHUS Hook. f. ex Planchon

1. Leaves elliptic-orbicular or oblong-obovate, apex retuse or obtuse; petals linear; ovary glabrous; capsule 2 cm dia., shining; seeds globose .... **C. collinus**
1. Leaves ovate-lanceolate, apex acuminate; petals obovate; ovary villous; capsule 0.8-1 cm dia., not shining; seeds cordate .... **C. patulus**

Cleistanthus collinus (Roxb.) Benth. ex Hook. f., Fl. Brit. India 5: 274. 1887; Gamble 2: 1282; N.P.Balakr. & Chakrab., Fam. Euphorb. India 324. 2007; Chakrb. *et al.*, in N.P.Balakr. *et al.*, Fl. India 23: 387. 2012. *Clutia collina* Roxb., Pl. Coromandel 2:37, t. 169. 1802 ('*Clutia*').

Deciduous, monoecious tree, up to 8 m high; bark dark brown, rough; wood dark reddish brown; branchlets terete, glabrous. Leaves coriaceous, elliptic, orbicular or oblong-obovate, 4-12 × 3-7.5, base obtuse, margin entire, apex obtuse or retuse, glabrous, bright green; secondary nerves 5-6 pairs; petiole to 0.7 cm. Flowers yellowish-green in small, axillary, silky clusters. Male flowers: pedicels 1 – 2 mm long; sepals occasionally 6, triangular-oblong or lanceolatee-oblong, pilose outisde; petals linear; disc annular; staminal column ca 11.5 mm long; pistillode ovoid. Female flowers: pedicels 0.5-1.5 mm long; calyx-tube ca 1 mm long; sepals triangular-lanceolate; petals subulatee; disc shortly cupular-annular; ovary subglobose, styles 3-4 mm long, almost free or basal column ca 1.5 mm long, shortly to deeply bifid above. Capsule globose, 2 cm across, 3-valved, dark brown, shining; seeds 3, globose, testa smooth, shiny.

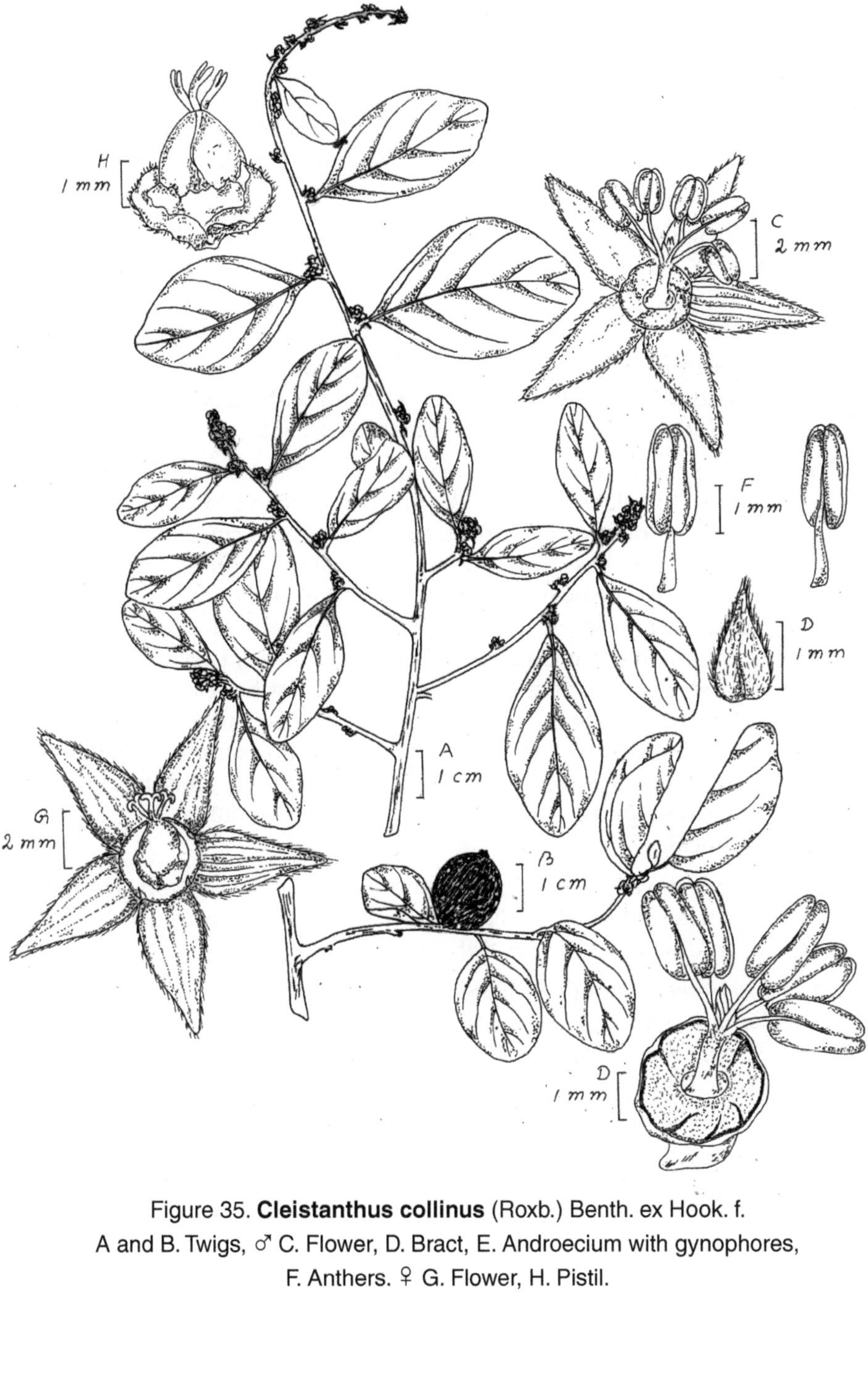

Figure 35. **Cleistanthus collinus** (Roxb.) Benth. ex Hook. f.
A and B. Twigs, ♂ C. Flower, D. Bract, E. Androecium with gynophores,
F. Anthers. ♀ G. Flower, H. Pistil.

Very common in dry forests, up to 800 m. Fl.: February, Fr.: May onwards dehiscing by next February. Vern.: Tel.: *Lodisa, Vodisa, Nalla-kordtha, Kodisi*; Ori.: *Karada;* Tam.: *Oduvan*. English: Star gooseberry.

Kalasamudram (ATP), *EC* 2885; Ahobilam (KNL), *TP* & *RVR* 2584; Bairluti (KNL), *SSR* 5213; Vinukonda (GNT), *VRK* 6902; Seethampet RF (SKLM), *SSR* 15576; Urlakonda (NLR), *MSRS* 1304 (CAL); Moulatta (PKM), *RKM* 406; Papi hills (WG), *DCSR* 73 (CAL); Addatigala (EG), *SKW* 2991 (BLAT); Rampachodavaram (EG), *MM* 101343; Madugala (VZN), *RKM* 406, *GVS* 47260 (CAL); Alagar hills (MDU), *SKS* 89 (SGH).

INDIA: West Bengal, Bihar, Madhya Pradesh, Odisha, Andhra Pradesh, Karnataka, Kerala and Tamil Nadu.

WORLD: Sri Lanka.

Cleistanthus patulus (Roxb.) Müll.Arg. in DC., Prodr. 15 505. 1866; FBI 5: 279. 1887; Gamble 2: 1282. 1925; Chakrbarty *et al.*, in N.P.Balakr. *et al.*, Fl. India 23: 389. 2012. *Clutia patula* Roxb., Pl. Coromandel 2: 37. t. 170. 1802 (*'Clutia'*).

Evergreen tree, up to 8 m high; monoecious; branches horizontal; bark smooth, grey, exfoliating in thin plates; wood pink or crimson, then yellowish, close-grained; branchlets terete, glabrous. Leaves coriaceous, 3.5-9.2 × 1.9-4 cm, glabrous, red when young, ovate-lanceolate, base acute or rounded, margin entire, apex acuminate, secondary nerves 5 pairs; petiole to 5 mm, thick. Flowers yellowish, small in clusters in axillary spikes; bracts ovate-deltoid, pilose without, often caducous. Male flowers: calyx glabrescent, up to 4 mm long, lobes 5 deltoid; petals 5, alternating with calyx lobes, deltoid; stamens 5-6 on a minute androphore, pistillode up to 1.5 mm long, 3-fid. Female flowers: sessile to subsessile; calyx tube cupular, sepals triangular, oblong-ovate to lanceolate; petals suborbicular; disc cupular, covering half of the ovary; ovary globose, tomentose, styles 1.5-2 mm long, free, bifid up to half way. Capsule 0.7 cm across, pilose-glabrous; seeds reniform, albumen thin, cotyledons large, fleshy, truncate.

Occasional, in Rayalaseema and Northern Eastern Ghats, chiefly in ravines or dry evergreen forests 600-1400 m. Fl.: January-March. Fr.: April-July. Vern.: Tel.: *Jegura, Jiguru, Tella vadisae*; Ori.: *Sonohat, Sanahati, Sar patria.*

Gundlabrahmeswaram (KNL), *SSR* 15529; Simhachalam (VSKP), *SSR* 15566; way to Talakona waterfall (CTR), *LR* & *MSRao* 2571 (BSID); Tadepally (WG), *NRR* & *TRS* 86119 (MH); Ramavaram (EG), *GVS* 27452 (MH); Minumuluru (VSKP), *GVS* 47347 (MH); Galikonda (VSKP), *GVS* 19624 (MH); Julka (Angul), *Haines* 2506 (DD); Bhanjabasa, Similipahar (MBJ), *Bal* 1005 (DD); Rayamatia (MBJ), *Bal* 120 (DD).

INDIA: Odisha, Andhra Pradesh, Karnataka, Kerala and Tamil Nadu.

WORLD: Sri Lanka.

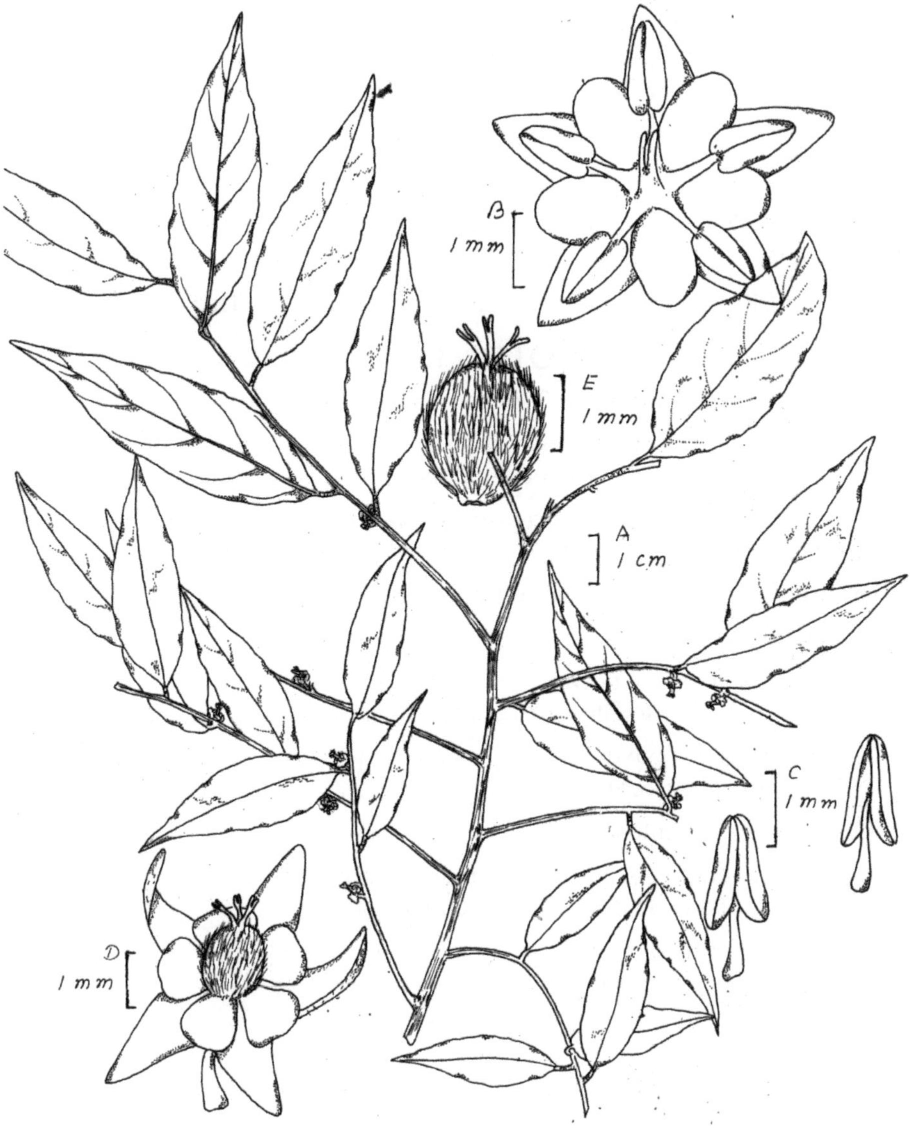

Figure 36. **Cleistanthus patulus** (Roxb.) Müll.Arg.
A. Twig, ♂ B. Flower, C. Outer tepals, ♀ D. Flower, E. Pistil.

CNESMONE Blume

Cnesmone javanica Blume, Bijdr. Fl. Ned. Ind. 630. 1826; FBI 5: 466. 1888; Saxena & Brahmam, Fl. Orissa 3: 1622. 1995; Chakrbarty & N.P. Balakr. in N.P.Balakr. *et al.*, Fl. India 23: 389. 2012.

Climbing undershrub, stem hairy. Leaves oblong, broadly oblong or obovate, 7.5- 16.5 × 4.5-9 cm, hairy on both sides, apex caudate or acuminate, margin irregularly crenate-dentate, base cordate; petiole 4-7 cm. Flowers in racemes, male flowers 6 mm diam., pedicelled; female flowers larger, subsessile; sepals broad, velvety, fruiting calyx as long as the capsule, rhomboid-ovate. Capsule hispid, of 3 cocci, each as large as a small pea; cocci velvety; seeds with mottled testa.

Rare in Similipahar forest in Odisha.

Baomukkabadi, Similipahar (MBJ), *SX, MB & Prabhakar Rao* 4582 (RRL-B).

INDIA: North-East India, West Bengal, Odisha.

WORLD: Bangladesh, Myanmar, Indonesia, Malaysia, Thailand,Vietnam.

CROTON Linnaeus

1. Herbs or subshrubs; capsule 3-lobed:
    2. Branchlets covered with glandular haris; leaves aromatic; fruits with redish disc-shaped glands .... **C. bonplandianus**
    2. Branchlets covered with strigose-stellate hairs; leaves non-aromatic; fruits with dense stellate haris .... **C. hirtus**
1. Shrubs of small trees; capsule subglobose:
    3. Branchlets glabrous:
        4. Tepals of female flower 5 + 5; stamens 11-14; capsule with stellate hairs .... **C. klotzschianus**
        4. Tepals of female flower 4 + 4; stamens 8-16; capsule pubescent .... **C. scabiosus**
    3. Branchlets tomentose:
        5. Leaves elliptic-ovate, base rounded, margin entire; tepals of female flower 5 + 5, stamens 14:
            6. Leaves hispid, scabrid-pubescent, pilose or sparsely pubescent above; style branches usually quadrifid; fruits 8-14 mm in diam., muricate, scattered pubescent .... **C. aromaticus**
            6. Leaves glabrous above; style branches usually bifid; fruits 13-25 mm in diam., smooth, subglabrous .... **C. tiglium**
        5. Leaves oblong-lanceolate, base acute, margin crenate-serrate; tepals of female flowers 3 + 3, stamens 12................ **C. persimilis**

Croton aromaticus L., Sp. Pl. 1005. 1753; FBI 5: 338. 1887; N.P.Balakr. & Chakrab., Fam. Euphorb. India 212. 2007; N.P.Balakr. in N.P.Balakr. *et al.*, Fl. India 23: 229. 2012. *C. laccifer* L., Sp. Pl. 1005. 1753; Gamble 2: 1315. 1925; Surya Prakash Babu in Pullaiah & Moulali, Fl. Andhra Pradesh, 2: 851. 1997.

Shrub, to 4 m high, monoecious; branchlets tomentose. Leaves 5-12 × 4-9 cm, elliptic-ovate, base rounded, margin entire, apex acute-apiculate; petiole to 3 cm, 2 stalked glands at top. Flowers in elongated, terminal racemes, pedicellate; bracts linear. Male flowers: 7 mm across; sepal 4 or 5, narrowly oblong to spathulate-oblanceolate; petals 4 or 5, narrowly oblong to spathulate-oblanceolate, 2.2 mm, recurved, obtuse, ciliate without; stamens 15-30, filaments free, 2.5 mm, bearded, anthers 1 × 0.6 mm, connectives broad, dotted. Female flowers: 5 mm across; sepals 5, 3 mm, ovate, truncate, acute, ciliate; petals 5, filiform, 2.5 mm, linear; ovary 2.5 mm across, globose, pubescent, styles 3, each 2-fid at halfway down. Capsule 8-14 mm in diam., subglobose, densely pubescent; seeds globose.

Rare in the hills on the bare, sunny slopes among rocks. Fl. & Fr.: April-August. Vern.: Tel.: *Aribilla.*

Anakapalli (VSKP), *JSG* 12519 (MH).

INDIA: Maharashtra, Karnataka, Kerala, Tamil Nadu and Andhra Pradesh

WORLD: Sri Lanka.

Croton bonplandianus Baill., Adansonia 4: 339. 1864; N.P.Balakr. & Chakrab., Fam. Euphorb. India 212. 2007; N.P.Balakr. in N.P.Balakr. *et al.*, Fl. India 23: 231. 2012. *C. sparsiflorus* Morong., Ann. New York Acad. Sci. 7: 221. 1893; Gamble 2: 1316. 1925.

Herbs or subshrubs, to 60 cm high, monoecious; much branched, branchlets terete. Leaves 2-5.5 × 1-2.5 cm, ovate-lanceolate, base acute-obtuse, margin sinuate, apex acute, dotted stellate hairs below; petiole to 1.5 cm, 2 glandular at top. Flowers in elongated terminal racemes, pedicellate, bracteate; male flowers many, from above the middle; female flowers few, at the lower half. Male flowers 4 mm across; tepals in two whorls, 5 + 5, free; outer (sepals) ovate, acute; inner (petals) oblong, villous; stamens many, anthers 0.5 mm across. Female flowers 2.5 mm across; tepals 5, 1.5 mm, margin villous; stamens many, inserted in a hairy receptacle, filaments free, 0.7 mm, anthers 0.7 mm across; disc of 5 glands, opposite to outer tepals. Female flowers 5.5 mm across; tepals in two whorls, 5 + 5; outer (sepals) 1-2 × 0.4-1 mm, connate at base, elliptic, acute; inner (petals) 1 × 0.3 mm, free, linear, acute; ovary 1.5 mm across, globose, tomentose stellate hairy, styles 3, each 2-fid, near the base; disc of 5 glands. Capsule 6 mm across, subglobose, stellate hairy.

Occasional in waste lands and road sides. Fl. & Fr.: August-January. Vern.: Tel.: *Galivana chettu, Nepalavemu*; Ori.: *Ban mirchi, Ban mircho;* Tam.: *Railpondu.*

Peravali (ATP), *TP* & *NY* 507; Gulemallebad (KNL), *RVR* 1388; Palakonda (CDP), *CS* 7762; Bhiravakona (NLR), *PSPB* 9109; Macherla (GNT), *VRK* 3256; Balapalli (CDP), *JLE* 14985 (MH); Salivendra kona (CDP), *JSG* 10787 (DD); Papannasanam

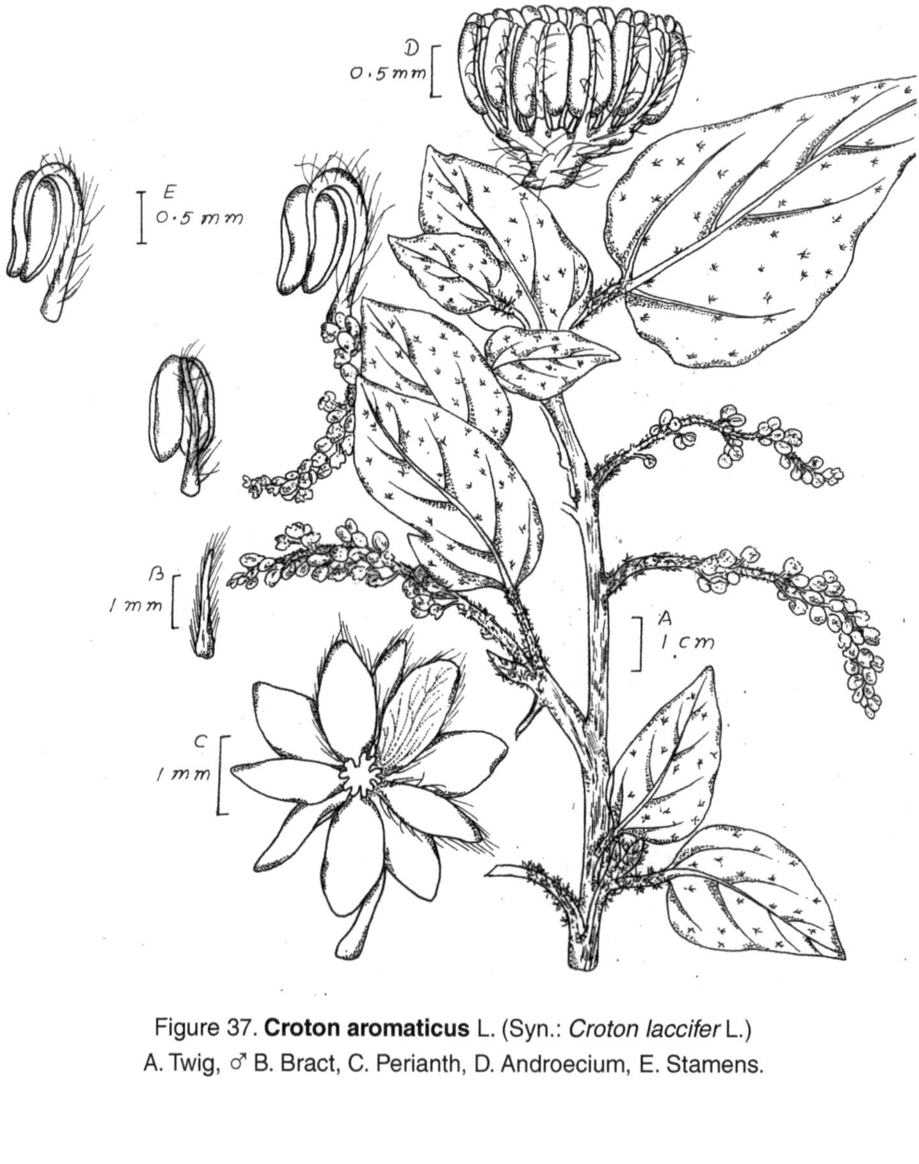

Figure 37. **Croton aromaticus** L. (Syn.: *Croton laccifer* L.)
A. Twig, ♂ B. Bract, C. Perianth, D. Androecium, E. Stamens.

(CTR), *KMS* 6955 (MH); Tirumala (CTR), *SKW* 5662, 8230 (BLAT); Attakanithippa (NLR), *PV* 111195 (BSID); Koringa (EG), *SKW* 1498 (BLAT).

INDIA: Almost through out India.

WORLD: Native of Tropical South America now widespread in S and E. Asia.

Croton caudatus Geiseler, Croton Monogr. 73. 1807; FBI 5: 388. 1887; Saxena & Brahmam, Fl. Orissa 3: 1624. 1995; N.P.Balakr. in N.P.Balakr. *et al.*, Fl. India 23: 232. 2012.

Scandent or straggling shrubs, all parts coarsely tomentellous or scattered scabrid-pubescent. Leaves ovate or orbicular, 5-25 x 3-12 cm, base cordate, margin dentate-serrate, apex acuminate to apiculate, stellate-tomentose; petiole 2- 5 cm. Flowers white, in racemes, monoecious, the lower 2-6 flowers being usually female; petals oblong, nearly as long as sepals, styles 5, bifid. Capsules globose or obovoid, scattered pubescent.

Frequent in scrub or open forest in Eastern Ghats of Odisha. Fl. & Fr.: April – August.

INDIA: Assam, Meghalaya, Arunachal Pradesh, Nagaland, West Bengal, Sikkim, Odisha, Karnataka, Kerala and Tamil Nadu.

WORLD: Sri Lanka, Nepal, Bhutan, Bangladesh, Myanmar, S. China, Thailand, Malaysia, Sumatra, Java, Philippines.

Croton hirtus L'Herit., Stirp. Nov. 1: 17. T.9. 1785; Kottaimuthu *et al.*, Ethnobotanical Leaflets 12: 300. 2008.

Erect herbs, up to 8 cm high; stem terete, dichotomously branched, branches clothed with pale strigose-stellate hairs. Leaves alternate, ovate-oblong, 2.5-5 x 1.5-4 cm, margin serrate, apex acute, base obliquely truncate; lateral nerves 3-5, obscure above, distinct below with 2 prominent glandular hairs at base of lamina; petioles 1-2 cm long; stipules linear, 0.2 cm long, caducous. Inflorescence a terminal, branched raceme with copious multibranched glandular hairs with globose tips, female flowers below, male flowers above, male flowers white; pedicels up to 0.5 cm long, stellate hairy; sepals 5, obovate, 0.2 cm long; petals 5, oblanceolate, 0.2 cm long, serrate along margin; stamens 9-11, filaments 0.1 cm long, glabrous, anthers oblong, pale brown. Female flowers green; pedicels 0.5-1.5 cm long, bracts linear, fringed with 2-5-stalked, capitates glands; sepals 5, unequal, linear-oblong, 0.3 cm long, dentate along margin; ovary ovoid, 0.1 cm across; styles 3, each divided into two, 0.2 cm long, apex reflexed. Capsule globose, 0.5 cm across, bracteate; sepals 3, trigonous, black.

Occasional in Eastern Ghats of Tamil Nadu. Fl. & Fr.: January – May.

On way to China Malaiyar, Karandamalai (DGL), *R.Kottaimuthu* & *M.Vimala* 370 (ATREE); on way to Mulaiyur, Perumalai (DGL), *R.Kottaimuthu* & *K. Natarajan* 3031 (ATREE); Puliyampallam, Sirumalai (DGL), *R. Kottaimuthu* & *R.Ganesan* 2014 (ATREE).

INDIA:Tamil Nadu, Kerala.

WORLD: West Indies, Central and Southen America,Tropical Asia.

Note: *Croton hirtus* L'Herit. is a native of West Indies, Central and South America, which has become an aggressive weed in Tropcal Asia and Africa.

Croton klotzschianus (Wight) Thwaites, Enum. Pl. Zeyl. 4: 276. 1861; FBI 5: 392. 1887; N.P.Balakr. & Chakrab., Fam. Euphorb. India 214. 2007; N.P.Balakr. in N.P.Balakr. *et al.*, Fl. India 23: 236. 2012. *Tiglium klotzschianum* Wight, Icon. Pl. Ind. Orient. t. 1914. 1852 (as '*Tiglium klotcheanum*').

Shrub, to 3 m high, monoecious. Leaves 3-7.5 × 1.5-3 cm, elliptic-oblong, base rounded-cuneate, margin serrate, apex obtuse-retuse, petiole to 1.5 cm, 2 stalked glandular at top; stipules minute, narrowly triangular, acute. Flowers in elongated terminal racemes, pedicellate, bracteate. Male flowers many, from above to more than half-way down; sepals oblong, elliptic to ovate, 1.5-3 × 0.5-2 mm; petals obovate to oblanceolate, stamens 11-14. Female florets few, at the base; pedicels 1.5-4 mm long; sepals 5, narrowly oblong to lanceolate; petals 0-5, filiform or subulate, up to 1.5 mm long; ovary subglobose, tomentose or hirsute, styles 3.5-6 mm long, free, bifid. Capsule 6 mm across, subglobose, stellate hairy.

Occasional in dry deciduous forests in Southern Eastern Ghats. Fl. & Fr.: August – January. Vern.: Tel.: *Nepalamu.*

Bhiravakona (NLR), *PSPB* 9109; Balapalli (CDP), *JLE* 14985 (MH); Salivendrakona (CDP), *JSG* 10787 (DD); Papanasanam (CTR), *KMS* 6955 (MH); Tirumala (CTR), *SKW* 5662. 8230 (BLAT).

INDIA: Andhra Pradesh, Tamil Nadu and Kerala.

WORLD: Sri Lanka.

Croton persimilis Müll.Arg., Linnaea 34: 116. 1865 & in DC., Prodr. 15(2): 618. 1866 (excl. var. *glabrata*) emend Philcox in Dassan. & Clayton, Rev. Handb. Fl. Ceylon 11: 92. 1997; N.P.Balakr. & Chakrab., Fam. Euphorb. India 215. 2007; N.P.Balakr. in N.P.Balakr. *et al.*, Fl. India 23: 240. 2012. *C. oblongifolius* Roxb., Fl. Ind. 3: 685. 1832; FBI 5: 386. 1887; Gamble 2: 1315. 1925. *C. roxburghii* N.P. Balakr., Bull. Bot. Surv. India 3: 39. 1962; Surya Prakash Babu in Pullaiah & Moulali, Fl. Andhra Pradesh 2: 851. 1997.

Small tree, to 8 m high, monoecious, branchlets tomentose. Leaves 12-24 × 6-11 cm, oblong-lanceolate, base acute, margin crenate-serrate, apex subacute, crowded at the ends of branchlets. Flowers in elongated terminal racemes. Male flowers: pedicels 2-5 mm long; sepals ovate-elliptic, to triangular; petals spathulate or narrowly oblong; stamens 10-15. Female flowers: pedicels 2-5 mm long; sepals ovate to elliptic; petals spathulatee to filiform; ovary subglobosee, styles 4-6 mm long, free, bifid. Capsule subglobose, thinly pubescent; seeds ellipsoid, rounded.

Rare in dry forests of Northern Eastern Ghats. Fl. & Fr.: March – December. Vern.: Tel.: *Bhutala bhairi, Bhuankusam, Peedachettu*; Ori.: *Desand, Masud, Putidi, Chandan, Dev chandan, Masundi.*

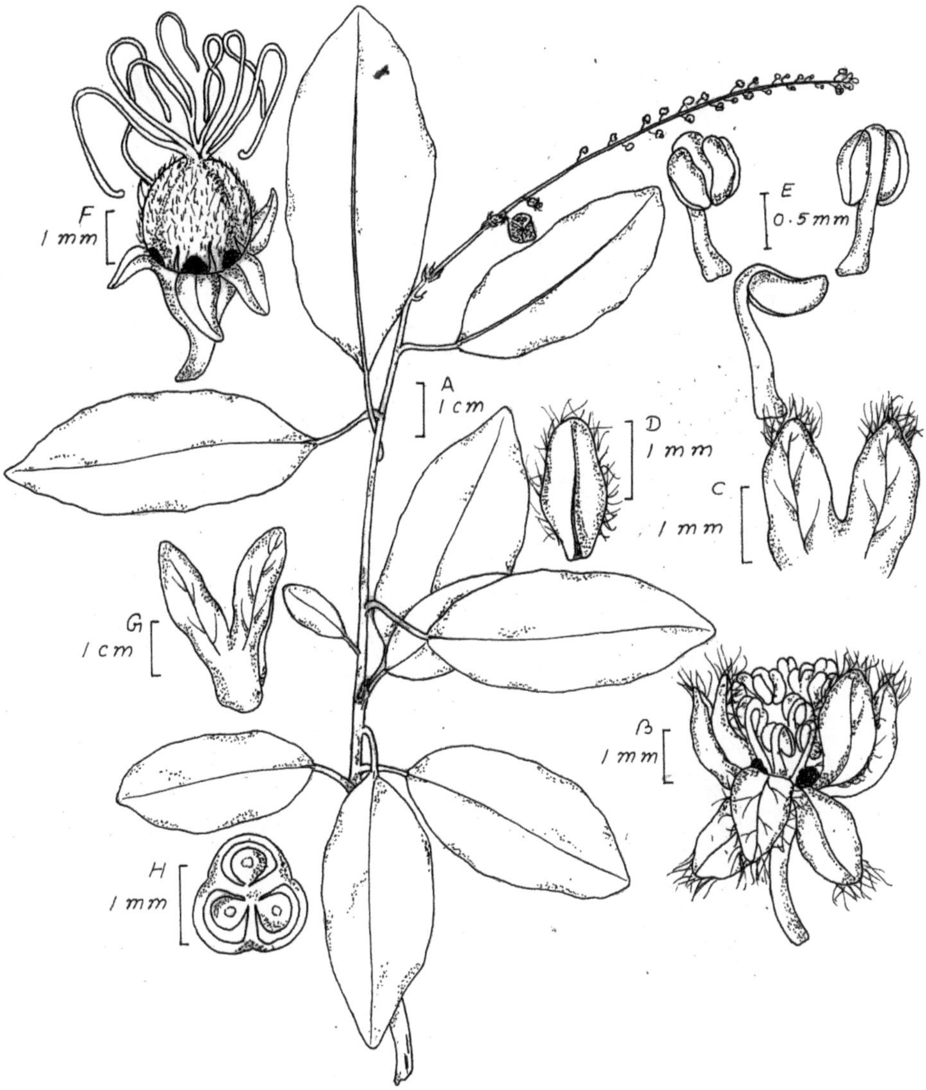

Figure 38. **Croton klotzschianus** (Wight) Thwaites
A. Twig, ♂ B. Flower, C. Outer tepals, D. Inner tepal, E. Stamens. ♀ F. Flower, G. Outer tepals, H. Ovary T.s.

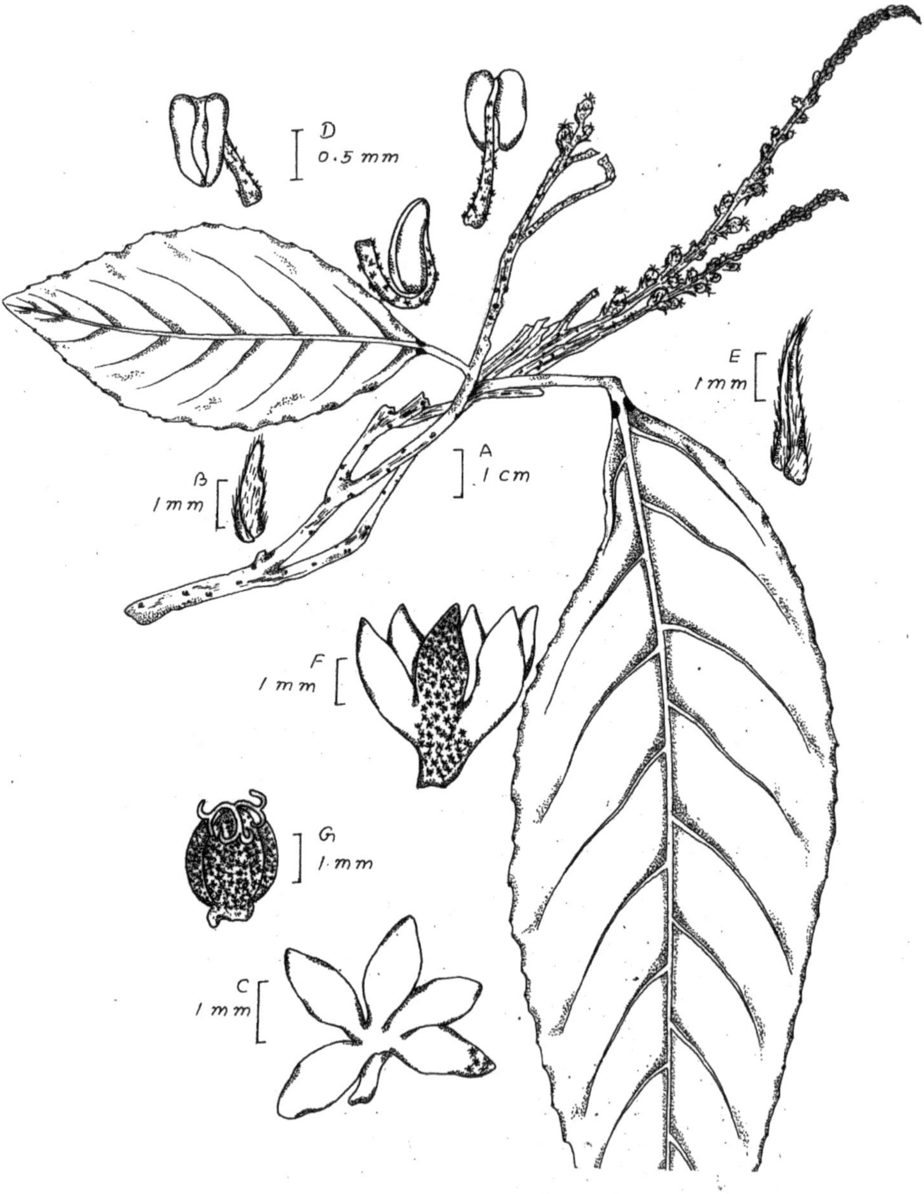

Figure 39. **Croton persimilis** Müll.Arg. (Syn.: *Croton roxburghii* N.P.Balakr.)
A. Twig, ♂ B. Bract, C. Tepals, D. Stamens, ♀ E. Bract, F. Tepals, G. Pistil.

Donubai (SKLM), *NRR* 83614; Suttiguda (VSKP), *NRR* 83795 (MH).

INDIA: Uttar Pradesh, Bihar, Jharkhand, West Bengal, Sikkim, Assam, Nagaland, Meghalaya, Odisha, Madhya Pradesh, Andhra Pradesh, Karnataka, Kerala and Tamil Nadu.

WORLD: Sri Lanka, Nepal, Bhutan, Bangladesh, Myanmar, SW. China, Indo-China and Thailand.

Croton scabiosus Bedd., Fl. Sylv. S. India 2. t. 283 B. 1872 & For. Man. 1873; FBI 5: 386. 1887; Gamble 2: 1314. 1925; N.P.Balakr. & Chakrab., Fam. Euphorb. India 215. 2007; N.P.Balakr. in N.P.Balakr. *et al.*, Fl. India 23: 240. 2012.

Deciduous tree, up to 6 m high; bark dark brown, tough, wood yellowish white, close grained, branchlets yellowish hairy. Leaves coriaceous, ovate, 4-10.5 × 7.8-8 cm, clothed with minute orbicular scales on both surfaces, pale and silvery beneath, green and very much rough above, base cordate, with 2 sessile glands, margin crenate, apex obtuse or slightly acute; petiole 3-5 cm. Flowers yellowish-white in terminal, 4-10 cm long racemes. Male flowers: many from above to more than half-way down; pedicels 2.5-3.5 mm long; sepals ovate to oblong-elliptic; petals oblong-elliptic to spathulate; stamens 8-16. Female flowers: few, at the base; pedicels 2-2.5 mm long; sepals as in male; ovary globose or ovoid, styles 3-5 mm long, free, bifid. Capsule sub-globose, 3-valved, 0.8-0.3 cm, yellowish; seeds oblong.

Common on dry rocky hills. Fl.: March-June; Fr.: July. Vern.: Tel.: *Chilla, Verri chilla.*

Pulivendula RF (CDP), *SSR* 10148; Guvvalacheruvu (CDP), *PSPB* 9197; Siddhout (CDP), *SKW* 5909, 5910 (BLAT); Janapal (CDP), *JSG* 11143 (DD); Thummalabailu (ATP), *SKW* 8009 (BLAT).

INDIA: Andhra Pradesh. Endemic to southern Andhra Pradesh.

Croton tiglium L., Sp. Pl. 1004. 1753; FBI 5: 393. 1887; N.P.Balakr. and Chakrab., Fam. Euphorb. India 215. 2007; N.P.Balakr. in N.P.Balakr. *et al.*, Fl. India 23: 243. 2012.

Trees, up to 7 m high; indumentum of stellate hairs. Stipules subulate, 1.5-4 cm, caducous, petiole 2.5-6 cm, subglabrous. Leaves ovate or ovate-elliptic, base rarely slightly cordate, with discoid glands, margin serrulate or subentire, apex acute or acuminate. Flowers unisexual in terminal racemes. Male flowers: bud subglobose, sparsely stellate-hairy, or glabrescent; sepals elliptic, oblong to ovate; petals oblong-elliptic to oblanceolate; stamens 14-20. Female flowers: pedicels 2-8 mm long; sepals oblong-lanceolate, glabrescent, petals 0-5, linear - filiform; ovary densely stellate-hairy, styles bipartite. Capsules ellipsoidal, oblong ovoid or subglobose; sparsely stelllate-hairy or glabrescent; seeds elliptic or oblong.

Evergreen or mixed forests or scrubs in Southern Eastern Ghats (N.P.Balakrishnan and Chakrabarty, 2007). Fl. & Fr.: January - December. Vern.: Tam.: *Nervalam.*

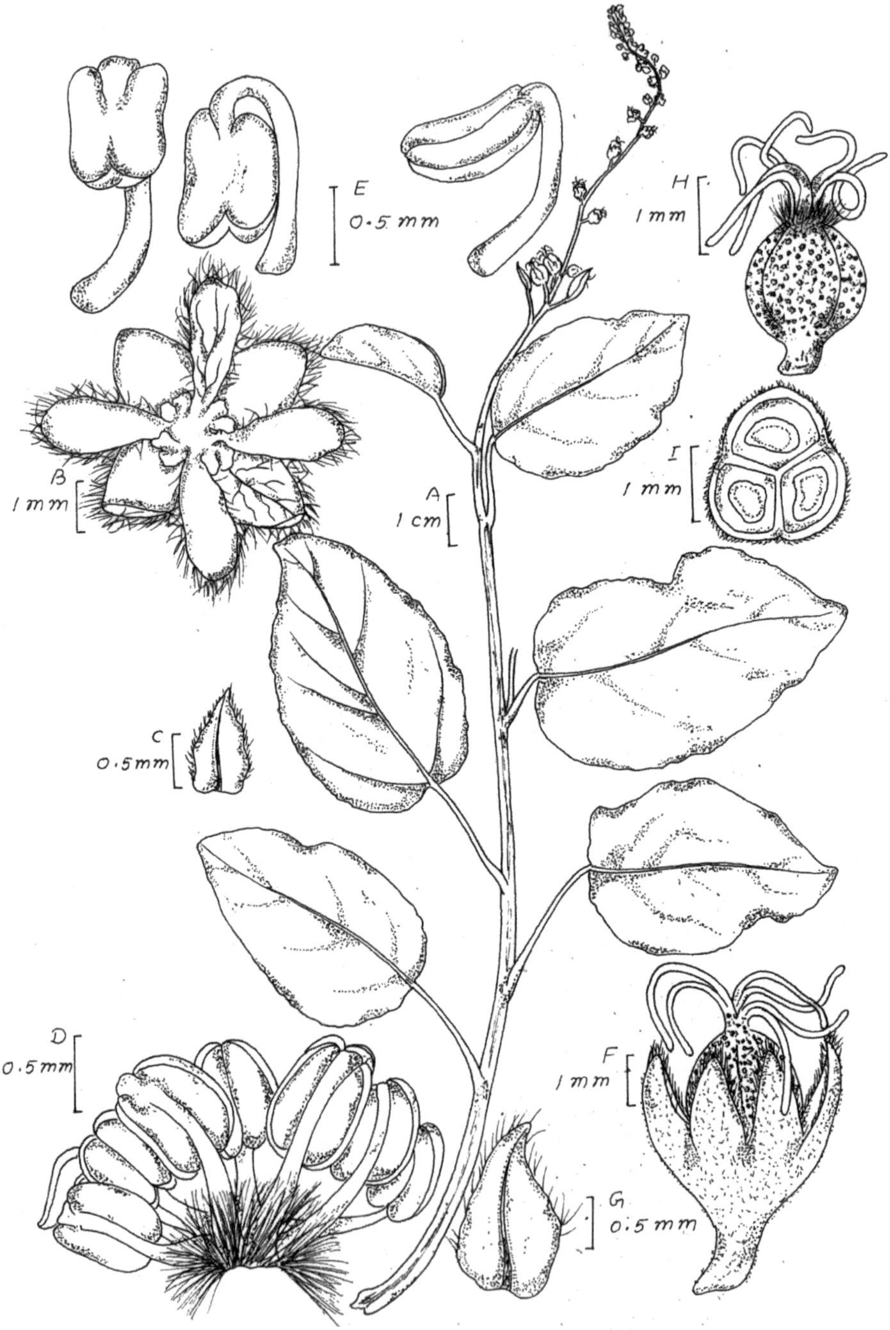

Figure 40. **Croton scabiosus** Bedd.

Twig, ♂ B. Tepals, C. Bract, D. Androecium, E. Stamens, ♀ F. Flower, G. Bract, H. Pistil, I. Ovary T.s.

INDIA: Arunachal Pradesh, Assam, Nagaland, Meghalaya, West Bengal, Uttar Pradesh, Madhya Pradesh, Bihar, Odisha, Adhra Pradesh, Maharashtra, Karnataka, Kerala and Tamil Nadu.

WORLD: Sri Lanka, Bhutan, Bangladesh, Myanmar, China, Thailand, Taiwan, Japan and Malesia.

DALECHAMPIA Linnaeus

1. Leaves 3-foliolate; tepals of female flowers lanceolate, glandular; anthers not apiculate; outer bract divided to below the middle .... **D. indica**

1. Leaves simple; tepals of female flowers pinnatifid, pilose; anthers apiculate; outer bract divided from above the middle .... **D. velutina**

Dalechampia indica Wight, Icon. Pl. Ind. Orient. 5: t. 1882. 1852; FBI 5: 467. 1887; Gamble 2: 1335. 1925; N.P.Balakr. & Chakrab., Fam. Euphorb. India 175. 2007; N.P.Balakr. in N.P.Balakr. *et al.*, Fl. India 23: 189. 2012.

Twiner, up to 4 m long, finely pubescent with stinging hairs except on upper surface of leaves. Leaves palmately tri-foliolate; petiole of 5 cm, central leaflet obovate, 5-10 × 2.5-4 cm; laterals smaller, oblique, subsessile, lanceolate, base acute-rounded, margin serrate, apex subacuminate, involucral bracts foliaceous. Male flowers: sepals 4-5, lanceolate, folded; stamens 20-25; filaments ca 0.8 mm long, anthers 1 mm across. Female flower: sepals 10-12, unequal, lanceolate, sericeous without, margin glandular hairy; ovary subglobose, styles simple, up to 1.5 mm long, stigma somewhat cupular. Capsule globose, fruiting calyx 6-8 mm long, glandular hairy; seeds globose, mottled dark grey or black.

Occasional on the hill slopes in southern Eastern Ghats. Fl. & Fr.: December-February. Vern.: Tam.: *Masukodi.*

Balepalli (CDP), *JLE* 15762 (CAL & MH); Penchalakona (NLR), *PV* 109951 (BSID); Rapur hills (NLR), *PV* 110033 (BSID).

INDIA: Andhra Pradesh and Tamil Nadu.

WORLD: Sri Lanka.

Dalechampia velutina Wight, Icon. Pl. Ind. Orient. 5: 1881. 1852; FBI 5: 467. 1888. *D. scandens* L., Sp. Pl. 1054. 1753 var. *velutina* (Wight) Müll.Arg., Prodr. 15(2): 1245. 1866.

Evergreen vine, to 6 m long, with stinging hairs. Leaves simple, 3(5)-lobed, coriaceous, velutinous beneath; petiole 3-8 cm long. Flowers in terminal or apparently axillary, head-like pleiochasium of cymules; peduncle to 8 cm long; involucral bracts 5 × 4 cm, outer one divided from above the middle; Male flowers: calyx 3-5-partite, stamens 8-12, anthers apiculate. Female flowers: Tepals ca 6, pinnatifid, pilose without, ovary hispdulous, stigma peltate. Fruit 3-loculed; seeds globose.

Hills 450-1300 m; dense mats on wayside thickets especially by road clearings (Matthew, 1991). Fl. & Fr.: December – Februry.

INDIA: Peninsula.

DIMORPHOCALYX Thwaites

1. Branchlets, petioles, peduncles and pedicels glabrous; leaves elliptic-ovate or obovate, rounded, obtuse to acute at base, acuminate at apex; female pedicels 3- 10 mm long; sepals glabrous; stamens 10-16, filaments of outer stamens free almost up to base .... **D. glabellus**

2. Branchlets, petioles, peduncles and pedicels pubescent; leaves obovate to oblanceolate, attenuate at base, obtuse to subacute at apex; sepals glabrous; female pedicels up to 2 mm long; sepals appressed-pilose; stamens 16-22, filaments of outer stamens shortly connate at base, free above .... **D. kurnoolensis**

Dimorphocalyx glabellus Thwaites, Enum. Pl. Zeyl. 278. 1861; FBI 5: 403.1887; Gamble 2: 1337. 1925; var. **glabellus**; N.P.Balakr. & Chakrab., Fam. Euphorb. India 205. 2007; N.P.Balakr. in N.P.Balakr. *et al.*, Fl. India 23: 223. 2012.

Small tree, to 5 m high, dioecious, sometimes monoecious; bark smooth or with large ridges of cork; new branchlets glabrous but with brown, shining, ovate scales. Leaves elliptic, ovate or obovate, 1.5-19 × 0.5-8 cm, base obtuse-acute, often cuneate, margin crenate or serrate-entire, apex shortly acuminate, acute or often obtuse-rounded or cuspidate; thinly coriaceous-chartaceous; secondary nerves ca. 8, slender; petiole 6.2-10 mm; stipules small, triangular or ovate. Flowers monoecious, the male and female inflorescences being on separate branches of the same tree. Male flowers axillary, solitary or umbellately 2 to 3-flowerd or in few branched cymes, white, 8.7-10 mm long; pedicels very short, calyx cupular or urceolate, 2.5 mm long, 5-toothed, nearly glabrous, corolla campanulate, 1.2 cm diam.; petals oblong, 1.2 cm long, nearly free, obtuse; stamens 10-16, outer series free, or connate to column, with longer filaments than the inner series on top of the column. Female flowers axillary and terminal, solitary to few in umbellate 4 mm long racemes, 1.8-2 cm diam.; peduncle 2-3.7 cm; sepals 5, spreading, obovate, rounded or retuse; petals free, broadly oblong, 1-1.2 cm long, overlapping to the right in bud, disc annular, ovary pubescent. Capsule globose, 7-15 × 10-18 mm, subglobose, 3-lobed and 6-grooved, thinly strigose-hairy, seated on the large spreading calyx; fruiting sepals very unequal, 0.5-3.3 × 0.3-1.8 cm, strongly nerved.

Occasional in evergreen forests. Fl. & Fr.: March-April.

Rapur (NLR), *BR, SKB* & *BSS* 30513; Tadepalli (EG), *NRR* & *TRR* 86117 (MH, BSID); Dolphin's nose (VSKP), *JSG* 16105 (CAL); Angul, *Haines* 2507 (DD).

INDIA: Odisha, Andhra Pradesh, Maharashtra, Karnataka, Kerala and Tamil Nadu.

WORLD: Sri Lanka

Dimorphocalyx kurnoolensis Venkataraju & Pullaiah, Bot. Bull. Acad. Sin. 35: 201. 1994; N.P.Balakr. & Chakrab., Fam. Euphorb. India 205. 2007; N.P.Balakr. in N.P.Balakr. *et al.*, Fl. India 23: 224. 2012.

Small tree, to 3 m high, dioecious branchlets terete, striate, pubescent. Leaves 5-19 × 2-6 cm, elliptic-lanceolate or obovate, base attenuate, margin entire-sinuate, apex subacute-obtuse; petiole to 2 cm, shallowly channeled above, stipules deltoid. Male flowers in terminal cymes; peduncle pubescent, bracts lanceolate, densely pubescent, acute-acuminate; flowers 4-5 mm across, sub-sessile; tepals in two whorls, 5 + 5; outer (sepals) cupular, 5-lobed, connate, adpressed-pilose without, lobes suborbicular, inner (petals) free, oblong, obtuse, often emarginated; stamens biseriate, 5 + 11-17, outer basally connate to the inner staminal column, filaments 1.8 mm, anthers 1.2 mm across, widely oblong, inner monadelphous, staminal column 7 mm, anther 0.8 mm across, orbicular, connectives 0.5 mm long, acute; disc of glands 5, free, ovate-oblate, hairy at top. Female flowers few, in short pedunculate racemes; flowers 5-8 mm across; pedicel 2 mm, pubescent; tepals in two whorls, 5 + 5; outer (sepals) cupular, shortly 5-lobed, connate, adpressed-pilose without, lobes suborbicular, 2 × 2.5 mm; inner (petals) free, oblong-obtuse, often emarginated; ovary 4 × 3 mm, adpressed-pilose, styles 3, connate at base, each 2-fid, from above the middle, papillose; disc of glands as in the male flower. Capsule 1-1.3 × 1.3-1.7 cm, subglobose, depressed, adpressed-pilose, 3-lobed, deeply furrowed, fruiting calyx (sepals) deeply 5-lobed, lobes 5.5 × 3.5 mm, adpressed-pilose without.

Rare along the sides of the streams at the foot hills of Erramalai hills near Owk in Kurnool district. Fl. & Fr.: Jan.-March. Vern.: Tel.: *Erra tandra.*

Erramalai hills near Owk (KNL), *RVR* 6661; *PSPB* 11847, 11848 (SKU).

INDIA: Andhra Pradesh. Endemic.

Note: This taxon is similar to *D. glabellus* but differs in having branchlets pubescent, leaves sparsely pilose without, sepals, ovary and capsule adpressed-pilose, stamens with connectives more in number (16-22) and styles with papillose.

N.P.Balakrishnan and Chakrabarty (2007) mentioned the authors as V.S.Raju and Pullaiah, it should be Venkataraju and Pullaiah.

DRYPETES Vahl

1. Leaves pubescent along nerves beneath, 6.7-18 × 4.5-7 cm; ovary 2-celled; drupe 2.5 × 1.5 cm; female flowers solitary, sessile .... **D. assamica**
1. Leave glabrous, 4-7 × 2.2-3 cm; ovary 1-celled; drupe 8 mm across; female flowers in fascicles, pedicellate:
    2. Fruit globose or fusiform, fleshy, fruting pedicel 4-10 mm long.. **D. sepiaria**
    2. Fruit obovoid, not fleshy, fruting pedicel 20-26 mm long............. **D. venusta**

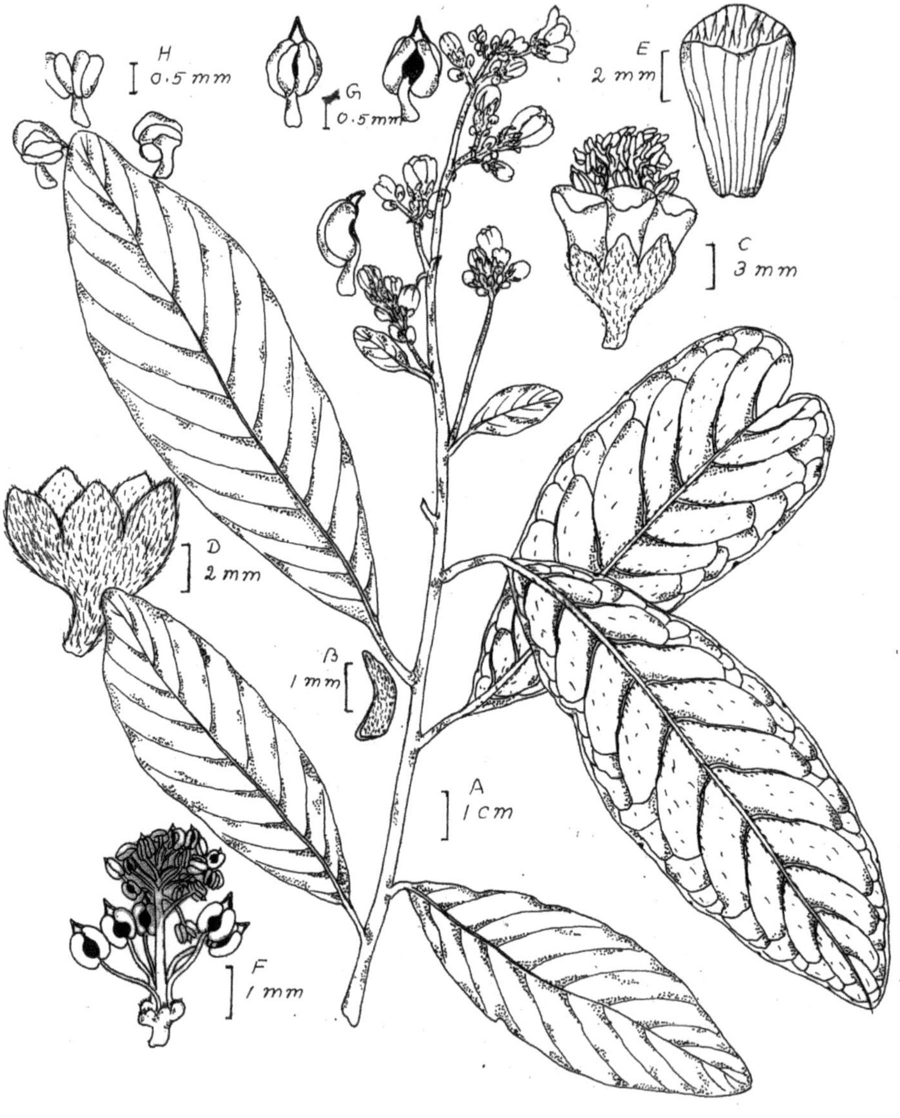

Figure 41. **Dimorphocalyx kurnoolensis** Venkataraju & Pullaiah
♂ A. Twig, B. Bract, C. Flower, D. Outer tepals (sepals), E. Inner tepal (petal), F. Androecium, G and H. Anthers.

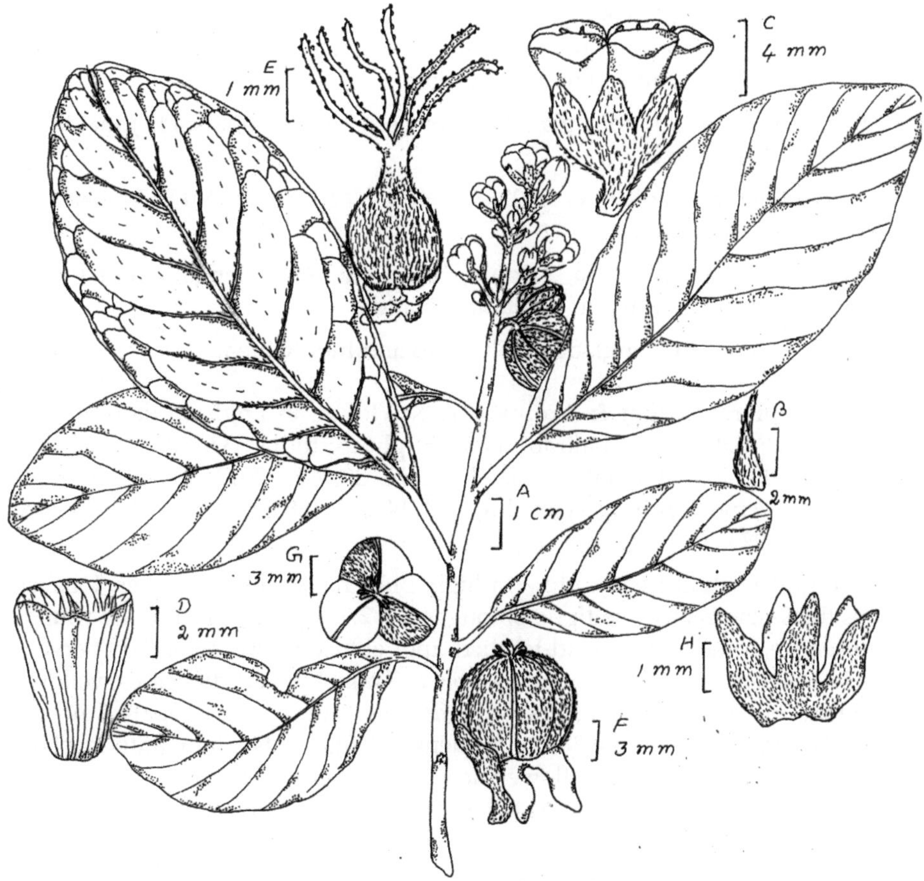

Figure 42. **Dimorphocalyx kurnoolensis** Venkataraju & Pullaiah

♀ A. Twig, B. Bract, C. Flower, D. Inner tepal (petal), E. Pistil, F. Capsule with calyx, G. Capsule (Upper view), H. Fruiting calyx.

Drypetes assamica (Hook.f.) Pax & Hoffm. in Engler, Pflanzenr. 5. 147 (15): 241. 1922; N.P.Balakr. & Chakrab., Fam. Euphorb. India 329. 2007; Chakrab. & N.P.Balakr. in N.P.Balakr. *et al.*, Fl. India 23: 394. 2012. *Cyclostemon assamica* Hook. f., Fl. Brit. India 5: 342. 1887; Gamble 2: 1301. 1925.

Small, evergreen tree, up to 8 m high, bark light; wood white, slightly yellow streaked; branchlets pubescent. Leaves subcoriaceous, elliptic, lanceolate or oblong, 6.7-18 × 4.5-7 cm, shining, pubescent along nerves beneath, base acute, margin entire, apex obtuse or shortly acuminate, secondary nerves 11-15 pairs, inter-costal reticulations between the nerves; petiole 8 mm. Flowers yellowish-green. Male flowers: clustered in axils; sepals 4, suborbicular; stamens 8-15; disc plicate with thee lobes mostly produced outwards embracing thee base of filaments, pubescent. Female flowers: solitary, sessile; sepals 4, suborbicular, puberulous; disc annular, pubescent; ovary globose or oblong-ellipsoid, stigmas sessile, 2, flabellate. Drupe scarlet, ovoid-oblong, bilocular. 2.5 × 1.5 cm.

Rare in the hills, in moist places of Northern Eastern Ghats. Fl.: June-August, Fr.: September-November.

Sunkarimemtta (VSKP), *GVS* 22667 (MH), *NPBK* 10936 (MH); Mahendragiri (GJM), *JSG s.n.* (MH)

INDIA: Assam, Mizoram, Tripura, Nagaland, West Bengal, Uttar Pradesh, Odhisha, Andhra Pradesh and Andaman Islands.

WORLD: Bhutan, Bangladesh, Thailand and Indo-China.

Drypetes sepiaria (Wight & Arn.) Pax & K. Hoffm. in Engler, Pflanzenr. 81: 2. 1922; N.P.Balakr. & Chakrab., Fam. Euphorb. India 333. 2007; Chakrab. & N.P.Balakr. in N.P.Balakr. *et al.*, Fl. India 23: 401. 2012. *Hemicyclia sepiaria* Wight & Arn., Edinburgh New Philos. J. 14: 297. 1883; FBI 5: 387. 1887; Gamble 2: 1299. 1925.

Evergreen, small tree, up to 7 m high; branchlets whitish, puberulous, stiff. Leaves thick- chartaceous, elliptic-oblong, 4-7 × 2.2-3 cm, glabrous, base rounded, margin entire, sometimes repand-toothed, apex obtuse-retuse, secondary nerves 5-6 pairs, impressed beneath, irregular; petiole to 7 mm. Flowers dioecious, yellowish, pedicellate, 6-8 in axillary clusters. Male flowers: 6 mm across, tepals 4, unequal, 2 + 2, 2 larger, membranous, 2 mm; stamens 4-8, filaments 2 mm, anthers oblong, 1.2 mm; disc glabrous. Female flowers in fascicles, pedicellate, 6 mm across; ovary ovoid, 2.5 mm, 1-locular, style 1, to 0.5 mm. Drupe red, globose, 8 mm across; seed 1 or 2, oblong, grooved on one face.

Occasional, along streams of moist forests to 600 m. Fl.: February-March. Fr.: April-July. Vern.: Tel.: *Bira, Manchi-Bira;* Tam.: *Virai.*

Siddulaiah kona (NLR), *SSR* 19367, 19376; Chintala forest (PKM), *SSR* 16150; Dornal (KNL), *JSG* 18685 (CAL); Peccheruvu (KNL), *SKW* 5098 (DD); Srisailam (KNL), *JLE* 23772; Kodur (CDP), *JLE* 15797 (MH), *JSG* 11241 (DD); Bakharapeta (CTR), *SKW* 5716 (BLAT); Panchalingalakona (NLR), *PV* 109943; Narsipatnam (VSKP), *GVS* 29650 (MH).

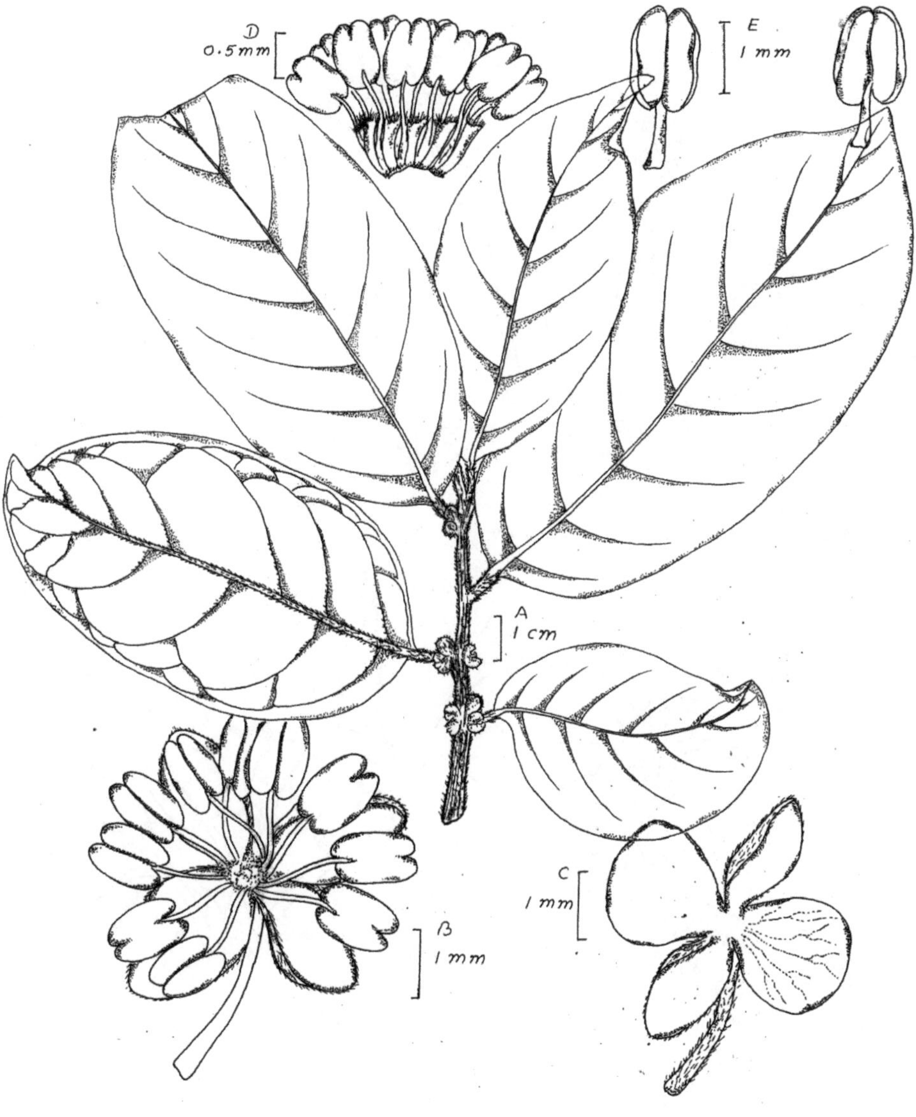

Figure 43. **Drypetes assamica** (Hook.f.) Pax & Hoffm.
♂ A. Twig, B. Flower, C. Tepals, D. Androecium, E. Anthers.

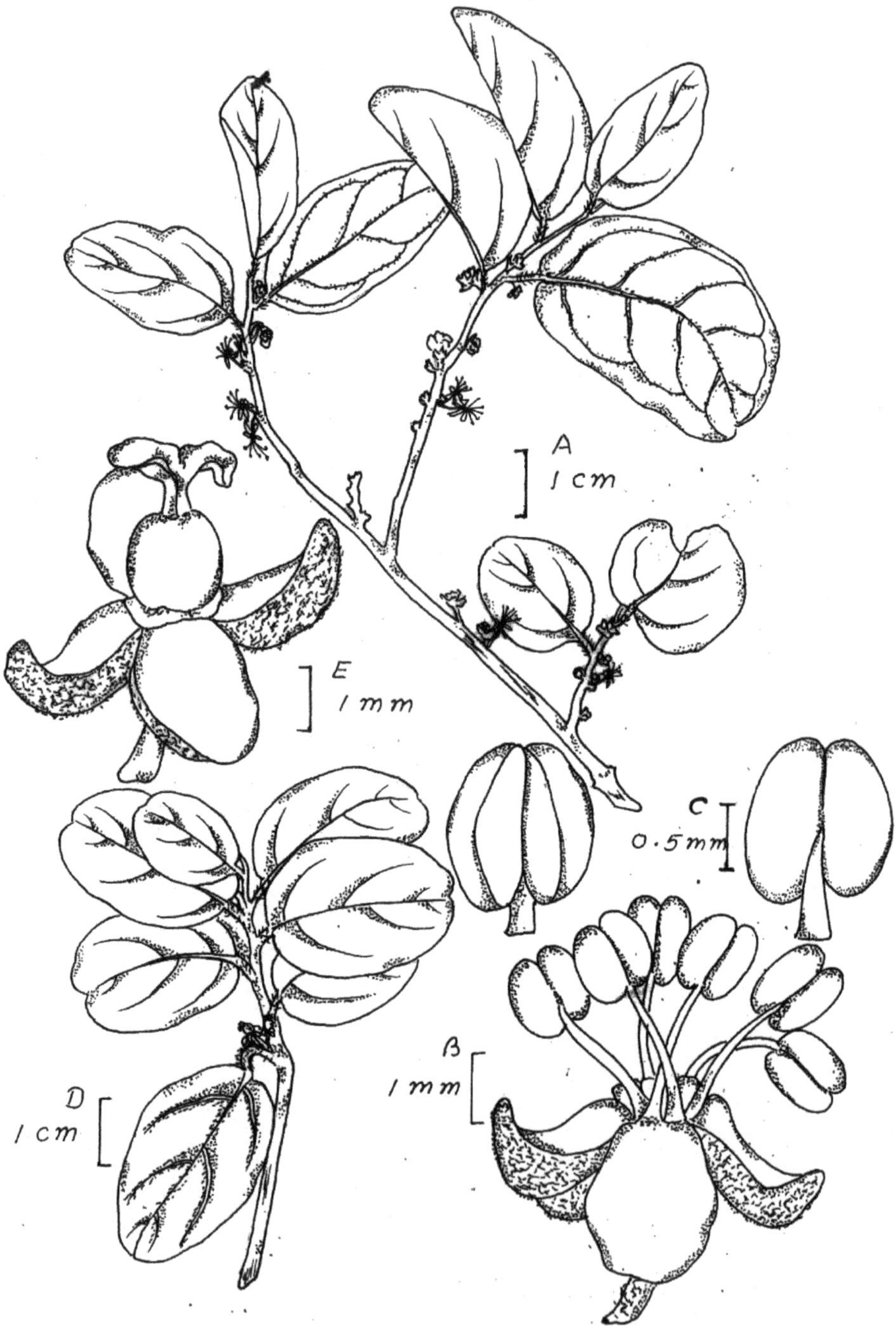

Figure 44. **Drypetes sepiaria** (Wight & Arn.) Pax & Hoffm.
♂ A. Twig, B. Flower, C. Anthers, ♀ D. Twig, E. Flower.

INDIA: Andhra Pradesh, Maharashtra, Karnataka, Kerala and Tamil Nadu.

WORLD: Sri Lanka.

Drypetes venusta (Wight) Pax & K. Hoffm. in Engl., Pflanzenr. IV.147. xv: 268. 1922. *Astylis venusta* Wight, Icon. Pl. Ind. Orient. 6: 13.t.1992. 1853. *Hemicyclia venusta* (Wight) Thwaites, Hooker's J. bot. Kew. Gard. Misc. 7: 272. 1855; FBI 5: 339. 1887; Gamble 2: 1300. 1925. *H. elata* Bedd., Fl. Sylv. S. India 2: t.279. 1872: Gamble 2: 1300. 1925. *Drypetes elata* (Bedd.) Pax & Hoffm., *loc.cit.* 268. 1922.

Trees, up to 30 m tall; branchlets drooping, glabrous. Leaves elliptic, oblong-elliptic to obovate, 4-15 x 1.5-5.5 cm, thinly coriaceous, glabrous, lateral nerves 7-16 pairs, base unequal or acute, margin entire, apex acuminate or sometimes acute to subacute; petiole to 10 mm long, glabrous. Inflorescence axillary. Male flowers: pedicles 3-10 mm long; sepals 4, elliptic-oblong, ovate to suborbicular, 2-4 x 1.5 -3 mm; stamens 6-10; anthers oblong, 1 mm long; disc lobulate, puberulous. Female flowers: pedicels to 15 mm long; sepals 4, ovate to obovate, 3-4 x 1.8-3.5 mm; ovary oblong, ovoid or subovoid, glabrous; style single, terminal; stigma orbicular or reniform. Fruits drooping on 2-4 cm long slender elongated pedicels, obovoid, 20-25 x 12-18 mm, smooth, unilocular, glabrous; seeds solitary.

Occasionally found in evergreen forest areas of Sirumalai hills. Fl. & Fr.: November – July. Vern.: Tam.: *Konamaram, Palakanni, Vellelambu.*

Sirumalai hills (DGL), *J.M.Pallithanam* 10014 (RHT).

INDIA: Maharshtra, Karnataka, Kerala and Tamil Nadu. Endemic.

EPIPRINUS Griffith

Epiprinus mallotiformis (Müll.Arg.) Croizat, J. Arnold Arbor. 23: 53. 1942. N.P.Balakr. & Chakrab., Fam. Euphorb. India 170. 2007; N.P.Balakr. in N.P.Balakr. *et al.*, Fl. India 23: 184. 2012. *Symphyllia mallotiformis* Müll.Arg., Linnaea 34: 156.1865; Gamble 2: 1323. 1925. *Adenochlaena indica* (Bedd.) Bedd. ex Hook.f., Fl. Brit. India 5: 418. 1887.

Tree, up to 10 m tall, monoecious, branchlets puberulous. Leaves broadly elliptic-ovate or orbicular, 7 – 17 × 3-8.5 cm, base rounded-acute, margin entire, apex subacute, coriacious. Flowers in interrupted subterminal spikes; male flowers many, clustered, calyx-lobes 4, ovate-lanceolate, up to 2 mm long; stamens 4, exserted; female flowers few, calyx lobes 6, unequal, ovate-lanceolate, up to 2 mm long, ovary globose, pubescent, styles 3, each bifid above with plumose stigmatic arms. Capsule depressed globose, rugose; seeds globose, crustaceous.

Occasional in forests in Southern Eastern Ghats. Fl and Fr.: November-January.

Papanasanam in Tirumala (CTR), *DRC* 1257 (SVU).

INDIA: Karnataka, Kerala and Tamil Nadu. Endemic.

EUPHORBIA Linnaeus

1. Stems not developed above the ground:
    2. Inflorescence synanthous .... **E. seshachalamensis**
    2. Inflorescence hysteranthous .... **E. fusiformis**
1. Stems well developed above the ground:
    3. Shrubs or small trees; stems and branches thick, fleshy:
        4. Branchlets unarmed, cylindric .... **E. tirucalli**
        4. Branchlets thick, fleshy, armed with stipular thorns:
            5. Branchlets cylindrical:
                6. Leaves 10-16 cm long; arborescent .... **E. nivulia**
                6. Leaves small, to 10 cm long; shrubs:
                    7. Involucre cupular; capsule slightly recurved ...... **E.venkatarajui**
                    7. Involucre turbinate; capsule triangular, truncate .... **E. caducifolia**
            5. Branchlets angled or twisted:
                8. Shrubs; leaves more than 2 cm long, early deciduous; podarium less than 1 cm long:
                    9. Leaves oblanceolate to spathulate, petiole absent, fruits sub globose .... **E. royleana**
                    9. Leaves obovate-oblong or broadly elliptic, petioles present 5-7 mm long, fruits ovoid- .... **E. lacei**
                8. Trees; leaves less than 1 cm long, early deciduous; podarium less than 1 cm long:
                    10. Branches crowded at apex; marbled white-coloured between the wings, flowers and fruits usually not developing .... **E. lactea**
                    10. Branches spreading; not marbled white-coloured between the wings; flowers and fruits developing in every season:
                        11. Styles simple .... **E. cattimandoo**
                        11. Styles 2-fid:
                            12. Branchlets not twisted, mature cyathia reddish .... **E. antiquorum**
                            12. Branchlets spirally twisted, mature cyathia greenish:
                                13. Cyathia solitary; capsule subglobose .... **E. susan-holmesiae**
                                13. Cyathia 3-4 in fascicle; capsule trigonous .... **E. tortilis**

3. Herbs, stems and branches not fleshy:

14. Appendages inconspicuous or 0:

15. Prostrate herbs:

16. Leaves subquadrate, apex retuse; one involucral lobe serpent hooded .... **E. serpens**

16. Leaves elliptic – oblong, apex subacute – obtuse:

17. Leaves of main stems distinctly larger than those elsewhere .... **E. clarkeana**

17. Leaves of stems and branches almost equal in size:

18. Leaf margin serrulate; involucre campanulate .... **E. thymifolia**

18. Leaf margin entire; involucre tubular:

19. Capsule prominently keeled .... **E. prostrata**

19. Capsule obtusely keeled .... **E. heyneana**

15. Erect herbs:

20. Involucral gland 1:

21. Shrubs or trees; upper leaves completely coloured; cyathia more than 6 mm long; gynophore long, protruding from involucre; seeds ca 10 × 8 mm, smooth .... **E. pulcherrima**

21. Herbs; upper leaves green or partially coloured; cyathiaa less than 5 mm long; gynophore least protruding from involucre; seeds 5 mm across, sculptured:

22. Upper leaves sinuate, yellow-red at above; gland bilabiate, transversely oblong .... **E. cyathophora**

22. Upper leaves entire, uniformly green; gland cupular .... **E. heterophylla**

20. Involucral glands 4 or 5:

23. Glands of involucre not horned, ciliate:

24. Leaves with gland tipped margin .... **E. agowensis**

24. Leaf margin not with glands .... **E. perbracteata**

23. Glands of involucre 2-horned .... **E. dracunculoides**

14. Appendages small or conspicuous:

25. Cyathia 1-3, within the distichously imbricate leaves:

26. Appendages sinuate .... **E. elegans**

26. Appendages laciniate:

27. Appendages glabrous or nearly so; seeds brown .... **E. laciniata**

27. Appendages ciliate; seeds white .... **E. cristata**

25. Floral leaves not distichous:

28. Appendages multifid:

29. Leaves obliquely rounded at base, apiculate at apex; cyathia few, terminal; styles filiform .... **E. longistyla**

29. Leaves obliquely truncate at base, acute at apex; cyathia solitary, axillary; styles simple:

30. Appendages of cyathial glands entire, minutely wavy .... **E. senguptae**

30. Appendages of cyathial glands 2-3-fid .... **E. kadapensis**

28. Appendages entire, not multifid:

31. Erect or ascending herbs:

32. Leaves linear-oblong, seeds smooth or pitted .... **E. deccanensis**

32. Leaves elliptic-ovate; seeds tuberculate:

33. Fruits globose, trigonous, obtusely keeled; seeds oblong or ovoid, obtusely quadrangular:

34. Limbs of glands indistinct or of same size as the glands:

35. Involucral glands 4; appendages small .... **E. hirta**

35. Involucral glands 5; appendages much larger than the glands .... **E. indica**

34. Limbs of glands distinct and a little to much larger than glands:

36. Cyathia many in each clusteer........ **E. hypericifolia**

36. Cyathia solitary, 2 or 3 in each cluster .... **E. pycnostegia**

33. Fruit ovoid, acutely keeled; seeds obscurely 2-3-furrowed:

37. Limbs of glands distinct and a little to much larger than the glands .... **E. sebastinei**

37. Limbs of glands indistinct or of the same size as the glands .... **E. hirta**

31. Prostrate herbs:

    38. Appendages pink or rose, the appendages of two upper glands obliquely oblong, lower two small .... **E. rosea**

    38. Appendages white, appendages subequal .... **E. corrigioloides**

Euphorbia agowensis Hochst. ex Boiss. in DC., Prodr. 15(2): 70. 1862; Kumari & Subba Rao, Bull. Bot. Surv. Ind. 15: 120. 1973 (1976); Binojk. & N.P. Balakr., Genus *Euphorbia* in India 207. 2010 & in N.P.Balakr. *et al.*, Fl. India 23: 301. 2012.

Key to Varieties

1. Leaves obovate, subovate, ovate-oblong to elliptic-lanceolate, rarely with gland-tipped marginal teeth at base .... var. **agowensis**
1. Leaves linear to linear-lanceolate, always with gland-tipped marginal teeth at base .... var. **pseudoholstii**

Euphorbia agowensis Hochst. ex Boiss. var. **agowensis.** *E. wightiana* Hook.f., Fl. Brit. India 5: 248. 1887, *p.p.*, non Boiss.1860. *Chamaesyce wightiana* V.S. Raju & P.N. Rao, Phytologia 37(5): 454. 1977.

Perennial or biennial herb, up to 1.5 m high; branches dichotomous or trichotomous, internodes long, grooved. Lower leaves alternate, upper opposite or ternate, young leaves puberulous, oblong lanceolate or obovate, 3-5 × 1-3 cm, base cuneate, margin entire or denticulate, apex acute-obtuse or emarginated; petiole short; stipulate; gland present at leaf blade. Cyathia terminal, subsessile, subtended by 2 or 3 ovate or rotund leaves; involucre turbinate, pubescent; glands 4 or 5, transversely oblong; appendages 0. Male florets 3, bracteolate; stalk hairy, anthers subglobose, yellow. Female floret laterally pendulous; ovary trigonus, hairy, styles 3, free, each 2-fid at above, hairy. Capsule trigonous, smooth, puberulous; seeds oblong, compressed longitudinally 5-ridged, wrinkled.

Rare especially on rocky hill slopes of Srisailam in Nallamalais. Fl. & Fr.: Throughout the year.

Near Srisailam dam (KNL) *NSD*, 31410; Srisailam (KNL), *VSR* 503, 511 (NU).

INDIA: Andhra Pradesh, Tamil Nadu.

WORLD: Arabian peninsula, Somalia, Tanzania and Kenya.

Euphorbia agowensis Hochst. ex Boiss. var. **pseudoholstii** (Pax) Bally & Carter, Kew Bull. 39(3): 650. 1984; Binojk. & N.P.Balakr., Genus *Euphorbia* in India 294. 2010. *Euphorbia pseudoholstii* Pax, Bot. Jahrb. Syst. 33: 287. 1903.

Moist forests, hill slopes, sporadic in southern Eastern Ghats. Fl. & Fr.: July - September.

INDIA: Andhra Pradesh, Tamil Nadu.

WORLD: Arabian peninsula, Ethiopia, Kenya and Tanzania.

Euphorbia antiquorum L., Sp. Pl. 450. 1753; FBI 5: 255. 1887; Gamble 2: 1277. 1925; Binojk. & N.P. Balakr., Genus *Euphorbia* in India 294. 2010 & in N.P. Balakr. *et al.*, Fl. India 23: 319. 2012.

Succulent shrub or small tree, up to 5 m high; branchlets 3 to 5-winged, with stipular divaricate persistent sharp spines. Leaves small, obovate-oblong, 4-12 × 3-6 mm, base cuneate, margin entire, apex rounded or subtruncate, deciduous, subsessile. Cyathia in axillary, lax, dichasial cymes of 3, the central one sessile, the lateral two pedunculate, unisexual or bisexual; involucre 6.5 × 9.5 mm, cupular; glands 5, suborbicular, thick; appendages 0. Male florets in 5 groups, each of 6-8, few sterile; bracteole laciniate; stalk 3 mm; anther 0.7 mm across. Female floret erect; gynophores to 2 mm long, ovary 2.5 mm, bluntly 3-angled, styles 3, each 2-fid at above, recurved, stigma papillose. Capsule 1 cm, trigonus; seeds 2 mm, globose, minutely carunculate, smooth.

Common in open plains usually in rocky places along foot hills in scrub jungles. Fl. & Fr.: December-February. Vern.: Tel.: *Bomma jemudu, Pedda jemudu;* Ori.: *Dokanasiju;* Tam.: *Sathurakkalli.*

Eswaramala RF (ATP), *TP* 827 (MH); Pennahobilam (ATP), *TP* & *NY* 687; Muchukota (ATP), *TP* & *NY* 208; Bugganipalli (KNL), *TP* & *RVR* 1815; Palakonda (CDP), *CS* 8259; Mypadu (NLR), *PSPB* 9101; Macherla (GNT), *VRK* 3816; Mellavagu (GNT), *VRK* 5847; Erramalais (KNL), *JSG* 10965 (DD); Srisailam (KNL), *JLE* 32692 (MH); Rampa (EG), *VNS* 675 (CAL); Rampa (EG), *VNS* 675 (CAL).

INDIA: Through out India.

WORLD: Pakistan, Bangladesh, Myanmar, China, Thiland, Sir Lanka, Indonesia and Malaysia.

Euphorbia caducifolia Haines, Indian Forester 40: 154. 1914; C. Fischer, Bull. Misc. Inform. Kew 1925: 341. 1925; Gamble 2: 1276. 1925; Binojk. & N.P.Balakr., Genus *Euphorbia* in India 298. 2010 & in N.P. Balakr. *et al.*, Fl. India 23: 320. 2012.

Succulent shrub, up to 2 m high; much branched, branches cylindrical, 3-5 cm in diam., smooth, pale green, with silver-white coating or peels when mature, with stipular divaricate persistent sharp spines. Leaves very small, oblong, 2-8 × 1-5 cm, fleshy, deciduous, subsessile. Cyathia in axillary cymes; solitary, paired or in triads, reddish pink, involucre 3 × 5 mm, cupular; glands 5, suborbicular, thick; appendages 0. Male florets in 5 groups, each of 4-6; bracteole laciniate; stalk 4 mm; anther 1mm across. Female floret laterally pendulous; gynophores to 1.3 mm long, perianth lobes 3, minute, triangular subacute, ovary 2.5 × 3.5 mm, sharply 3-angled; styles 3, connate at base, simple, recurved. Capsule 1 cm, trigonus, keeled; seeds ovoid to spherical, 2 mm, smooth, brownish.

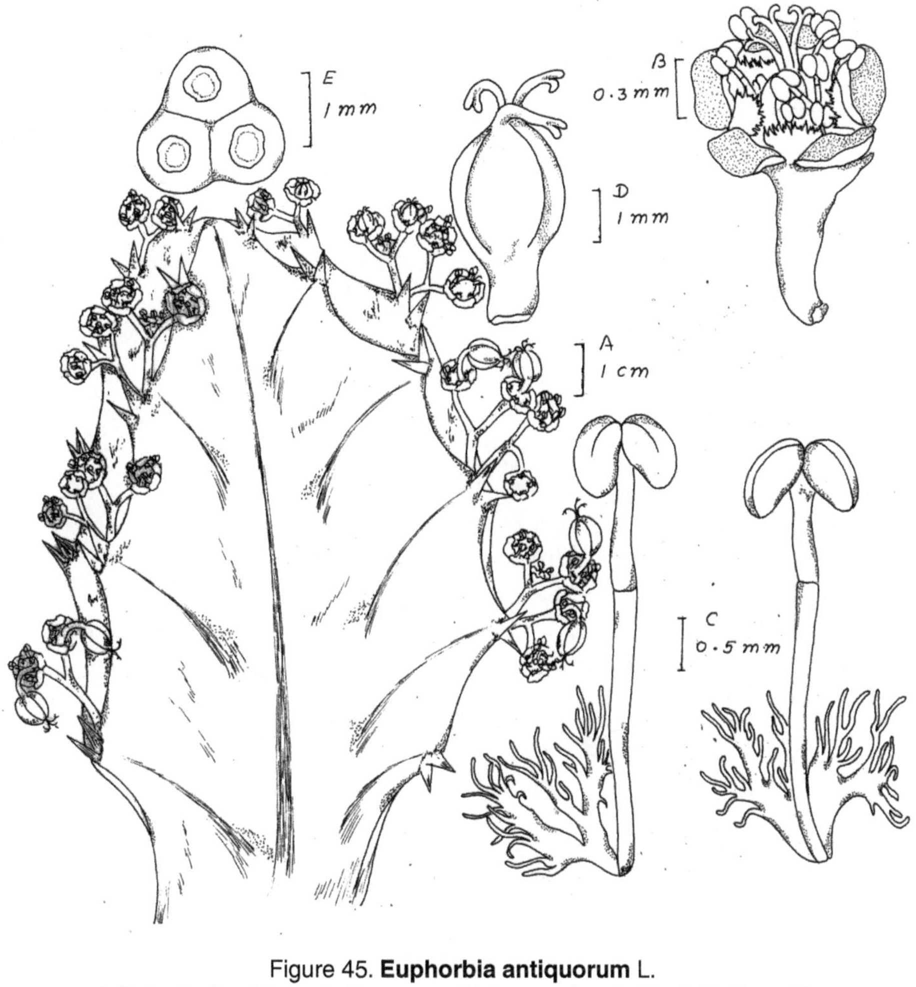

Figure 45. **Euphorbia antiquorum** L.
A. Twig, B. Cyathium, C. Stamens with bracteoles, D. Pistil, E. Ovary T.s.

Figure 46. **Euphorbia caducifolia** Haines
A. Twig, B. Inflorescence, C. Cyathium, D. Stamens with bracteoles, E. Ovary T.s.

Common in scrub jungles, waste lands and rocky hills. Fl. & Fr.: March – May. Vern.: Tel.: *Katte jemudu, Kola jamudu*; Ori.: *Khirsiju.*

Muchukota (ATP), *TP* & *NY* 964; Singanamala (ATP) *PSPB* 1184; Palakonda (CDP), *CS* 9322; V.P. South (GNT), *VRK* 3954; Owk (KNL), *RVR* 3048; Srisailam (KNL), *JLE* 23785; Chirukonda (VSKP), *GVS* 29730 (MH).

INDIA: Rajasthan, Gujarat, Madhya Pradesh, Bihar, Odisha, Chhattisgarh, Andhra Pradesh and Maharashtra.

WORLD: Pakistan.

Euphorbia cattimandoo Elliot ex Wight, Icon. Pl. Ind. Orient. 6: t. 1993. 1853; Gamble 2: 1277. 1925. *E. trigona* Hook.f., Fl. Brit. India 5: 256. 1887 *p.p.;* Binojk. & N.P. Balakr., Genus *Euphorbia* in India 302. 2010 & in N.P. Balakr. *et al.*, Fl. India 23: 321. 2012.

Succulent shrub or small tree, to 3 m high; branchlets 5-winged, with stipular divaricate persistent sharp spines. Leaves petiolate, 2-5 × 0.6-1.5 cm, oblanceolate, base acute, margin entire, apex mucronate, subsessile-sessile. Cyathia solitary, in diads or triads, involucres cupular, 4 × 5 mm, bracts opposite, below the involucres, oblong, fimbriate at apex, lobes sheathing at base, laciniate, glands 5. Male florets in small fascicles or dispersed, anthers subglobse, bracteoles laciniate. Female florets: gynophores to 1 cm long, protruding out of the involucres cup, ovary subglobose, to 4 mm across, glabrous, styles 3, connate at base, free above, stigma papillose. Capsule subglobose, 7 mm across; seeds ovoid, blackish brown.

Occasional along the forest paths and the rocky hills. Fl. & Fr.: January – February. Vern.: Tel.: *Cattimandoo.*

Pinjari konda (EG), *PSPB* 11843; Y. Ramavaram (EG), *PSPB* 11844; Addatheegala (EG), *NSD* 31429; Domial (KNL), *JSG* 18718 (MH); Talantrakota (SKLM), *GVS* 62307 (MH).

INDIA: Andhra Pradesh. Endemic.

Euphorbia clarkeana Hook.f., Fl. Brit. India 5: 253. 1887; Cooke, Fl. Pres. Bombay 2: 578. 1906; Radcliff-Smith in Fl. Pakistan 172: 102, f. 19 A – C. 1986; N. P. Balakr. & Chakrab., Fam. Euphorbiaceae India 252. 2007; Binojk. & N. P. Balakr., Genus *Euphorbia* in India 106. 2010 & in N.P.Balakr. *et al.*, Fl. India 23: 280. 2012.

Annual prostrate herbs, up to 25 cm long; stems several to many from a woody crown; branches pink to purplish, pubescent, nodes thickened, internodes 0.5 – 0.7 cm (floral branches 0.2–0.5 mm) long. Leaves two types, sub sessile, elliptic – oblong, base oblique, margin entire, apex apiculate, coriaceous, 0.4 – 0.5 × 0.2 – 0.3 cm, (microphyll 1.21 – 2.96 × 0.6 – 1.4 mm), glabrous, midnerve prominent below; petioles ca. 0.36 mm (microphyll 0.32 mm), glabrous; stipules interpetiolar, linear lanceolate, 0.96 mm, (microphyll 0.6 mm), with glandular hairs; oval-shaped glands present at nodes, ca. 0.45 – 0.3 mm (microphyll 0.15 mm), pinkish, puberulous. Cyathia solitary,

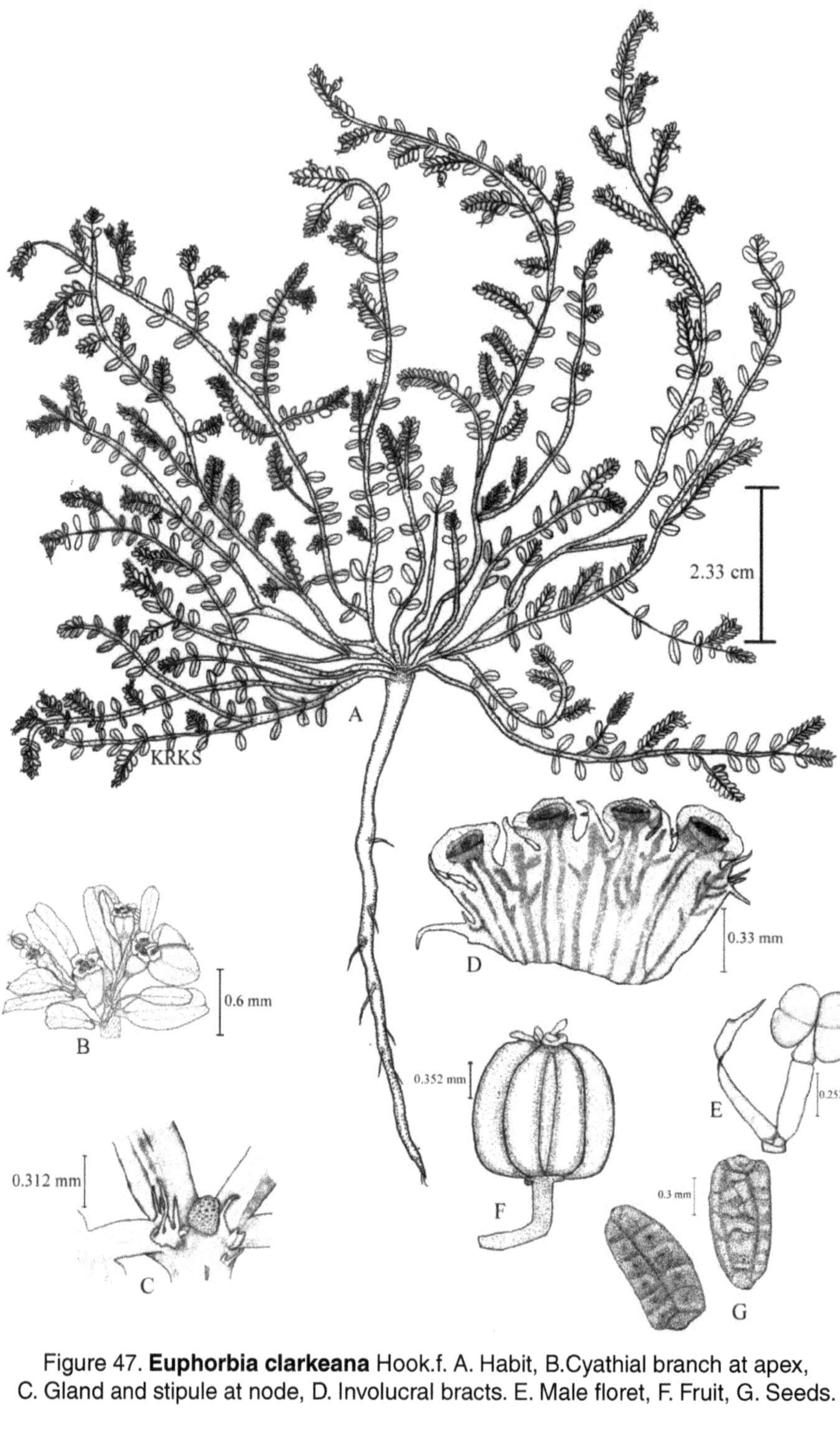

Figure 47. **Euphorbia clarkeana** Hook.f. A. Habit, B.Cyathial branch at apex, C. Gland and stipule at node, D. Involucral bracts. E. Male floret, F. Fruit, G. Seeds.

axillary, 3 – 5 on congested by reduction of subtending leaves, often arranged in microphyllous branches or lateral branchlets, 1.26 – 1.89 mm, glabrous, peduncle 0.45 – 0.36 mm; involucre campanulate, 1.26 × 0.81 mm, glabrous, lobes 5, oblong, laciniate, glabrous, glands 4, shortly stalked, 0.11 × 0.23 mm, oblong, cupular, pink, limb of gland inconspicuous, 0.32 – 0.51 mm dia., white, margin wavy. Male florets: few, up to 4, 0.52 - 1 mm long, pedicels ca 0.32 – 0.54 mm long, glabrous, anther subglobose, 0.23 × 0.3 mm, transversely dehiscing, bracteoles lanceolate, simple, 0.3 – 0.54 mm long. Female florets: gynophore 1.26 – 1.51 mm long, glabrous, ovary 0.53 – 0.63 mm long, glabrous, style 3, 0.36 – 0.583 mm long, pink, bifid from the middle. Capsule subglobose, 3-angled, 1.36 × 1.06 mm; seeds quadrangular, subovoid – oblong, ovoid, ca. 1.15 – 1.21 × 0.69 – 0.72 mm, transversely 4-furrowed, reddish brown, white coated, all facets with 5 - 7 transverse ridges.

Rare in cultivated fields, roadsides and open waste lands in Kadapa and Anantapuram districts. Fl. & Fr.: April – December.

Thalupula (ATP), *KRKS* 41855; Daduluru near Mamillapalli (ATP), *NSD* 31414; Boggudupalle (CDP), *KRKS* 41960.

INDIA: Jammu & Kashmir, Himachal Pradesh, Punjab, Hryana, Delhi, Rajasthan, Gujarat, Uttar Pradesh, Uttarakhand, Maharashtra, Karnataka and Andhra Pradesh.

WORLD: Afghanistan and Pakistan.

Euphorbia corrigioloides Boiss. Cent. Euphrob. 11. 1860 & in DC., Prodr. 15: 32. 1862; FBI 5: 251. 1887; Gamble 2: 1275. 1925; Binojk. & N.P. Balakr., Genus *Euphorbia* in India 153. 2010 & in N.P.Balakr. *et al.*, Fl. India 23: 282. 2012.

Prostrate herb; branches from a stout root stock. Leaves 0.6-2 × 0.4-0.7 cm, base obliquely truncate, margin serrulate, apex obtuse, opposite, single-nerved, blotched with red; petiole 1 mm; stipules triangular, laciniate. Cyathia 3-10, solitary in the upper axils or in terminal clusters; involucre 2.2 × 2 mm, broadly campanulate; lobes laciniate, thinly pubescent; glands 4, equal, suborbicular; appendages orbicular, subequal, white-pinkish. Male florets 6-10; bracteole laciniate; stalk 0.9 mm; anther 0.4 mm across. Female floret laterally pendulous; ovary 1.2 × 1mm, thinly pubescent, styles 3, each 2-fid from the base. Capsule 2.5 mm, thinly pubescent; seeds 1.5 mm, 4-angled, smooth

A common weed of cultivated fields especially in black cotton soils. Fl. & Fr.: July – February. Vern.: Tel.: *Errapala alumu.*

Velugodu (KNL), *SKW* 4780, 4799 (BLAT); Regadi stream (SKLM), *GVS* 62481 (CAL).

INDIA: Andhra Pradesh, Maharashtra, Karnataka, Kerala and Tamil Nadu. Endemic.

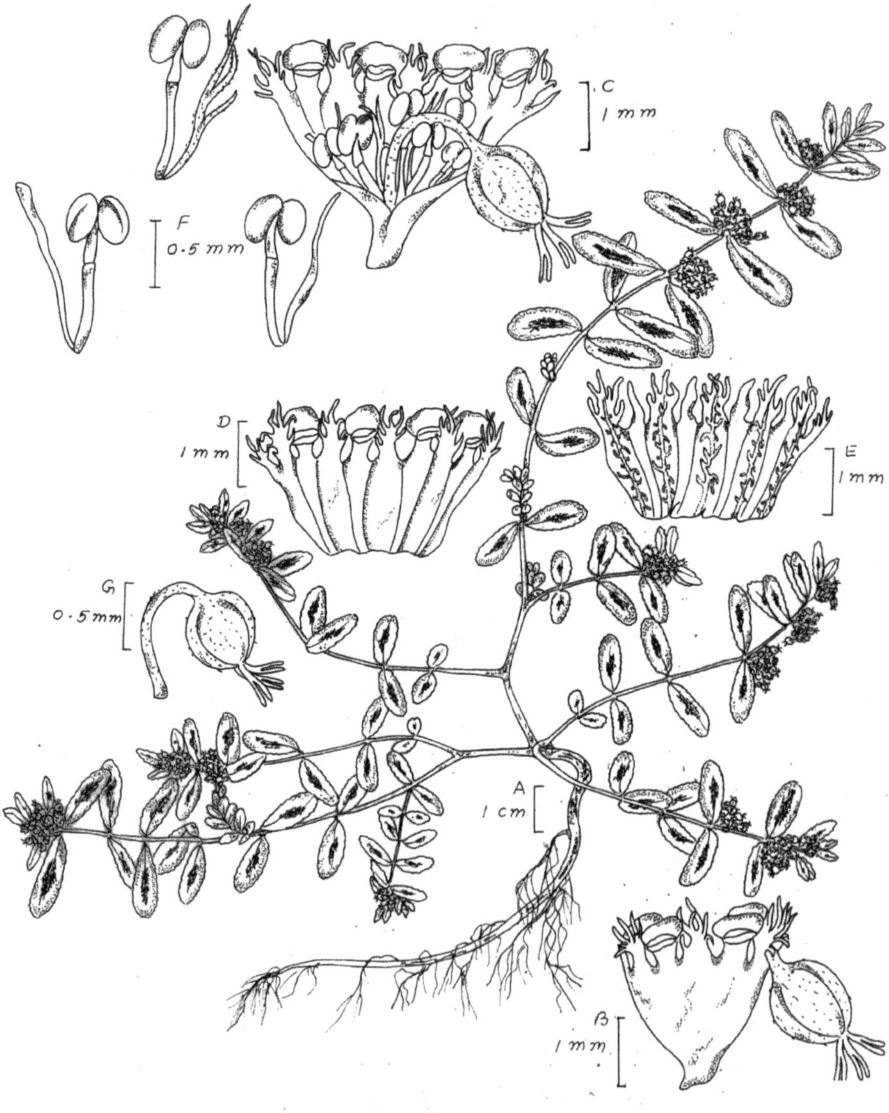

Figure 48. **Euphorbia corrigioloides** Boiss.
A. Habit, B. Cyathium, C. Cyathium split open, D. Involucre split open (with glands), E. Involucre split open (without glands), F. Stamens with bracteoles, G. Pistil.

Euphorbia cristata B. Heyne ex Roth, Nov. Pl. Sp. 226. 1821; FBI 5: 247. 1887; Gamble 2: 1274. 1925; Binojk. & N.P. Balakr., Genus *Euphorbia* in India 136. 2010 & in N.P.Balakr. *et al.*, Fl. India 23: 282. 2012. *Chamaesyce cristata* (Heyne ex Roth) Webster, J. Arnold Arbor.48: 424. 1967.

Annual, erect, ascending or prostrate and spreading herb, up to 35cm high, dichotomously branched; branches striate, pubescent. Leaves 0.5-1 × 0.4-0.8 cm, broadly ovate-oblong, base obliquely cordate, margin serrulate, apex acute, opposite, densely crowded at apex, sparsely hairy; petiole 1 mm; stipules ovate, obtuse; floral leaves distichous. Cyathia 1-3, axillary and subterminal; involucre 2.2 × 4 mm, campanulate; lobes triangular, ciliate; glands 4, unequal, laciniate and ciliate. Male florets 6-8; bracteole ciliate; stalk 1.7 mm; anthers 0.3 × 0.5 mm. Female floret laterally pendulous; ovary 2 × 1.5 mm, pubescent; styles 3, each 2-fid, from near the base. Capsule 4 mm, pubescent; seeds 1.5 mm, 4-angled, furrowed, white.

Occasional in fallow fields, in crevices of exposed rocks and in forest areas. Fl. & Fr.: August-January.

Bukkapatnam (ATP), *NY* 631; Muchukota (ATP), *TP* & *NY* 978 (MH); Gavisiddeswara hills (ATP), *PSPB* 11857; Gooty hills (ATP), *JSG* 16401 (CAL); Balaparallagutta (CTR), *GVS* 46802; Talakona RF (CTR), *GVS* 46947 (MH); Balapalli (CDP), *SKW* 7635 (BLAT), *JLE* 14991 (MH); Vandalur RF (CGP), *ANH* 91813 (MH), *ANH* 47129 (CAL); Sirumalai hills (DGL), *KCB* 53403 (MH).

INDIA: Jammu & Kashmir, Uttar Pradesh, Madhya Pradesh, Maharashtra, Andhra Pradesh, Karnataka, Kerala and Tamil Nadu.

WORLD: Sri Lanka, Bangladesh, Myanmar and Thailand.

Euphorbia cyathophora Murray, Comment, Soc. Regiae. Sci. Gott. 7: 81. 1786; Binojk. & N.P. Balakr., Genus *Euphorbia* in India 355. 2010 & in N.P.Balakr. *et al.*, Fl. India 23: 329. 2012. *Poinsettia cyathophora* (Murr.) Klotzsch & Garcke, Monastber. Deutsch. Akad. Wiss. Berin 1859: 253. 1859. *Euphorbia heterophylla* auct. non L.; Gamble 2: 1278. 1925.

Annual or biennial, erect herb, up to 1.5 m high, dichotomously branched; branchlets terete, striate. Leaves 5-7 × 1.5.3 cm, obovate-oblanceolate or panduriform, acute-attenuate at base, entire-serrate, acute, alternate below, opposite-whorled above, floral leaves with scarlet red or yellow patch at basal half; petiole to 1 cm. Cyathia 12-15, in terminal clustered pseudodichasia, each forking ca 4 times; involucre 3.5 × 3.8 mm, campanulate; lobes broadly triangular, laciniate; gland 1, bilabiate, transversely oblong; appendages 0. Male florets 20-30; bracteole laciniate, ciliate; stalk 3 mm; anther 0.8 × 1 mm. Female floret laterally subpendulous, ovary 2 mm; styles 3, each 2-fid, at above. Capsule trigonous to subglobose, deeply 3-lobed, 3.5 mm, glabrous or sparsely puberulous; seeds 2.5 mm, globose, tuberculate, greyish black.

Occasional weed in moist places, sometimes in the cultivated fields and gardens. Fl. & Fr.: June-January.

INDIA: Usually cultivated, some times as an escape through out India.

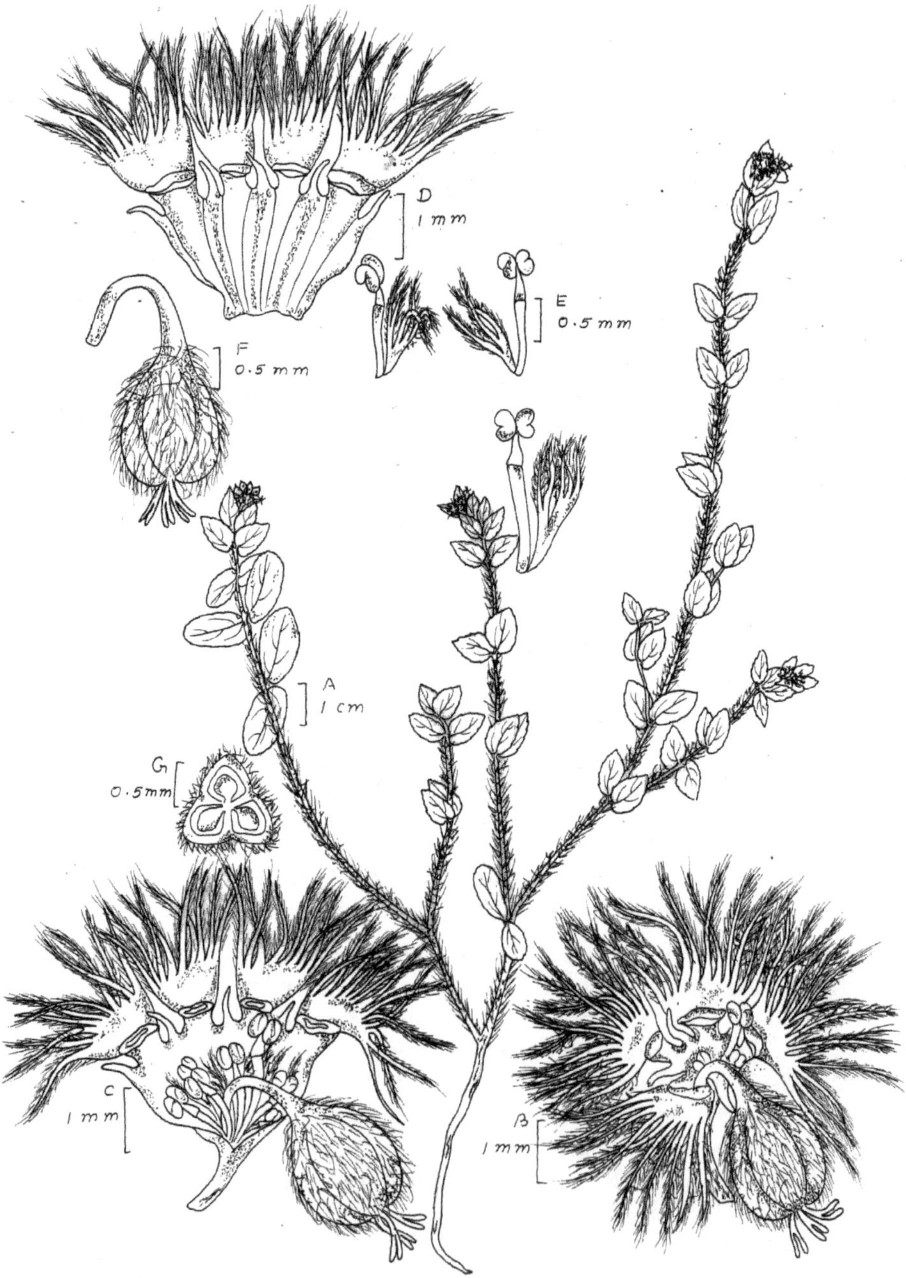

Figure 49. **Euphorbia cristata** Heyne ex Roth
A. Twig, B. Cyathium, C. Cyathium split open, D. Involucre split open, E. Stamens with bracteoles, E. Pistil, F. Ovary T.s.

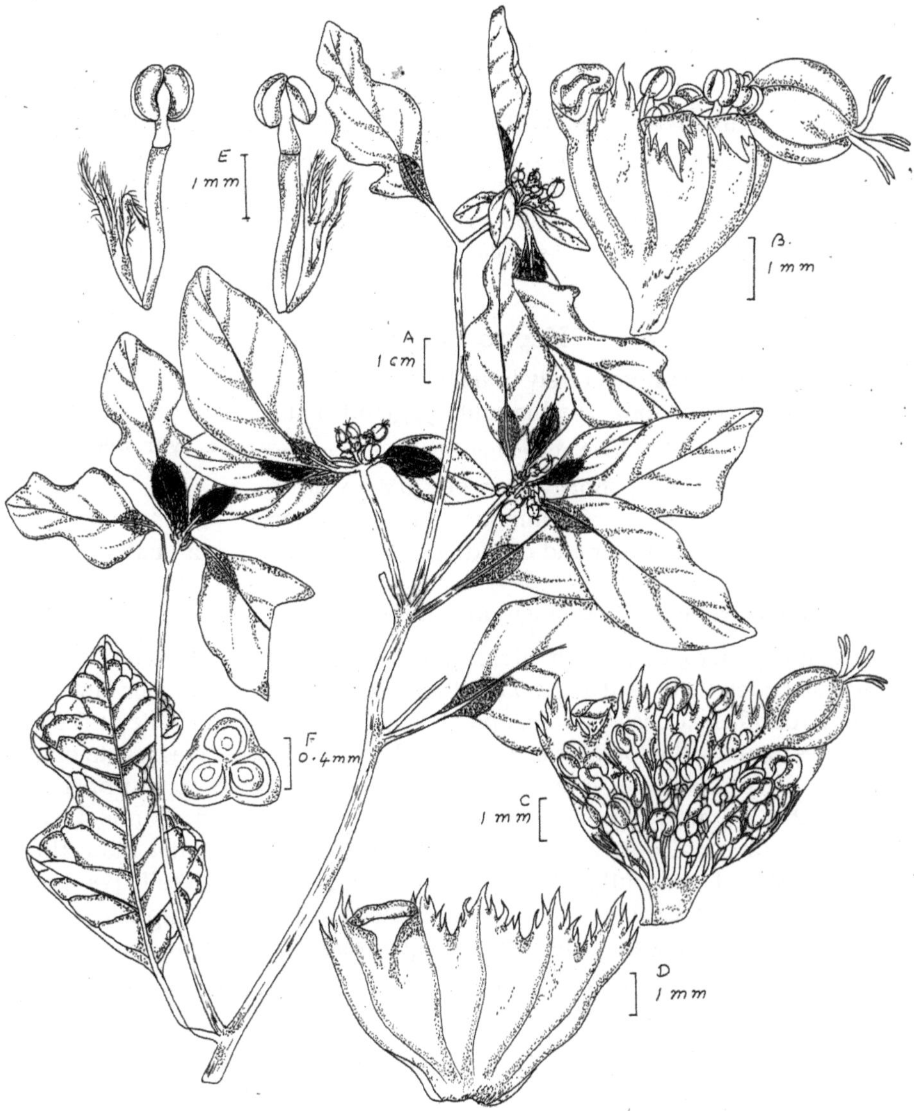

Figure 50. **Euphorbia cyathophora** Murr.
A. Twig, b. Cyathium, C. Cyathium split open, D. Involucre split open (without), E. Stamens with bracteoles, F. Ovary T.s.

WORLD: Native of Tropical and Subtropical America. Widely seen in Old World Tropics.

Note: *Euphorbia cyathophora* Murray superficially resembles *E. heterophylla* L. However, the two are distinguished by the former having floral leaves brightly coloured, involucral glands bilabiate and seeds sharply tuberculate and non-glandular, while the later has floral leaves green, involucral glands circular and seeds bluntly tuberculate and angular.

Euphorbia deccanensis V.S.Raju, Taxon 34: 519. 1985; Binojk. & N.P. Balakr., Genus *Euphorbia* in India 181. 2010 & in N.P.Balakr. *et al.*, Fl. India 23: 283. 2012. *E. linearifolia* Roth, Nov. Pl. Sp. 224. 1821; FBI 5: 249. 1887; Gamble 2: 1276. 1925; Surya Prakash Babu in Pullaiah & Moulali, Fl. Andhra Pradesh 2: 864. 1997. *Chamaesyce linearifolia* (Roth) J. Sojak, Cas. Nar. Muz. (Prague) 140(3-4): 169. 1972.

Key to Varieties

1. Rootstock slender; stems rarely branched; internodes 3-6 mm long; cyathia solitary, glands flat and transversely oblong .... var. **deccanensis**
1. Rootstock stout; stems shortly branched at apex or unbranched; internodes 5-20 mm long; cyathia solitary or paired; glands cupular and rounded .... var. **nallamalayana**

Euphorbia deccanensis V.S.Raju var. **deccanensis**

Decumbent herb, up to 20 cm high; many branches from a stout root, branches swollen at nodes. Leaves 0.8-2 × 0.2-0.4 cm, linear-oblong or linear-lanceolate, obliquely rounded at base, serrate, apiculate, opposite, subimbricate at the ends of branches, usually the lower leaves shorter than the upper leaves, ovate-oblong; petiole to 1.2 mm; stipules interpetiolar, ovate. Cyathia solitary in upper axils; involucre 2 × 2.5 mm, campanulate; lobes triangular, fringed; glands 4, small; appendages unequal obovate, cunenate, rounded or crenulate. Male florets bracteolate; stalk 1 mm; anther 0.25 × 0.35 mm. Female floret laterally pendulous; ovary 0.8 mm; styles 3, each 2-fid, from above the middle, stigma capitate. Capsule subglobose, 2 mm, smooth, glabrous; seeds quadrangular, pale brown.

Occasional in hill slopes and plains of Southern Eastern Ghats. Fl. & Fr.: August -November.

Hindupur (ATP), *PSPB* 11849; Guvvalacheruvu (CDP), *NSD* 31426; Palamaneru (CTR), *CECF* 4311 (CAL).

INDIA: Andhra Pradesh, Karnataka, Keraala and Tamil Nadu. Endemic.

Euphorbia deccanensis V.S.Raju var. **nallamalayana** (Ellis) V.S. Raju, Taxon 34: 520. 1985; Binojk. & N.P. Balakr. in N.P.Balakr. *et al.*, Fl. India 23: 284. 2012. *E. linearifolia* Roth var. *nallamalayana* Ellis, Bull. Bot. Surv. Ind. 8: 341-346. 1967.

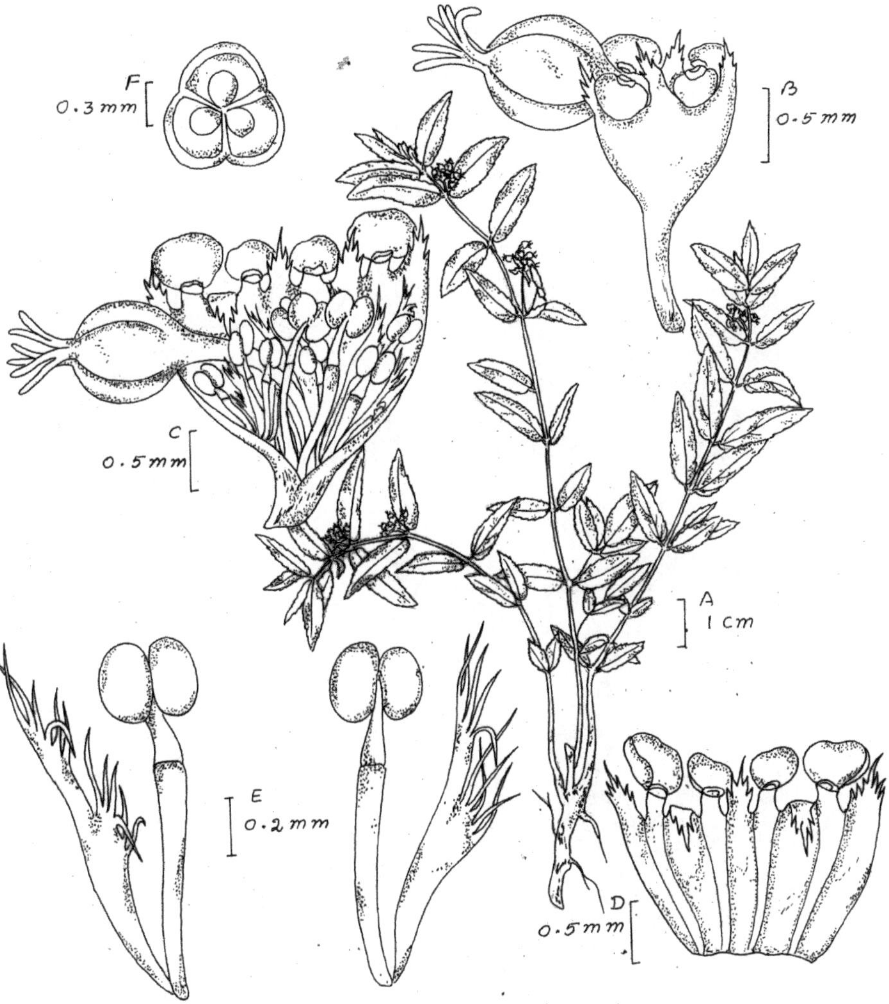

Figure 51. **Euphorbia deccanensis** V.S.Raju var. **deccanensis**
(Syn.: *Euphorbia linearifolia* var. *linearifolia*)
A. Twig, B. Cyathium, C. Cyathium split open, D. Involucre split open,
E. Stamens with bracteoles, F. Ovary T.s.

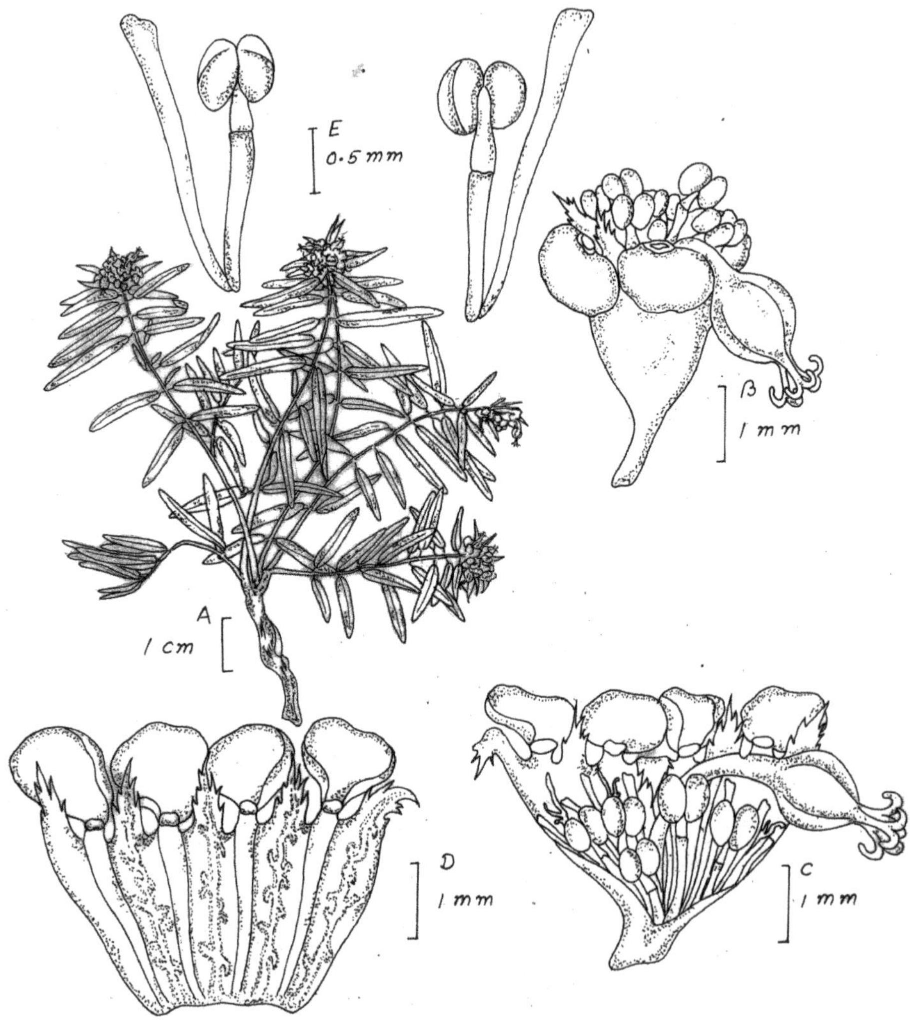

Figure 52. **Euphorbia deccanensis** V.S.Raju var. **nallamalayana** (Ellis) V.S. Raju (Syn.: *Euphorbia linearifolia* Roth var. *nalamalayana* Ellis)
A. Twig, B. Cyathium, C. Cyathium split open, D. Involucre split open, E. Stamens with bracteoles.

Chamaesyce linearifolia (Roth) J. Sojak var. *nallamalayana* (Ellis) V.S.Raju & P.N. Rao, Phytologia 37: 454. 1977.

Decumbent or spreading herb, to 15 cm high; many branches from a stout root; branches swollen at nodes. Leaves 1-3 × 0.15-0.3 cm, linear-linear lanceolate, base obliquely subcordate, margin inconspicuously serrulate, apex apiculate, opposite, subimbricate at the ends of branches; petiole to 1 mm; stipules interpetiolar, slightly laciniate. Cyathia solitary at upper axils; involucre 3.5 × 4 cm, campanulate; lobes triangular, fimbriate; glands 4, equal, suborbicular; appendages equal, orbicular. Male florets: bracteole laciniate; stalk 1.3 mm; anther 0.5 mm across. Female floret: laterally pendulous; ovary 1 mm, styles 3, each 2-fid, from above the middle. Capsule 2 mm; seeds obtusely 4-angled, smooth, pale brown.

Rare in the forests. Fl. & Fr.: January – November.

Chinnagunitam (KNL), *PSPB* 9141; Guvvalacheruvu (CDP), *RVR* 1589; Chelama (KNL), *JLE* 16774, 23711 (MH); way to Balugram, Nallamalais (KNL), *JLE* 32591 (MH); Bairanidokku (KNL), *JLE* 42238 (CAL); East of Nerabylu (CTR), *GVS* 46978 (CAL, MH).

INDIA: Andhra Pradesh. Endemic.

Euphorbia dracunculoides Lam., Encycl. 2: 428. 1788; FBI 5: 262. 1887; Gamble 2:1278. 1925; Binojk. & N.P. Balakr., Genus *Euphorbia* in India 355. 2010 & in N.P.Balakr. *et al.*, Fl. India 23: 305. 2012.

Annual or sometimes perennial herb, to 50 cm high, much dichotomously branched from root stock; branchlets terete, striate. Leaves 1-5 × 0.2-0.5 cm, linear-linear lanceolate, base acute, margin entire, apex subacute, alternate below, opposite above, sessile, palminerved; floral leaves opposite, 2-3 × 0.3-0.6 cm. Cyathia solitary, terminal, subsessile; involucre 2.2 × 3.5 mm, broadly campanulate, thinly pubescent within; lobes short, triangular, ciliate; glands 5, semilunate, 2-horned; appendages 0. Male florets 6-8; bracteole laciniate, ciliate; stalk 2.5 mm; anthers 0.5 × 0.6 mm. Female floret erect; ovary 2.3 mm, styles 3, each 2-fid, from above the middle. Capsule subglobose, roundly trigonous, 3.5 mm, keeled, smooth, glabrous; seeds 2.5 mm, ellipsoid, minutely tuberculate.

Common in wastelands and in cultivated fields. Fl. & Fr.: December-March. Vern.: Tel.: *Tillakada.*

Kondapudi (EG), *CAB* 5207 (MH).

INDIA: Punjab, Haryana, Rajasthan, Uttar Pradesh, Bihar, West Bengal, Odisha, Chhattisgarh, Madhya Pradesh, Gujarat, Maharashtra, Andhra Pradesh, Karnataka, Kerala and Tamil Nadu.

Euphorbia elegans Spreng., Syst. Veg. (ed. 16) 3: 794. 1826; FBI 5: 246. 1887; Gamble 2: 1274. 1925; Binojk. & N.P. Balakr., Genus *Euphorbia* in India 355. 2010 & in N.P.Balakr. *et al.*, Fl. India 23: 284. 2012. *Chamaesyce elegans* (Spreng.) Sojak, Cas. Nar. Muz. (Prague) 140: 169. 1972.

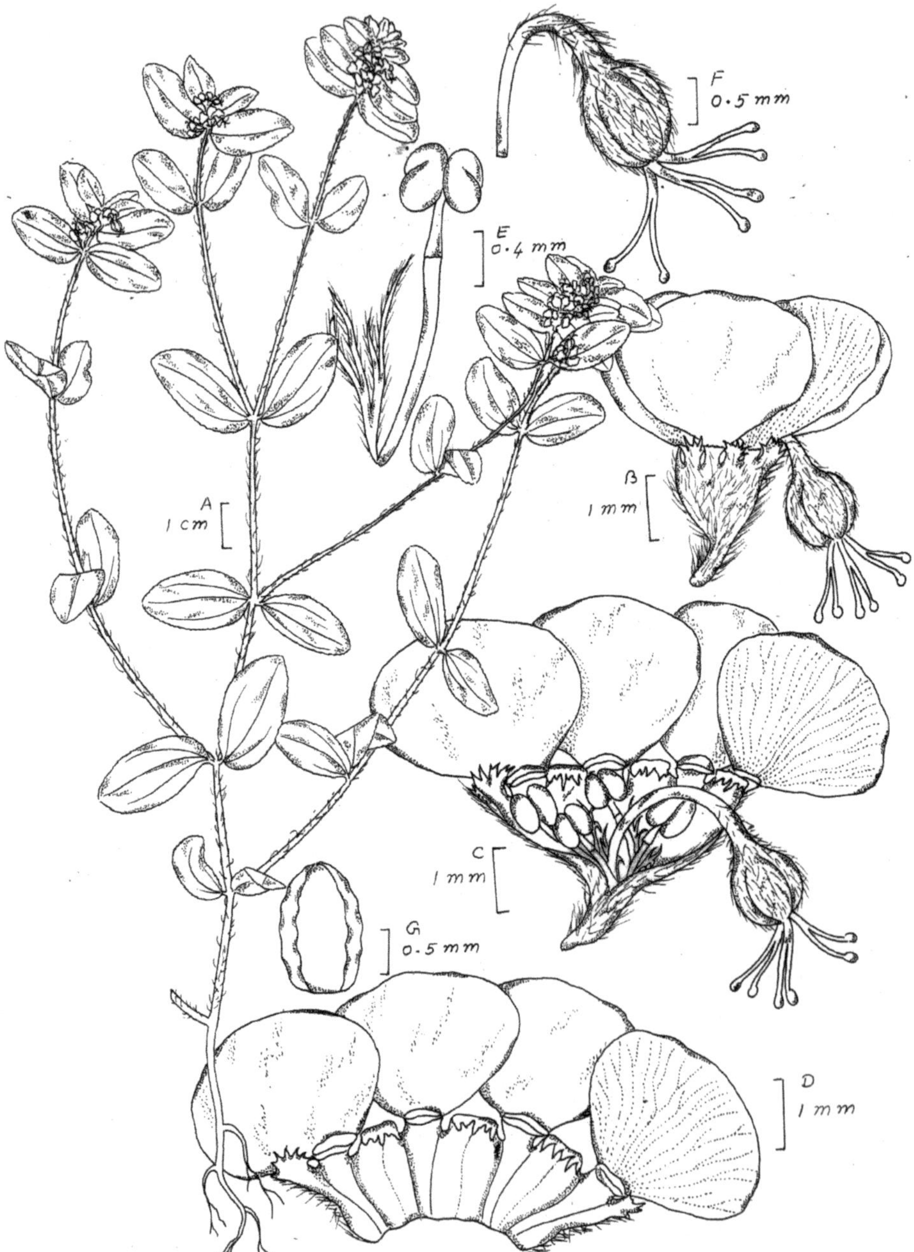

Figure 53. **Euphorbia elegans** Spreng.
A. Twig, B. Cyathium, C. Cyathium split open, D. Involucre split open, E. Stamen with bracteole, F. Pistil, G. Seed.

Ascending herb, to 40 cm high, dichotomously branched; branchlets clothed with long scattered soft hairs. Leaves 1.5-2.8 × 0.7-1.5 cm, elliptic-oblong, base obliquely truncate-cordate, margin serrulate, apex obtuse, opposite, subsessile; floral leaves distichously imbricate, broadly elliptic or obovate, serrulate, hairy below, forming short spikes at the ends of the branchlets. Cyathia 1-3, axillary or subterminal within the imbricate leaves; involucre campanulate-turbinate, hairy; lobes forked, hairy; glands 4, equal, transversely oblong; appendages equal, larger than the glands, hairy within, near the base, sinuate at apex. Male florets 18-20; bracteole laciniate, ciliate; stalk 2.4 mm; anthers 0.4 × 0.6 mm. Female floret laterally pendulous; ovary 0.8 × 1 mm, subglobose, hairy; styles 3, each 2-fid, from near the base, capitate. Capsule 3.5 mm, hairy; seeds 1.5 mm, 4-angled, furrowed.

Occasional in foothills. & Fr.: December- March.

Balugram (KNL), *TP* & *RVR* 2739; Chalama (KNL), *PSPB* 9996, 11820, *JLE* 17943 (MH); Diguvametta (KNL), *SKW* 5372, 5374 (BLAT), *CAB* 8009 (MH); Palakonda (CDP), *EKK* 77826 (DD); Thummalabailu (CDP), *SKW* 7784 (BLAT); Sambalpur (SBP), *HFM* 1456 (DD); Peria Kalrayans (SM), *KM* 5403 (RHT).

INDIA: Odisha, Chhattisgarh, Madhya Pradesh, Gujarat, Maharahtra, Karnataka and Tamil Nadu. Endemic.

Euphorbia fusiformis Buchn.-Ham. ex D. Don, Prodr. Fl. Nepal 62. 1825; Boiss. in DC., Prodr. 15: 93. 1866; FBI 5: 257. 1887; Binojk. & N.P. Balakr., Genus *Euphorbia* in India 364. 2010 & in N.P.Balakr. *et al.*, Fl. India 23: 332. 2012. *E. acaulis* Roxb., Fl. Ind. 2: 472. 1824; Gamble 2: 1276. 1925; Surya Prakash Babu in Pullaiah & Moulali, Fl. Andhra Pradesh 2: 856. 1997.

Herb, stem reduced to a stout underground rootstock, fusiform, 10-15 × 4-6 cm. Leaves in apical rosettes, obovate or oblanceolate, base cuneate, margin crenulate, apex obtuse-subacute, fleshy, subsessile. Cyathia in cymes, few, on long peduncles, appears after the leaves have fallen; bracts at the forks, 2 × 2.5 mm, ovate – lanceolate, scarious; involucre 3 × 3.5 mm, broadly cupular; lobes triangular, fringed; glands 4-5, oblong. Male florets many, bracteolate; stalk 2 mm; anther 1 mm across. Female floret erect - subpendulous; ovary 1.5 mm, styles 3, connate at base, each 2-fid at above. Capsule 6 mm; seeds 2 mm, broadly ovoid, smooth.

Rare in the forests. Fl. & Fr.: Feb. – Sept. Vern.: Tel.: *Barresepu gadda*; Ori.: *Khirkanchan*.

Batrepalle (ATP), *NSD* 31433; Nigidi (ATP), *B.Aswarthappa* 210 (BSID); Seshachalam (CDP), *PSPB* 11852; Dornala (PKM), *BSS* & *SKB* 29404; Lukke (KNL), *JLE* 2218 (MH); Balepalli (CDP), *JLE* 14266 (MH), *SKW* 7639, 7640 (BLAT).

INDIA: Odisha, Bihar, Uttar Pradesh, Rajasthan, Gujarat, Maharashtra, Andhra Pradesh, Karnataka and Tamil Nadu.

WORLD: Nepal.

Note: Boissier (1862) treated *E. acaulis* and *E. fusiformis* as distinct species and E. *humilis* Royle as a synonym of *E. fusiformis*. J. D. Hooker (1887), on the contrary, opined that *E. fusiformis* and *E. acaulis* were conspecific; he did not

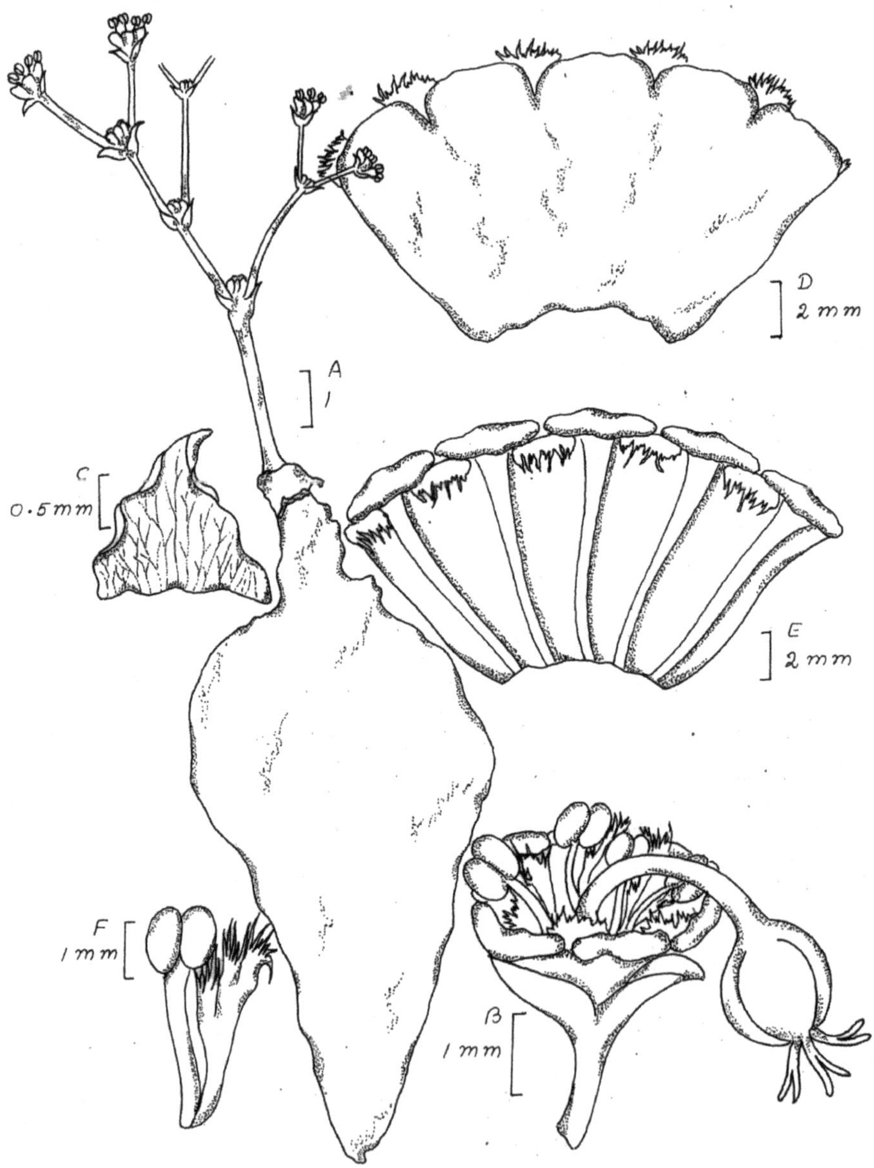

Figure 54. **Euphorbia fusiformis** Buch.-Ham. ex D. Don (Syn.: *E. acaulis* Roxb.)
A. Twig, B. Cyathium, C. Bract, D. Involucre split open (without), E. Involucre split open (within), F. Stamen with bracteole.

mention *E. humilis,* but also included *E. nana* in the synonymy. Cooke (1906) and Gamble (1925) also followed this treatment. Comparison of the original descriptions and type photographs confirms that *E. fusiformis, E. acaulis* and *E. humilis* are conspecific and that *E. fusiformis* Buch.-Ham. ex D.Don is the earliest validly published name for the species.

Euphorbia heterophylla L., Sp. Pl. 453. 1753; Binojk. & N.P. Balakr., Genus *Euphorbia* in India 355. 2010 & in N.P.Balakr. *et al.,* Fl. India 23: 330. 2012. *E. geniculata* Orteg., Nov. Rar. Pl. Matrit. 18. 1797; FBI 5: 239. 1887; Gamble 3: 1884. 1936.

Annual or biennial, erect herb, to 60 cm high, with white latex, dichotomously branched; branchlets terete, striate. Leaves 5-10 × 1.5-5 cm, obovate-rhomboid, base acute, margin serrate, apex acute, sparsely hairy, alternate below, opposite above; petiole to 2.5 cm. Cyathia 6-10, in terminal clusters; involucre 2.5 × 3 mm, campanulate; lobes laciniate, ciliate; gland 1, cupular; appendages 0. Male florets 12-15; bracteole ciliate; stalk 1-2 mm; anther 0.7 mm across. Female floret laterally pendulous; ovary 1.8 mm across; styles 3, each 2-fid, from above the middle. Capsule trigonous-subglobose, 5 mm; seeds 2 mm, 3-angled, tuberculate, brownish-black.

Common weed of wastelands and cultivated fields. Fl. & Fr.: Throughout the year.

Penakacherla (ATP), *NY* 594; Thummalapalle (CDP), *TP & Party* 33668; Tirumala (CTR), *PSPB* 9119; Mellavagu (GNT), *VRK* 5730; Maddileti stream (KNL), *RVR* 3065; Horseley hills (CTR), *SKW* 6313 (BLAT); Anantagiri (VSKP), *NPBK* 833 (CAL); way to Upper Ahobilam (KNL), *JLE* 25571 (MH); Araku (VSKP), *HS* 20446 (BLAT); Below Deomali (KPT), *SX & MB* 6580 (RRL-B); Lanjigarh (KHD), *MB & Dhal* 7392 (RRL-B); Pottangi (KPT), *HFM* 3884 (DD); Brooks hill (SBP), *SP & APD* 1162 (CAL); Below Deomali (KPT), *SX & MB* 6580 (RRL-B); Kollimalai (TP), *KCJ* 11354 (MH).

INDIA: Through out the country.

WORLD: Native of Tropical and Subtropical America. Widely seen in Old World Tropics.

Euphorbia heyneana Spreng., Syst. Veg. (ed.16) 3: 791. 1831. (as '*heyniana*') subsp. **heyneana;** Binojk. & N.P. Balakr., Genus *Euphorbia* in India 115. 2010 & in N.P.Balakr. *et al.,* Fl. India 23: 286. 2012. *E. microphylla* Heyne ex Roth, Nov. Pl. Sp. 229. 1821, auct. non Lam. 1788; FBI 5: 252. 1887; Gamble 2: 1276. 1925. *E. orbiculata* Miq., Fl. Ind. Bot. 1: 421. 1859, (non H.B.K. 1817). *E. bombaiensis* Santapau, Bull. Bot. Soc. Bengal 8:17. 1955, excl. syn. *Chamaesyce heyneana* (Spreng.) Sojak, Cas. Nar. Muz. (Prague) 140: 169. 1972.

Postrate herb, up to 15 cm long, stems many, spreading, slender, filiform, glabrous, internodes grey or reddish brown. Leaves opposite, subsessile, 3.5 × 1.5-3 mm, obovate-rounded or oblong, base obliquely truncate-cordate, margin entire, apex subacute-retuse, stipules 2-partite, laciniate. Cyathia solitary, axillary; involucre 0.9 × 1 mm, campanulate; lobes triangular, apiculate; glands 4; appendages minute,

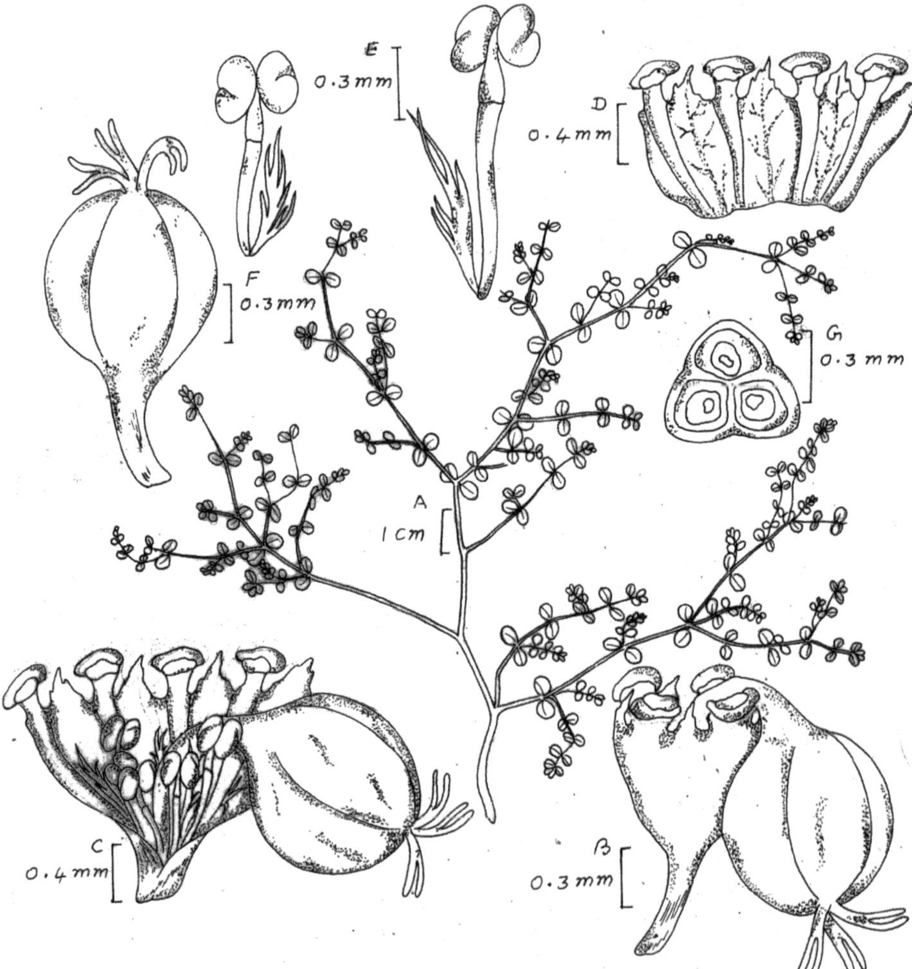

Figure 55. **Euphorbia heyneana** Spreng.
A. Twig, B. Cyathium, C. Cyathium split open, D. Involucre split open,
E. Stamens with bracteoles, F. Pistil, G. Ovary T.s.

inconspicuous. Male florets: 1-4, ebracteolate; stalk 1 mm; anthers 0.3 mm across. Female florets: laterally pendulous; ovary 0.4 mm, styles 3, each 2-fid, from above the middle. Capsule 1.5 mm, obtusely keeled; seeds 1 mm, 4-angled, faintly furrowed.

Common weed in moist localities. Fl. & Fr. December – February. Vern.: Tel.: *Alumu.*

Gutturkona (ATP), *NY* 995 (MH); Yaganti (KNL), *RVR* 1444; Chinnagunitam (KNL), *PSPB* 9142; Thungabhadra (KNL), *PSPB* 9183; Vijayapuri (GNT), *VRK* 5786; Nandyal (KNL), *SKW* 5230, 5231 (BLAT); Nellimalai RF (CBE), *KRM* 14777 (MH).

INDIA: Jammu & Kashmir, Himachal Pradesh, Punjab, Bihar, West Bengal, Odisha, Chhattisgarh, Madhya Pradesh, Gujarat, Maharashtra, Goa, Andhra Pradesh, Karnataka, Keral and Tamil Nadu.

WORLD: Africa, Pakistan, Bangladesh, Myanmar, China and Java.

Note: In 1821, Roth established this species basing on Heyne's collection from 'India Orientalis', as *E. microphylla.* As it happened to be a later homonym of *E. microphylla* Lam. (1788), Spreng. (1831) changed the specific epithet to *E. heyneana.* Santapau (1955) proposed a new name *E. bombaiensis* for this species observing that *E. heyneana* Spreng. doubtfully applies to Indian Plants. Rajagopal and Panigrahi (1968) suggested that *E. orbiculata* Kunth as the correct name for this species. However, *E. orbiculata* Kunth is a distinct species, restricted to the New World. Panigrahi (1974) cleared the confusion and recognized *E. heyneana* as the correct name for the Indian plants.

Euphorbia hirta L., Sp. Pl. 454. 1753; Gamble 2: 1275. 1925; Binojk. & N.P. Balakr., Genus *Euphorbia* in India 156. 2010 & in N.P.Balakr. *et al.,* Fl. India 23: 287. 2012. *E. pilulifera* sensu Hook.f., Fl. Brit. India 5: 250. 1887, non L. 1753.

Ascending hispid herb, to 60 cm high; branchlets often 4-angled, densely patently strigose with multicellular yellowish or brownish yellow hairs and also minutely appressed-pubescent with whitish crisped hairs. Leaves opposite, 0.5-3 × 0.6-1.5 cm, elliptic-lanceolate, base oblilquely acute-rounded, margin serrulate-serrate, apex subacute-acute; petiole to 2 mm. Cyathia clustered in dense cymose glomerate heads, axillary, at upper nodes or terminal, often tinged reddish; heads with 20-30 cyathia, single or often paired on ca 1 cm long, simple or dichotomous leafless branches, subsessile or on 2-6 mm long leafless peduncles, male or bisexual; involucre 1 × 0.8 mm, campanulate; lobes triangular, laciniate; glands 4, minute; appendages small. Male florets 4-6, ebracteolate; stalk 1mm; anthers 0.3 mm across. Female floret laterally pendulous; ovary 1 × 0.9 mm; pubescent; styles 3, each 2-fid, from base. Capsule subglobose, sharply trigonous, 1.2-1.5 mm across, pubescent; seeds oblong, 1mm, 4-angled, faintly furrowed, purplish or reddish grey, grey when mature.

Common weed along roadsides and in open forests. Fl. & Fr: Throughout the year. Vern.: Tel.: *Reddyavari nanabalu;* Ori.: *Chelduddhi, Chitakuli, Chita-kuteai, Harahari, Hariharik;* Tam.: *Ammanpacharisi.*

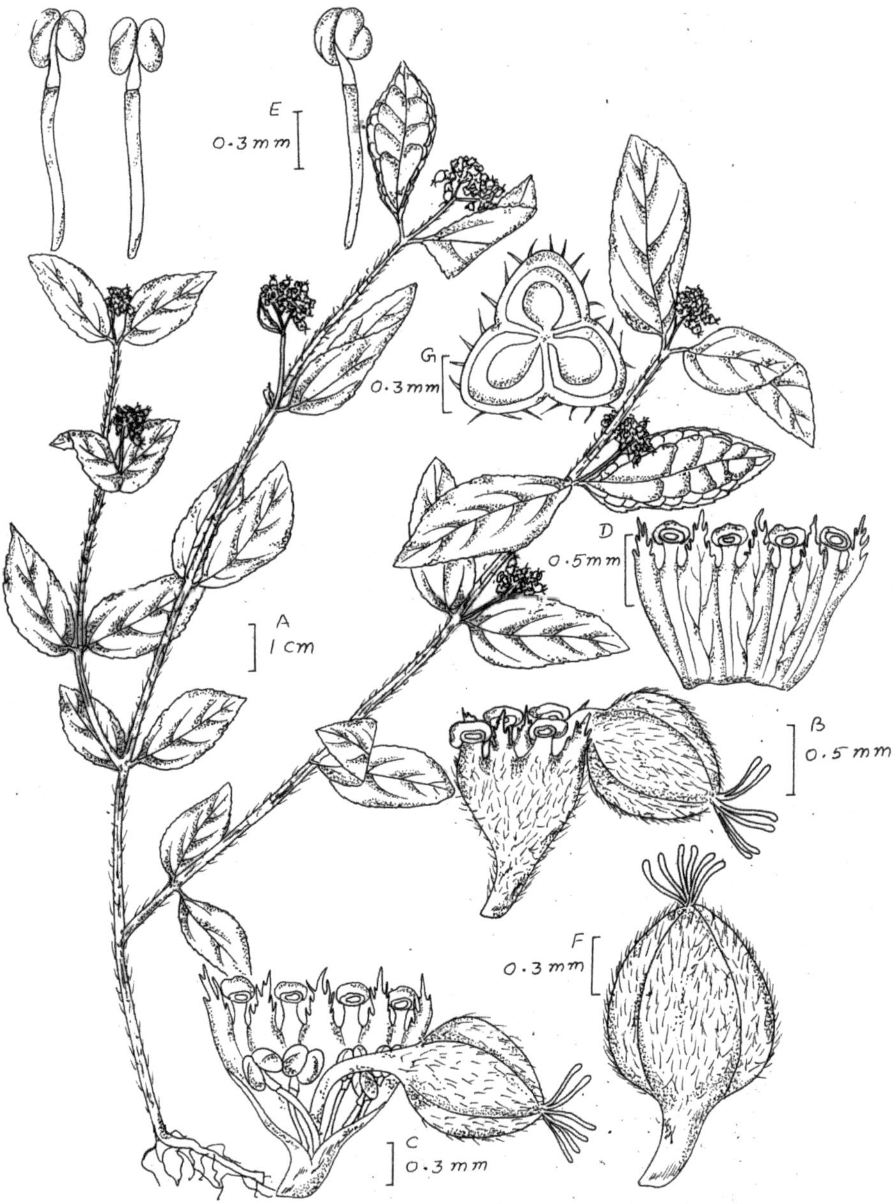

Figure 56. **Euphorbia hirta** L.
A. Twig, B. Cyathium, C. Cyathium split open, D. Involucre split open, E. Stamens, F. Pistil, G. Ovary T.s.

Maddileti stream (KNL), *RVR* 3067; Palakonda (CDP), *CS* 7796; Chintala tanda (GNT), *VRK* 3224; Balapalle (CDP), *JLE* 14297 (MH); Addatigala (EG), *VNS* 680; Kota (EG), *VNS* 524 (CAL); Maredumilli (EG), *MM* 101383 (BSID); Simhachalam (VSKP), *DCSR* 1876 (CAL); Araku valley (VSKP), *NPBK* 638 (CAL); Salur (VZN), *NPBK* 992 (CAL); Vepampatti RF (DMP), *EV* 51841 (MH); Tippucadu RF (NA), *KRM* 16644 (MH); Rangappanur RF (SA), *KRM* 77465 (MH).

INDIA: Almost through out India.

WORLD: Tropics and subtropics of Old and New World.

Note: Several floras treat this as either *E. hirta* L. or *E. pilulifera* L., or the latter as a synonym of the former. Boisser (1862), J.D.Hooker (1887), Prain (1903) and others called this species as *E. pilulifera*. Burman (1768), Duthie (1911), Haines (1925), S. Carter (1984), Radcliffe-Smith (1986) and others treat this species as *E. hirta*. Study of the Linnean protologue of *E. pilulifera* reveals that it consists of more than one species, as the reference given by him, 'Burm. Zeyl. 224.t.105.f.1' points to *E. hirta*. Neverthless, the specimen No. 630.8 Linn (LINN) designated as type of *E. pilulifera* differs in several ways from *E. hirta* and from Burman's illustration.

E. hirta resembles *E. indica* and *E.hypericifolia* in the presence of glomerulate cyathia, but differs in its characteristic yellowish hairs.

Euphorbia hypericifolia L., Sp. Pl. 454. 1753; FBI 5: 250. 1887 *p.p.*; Binojk. & N.P.Balakr. Genus *Euphorbia* in India 156. 2010 & in N.P.Balakr. *et al.*, Fl. India 23: 289. 2012.

Herbs, erect, up to 60 cm high, glabrous; stems woody, dichotomously branched, branches terete; nodes thickened; internodes 2-2.5 cm long. Leaves oblong-lanceolate, oblique at base, serrulate above the middle on margins, acute at apex, 1.5- 3.5×0.7-1 cm; lateral nerves obscure; petioles 1-6 mm long; stipules triangular. Cyathia in axillary glomerules, often with a few floral leaves; involucre campanulate, 1.6 × 0.9 mm, glabrous; lobes lanceolate 1.3 mm long, laciniate; glands 4, stalked, orbicular, ca 1 mm across, sparsely puberulous at base; limbs orbicular, entire, ca 5×4 mm, white, or pale pink. Male florets: pedicels ca 1 mm long, anthers sub-globose; bracteoles filiform, ca 1.3 mm long. Female floret: gynophores 1.8 mm long; ovary ca 2 mm long, glabrous; styles 0.6 mm long, free, deeply bifid, stigma capitate. Fruits obtusely keeled, sub- spherical, 3.2 × 3.5 mm, glabrous; seeds ovoid, 1.3 mm long, wrinkled, brown.

Weed in cultivated fields in Anantapur district. Introduced into India. Fl. & Fr.: July - November.

Mamillapalli (ATP) *NSD*, 31403.

INDIA: Jammu & Kashmir, Uttarakhand, Maharashtra, Andhra Pradesh, Karnataka, Kerala and Tamil Nadu.

WORLD: Centrl Tropical America, Mexico, Venezuela, Pakistan, Bangladesh, Sri Lanka and Myanmar.

Euphorbia indica Lam., Encycl. 2: 423. 1786; V.S.Raju & P.N. Rao, Indian J. Bot. 2(2): 205. 1979; Binojk. & N.P. Balakr., Genus *Euphorbia* in India 168. 2010 & in N.P.Balakr. *et al.*, Fl. India 23: 290. 2012. *E. hypercifolia* auct. non L. 1753; FBI 5: 249. 1887 *p.p.*; Gamble 2: 1275. 1925. *E. parviflora* auct. non L. 1759; FBI 5: 250. 1887; Surya Prakash Babu in Pullaiah & Moulali, Fl. Andhra Pradesh 2: 865. 1997.

Ascending or erect herb, to 45 cm high, sparsely pubescent; branches arising from rootstock, slender, filiform, fleshy, flexuous, branchlets often quadrangular. Leaves opposite, 1-2.8 × 0.5-1.5 cm, elliptic-broadly oblong, obliquely truncate at base, serrulate, acute-obtuse, opposite, sparsely pubescent below; petiole to 4 mm; stipulate. Cyathia 13-15, aggregated in axillary and terminal clusters; involucre 1.3 × 1.2 mm, campanulate; lobes triangular, laciniate; glands 5, minute; appendages much larger than glands. Male floret bracteole laciniate or simple; stalk 1mm; anther 0.35 mm across. Female floret laterally pendulous; ovary 1.5 × 2 mm, hairy; styles 3, each 2-fid, near the base. Capsule 1.5 mm, subglobose, hairy; seeds 1 mm, 4-angled, minutely furrose or smooth.

Common along the forest paths, foothills and in cultivated fields. Fl. & Fr.: July-March.

Kalasamudram RF (ATP), *EC* 2913; Gani fields (KNL), *TP* & *AA* 1931; Thummalapalli (CDP), *KRKS* 36916; Mellavagu (GNT), *VRK* 5733; Veldurthi RF (KNL), *RVR* 1758; Kalasamudram RF (ATP), *EC* 2913; Balapalle (CDP), *JLE* 14944 (MH); near Dongalacheruvu lake (CDP), *KS* 6320 (MH); Lakshmipuram (VSKP), *CAB* 1725 (MH); Pathalaganga, Srisailam (KNL), *JLE* 16840(MH); Melpat (SA), *CAB* 1067 (MH).

INDIA: Through out India except North-East.

WORLD: Africa and Asia.

Note: The identity of this species, since long, has been a taxonomic problem. Various authors have applied the names *E. hypericifolia, E. indica, E. bracteolaris* and *E. parviflora* to the specimens belonging to *E. indica*. Having failed to solve the problem J.D.Hooker (1887) brought these four names together under *E. hypericifolia* giving Boissier's diagnostic characters for each and left it to Indian Botanists to test their value and make the final decision. According to V.S.Raju and P.N.Rao (1979a), there is only one species in India and it is *E. indica.*

Euphorbia hypericifolia L. is a plant of New World Tropics, characterized by glabrous capsules now introduced into India (Binojkumar and Balakrishnan, 2012).

Euphorbia kadapensis Sarojinidevi & Venkataraju, Phytotaxa 181: 179. 2014.

Erect, decumbent to ascending herb (hemicryptophyte) up to 30 cm tall; latex milky, stem sympodial, dichotomously branched; branches terete, glabrous, pink, filiform towards apices, sparsely hairy, hairs hyaline; internodes 1–2.5 cm long, thickened at nodes; stipules scaly, connate, 0.1 cm long, shortly laciniate. Leaves simple, opposite, equal, inequilateral, subsessile; petiole 0.1 cm long, sparsely hairy;

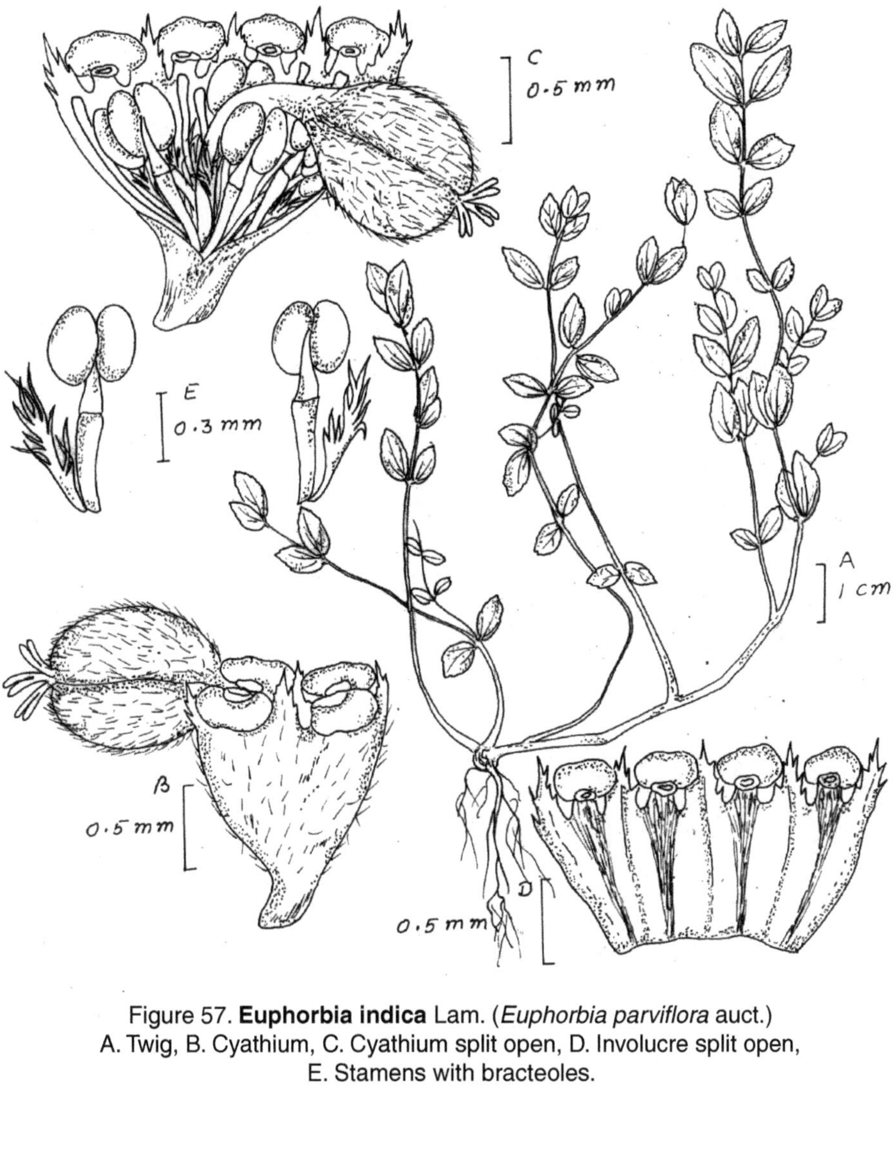

Figure 57. **Euphorbia indica** Lam. (*Euphorbia parviflora* auct.)
A. Twig, B. Cyathium, C. Cyathium split open, D. Involucre split open,
E. Stamens with bracteoles.

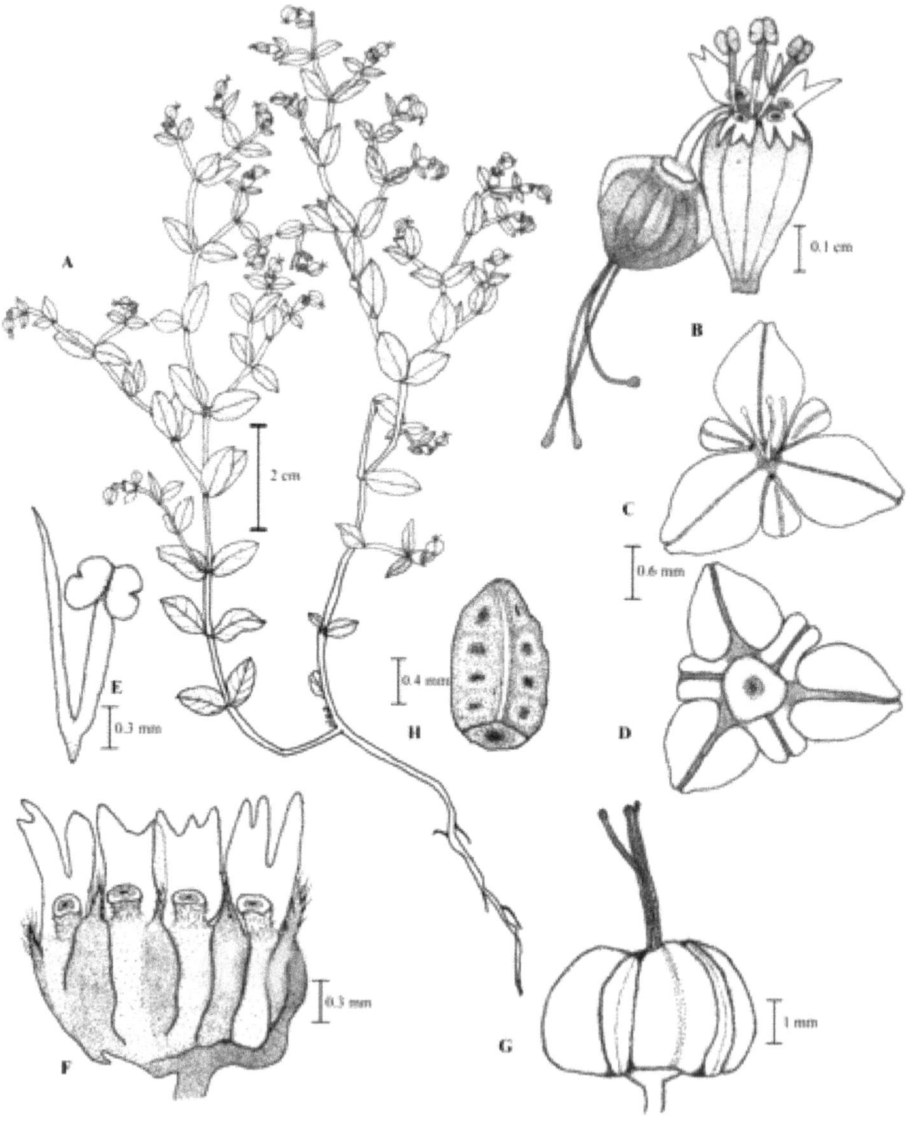

Figure 58. **Euphorbia kadapensis** Sarojinidevi & Venkataraju.
A. Habit. B. Cyathium. C. Ovary, top view. D. Capsule, bottom view.
E. Staminate flower, with bracteole. F. Involucre split open, revealing the four glands with their petaloid limbs (note the hispid involucral lobes and stipitate of glands).
G. Capsule with persistent styles and conspicuous ridges in between cocci. H. Seed.
(After Sarojinidevi and Venkataraju 2014).

lamina ovate to elliptic, 1.0–2.5 × 0.4–0.6 cm, glabrous, oblique at base, margin serrulate and more so distally, apiculate, midrib prominent, nerves slightly distinct on adaxial surface, in 6–8 pairs. *Cyathia* terminal, at times axillary, 1–2, 0.6–0.9 cm long; peduncles 0.2 cm long; involucre turbinate, 4 × 2 mm, glabrous, alternately pink and green; involucral lobes 5, laciniate; appendages of glands 4, 2–3-fid up to half of its length and occasionally to the base, 2 × 1 mm, pink; glands 4, oblong, yellowish green to brown, stipitate, stalks hairy; staminate florets 15–18, exserted, 1.5 mm long, pink, bracteate; bract simple, 2 mm long; anther lobes subglobose, dehiscing transversely, yellow; pistillate flowers laterally pendulous, 0.5 × 0.3 cm, glabrous, pedicel 0.5 cm long, knobbed distally; ovary tricarpellate; styles 3, free from base, 0.2 cm long; stigmas 3, simple, pink. Fruiting pedicels recurved, capsules schizocarpic, trigonous, distinctly keeled, prominently ridged along sutures, glabrous; cocci obscurely venulose. Fruits 3.5 mm long; seeds 3 per capsule, ecarunculate, 1.5 × 1.0 mm oblong-ovoid, tetragonal, transversely and obscurely furrowed, truncate at base, not tuberculated, brown.

Rare in Palakonda hills in Kadapa district. Endemic. Fl. & Fr.: September – December.

Mabbuchintalapalle (CDP), *NSD* & *RVR* 31413 (holotype SKU; isotypes BSID, MH).

INDIA: Andhra Pradesh. Endemic.

Euphorbia lacei Craib, Bull. Misc. Inf. Kew 1911: 456. 1911 & Contr. Fl. Siam. Dicot. 182. 1912; Esser & Chayamarit, Harvard Papers in Bot. 6(1): 262. 2001; Binojk. & N.P. Balakr. Genus *Euphorbia* in India 168. 2010 & in N.P.Balakr. *et al.*, Fl. India 23: 321. 2012. *E. barnhartii* Croizat Euphorb. Antiq. 54. 1934; Surya Prakash Babu in Pullaiah & Moulali, Fl. Andhra Pradesh 2: 857. 1997. *E. trigona* Roxb., Fl. Ind. 2: 468. 1832 *nom. illeg.*; Hook.f., Fl. Brit. India 5: 256. 1887. *p.p;* Gamble 2: 1276. 1925.

Large succulent shrub or small tree, to 7 m high, with ascending branches, usually leafless; brachlets exclusively 3-winged, wings less broad in proportion to the length of the joints, with stipular divaricate persistent sharp spines. Leaves alternate, sessile, obovate-spathulate, base attenuate, margin entire, apex mucronate, deciduous. Cyathia in axillary lax cymes; involucre cupular; glands 5, suborbicular, thick; appendages O; Male florets in 5 groups, each of 6-8, few sterile; bracteole laciniate, anthers subglobose, red. Female floret erect; ovary glabrous, styles 3, simple, slightly connate, stigmas emarginated. Capsule ovoid, trigonus, deeply sulcate and sharply keeled; seeds light brown to white, smooth.

Occasional in open plains usually in rocky places and along foothills in scrub jungles. Fl. & Fr.: December-February. Vern.: Tel.: *Aavijamuku.*

Gavisiddeswara hills (ATP), *PSPB* 11845; Srisailam (KNL), *TRS* & *PVS* 79535 (MH).

INDIA: Uttar Pradesh, Bihar, West Bengal, Nagaland, Meghalaya, Andhra Pradesh, Maharashtra, Karnataka, Kerala and Tamil Nadu.

WORLD: Native of Moluccas, naturalized in S and SE Asia.

Euphorbia laciniata Panigrahi, Kew Bull. 30: 531. 1975; Binojk. & N.P. Balakr. Genus *Euphorbia* in India 168. 2010 & in N.P.Balakr. *et al.*, Fl. India 23: 292. 2012. *E. fimbriata* Heyne ex Roth, Nov. Pl. Sp. 227. 1821, non Scop. 1788; Gamble 1273. 1925.

Dwarf undershrubs, up to 15 cm high; branchlets many, short; rootstock thick, woody, black. Leaves oblong-ovate, base subcordate, margin sinuate-serrate, apex apiculate or mucronate, opposite, crowded towards the tips of the branches, floral leaves distichously imbricate. Cyathia solitary, axillary, short pedunculate, involucres densely villous; appendages palmate multifid, laciniate, glabrous or subciliate. Male florets in 5 fascicles of 6-8 flowers each, bracteoles setaceous, sometimes bifid, peripheral ones adnate to involucral wall, laciniate and hairy. Female floret: gynophores ca 2 mm long, projecting outside the cyathial cup and recurved, hairy; ovary globose, white-hirsute, styles ca 1.5 mm long, free from base, each bifid, recurved, stigma subcapitate. Fruits globose to subglobose, obtusely keeled, villous-pubescent; seeds 4-angled, with concave surfaces, horizontally furrowed, brown.

Occasional along roadsides and dry grassy slopes in Dharmapuri and Salem districts in Tamil Nadu. Fl. & Fr.: October – March.

INDIA: Maharashtra, Karnataka and Tamil Nadu. Endemic.

Note: *E. fimbriata* Roth (1821) is a later homonym of *E. fimbriata* Scopoli (1788). Roth bases his description on Heyne's collection from 'India Orientalis'. There is no type specimen in Berlin Herbarium, probably destroyed during Second World War. Panigrahi (1975) redescribed this species as *E. laciniata* Panigr., selecting Beddome's collection from Tirunelveli, Tamil Nadu as the Holotype (Binojkumar and Balakrishnan, 2010).

Euphorbia lactea Haw., Syn. Pl. Succ. 127. 1812; Binojk. & N.P. Balakr. Genus *Euphorbia* in India 307. 2010 & in N.P.Balakr. *et al.*, Fl. India 23: 322. 2012.

Succulent spiny shrubs or small trees, up to 10 m high; stems woody, up to 20 cm in diam. at base, older parts terete; branches many, crowded at apex, erect, bushy; young branches 3 (or 4)-winged, ascending; wings ca 1 cm thick, wavy-margined, white-marbled or whitish-greyish at the space between the wings; spines paired, ca 3 mm long, brownish black, persistent. Leaves rarely present, alternate, sessile, scale like, ovate to orbicular, base rounded to cuneate, margin entire, apex obtuse, 3-5 × 2-3 mm, obscurely nerved, fleshy, caducous. Flowers and fruits not seen.

Usually cultivated in gardens as a fence, sometimes seen as an escape (Binojkumar and Balakrishnan *loc. cit.*).

Varlavaripalli (CDP), *BSS* & *P.Kiran Babu* 34318.

INDIA: Uttar Pradesh, West Bengal, Andhra Pradesh, Maharashtra, Karnataka, Kerala and Tamil Nadu.

WORLD: Widely cultivated in Tropical Africa and Asia.

Euphorbia lawsonii Binojk. & Dwarakan, Rheedea 4: 60. 1994; Binojk. & Balakr. Genus *Euphorbia* in India 120. 2010 & in N.P.Balakr. *et al.*, Fl. India 23: 292. 2012. *Chamaesyce lawsonii* (Binojk. & Dwarakan) V.S.Raju, J. Econ. Tax. Bot. 28: 92. 2004.

Erect, suberect or rarely decumbent herb, to 20 cm tall, puberulous to pubescent, stems pseudo-dichtomously branched; nodes annular, slightly thickened; internodes 1-2 cm long, ribbed. Leaves ovate to orbicular, hirsute, petiole to 1 cm long. Cyathia terminal, solitary or paired, without or rarely with microphyllous leaves; involucres turbinate, to 1 x 1.2 mm, lobes triangular, laciniate, glands 4, tiny, pink. Male flowers: pedicels to 1 mm long, anthers subglobose, transversely dehiscing, bracteoles filiform, to 1 mm long. Female flowers: gynophore to 0.5 mm long, ovary globose, 1 mm across, pubescent, styles free from base, each bifid up to half of the length, stigma capitate. Capsule subglobose, obtusely keeled, to 2 mm across, densely hairy; seeds tetragonous, oblong-ovoid, all faces with 2 or 3 clear transverse irregular grooves.

Rare in moist areas and in evergreen forests, 500-1200 m altitude in Kolli hills. Fl. & Fr.: Nov.-April.

Kolli hills (SM), *Dwarakan* 513 (MH & K).

INDIA: Karnataka and Tamil Nadu, endemic.

Euphorbia longistyla Boiss., Cent. Euph. 9. 1860 & in DC., Prodr. 15(2): 20. 1862; FBI 5: 247. 1887; Gamble 2: 1275. 1921; Binojk. & N.P. Balakr., Genus *Euphorbia* in India 194. 2010 & in N.P.Balakr. *et al.*, Fl. India 23: 293. 2012.

Erect herbs, up to 30 cm high, stems pseudodichotomously branched, branches filiform towards the apex, base minutely velutinous at base. Leaves elliptic to oblong-elliptic, 1-2 × 0.5-1 cm, glabrous, oblique at base, entire or serrate, obtuse-apiculate at apex, stipules linear-lanceolate. Cyathia terminal, solitary, peduncle ca 2 mm long, involucres turbinate, 1-2 × 1-1.5 mm, glabrous, lobes 5, laciniate, pubescent towards margins, glands 4, shortly stalked, rounded. Male florets few, anthers subglobose, bracteoles filiform, laciniate. Female floret: ovary subovoid, 2 mm across, styles filiform, free from base, stigma capitate. Capsule subglobose, keeled, glabrous; seeds quadrangular, subovoid to oblong-ovoid, reddish brown.

Rare in moist slopes, grasslands, under shade in forests of Kadapa and Kurnool districts. Fl. & Fr.: July- November.

Kalasamudram (ATP) *NSD*, 31406; Batrepalli (ATP, *KRKS* 37890; Somulavandlapalli (CDP), *KRKS* 41902; Pulivendula ghat (KD), *KRKS* 39665; Mudsumpad (CDP), *JSG* 21266 (K, CAL); Guvvalacheruvu, Palakonda RF (CDP), *KS* 6378 (MH); Srisailam (KNL), *JLE* 2209 (CAL).

INDIA: Andhra Pradesh, Karnataka. Endemic.

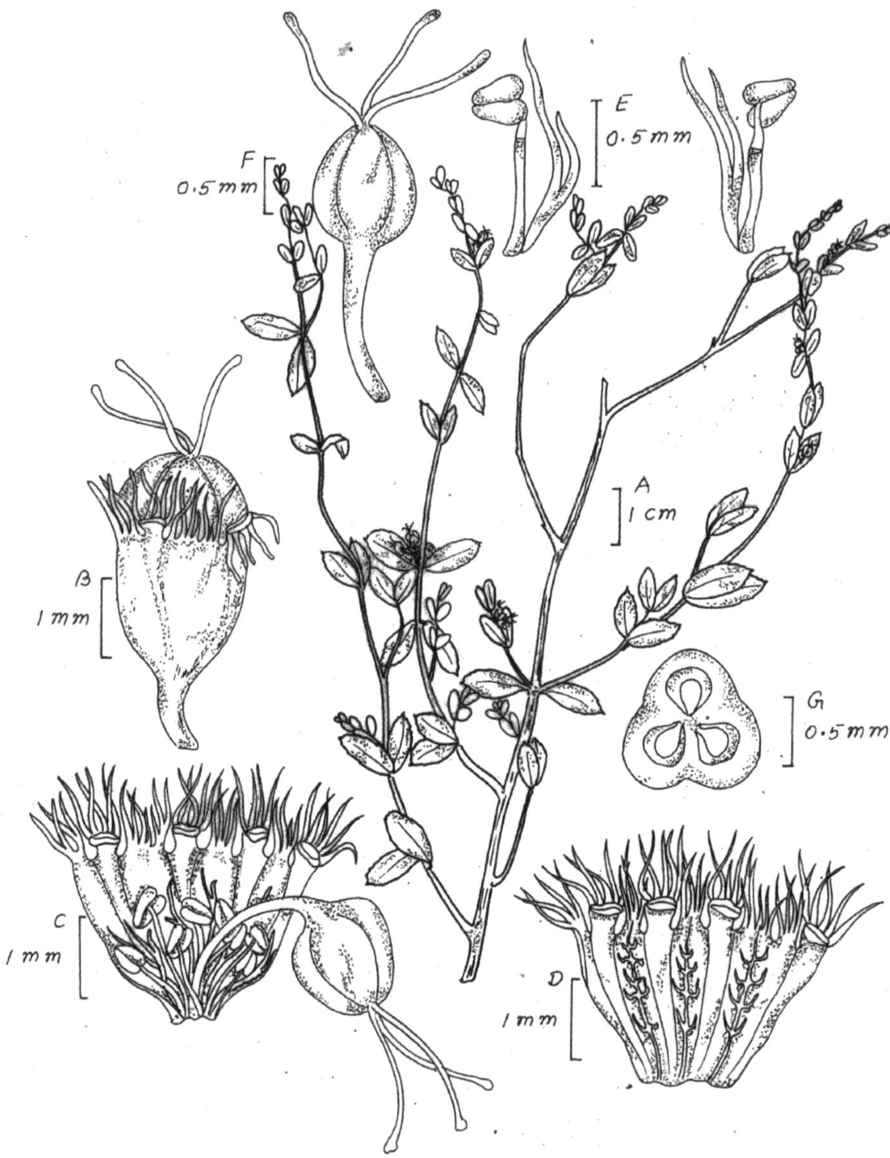

Figure 59. **Euphorbia longistyla** Boiss.
A. Twig, B. Cyathium, C. Cyathium split open, D. Involucre split open, E. Stamens with bracteoles, F. Pistil, G, Ovary T.s.

Euphorbia neriifolia L., Sp.Pl. 451. 1753, *p.p.* FBI 5: 255. 1887; Gamble 2: 1277. 1925. *E. ligularia* Roxb. ex Buch.-Ham., Trans. Linn. Soc. London 14: 285. 1825, *nom. illegit.*

Shrub or small tree, up to 5 m tall with pairs of stipular spines on tubercles or swellings of the branchlets; tubercles more or less confluent in five vertical or slightly spiral lines so that the branch is more or less obtusely 5-gonous. Leaves alternate, crowded at ends of branches, sessile or subsessile, obovate-oblong or oblanceolate, 15-23 × 4-6.5 cm, apex obtuse or acute, margin entire, base cuneate. Cyathia yellowish, solitary, in diads or triads, the central and oldest cyathia sessile and male, the lateral ones bisexual and with short thick peduncle; involucral cup subtended by 2, ovate bracts; lobes 5, suborbicular, glands 5, transversely oblong. Male florets: in 4 or 5 fascicles of 3 or 4 flowers each. Female florest: gynophores c. 3 mm long, styles connate up to middle, stigma papillose. Capsule deeply 3-lobed; seeds globoose.

Planted in village hedges or wild in rocky places in Odisha (Saxena and Brahmam, 1995). Fl. & Fr. December – June.

INDIA: Punjab, Haryana U.P., West Bengal, Tripura, M.P., Rajasthan, Gujarat, Maharashtra, Karnataka, Tamil Nadu and Odisha.

WORLD: Tropical Asia.

Euphorbia nivulia Buch.-Ham., Trans. Linn. Soc. London 14: 286. 1825; FBI 5: 255; Gamble 2: 1277. 1925; Binojk. & N.P. Balakr., Genus *Euphorbia* in India 315. 2010 & in N.P.Balakr. *et al.*, Fl. India 23: 324. 2012.

Armed, deciduous, succulent tree, up to 10 m high; trunk straight; branches whorled, branchlets cylindrical with stipular, well developed, divaricate, persistent, sharp spines on small corky bases arranged in regular rows; wood white, soft, even-grained. Leaves crowded at ends of branches, obovate to oblanceolate, 8-20 × 3-4 cm, subsessile, fleshy, subspathulate, glabrous, apex obtuse, margin entire, base cuneate, nerves not conspicuous. Cyathia in 2-4 cm long, terminal, lax cymes, involucres cup-shaped, transversely oblong, greenish yellow, bracts 2, broadly ovate, scarious, involucral lobes 5, broadly orbicular, glands 5, male florest in 5 fascicles of 6 to 7 flowers each, anthers subglobose, longitudinally dehiscing; female florets: ovary subglobose, styles connate up to middle, each bifid at tip, stigma papillose. Capsules trigonous, 6 × 8 mm, glabrous; seeds 4 mm, 4 angular, smooth.

Common along the rocky areas in open dry forests. Fl.: January-March. Fr.: April-May. Vern.: Tel.: *Bontha jemudu, Aku-jemudu, Erra-kalli, Modu-jemudu, Jemudu*; Ori.: *Siju, Kath siju, Bad siju;* Tam.: *Ilaikalli.*

Upper Ahobilam (KNL), *SSR* 15286; Peccheruvu (KNL), *PSPB* 9136, *SKW* 5138, 5240 (BLAT); Macherla RF (GNT), *VRK* 3819; Annavaram (EG), *SKW* 148 (BLAT); Dummagudem (EG), *JSG* 15796 (CAL); Pachamalais, (Madurai), *KMM* 121376 (RHT); Foothills of Yercaud (SM), *KCJ* 18080 (MH); Yercaud (SM), *KMM* 26846 (RHT).

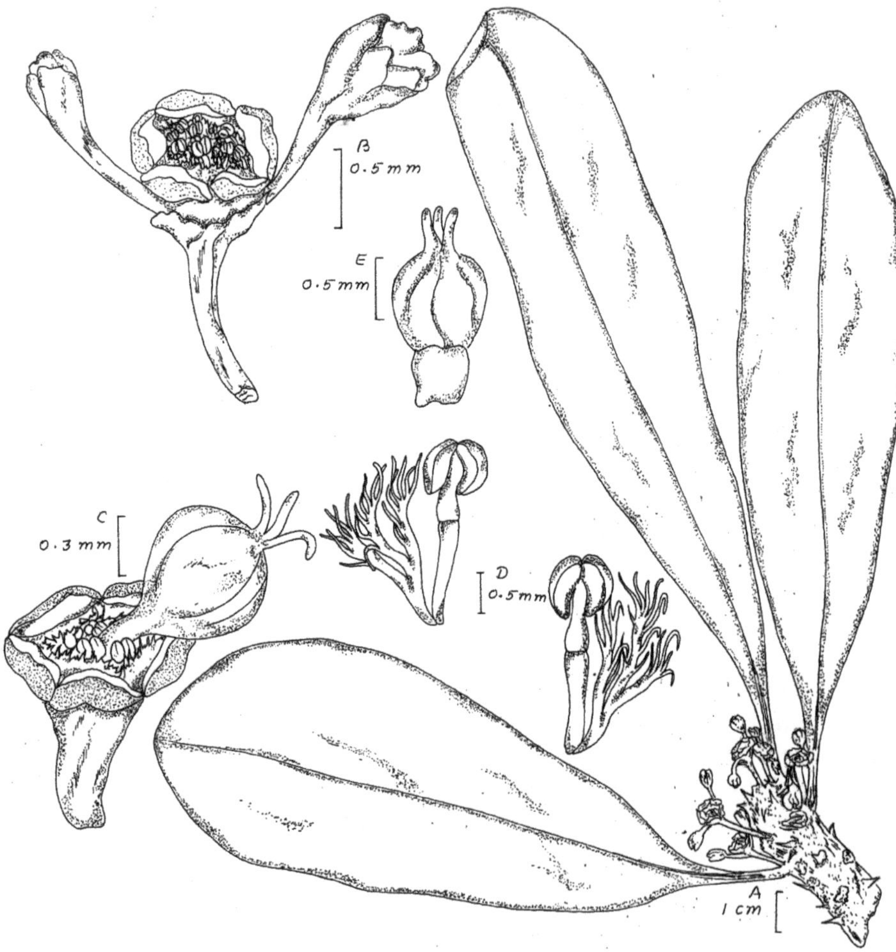

Figure 60. **Euphorbia nivulia** Buch.-Ham.
A. Twig, B. Inflorescence, C. Cyathium, D. Stamens with bracteoles, e. Pistil.

INDIA: Punjab, Haryana, Delhi, Rajasthan, Gujarat, Madhya Pradesh, Uttar Pradesh, Bihar, West Bengal, Odisha, Andhra Pradesh, Maharashtra, Karnataka, Kerala and Tamil Nadu.

WORLD: Pakistan, Bangladesh Myanmar and Sri Lanka.

Euphorbia perbracteata Gage, Bull. Misc. Inform. Kew 1914: 238. 1914; Binojk. & N.P. Balakr., Genus *Euphorbia* in India 230. 2010 & in N.P.Balakr. *et al.*, Fl. India 23: 313. 2012.

Erect herb, up to 50 cm high, glabrous. Leaves alternate, sessile, linear-lanceolate or linear-oblong, 2-6 × 0.2-0.5 cm, pale green, base cuneate, margin entire, apex acute. Pseudoumbels 2-4-rayed, pseudoumbel leaves broader than stem leaves, base cuneate, margin entire, apex obtuse or subacute, 3-3.5 × 0.5-1 cm, rays 5-10 cm long, 2-4 times dichotomously branched; ray leaves opposite, connate at base, broadly orbicular, 1-2 × 1-1.5 cm. Cyathia solitary, terminal, sparsely hairy, bracts surrounding cyathia foliaceous, conspicuously veined and often connate at base, involucre turbinate, pubescent outside, involucral lobes 5, triangular, hairy along margin, glands 4. Male florets pedicels ca 2 mm long, glabrous, anthers subglobose. Female floret: gynophores ca 4 mm long, glabrous, ovary subglobose, trigonous, styles to 1 mm long, free, each bifid, stigma capitate. Capsule ovoid, subglobose; seeds subglobose, ovoid.

Weed in wastelands and moist areas, up to 1000 m altitude in East Godavari district. Fl. & Fr.: January-March. Vern.: Tel.: *Vathani.*

Sanjamala (KNL) *NSD* 31411; Kontrukota near Vadapalli (EG), *DCSR* 149 (CAL).

INDIA: Bihar, Madhya Pradesh, Chhattisgarh, Andhra Pradesh, Maharashtra, Goa and Karnataka. Endemic.

Euporbia prostrata Aiton, Hort. Kew. 2: 139. 1789; FBI 5: 266. 1887; Gamble 2: 1276. 1925; Binojk. & N.P. Balakr., Genus *Euphorbia* in India 121. 2010 & in N.P.Balakr. *et al.*, Fl. India 23: 295. 2012. *Chamaesyce prostrata* (Aiton) Small, Fl. SE U.S. 713. 1903. *Euphorbia chamaesyce* auct. plur. non L. 1753; Surya Prakash Babu in Pullaiah & Moulali, Fl. Andhra Pradesh 2: 858. 1997.

Perennial, spreading, prostrate herb, up to 20 cm high, branchlets hispid-pubescent. Leaves opposite, distichous, elliptic to obovate, 3-9 × 1-4 mm, prominently 3-nerved, base obliquely truncate to rounded, margin crenulate, apex obtuse, always with small, axillary leaves, subsessile, stipulate. Cyathia axillary, solitary or 2 or 3 together in microphyllous branches, involucre tubular, 1 × 0.7 mm; glands 4, equal; appendage inconspicuous. Male florets 1-4, ebracteolate; stalk 0.5 mm; anther 0.1 mm, dehiscence transverse. Female florets: laterally pendulous, ovary subglobose, 0.5 mm across; styles 3, forked from base, 0.4 mm, stigma obtuse. Capsule prominently beaked, angles with stiff, spreading hairs; seeds obtusely 4-angled, 1 mm, faintly furrowed, reddish green, pale brown to pale grey.

A common weed of cultivated fields in black cotton soils. Fl. & Fr.: July – February. Vern.: Tel.: *Pala alumu;* Tam.: *Chinna-amman-patcharisi.*

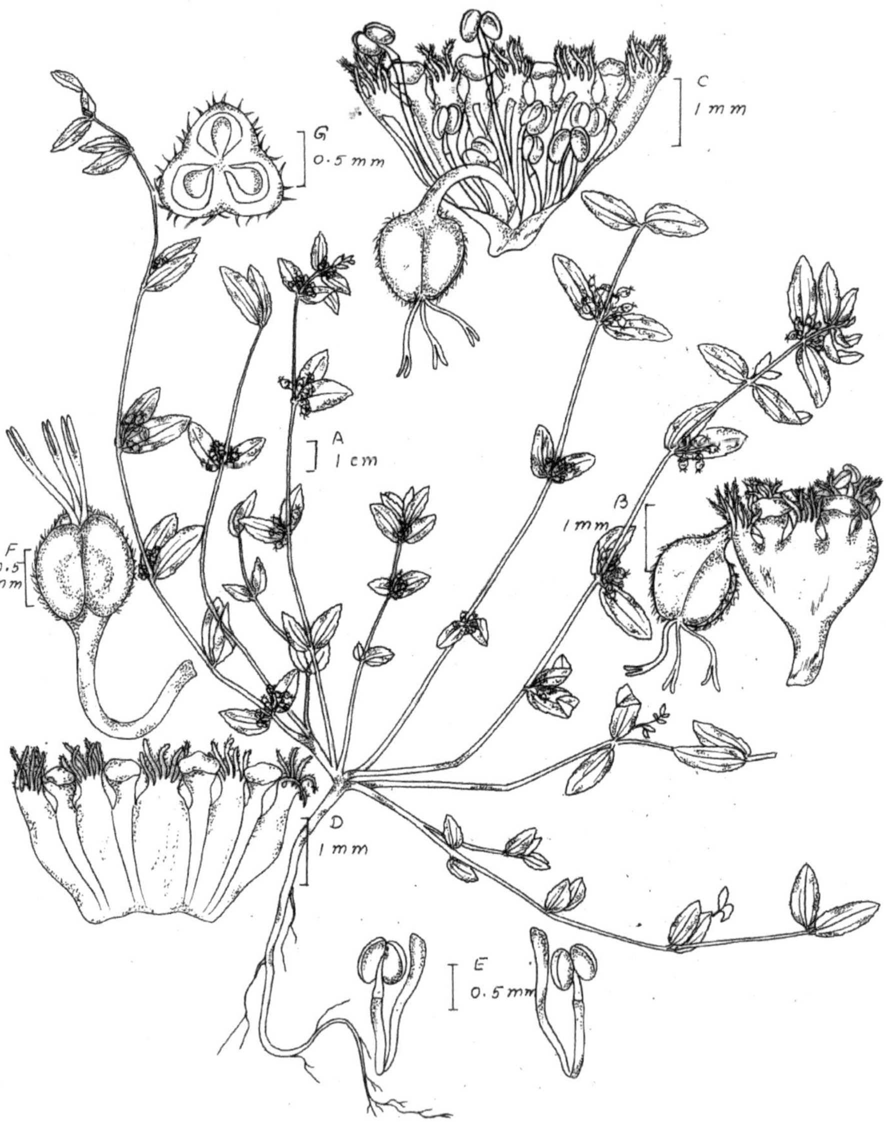

Figure 61. **Euphorbia prostrata** Aiton (*Euphorbia chamaesyce* auct plur non L.)
A. Twig, B. Cyathium, C. Cyathium split open, D. Involucre split open,
E. Stamens with bracteoles, F. Pistil, G. Ovary T.s.

Nayakunipalli (ATP), *TP* 463; Kalasamudram (ATP), *NSD* 31422; Kammavaripalli (KNL), *TP & PVP* 3368; Maredumilli (EG), *GVS* 24255 (MH); Araku (VSKP), *NPBK* 10801 (MH), Lakshmipuram (VSKP), *CAB* 1726 (MH); Lammasingi (VSKP), *HS* 20765 (BLAT); Simhachalam (VSKP), *GVS* 21506 (MH); Similipahar (MBJ), *SX* 447 (RRL-B), *SX & MB* 5548 (RRL-B); Pottangi, *Subba Rao* 29704 (ASSAM).

INDIA: Almost through out India.

WORLD: Native of Tropical and Subtropical America, introduced into many parts of Old World.

Note: In certain Indian Floras, the specific epithet used for this species is *E. chamaesyce,* with *E. prostrata* treated as a synonym. The two species are distinct and the former does not occur in India. *E. chamaesyce,* with green or pale green leaves and the cocci of fruits obtusely keeled and hairy throughout is a very common invasive weed found throughout North America and the Mediterranean parts of Europe and so far not been seen in Asia. *E. prostrata* differs from *E. chamaesyce* by the blackish green leaves and the acutely keeled fruits, which are hairy on the keels only. *E. prostrata* is also of American origin, but introduced to Asia long ago and is now widespread in tropical Asia (Mitra, 1969).

Euphorbia pulcherrima Willd. ex Klotzsch, Allg. Gartenzeitung 2: 27. 1834; Gamble 2: 1278. 1925; Binojk. & N.P. Balakr., Genus *Euphorbia* in India 360. 2010 & in N.P.Balakr. *et al.,* Fl. India 23: 331. 2012. *Poinsettia pulcherrima* (Willd.) Graham, Edinb. New Philos. J. 20: 412. 1836 & Bot. Mag. 63: t. 3493. Dec. 1836; FBI 5: 239. 1886.

Shrubs, up to 3 m high, somewhat woody, branchlets hollow. Lower leaves alternate, elliptic-ovate, 5-12 × 3-7 cm, base cuneate and subobtuse, margin entire, undulating, apex acute to acuminate, sparsely hairy; petioles 3-6 cm long, reddish green, canaliculated on upper side. Upper floral leaves alternate or whorled, elliptic, 3-11 × 1-5 cm, bright red or yellow or pale yellow, usually glabrous, petioles 1-3 cm long. Cyathia in leaf-opposed condensed pseudodichasia; peduncles 2-5 mm long, involucres turbinate, glabrous outside, hairy inside; gland solitary, cupular, yellowish green. Male florets in 5 fascicles of 8-12 flowers each; pedicels 3-4 mm long; filaments red; bracteoles 4-6 mm long, laciniate. Female floret: gynophores 5-8 mm long, pale red; ovary subobovate, hidden within the cyathium, green; styles ca 4 mm long, connate up to middle, each bipartite, recurved. Fruit on deflexed red stalk, trilobate-subglobose, obtuse; seeds ovoid, smooth, pale grey.

Cultivated for its showy foliage as ornamental garden plant; rarely an escape along roadsides. Fl. & Fr.: December – March.

INDIA: Cultivated through out India as ornamental garden plant.

WORLD: Native of Mexico, cultivated in tropics and subtropics.

Euphorbia pycnostegia Boiss., Cent. Euphorb. 9. 1860 & in DC., Prodr. 15(2): 18. 1862; FBI 5: 216. 1887; Gamble 2: 1274. 1925. var. **zornioides** (Boiss.) Santapau, Bull. Bot. Soc. Bengal 8: 11. 1955 & in Rec. Bot. Surv. India ed. 3. 16: 241. 1967;

Binojk. & N.P. Balakr., Genus *Euphorbia* in India 148. 2010 & in N.P.Balakr. *et al.*, Fl. India 23: 296. 2012. *E. zornioides* Boiss. in DC., Prodr. 15(2). 19. 1862; FBI 5: 246. 1887; Gamble 2: 1274. 1925.

Slender, erect, glabrous herb, up to 70 cm high, woody at the base, with long internodes. Leaves linear-oblong 1.5-2.5 × 1.1-1.5 cm, base subcordate, margin serrulate, apex apiculate or obtuse, floral leaves distichously imbricating, ovate, gradually smaller upwards, pinkish and ciliate along margins. Cyathia solitary or paired, involucre campanulate, lobe 5, oblong, fimbriate, glands 4, limbs of glands ovate, rounded to truncate at apex, 1 – 1.5 mm, white, rosy or pink. Male floretss: pedicels ca 1 mm long; anthers subglobose, bracteoles filiform. Female florets: gynophore ca 2 mm long, glabrous; ovary globose, styles free, bifid, stigma capitate. Capsule subglobose, glabrous, cocci obtusely angled; seeds ovoid, 4-angled, papillose.

Rare in hilly forest areas in Dummakonda in Rampa hills (Rolla S. Rao *et al.*, 1999). Fl. & Fr.: February.

Dumakonda, Rampa (EG), *VNS* 665 (CAL).

INDIA: Madhya Pradesh, Andhra Pradesh, Maharashtra, Goa, Karnataka, Kerala and Tamil Nadu. Endemic.

Euphorbia rosea Retz., Observ. Bot. 4: 26. 1786; FBI 5: 251. 1887; Gamble 2: 1275. 1925; Binojk. & N.P. Balakr., Genus *Euphorbia* in India 174. 2010 & in N.P.Balakr. *et al.*, Fl. India 23: 296. 2012. *Chamaesyce rosea* (Retz.) Webster, J. Arnold Arbor. 48: 423. 1967.

Prostrate herb, dichotomously branched; branches brittle, wavy. Leaves 3-12 × 1.5-5 mm, oblong – obovate or linear spathulate, base obliquely cuneate – truncate, margin entire – serrulate, apex obtuse – retuse or rounded, opposite, subsessile, stipules triangular. Cyathia solitary or to 10 in axillary and subterminal lax clusters; involucre 2 × 1.8 mm, subcampanulate, puberulous; lobes triangular, ciliate; glands 4, transversely oblong; appendages of two upper glands obliquely oblong, crenulate; two lower glands suborbicular, short, crenulate. Male florets: 4-6; bracteole laciniate; stalk 1 mm; anthers 0.45 mm across. Female floret: laterally pendulous; ovary 1.2 mm, puberulous, styles 3, connate at base, each 2-fid, from above to quarter down. Capsule 2.5 mm, puberulous, bluntly keeled; seeds 1.5 mm, 4-angled, transversely furrowed.

Occasional in hilly areas. Fl. & Fr.: June – November. Vern.: Tam.: *Chinna-aman-patcharisi, Tippiyavanitam, Utarappalatai.*

Alurukona (ATP), *TP* & *NY* 915, 1132; Thungabhadra (KNL), *PSPB* 9184; Gunjana river bed (CDP), *JLE* 14359 (CAL, MH); Palakonda (CDP), *EKK* 80360 (DD); Kambakkam to Thandipandal (CTR), *MCB* 45104; Ipur RF (NLR), *SKW* 6638 (BLAT); Balukhand RF, *Abraham* 32(CAL); Kambakkam (CPT), *Bourne* 2196 (PCM).

INDIA: Uttarakhand, Uttar Pradesh, Madhya Pradesh, Odisha, Andhra Pradesh, Maharashtra, Karnatak, Kerala and Tamil Nadu.

WORLD: Iran, Afghanistan, Pakistan, Nepal and Sri Lanka.

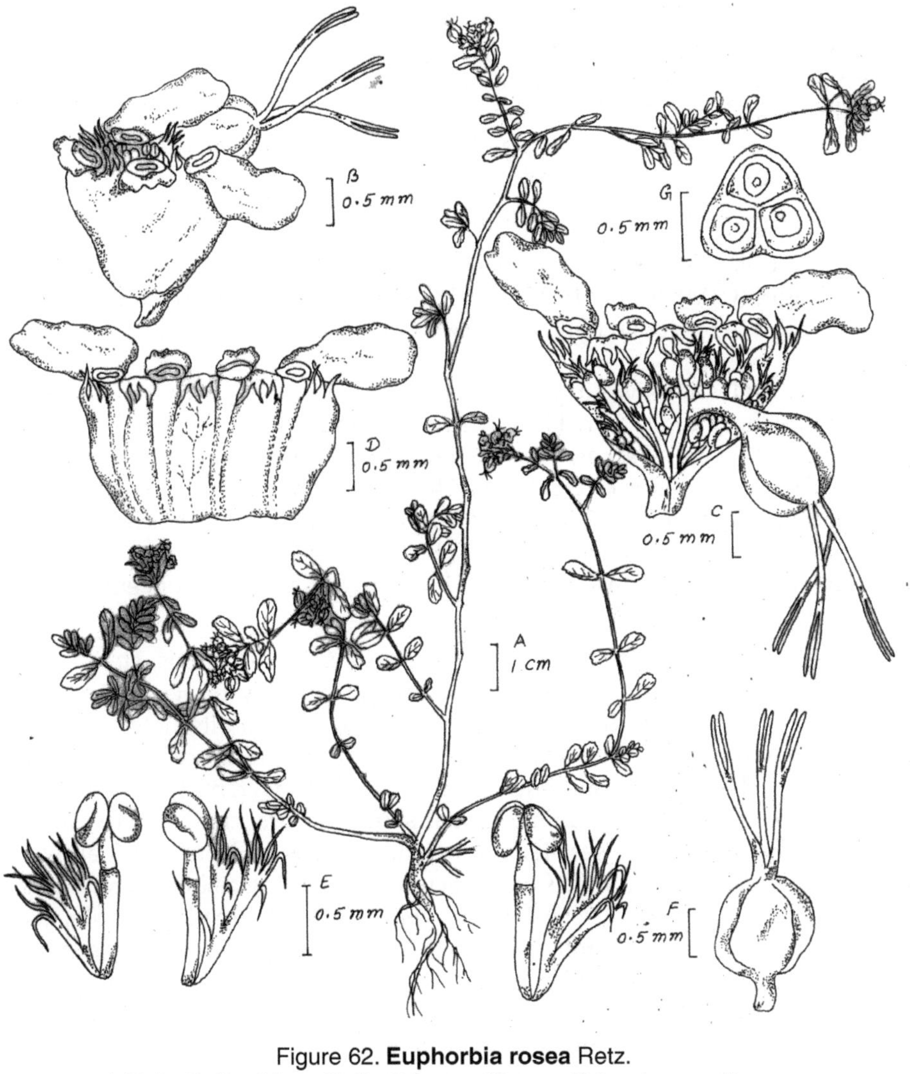

Figure 62. **Euphorbia rosea** Retz.
A. Twig, B. Cyathium, C. Cyathium split open, D. Involucre split open, E. Stamens with bracteoles, F. Pistil, G. Ovary T.s.

Euphorbia rothiana Spreng., Syst. Veg. 3: 796. 1826; FBI 5: 273. 1887; Gamble 2: 1278. 1925; Binojk. & N.P. Balakr., Genus *Euphorbia* in India 125. 2010 & in N.P.Balakr. *et al.*, Fl. India 23: 314. 2012. *E. laeta* B. Heyne ex Roth, Nov. Pl. Sp. 230. 1821 non Aiton 1789. *E. rothiana* var. *pubescens* (Boiss.) Gamble, Fl. Madras 2: 1278. 1925.

Perennial, erect herb, up to 1.5 m high. Leaves alternate, linear, linear-lanceolate, 5-11 × 0.6-1.8 cm, glabrous, apex acute to subacute, margin entire, base narrowed; petioles 2-6 mm long. Cyathia solitary, terminal or axillary; involucres campanulate, glabrous outside, villous inside, lobes 5, subquadrate, glands 4-5, equal, 2-horned. Male florets: anthers subglobose. Female floret: gynophores ca 5 mm long, glabrous, styles 1-2 mm, deeply bifid, stigma capitate. Capsule subglobose, with rounded cocci, glabrous; seeds oblong-ovoid, bluish-grey to black.

In open grasslands in Kolli hills and Mahendrgiri hills. Fl. & Fr.: July- October.

Kolli Hills (SM), *Arokiaswamy* 7690 (RHT), *Mohanan* 13186 (CAL, RHT); Mahendragiri (GJM), *SX* & *MB* 3721, 4018 (RRL-B), *G.Saran* & *Party* 38356 (LWG), *VNS* 5752 (MH); Singaraju (GJM), *SM* 2952 (RRL-B).

Euphorbia royleana Boiss. in DC., Prodr.15(2): 83.1862, *E. pentagona* Royle, non Haw.1828; Brandis, For. Fl. northwestern India 438.1874; Hook.f., Fl. Brit. India 5:247.1887; Parker, For. Fl. Punjab 445.1918; Duthie, Fl.U. Gangetic Pl.3:76.1903; Collett, Fl. Simlensis 446.1921; Brandis, Indian Trees 558.1921; Gamble, Man. Indian Timb.591.1922; Bor, Man. Indian For.Bot.172. 1953; Babu, Herb. Fl. Dehra Dun 459.1977; Bir & Sharma in Indian For.106:744.1980; Almeida, Fl. Maharashtra IVB: 319.2003; Sarojinidevi & Venkataraju, J. Threatened Taxa 9: 10233. 2017.

Shrubs or small trees, 2–4 m tall with dense crown, glabrous, stems woody, branched; branches whorled above, 5–7-angled, slightly projected 5mm long, blackish; spines paired, divaricate 2–4 mm long. Leaves alternate, sub-sessile, obovate-oblanceolate attenuate at base, margin entire, apex mucronate, 4–10 × 2–4 cm, glabrous, fleshy, caducous, often absent when in flower, midnerve prominent. Cyathia sub-terminal and from internodal sinuses of branches, solitary, in diads or triads, the central one sessile, yellowish, primary peduncles 1-3 above the podarium, ca. 5 mm long. Bracts ovate, acuminate, as long as involucre, membranous. Involucre turbinate, 3–4 by 4–5 mm, yellow, lobes 4, broad, fimbriate; glands 5, transversely oblong, ca. 3 mm wide, ca. 1.5 mm long. Staminate florets: pedicels ca. 1 mm long, bracteoles laciniate, ca. 1.8 mm long, an thers sub-globose. Pistillate floret: gynophores 15 mm long, ovary sub-ovoid, styles free from the middle, stigma shortly bifuricated. Fruits subglobose, trigonous with three compressed, obtuse cocci, 6–9 mm in diam., glabrous, greyish-green. Seeds sub-globose, 2–2.5 mm in diam., brown to pale black, ecarunculate.

Rare in scrub forests and dry rocky hill slopes. Fl. & Fr.: February – June.

Near Balapur (Khurda District, Odisha), *N. Sarojini Devi,* 31440 (SKU).

INDIA: Himachal Pradesh, Punjab, Uttarakhand, Arunachal Pradesh, Nagaland, Meghalaya, Tripura, Assam and Odisha.

Euphorbia sebastinei Binojk. & N.P. Balakr., Rheedea 3:26. 1993; Binojk. & N.P. Balakr., Genus *Euphorbia* in India 125. 2010 & in N.P.Balakr. *et al.*, Fl. India 23: 297. 2012. *Chamaesyce sebastinei* (Binojk. & N.P. Balakr.) V.S. Raju, J. Econ. Taxon. Bot. 28:92.2004.

Herbs, erect or decumbent, up to 25 cm tall, pubescent; stems many from thick rootstock; branches 2-6 together; nodes thickened, internodes 1.5-2 cm long, terete. Leaves petiolate, oblong to elliptic-oblong, oblique at base, serrate, acute or sub-acute at apex, 6–10×4–6 mm, pubescent on both the surfaces; nerves obscure; stipules triangular, laciniate, ca 2 mm long. Cyathia in axillary and terminal clusters; peduncles ca 1 mm long; involucres turbinate, 0.8-1 mm across; lobes triangular, toothed, exceeding the glands 4, tiny, orbicular; limbs obovoid, wavy at apical margin, ca 0.5×0.9 mm white or pink. Male florets: pedicels ca 0.5 mm long; anthers sub-globose, bracteoles filiform, ca 0.8 mm long. Female floret: gynophores ca 1.5 mm long; ovary sub-globose, ca 1mm in diam., hairy; styles free, 0.4 mm long, each bifid at apex. Fruits ovoid, 2 × 2 mm, acutely keeled, hairy; seeds oblong, tetragonous, 2 or 3-grooved on faces, 1 × 0.5 mm.

Occasionally seen in waste lands and also as garden weed. Fl. & Fr.: June-February.

INDIA: Gujarat, Maharashtra and Andhra Pradesh. Endemic.

Euphorbia senguptae N.P.Balakr. & Subr., Bull. Bot. Surv. Ind. 2: 175. 1960; Binojk. & N.P. Balakr., Genus *Euphorbia* in India 126. 2010 & in N.P. Balakr. *et al.*, Fl. India 23: 298. 2012. *Chamaesyce senguptae* (N.P. Balakr. & Subr.) V.S. Raju & P.N. Rao, Phytologia 37: 454.1977.

Herbs, erect, up to 40 cm tall, entirely glabrous; stems woody, terete, pseudo dichotomously, profusely branched; branches slender, thickened at nodes; internodes 1-3.5 cm long. Leaves oblong to elliptic, oblique at base, serrulate, acute at apex, 0.8–2.5 × 0.4–0.8 cm, glabrous; lateral nerves obscure; petioles 2 mm long; stipules scaly, laciniate. Cyathia terminal, solitary; peduncle ca 2 mm long; involucres turbinate, 2×2.5 mm; lobes 5, triangular; glands 4, oblong, shortly stalked; limbs sub-ovate, entire, minutely sinuate, hyaline, pale pink. Male florets: pedicels ca 1mm long; anthers sub-globose; bracteoles filiform, 2 mm long. Female florets: gynophores ca 2 mm long; ovary oblong-ovoid, ca 1.5× 1mm; styles up to 2 mm long, free, simple, stigma minute, capitate. Fruits sub-ovoid, deeply keeled, 3×3 mm, glabrous; seeds oblong, tetragonous, 2×1 mm, reddish brown.

Occasional on grassy hill slopes, stream sides and partially shaded places in Southern Eastern Ghats. Fl. & Fr.: September-December.

Amagondapalem RF (ATP), *NY* 1048; Muchukota RF (ATP), *TP* & *NY* 979 (MH); Kalasamudram RF (ATP), *EC* 2993; Dhone (KNL) *NSD*, 31405; Mahanandi (KNL), *JLE* 25462, 42331; Srisailam (KNL), *JLE* 22095 (MH); Palakonda RF, Guvvalacheruvu (CDP), *KS* 6820 (CAL, MH, K).

INDIA: Andhra Pradesh. Endemic.

Euphorbia serpens Kunth, Nov. Gen. Sp. 2: 52. 1817; Mitra, J. Bombay Nat. Hist. Soc. 68: 854. 1971; Binojk. & N.P. Balakr., Genus *Euphorbia* in India 126. 2010 & in N.P. Balakr. *et al.*, Fl. India 23: 298. 2012. *E. microphylla* auct. non Heyne ex Roth 1821; FBI 5: 252. 1887 *p.p.*, *E. orbiculata* Kunth. var. *jawaharii* Rajagopal & Panigrahi, Taxon 17: 547. 1968.

Annual prostrate herb, many branches spreading from the root, dichotomously branched; branches striate, rooting at nodes. Leaves opposite, subsessile 3-5 × 1-4 mm, subquadrate, base obliquely truncate, margin entire, apex retuse; stipules 2-partite. Cyathia solitary, axillary, terminating in axillary shoots in upper leaf-axils; involucre 0.8 × 1 mm, broadly campanulate; lobes triangular, apiculate, one lobe serpent hooded; glands 4, equal; appendages minute. Male florets 1-4; bracteole laciniate, ciliate; stalk 0.6 mm; anther 0.25 mm across. Female floret laterally pendulous; ovary 0.7 × 0.6 mm; styles 3, each 2-fid, from above the middle. Capsule 1.5 mm; seeds 1 mm, 4-angled, smooth, white.

Common weed in open places, along roadsides, waste lands and other moist areas in fields, dried ponds and low lying banks of rivers. Fl. & Fr.: June – December.

Dhaniyanicheruvu (ATP), *TP* & *NY* 842 (MH); Chinnapalle (ATP), *KRKS* 37998; Nagarjunakonda valley (GNT), *KT* 9831 (CAL); Gobanapalem (WG), *KS* 5152 (MH).

INDIA: Punjab, Haryana, Delhi, Rajasthan, Uttar Pradesh, West Bengal, Madhya Pradesh, Andhra Pradesh, Maharashtra, Goa, Karnataka, Tamil Nadu and Andamn Islands.

WORLD: Native of Tropical America introduced into the Mediterranean region, Asia and Africa.

Note: *Euphorbia serpens* Kunth is often confused with *E. microphylla* Heyne but is distinguished by leaves subquadrate to ovate with retuse apex, stipules incised and branches rooting at nodes, while the latter has leaves obovate to oblanceolate with acute apex, stipules laciniate and branches not rooting at nodes.

Euphorbia seshachalamensis Prasad & Prasanna, Ann. Bot. Fennici 53: 73. 2016.

Geophytic herbs, tuberous; main tuber subterranean, elongate, variously shaped, ca 14 × 8 cm, with milky latex; stems 2-6, unbranched, subterranean, near tip of main tuber, ca 4.5 cm long, ca 1 cm thick. Leaves in rosettes from stem apices, succulent, petiolate, petiole ca. 2 cm long, lamina narrowly lanceolate, 4-7 × 1 cm, apex acute, margins crispate and rarely purple, base narrowed, veins immersed in lamina and indistinct, greenish-yellow. Inflorescence with leaves (synanthous), axillary, usually in branched cymes, rarely simple, ca 10 cm. long, greenish yellow or rarely greenish purple, pedunculate; peduncles ca. 4 cm long; bracts broadly oblong-obovate, 3-4 × 2.2 – 2.5 mm, acuminate, sparsely dentate on margins, greenish-yellow. Cyathia pedunculate; peduncles 5-6 mm long; involucres broadly cup-shaped, on 0.5-0.6 mm long cyathial pedicels, greenish-yellow, rarely purple; involucral lobes 5, rectangular, ca 0.5 × 0.75 mm, fringed, white, glands

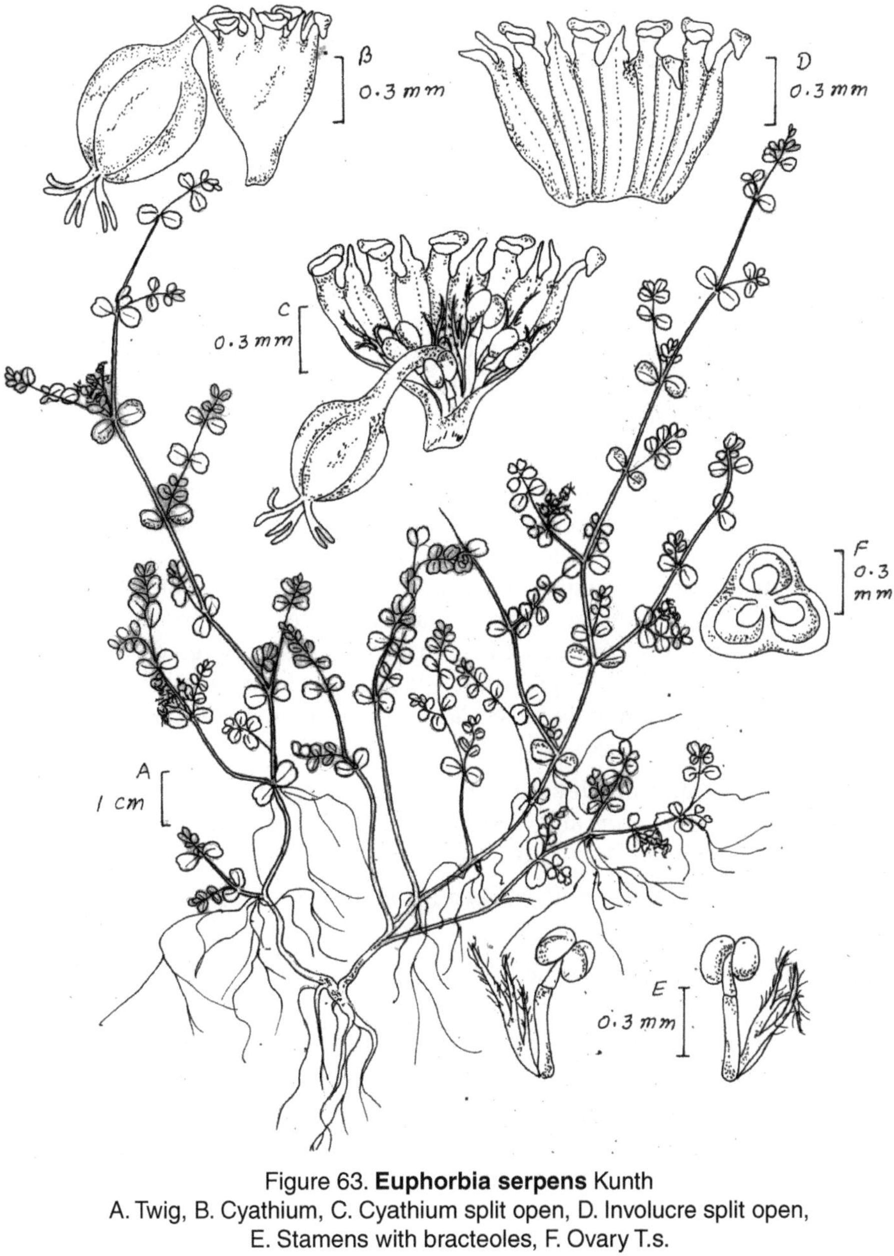

Figure 63. **Euphorbia serpens** Kunth
A. Twig, B. Cyathium, C. Cyathium split open, D. Involucre split open, E. Stamens with bracteoles, F. Ovary T.s.

5, transversely oblong, 2.1-2.4 mm long, greenish-yellow; appendages absent. Staminate florets in 10-12 fascicles, each fascicle 3-4-flowered; bracteoes laciniate, 1.8-2.3 mm long; florets on 1.9-2 mm long pedicels; filaments 0.5 mm long; anthers yellow, bilobed, lobes subglobose, 0.6-0.64 × 0.42 mm. Pistillate floret: ovary smooth, tricarpellate on 2 mm long gynophores; styles 3, ca 1.1 mm long, connate below for 0.4-0.5 mm, free above; stigmas 3, ca 0.4 x 0.6 mm, slightly bilobed. Capsules deeply 3-lobed, rarely 4-lobed, 3.8-4.25 × 6-7.6 mm, with compressed smooth cocci; pedicels 8.5-9 mm long, recurved; seeds globose, 2.3-2.6 × 2.2-2.4 mm, hairy, variegated, pale with brownish lines, ecarunculate.

In crevices on rocky ground in grass dominated dry deciduous forests at an altitude rage of 300-500 m a.s.l. in Seshachalam hills. Fl. & Fr.: August – September.

Jandrapenta RF (CDP), *PVP, KP, M. Sankara Rao* & *T.Thulasaiah* 4538 (holotype BSID, isotype CAL).

Euphorbia susan-holmesiae Binojk. & Gopalan in Cact. Succ. J. (U.S.) 65: 277-278. f1.1993; Binojk. & Balakr. Genus *Euphorbia* in India 325. 2010 & in N.P.Balakr. *et al.*, Fl. India 23: 326. 2012.

Shrubs or small trees, up to 4 m tall, main stem terete, green, branches crowded at apex, slender, twisted, usually 4-winged, rarely 3- or 5-winged, with a flat intervening space between the wings, spines 2, divaricate. Leaves sessile in the axil of spines, orbicular, less than 1 cm long, fleshy, entire, obtuse at apex, early caducous. Cyathia solitary, above the spines, bracts 2, triangular, fimbriate at apical margin, pale yellow, involucres cupular, ca 3 mm diameter, lobes 5, suborbicular, glands 5, yellow. Staminate florets in 5 fascicles of 3-6 florets in each, covered by bracteoles. Pistillate floret: ovary globose, glabrous, styles 1 mm long, connate nearly up to middle, bifid at apex, stigma plumose. Capsule 3-4 mm in diam., cocci 3, glabrous, slightly pinkish, bivalve; seeds subglobose, smooth.

Scrub jungles and semievergreen forests on shallow soil over bedrock, associated with *E. antiquorum* and *E. tortilis*, at 250-1100 m. altitude in Dharmapuri district of Tamil Nadu. Fl. & Fr.: July-Sep. and Dec. – Feb.

Tirthamalai (DP), *KMM* & *Venugopal* 14196 (RHT), *KMM, Britto* & *Rani* 29193 (RHT).

INDIA: Tamil Nadu. Endemic.

Euphorbia thymifolia L., Sp. Pl. 454. 1753; FBI 5: 252. 1887; Gamble 2: 1276.1925; Binojk. & N.P. Balakr. Genus *Euphorbia* in India 345. 2010 & in N.P. Balakr. *et al.*, Fl. India 23: 299. 2012. *Chamaesyce thymifolia* (L.) Millsp., Pub. Field Columb. Bot. Mus. Ser. 2: 412. 1916.

Prostrate herb, forming dense tufts or mats of about 25 cm in diam., partially erect, up to 10 cm high, stems arising from a woody rootstock, slender, dichotomously branched, branches thinly pubescent. Leaves opposite, 3-7 × 1.5-4 mm, elliptic – oblong, base rounded – obliquely cordate, margin crenulate – serrate, apex rounded – obtuse, subsessile; stipules fimbriate. Cyathia solitary or 2 – 3 in axils; involucre 1 × 0.8 mm, campanulate; lobes triangular, laciniate; glands 4; appendages minute.

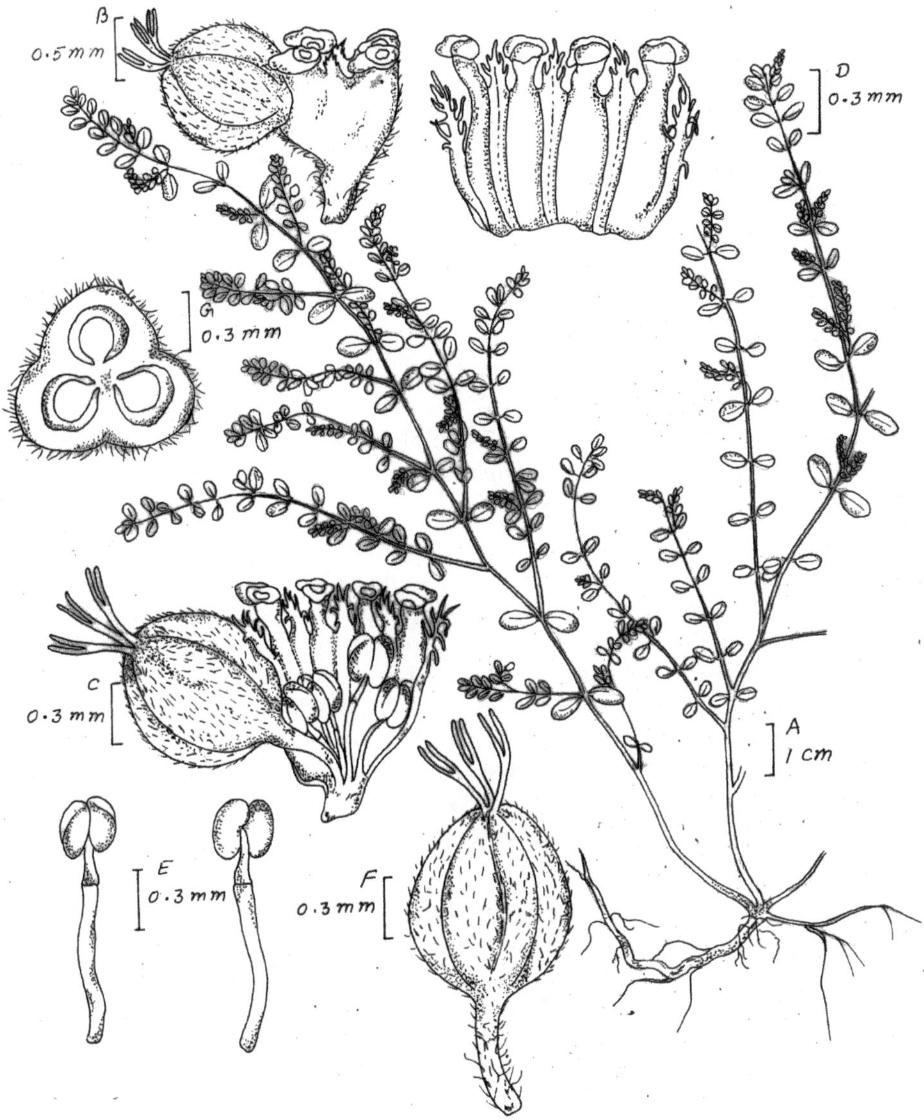

Figure 64. **Euphorbia thymifolia** L.
A. Twig, B. Cyathium, C. Cyathium split open, D. Involucre split open, E. Stamens, F. Pistil, G. Ovary T.s.

Male florets 1-4, anthers oblong-obovte to subglobose, bracteoles setaceous. Female floret: gynophore to 0.5 mm long, ebracteolate, stalk 1 mm, tomentose; styles 3, each 2-fid, from above the middle. Capsule 1.5 mm, obtusely keeled, keels tomentose; seeds 1 mm, 4-angled, minutely tuberculate.

Common in dry areas along roadsides, rocky areas and as a weed in gardens. Fl. & Fr.: August – March. Vern.: Tam.: *Chinnamman Paccharisi.* Tel.: *Erra – chittipayalaku;* Ori: *Kalijati, Patra-siju.*

Kalasamudram (ATP), *NY* 1169, Rayadurg (ATP), *PSPB* 9189; Palakonda (CDP), *CS* 9397; Palamaner (CTR), *CECF* 4316 (CAL); Duttalur (KNL), *CECF* 4134 (CAL); Kambakkam (CTR), *MCB* 45182 (MH); Gobanapalem (WG), *KS* 5157 (CAL); Gokavaram (EG), *SKW* 1794 (BLAT); Araku (VSKP), *HS* 20409 (BLAT), *VSR* 2290; Nattavaram (VSKP), *CAB* 10172 (MH); Chittarkonda (KPT), *DCSR* 897 (CAL); Hogainakkal (SM), *KCJ* 18002 (MH).

INDIA: Almost through out the country.

WORLD: Pantropical in Asia, Africa and America.

Euphorbia tirucalli L., Sp. Pl. 452. 1753; FBI 5: 254. 1887; Gamble 2: 1276. 1925; Binojk. & N.P. Balakr., Genus *Euphorbia* in India 345. 2010 & in N.P. Balakr. *et al.*, Fl. India 23: 335. 2012.

Succulent shrub or small tree, to 5 m high; unarmed, branchlets terete glabrous, articulate. Leaves 0.4-1.5 × 1-0.2 cm, linear, cuneate at base, entire, obtuse – acute, caducous, subsessile, early deciduous. Cyathia subsessile, solitary or in small capitate glomerules, congested at the apices of the branchlets, yellow; bracts rounded, involucre 2 × 2.5 mm, cupular, coriaceous; lobes triangular, fringed and ciliate; glands 4-5, orbicular, thick; appendages 0. Male florets 6-8; bracteole laciniate; stalk 1.5 mm; anther 0.45 mm across. Female floret erect; perianth distinctly 3-lobed; ovary 2 × 1.7 mm, pubescent, styles 3, spreading, each 2-fid, from above the middle, recurved. Capsule 0.7 mm, globose, pubescent; seeds 3.5 mm, globose, smooth, buff, speckled brown and with dark brown ventral line.

Common in field hedges near villages in dry areas. Fl. & Fr.; January – April. Vern.: Tam.: *Kombukalli, Tirugu kalli, Tirukalli.* Tel.: *Tirukalli;* Ori.: *Kharseju, Jadengachha, Hadabhongam, Dangulisiju.*

Bukkapatnam (ATP), *TP* & *NY* 2102; Rayadurg (ATP), *PSPB* 9192; Kalasamudram (ATP), *KRKS* 39124; Palakonda (CDP), *CS* 8280; Anchetty-Kundbukottai forest (DMP), EV 57918 (CAL, MH).

INDIA: Through out the country.

WORLD: Africa, Pakistan, Sri Lanka, Bangladesh, Myanmr, Vietnam, Thailand and Malesia.

Euphorbia tortilis Rottl. ex Ainslie, Mat. Ind. 2: 424. 1826; FBI 5: 256. 1887; Gamble 2: 1277. 1925; Binojk. & N.P. Balakr. Genus *Euphorbia* in India 328. 2010 & in N.P.Balakr. *et al.*, Fl. India 23: 326. 2012.

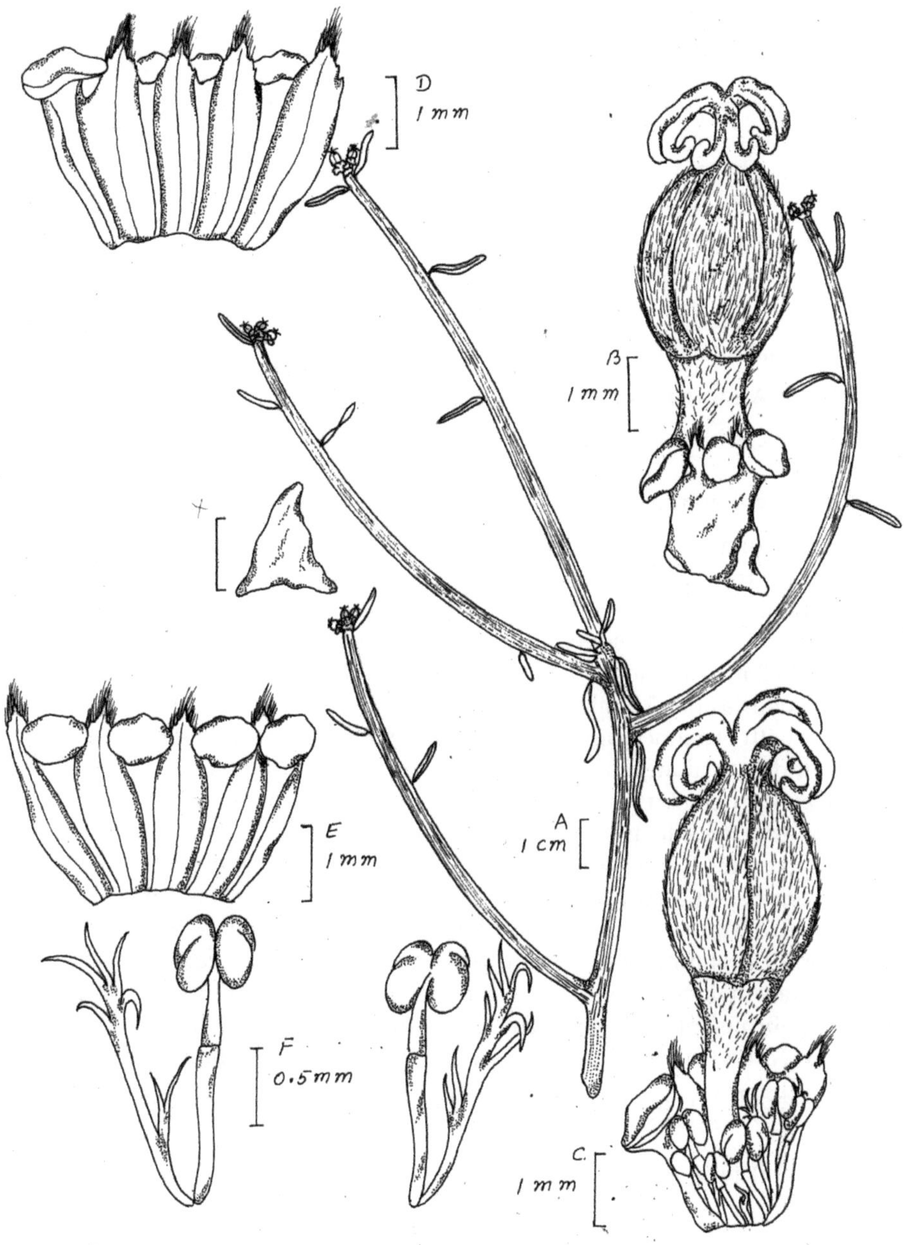

Figure 65. **Euphorbia tirucalli** L.
A. Twig, B. Cyathium, C. Cyathium split open, D and E. Involucre split open, F. Stamens with bracteoles.

Succulent shrub or small tree; up to 3 m high, branches jointed, branchlets 3-winged, twisted, with stipular divaricate persistent sharp spines. Leaves small, 4-6 × 4-5 mm, glabrous, fleshy, obovate – oblong, base cuneate, margin entire, apex subtruncate, early deciduous, subsessile. Cyathia 3-4, in fascicles, axillary, pale yellowish green, bracts broadly triangular, obtuse at apex, involucres cup-shaped, 4 × 3 mm, lobes 5, laciniate along the margins, glands 5, fleshy, yellow. Male florets in 5 fascicles of 5-9 flowers each, anthers subglobose, transversely dehiscing. Female florets: ovary sharply 3-angled, styles 3, each 2-fid at above. Capsule trigonus; seeds 4-angled, smooth.

Occasional in the plains of scrub jungles in Southern Eastern Ghats. Fl. & Fr.: February – May. Vern.: Tel.: *Tirugu jemudu, Naga jemudu*; Tam.: *Thirukukalli*

Rachakuntapalli (CDP), *KRKS* 39074, *NSD*, 31424.; Alagarkoil (MDU), *EV* 25294 (MH); Narthamalai (TCP), *KMM* & *Rani* 29349(RHT).

INDIA: Andhra Pradesh, Karnataka, Kerala and Tamil Nadu.

WORLD: Sri Lanka.

Euphorbia venkatarajui Sarojin., Nordic J. Bot. 35: 359. 2017.

Fleshy, compact cushion-like, up to 1 m tall shrub; branches cylindrical, 3 cm in diameter, green; podaria conical 1.0– 1.2 cm long, closely and spirally arranged on stem, with spines at the tip; spines paired, divaricated, 5–8 mm long, often with two 0.5 mm long spines above, rigid, reddish when young, blackish when mature; spiny shield 3 mm in diameter, discoid. Leaves variable, positioned just above the spines at podaria apex, sessile, spherical, ovate or slightly obovate, 0.6–0.8 × 0.4–0.6 cm, fleshy with prominent mid-nerve and indistinct lateral nerves, rounded at base, slightly undulate along margin, rounded to acute at apex, caducous. Cyathia in triads of axillary clusters or fascicles, yellowish to pink; peduncles 0.5–1.0 cm long; bracts opposite, broadly triangular; involucre cupular, 3–5, 2–3 mm; lobes 5, broadly cuneate, fimbriate at apex; glands 5, transversely oblong, yellow. Staminate florets: in 5 fascicles of 6 or 8 flowers in each; pedicels ca 1 mm long; anthers sub-globose; bracteoles ca 1.5 mm long, broad, laciniate at apex, covering florets. Pistillate florets: gynophores ca 1.4 mm long, curved; perianth lobes 3, triangular; ovary trigonous, styles connate up to middle; stigma papillose. Capsule trilobed, 3–9 mm in diameter, smooth, glabrous, pinkish; cocci curved, obtusely keeled. Seeds ovoid to globose, 2.5-3.0 mm, brownish black, glabrous, smooth with a ventral line; caruncle minute.

Occasional in Yaganti (Yerramalais) in Kurnool district. Fl. & Fr.: February – May.

Yaganti (KNL), *Naidu Sarojinidevi* 31438 (SKU, BSID, CAL, MH).

EXOECARIA Linnaeus

Excoecaria crenulata Wight, Icon. Pl. Ind. Orient. 5: 1865. 1852; FBI 5: 473. 1888; Gamble 2: 1345. 1925; Chakrab. & N.P.Balakr. in N.P.Balakr. *et al.*, Fl. India 23: 344. 2012. *E. oppositifolia* Griff. var. *crenulata* (Wight) Chakrab. & M.Gangop., J. Econ. Taxon. Bot. 18: 208. 1994. *E. robusta* Hook.f., Fl. Brit. India 474. 1888.

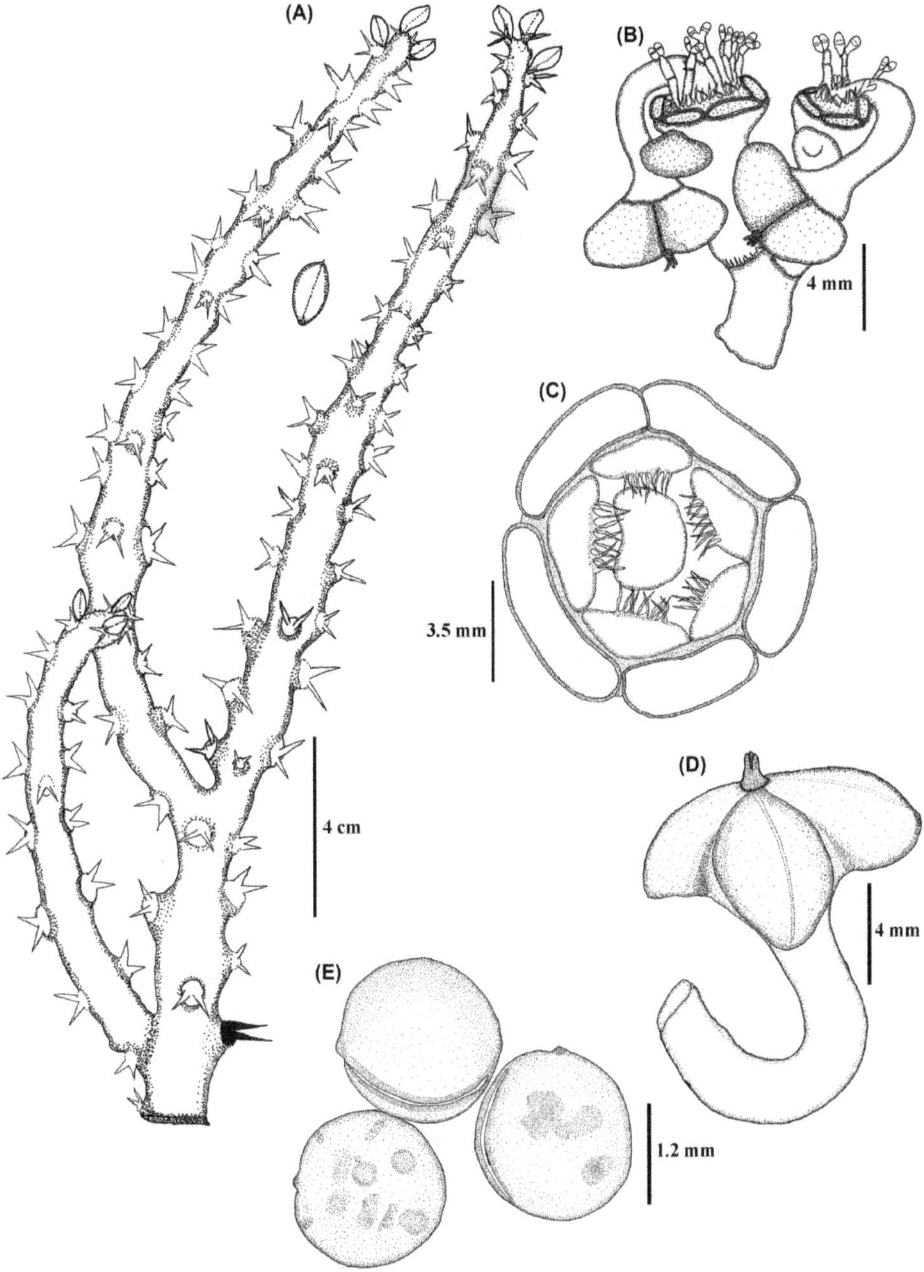

Figure 66. **Euphorbia venkatarajui** Sarojin. (A) habit, (B) cyathium, (C) involucral glands, (D) capsule, (E) seeds. (Sarojinidevi, 2017).

Shrub or tree, 5-10 m tall, latex milky, poisonous. Leaves opposite, obovate to oblanceolate, 12-18 × 3-6 cm, apex shortly acuminate, margin entire, base acute or obliquely rounded; petiole 1-2 cm long; stipules subulate. Male spikes to 8 cm, axillary and terminal; sepals 3; stamens 3. Female spikes to 3 cm, axillary or terminal, sepals 3. Capsule pendulous, 3-lobed, 1-3 cm across; seeds broadly ovoid, mottled.

Deciduous forests. Occasional in Pacchamalais and Kolli hills, abundant in Guthirayans. Fl.: March – July; Fr.: August – September.

INDIA: Karnataka, Kerala, Tamil Nadu.

WORLD: Sri Lanka.

FALCONERIA Royle

Falconeria insignis Royle, Ill. Bot. Himal. Mts. 354. 1839; N.P.Balakr. & Chakrab., Fam. Euphorb. India 292. 2007 & in N.P.Balakr. *et al.*, Fl. India 23: 347. 2012. *Sapium insigne* (Royle) Trimen, Cat. Pl. Ceylon 83. 1885; FBI 5: 471. 1888; Gamble 2:1346. 1925; Surya Prakash Babu in Pullaiah & Moulali, Fl. Andhra Pradesh 2: 885. 1997.

Deciduous tree, up to 15 m high; monoecious; bark grey, smooth, shining; wood white to whitish yellow, grey, straight- grained, medium textured; branches thick, milky poisonous juice. Leaves crowded towards the ends of the branches, thick, elliptic-lanceolate or obovate, 6.5-22 × 4-8 cm, glabrous, bright green, base attenuate, margin crenate-serrate, often with glands on the margin, apex acuminate; secondary nerves 10-16 pairs, slender; petiole to 5 cm, with 2-conspicuous glands at the top. Flowers reddish in terminal, large (ca.12.5 cm long) spikes. Male flowers: above, sessile; calyx lobes 2, orbicular, concave; stamens 2, free, alternating with sepals, exserted. Female flowers: solitary, subsessile; calyx 2-lipped, with 2 or 3 ovate, long pointed sepals; ovary ovoid to obovoid, glabrous, styles 3, shortly recurved; stigma subsessile. Capsule globose, 1 cm dia., fleshy at first, dry and irregularly dehiscent when ripe, packed closely together along the rachis; seeds 3, ovoid, 0.6 cm in dia., smooth, pale.

Rare, in dry rocky hills of dry-deciduous forests above 800 m. Fl.: January-February. Fr.: March. Vern.: Tel.: *Devaswroopi, Garbhasula, Kadura.*

Tirumala (CTR), *SSR* 15271; Mogilipenta (CDP), *JSG* 21287 (MH); Seshachalam forest (CDP), *PSPB* 11851; Veligonda (NLR), *MSR* 1414 (CAL).

INDIA: Himachal Pradesh, Uttarakhand, Uttar Pradesh, Bihar, West Bengal, Sikkim, Arunachal Pradesh, Assam, Meghalaya, Odisha, Andhra Pradesh, Maharashtra, Goa, Karnataka, Kerala and Tamil Nadu.

WORLD: Sri Lanka, Nepal, Bhutan, Myanmar, Laos, Cambodia, Vietnam, China, Thailand and Malaysia.

FLUEGGEA Willdenow

1. Branchlets arrested, thorn like (armed); pistillodes 2, free .... **F. leucopyrus**
1. Branchlets unarmed; pistillodes 3, connate at base .... **F. virosa**

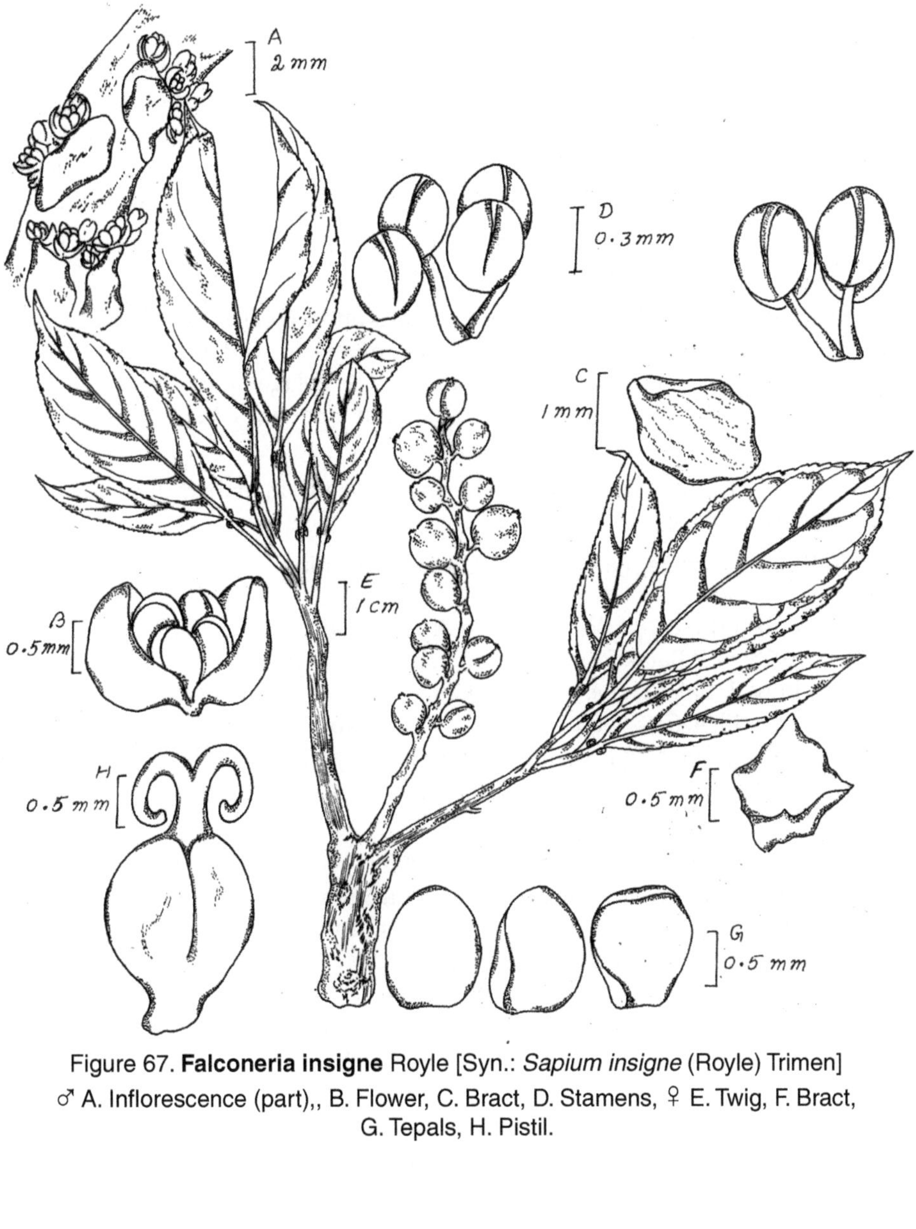

Figure 67. **Falconeria insigne** Royle [Syn.: *Sapium insigne* (Royle) Trimen]
♂ A. Inflorescence (part),, B. Flower, C. Bract, D. Stamens, ♀ E. Twig, F. Bract, G. Tepals, H. Pistil.

Flueggea leucopyrus Willd., Sp. Pl. 4: 757. 1805; FBI 5: 328. 1887; Gamble 2: 1296. 1925; N.P. Balakr. & Chakrab., Fam. Euphorb. India 342. 2007; Chakrab. *et al.*, in N.P. Balakr. *et al.*, Fl. India 23: 412. 2012. *Securinega leucopyrus* (Willd.) Müll. Arg. in DC., Prodr. 15: 451. 1866; Surya Prakash Babu in Pullaiah & Moulali, Fl. Andhra Pradesh 2: 887. 1997.

Shrub, up to 5 m high, dioecious; branchlets terete, long thorn like, with leaves and flowers. Leaves 1-2 × 0.4-1.2 cm, obovate, base acute, margin entire, apex acute; petiole to 5 mm; stipules triangulr, persistent. Flowers in axillary clusters; pedicels to 5 mm, filiform; bracts minute. Male flowers: 2 mm across; sepals 5, 0.6-1 × 0.5-1 mm, obovate, concave, obtuse; disc glands 5, angular; stamens 5, free, filaments 2 mm, anthers 0.5 mm across; pistillodes 2, 1.3 mm. Female flowers 2.7 mm across; sepals 6, 1.7 × 1 mm, free, obovate, concave, obtuse; disc flat; ovary 1.8 × 1.5 mm, subglobose, styles 3, each 2-fid. Capsule 5 mm, globose; seeds plano-convex, smooth, brown.

Common along roadsides and also in open places. Fl. & Fr.: June – October. Vern.: Tel.: *Tella purugudu*; Tam.: *Vellaipoolan*.

Penchalakona (NLR), *PSPB* 11864; Penukonda (ATP), *SKW* 6010 (BLAT); Guvvalacheruvu (CDP), *KS* 6408 (MH); Rajampeta to Kadapa (CDP), *A.M.Rao* 1920 (BSID); Papanasanam (CTR), *DRC* 455 (SVU); Velagalapalli (NLR), *SWL* 1231 (DD); Koradi RF (NLR), *PV* 112133 (BSID); Diguvametta (PKM), *KMS* 6342 (MH); Suryalanka (GNT), *SKW* 3840 (BLAT).

INDIA: Rajasthan, Madhya Pradesh, Andhra Pradesh, Maharashtra, Karnataka, Kerala and Tamil Nadu.

WORLD: Sri Lanka, Pakistan, Myanmar and China.

Flueggea virosa (Roxb. ex Willd.) Royle, Ill. Bot. Himal. Mts. 328. 1836; Gamble 3: 1296. 1925; N.P.Balakr. & Chakrab., Fam. Euphorb. India 342. 2007; Chakrab. *et al.*, in N.P. Balakr. *et al.*, Fl. India 23: 413. 2012. *Phyllanthus virosus* Roxb. ex Willd., Sp. Pl. 4: 578. 1805. *Securinega virosa* (Roxb. ex Willd.) Bail., Adansonia 6: 334. 1866, quod. syn. tantum. Emend; Surya Prakash Babu in Pullaiah & Moulali, Fl. Andhra Pradesh 2: 887. 1997. *Flueggea microcarpa* Blume, Bijdr. Fl. Ned. Ind. 580. 1826; FBI 5: 328. 1887.

Shrub, up to 5 m high, dioecious; branchlets terete. Leaves 3-10 × 1.5-5.5 cm, variable, elliptic – obovate, base cuneate – acute, margin entire, apex acute; petiole to 7 mm. Flowers in axillary clusters; pedicels of male to 5 mm, filiform; female to 3 mm. Male flowers: 2 mm across; sepals 5, 1.3 × 1 mm, ovate, obtuse, concave; stamens 5, filaments 1.5 mm, free; anthers 0.6 × 0.5 mm; pistillodes 3, 1.2 mm; disc of glands 5, free. Female flowers: 1.7 mm across; sepals 5, 1.4 × 1 mm, ovate, obtuse, concave; disc annular, entire; ovary subglobose; styles 3, each 2-fid. Capsule 4 mm, globose; seeds plano-convex, faintly reticulate, brown.

Occasional in scrub and deciduous forests on bare slopes. Fl. & Fr.: June – December. Vern.: Tel.: *Sulamunata, Patangi karji*; Ori.: *Janjingi, Kanilehya, Bhoji bhaji, Patri*; Tam.: *Karumpolan*.

Figure 68. **Flueggea leucopyrus** Willd.
[Syn.: *Securinega leucopyrus* (Willd.) Müll.Arg.]
♂ A. Twig, B. Flower, C. Androecium, D. Anthers, ♀ E. Twig, F. Flower, G. Pistil, H. Ovary T.s.

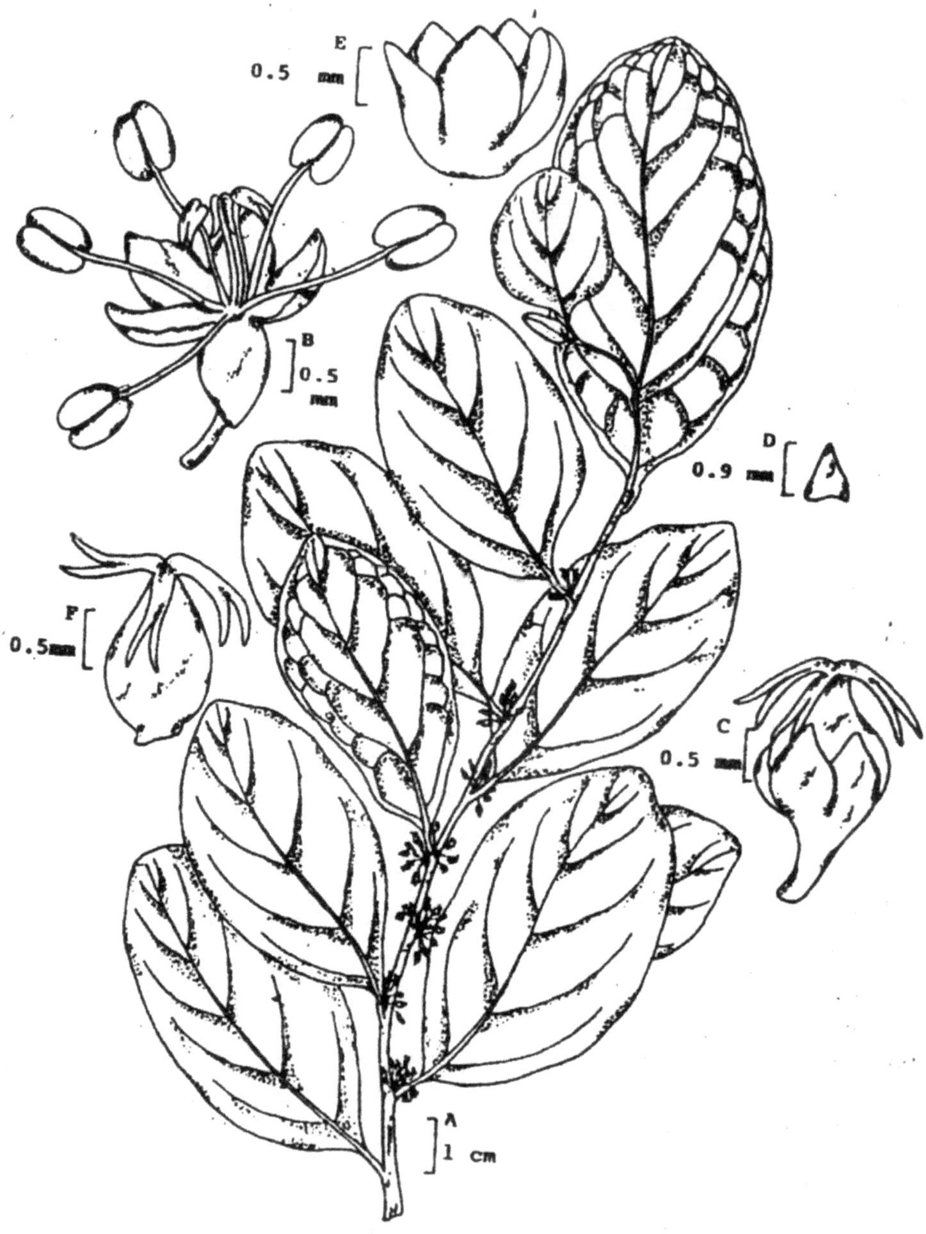

Figure 69. **Flueggea virosa** (Roxb. ex Willd.) Voigt
[*Securinega virosa* (Roxb. ex Willd.) Bail.]
♂ A. Twig, B. Flower, ♀ C. Flower, D. Bract, E. Tepals, F. Pistil.

Seshachalam forest (CDP), *PSPB* 11853; Muskambavi (KNL), *VBH* 83940 (BSID); Durgam (NLR), *PV* 111141 (BSID); Simhachalam hills (VSKP), *NRR* 86129 (BSID).

INDIA: Through out India.

WORLD: Tropical Asia, Africa, Australia and Polynesia.

GIVOTIA W. Griffith

Givotia moluccana (L.) Sreem., Taxon 24: 696. 1975. *Croton moluccanus* L., Sp. Pl 1005. 1753. *p.p. Givota rottleriformis* Griff., Calcutta. J. Nat. Hist. 4: 388. 1844; FBI 5: 395. 1888; Gamble 2: 1342. 1925; N.P.Balakr. & Chakrab., Fam. Euphorb. India 232. 2007; Chakrab. & N.P. Balakr. in N.P.Balakr. *et al.*, Fl. India 23: 264. 2012.

Deciduous tree, up to 20 m high; bark smooth, peeling off in circular thick bosses; wood white; branchlets fulvous tomentose. Leaves chartaceous, broadly ovate or orbicular, 5-15 × 5-20 cm, glabrous above, yellowish tomentose beneath, base cordate, margin irregulary dentate, apex acuminate, glandular, 5-nerved; petiole to 10 cm with 2 glands. Flowers yellow in 10-15 cm long, axillary or sub terminal racemose or paniculate cymes. Male flowers: sepals oblong to obovate, cucullate, densely stellate pubescent; petals campanulate, 5-lobed with cuneate and truncate lobes, longer than sepals; disc of 5, free orbicular glands; stamens 8-20, outer ones ca 6 with free filaments and inner ones ca 10 with filaments connate at base and exserted, filaments unequal. Female flowers: sepals and petals as in male; ovary globose, stellate hairy; styles 2, forked from base. Drupe oblong or subglobose, to 3 x 2.5 cm, brown, smooth; seed one, globose, hairy.

Occasional, in dry deciduous forests at 400-1000 m. Fl.: May-July. Fr.: March. Vern.: Tel.: *Konda ponaku, Tella poliki, Yella dabba*; Tam.: *Kottaithanakku.*

Kalasamudram RF (ATP), *SSR* 13752, *KRKS* 39193; Palakonda (CDP), *DAM* 5959; Diguvametta (PKM), *SSR* 16103; Chelama (KNL), *JSG* 78056 (DD); Jathararevu (KNL), *VBH* 84971 (BSID); Balapalli (CDP), *JSG* 1713 (DD), *JLE* 14267 (MH).

INDIA: Andhra Pradesh, Maharashtra, Karnataka, Kerala and Tamil Nadu.

WORLD: Sri Lanka.

GLOCHIDION J.R.Forster & J.G.A. Forster nom. cons.

1. Anthers 4 or more:
   2. Inflorescence supra axillary and pedunculate .... **G. zeylanicum**
   2. Inflorescence axillary and sessile .... **G. talakonense**
1. Anthers 3:
   3. Female calyx cupular, campanulate or obconic, with 4-6 irregular lobes.. **G. candolleanum**
   3. Female calyx not as above, comprising of (5-) 6 more or less free sepals:
      4. Female flowers pedicellate; capsule 4-locular.....................**G. heyneanum**

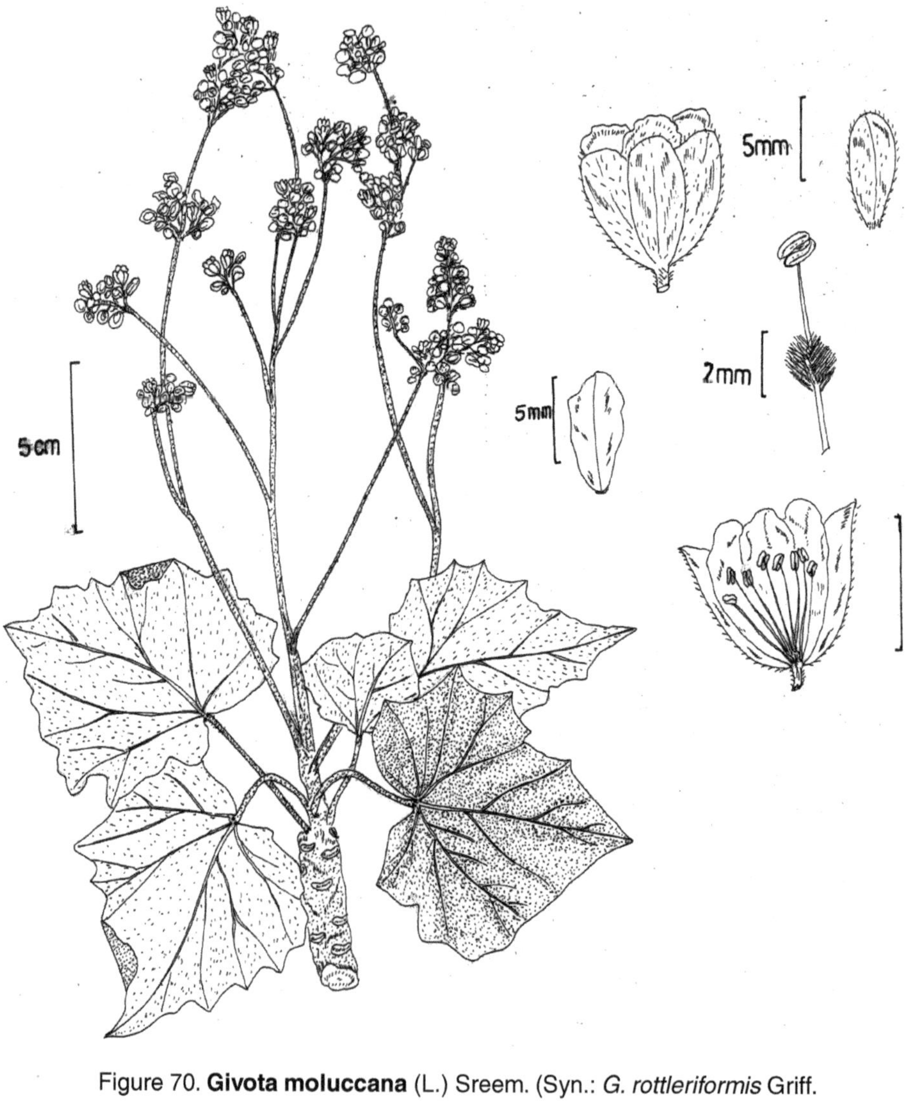

Figure 70. **Givota moluccana** (L.) Sreem. (Syn.: *G. rottleriformis* Griff.
♂ A. Twig, B. Flower entire, C. Sepal, D. Petal, E. Stamen, F. L.s. of flower.

4. Female flowers sessile; capsule 5-locular:

 5. Inflorescence densely flowered; ovary subglobose, tomentellous to glabrous; fruits unlobed or shallowly lobed, crustaceous .... **G. ellipticum**

 5. Inflorescence 3-6 flowered; ovary urn-shaped, pubescent; fruits deeply lobed, lobes bilobulate, coriaceous .... **G. tirupathiense**

Glochidion candolleanum (Wight & Arn.) Chakrab. & M. Gangop., J. Econ. Taxon. Bot. 19: 191. 1995; Chakrab. *et al.*, in N.P. Balakr. *et al.*, Fl. India 23: 418. 2012. *Lobocarpus candolleanus* Wight & Arn., Prodr. Fl. Ind. Orient. 1: 7. 1834. *Glochidion neilgherrnse* Wight, Icon. Pl. Ind. Orient. 5(2): 29. 1852; FBI 5: 316. 1887; Gamble 2: 1307. 1925. *G. arboreum* Wight, *loc.cit.* 29, t. 1907. 1852; FBI: 5: 316. 1887; Gamble 2: 1307. 1925, *p.p. Phyllanthus pycnocarpus* Müll.Arg., Flora 48: 386.1865 & *loc. cit.* 304. 1866. *Glochidion pycnocarpum* (Müll.Arg.) Bedd., Forester's Man. 194. 1873; FBI 5:. 315. 1887. *G. pauciflorum* Gamble, Fl. Madras 2:. 1307. 1925. *G. sisparaense* Gamble, Fl. Madras 2: 1307. 1925.

Trees, 3-12 m high. Leaves elliptic, oblong-elliptic, ovate-lanceolate to lanceolatee or subfalcate, base acute to rounded or often unequal, curling upwards along margins on drying, apex acute, apiculate, attenuate to acuminate, 5-14 × 22-6 cm, glabrous; petioles 2-7 mm long. Flowers in racemes. Male flowers: pedicels 4-10 mm long; sepals oblong, elliptic to obovate, anthers 3, 1.5-3 mm long. Female flowers: sessile or pedicelled; calyx usually campanulate, sometimes cupular; ovary subglobose, style clavate, turbinate or obconic, lobes 4-6. Capsule depressed-subglobose.

Rare in Mixed forests or sholas, at 600-1500 m altitudes (Chakrab. *et al., loc. cit.*). Fl. & Fr.: January - October.

INDIA: Andhra Pradesh, Karnataka, Maharashtra, Kerala and Tamil Nadu.

WORLD: Sri Lanka.

Glochidion ellipticum Wight, Icon. Pl. Ind. Orient. 5(2): t. 1906. 1852; FBI 5: 321. 1887; Gamble 2: 1308. 1925; N.P.Balakr. & Chakrab., Fam. Euphorb. India 348. 2007; Chakrab. *et al.*, in N.P.Balakr. *et al.*, Fl. India 23: 420. 2012. *Phyllanthus malabaricus* Müll.Arg., Linnaea 34: 69. 1865. *Glochidion malabaricum* (Müll. Arg.) Bedd., For. Man. Bot. 194. 1873; FBI 5: 319. 1887. *Phyllanthus assamicus* Müll.Arg., Flora 48: 378. 1865. *Glochidion assamicum* (Müll.Arg.) Hook.f., Fl. Brit. India 5: 319. 1887.

Evergreen, monoecious tree, up to 8 m high; branchlets slender, 4-angular, glabrous. Leaves coriaceous, elliptic-oblong, 3-11 × 1.5-3.5 cm, glabrous on both surfaces, base acute, margin entire, apex acuminate, often oblique; secondary nerves 4-6 pairs; petiole to 4 mm. Flowers in axillary fascicles, male and female flowers mixed, greenish yellow; male flowers pedicellate, female flowers sessile. Male flowers: up to 4 mm across, pedicels up to 1.1 cm long, slender, glabrous; tepals 6 in two series, broadly ovate, minutely pubescent on both sides; stamens

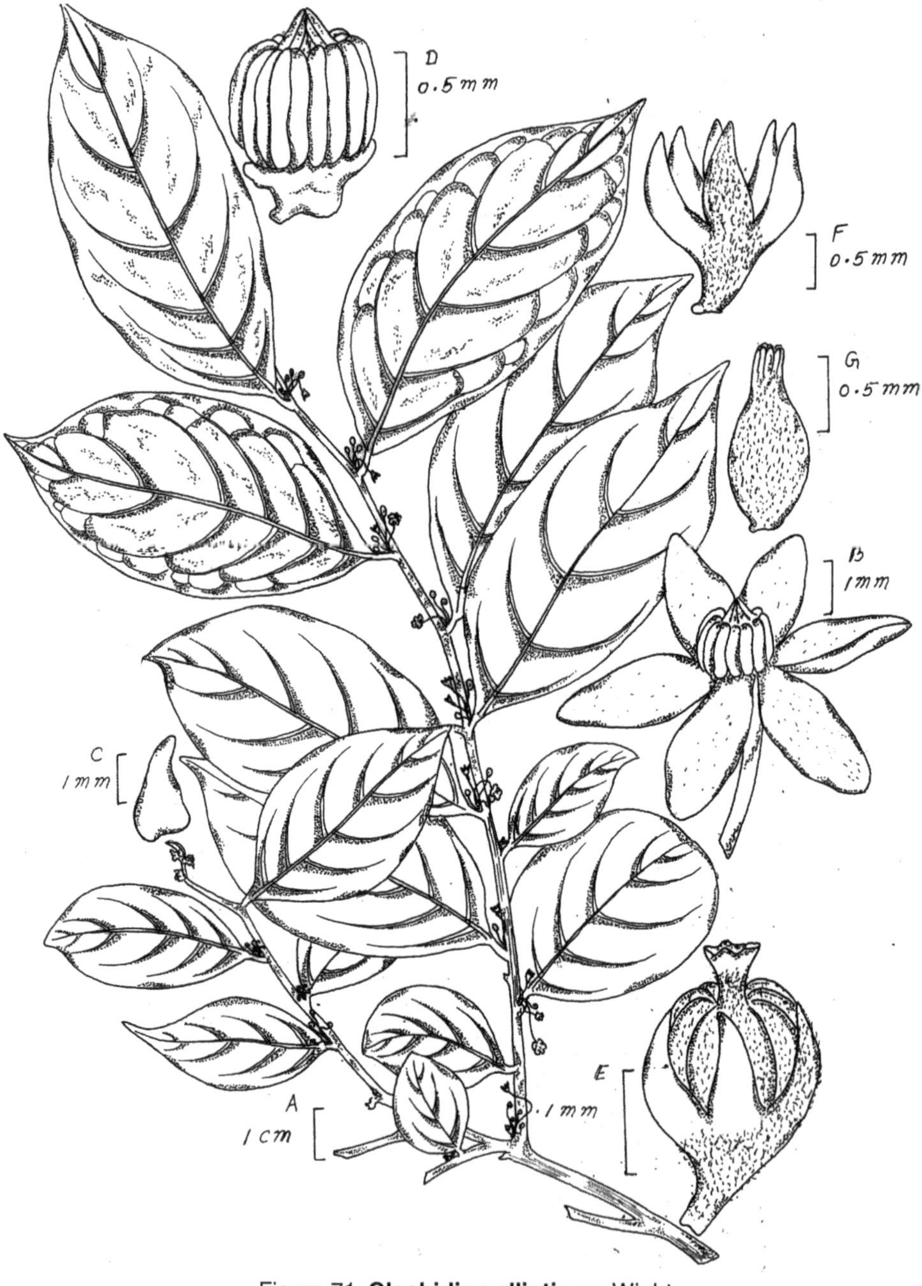

Figure 71. **Glochidion ellipticum** Wight
A. Twig, ♂ B. Flower, C. Bract, D. Androecium, ♀ E. Flower, F. Perianth, G. Pistil.

3, connectives produced. Female flowers: pedicels up to 0.5 mm long, villous; tepals 6, unequal, lanceolate, pubescent on both sides; ovary globose, pubescent, stylar column up to 0.7 mm long, 6-lobed. Capsule depressed globose, 1 cm across, 4-5-lobed, pubescent; seeds red.

Occasional on the slopes of moist dry forests. Fl. & Fr.: Throughout the year. Vern.: Tam.: *Thinakkolli, Panippadukkaanai.*

Anantagiri (VSKP), *PSPB* 11805; Birmisala (VSKP), *GVS* 47334 (MH).

INDIA: Uttar Pradesh, Bihar, West Bengal, Sikkim, Assam, Arunachal Pradesh, Meghalaya, Nagaland, Tripura, Odisha, Andhra Pradesh, Maharashtra, Karnataka, Kerala and Tamil Nadu.

WORLD: Nepal, Bhutan, Bangladesh, Myanmar, Thailand, China, Taiwan and Vietnam.

Glochidion heyneanum (Wight & Arn.) Wight, Icon. Pl. Ind. Orient. 5: t. 1908. 1852; FBI 5: 323. 1887; N.P.Balakr. & Chakrab., Fam. Euphorb. India 348. 2007; Chakrab. *et al.*, in N.P.Balakr. *et al.*, Fl. India 23: 421. 2012. *Gynoon heyneanum* Wight & Arn., Edinb. New Phil. J. 14: 300. 1833. *Glochidion velutinum* Wight, Icon. Pl. Ind. Orient. t. 1907. 1852; FBI 5: 322. 1887; Gamble 2: 1307. 1925; Surya Prakash Babu in Pullaiah & Moulali, Fl. Andhra Pradesh 2: 870. 1997. *Phyllanthus asperus* Müll.Arg., Flora 48: 377. 1865 and in DC., Prodr. 15(2): 297. 1866; FBI 5: 327. 1887.

Small tree, up to 8 m high; monoecious, bark brown and rough, wood crimson; branchlets zig-zag, velutinous. Leaves thinly coriaceous, elliptic-oblong or elliptic-lanceolate, 3.5-10 × 2-4 cm, base unequal, acute, margin entire, undulate, apex acute, thinly pubescent above, densely pubescent beneath; secondary nerves 6-8 pairs; petiole to 3 mm. Male flowers: yellow; pedicels up to 6 mm long, slender, pubescent; tepals 6, subequal, ovate, keeled, pubescent without; stamens 3, connectives connate, produced. Female flowers: green; tepals 6, subequal, lanceolate, pubescent on both sides, keeled without, persistent in fruit; ovary globose, hirsute, styles connate, column up to 3 mm long, pubescent, minutely 6-lobed. Capsule orbicular, depressed, 8-lobed, 0.4-0.7 cm, pubscent; seeds bright red in colour, flattened.

Ocasional, along the streams in moist deciduous forests at 600-1000 m. Fl.: April-May. Fr.: June-August.

Jyothi (CDP), *BR* & *SKB* 30543; Tirumala (CTR), *SSR* 13776; Tyda (VSKP), *PSPB* 11809; Gundlabrahmeswaram (KNL), *JLE* 16941 (MH); Tiger camp, Valamuru (EG), *MM* 105040 (BSID); Chintapalli (VSKP), *SKW* 2844 (BLAT); Galikonda (VSKP), *GVS* 11596 (MH); Sunkarimetta (VSKP), *HS* 20531 (BLAT); Similipahar (MBJ), *SX* 548 (RRL-B); Mahendragiri (GJM), *SX* & *MB* 211, *VNS* 5595 (MH); Laxmipur-Rayagada (KPT), *MB* & *Raut* 6981 (RRL-B).

INDIA: Almost through out the country.

WORLD: Pakistan, Nepal, Bhutan, Bangladesh, Myanmar and China.

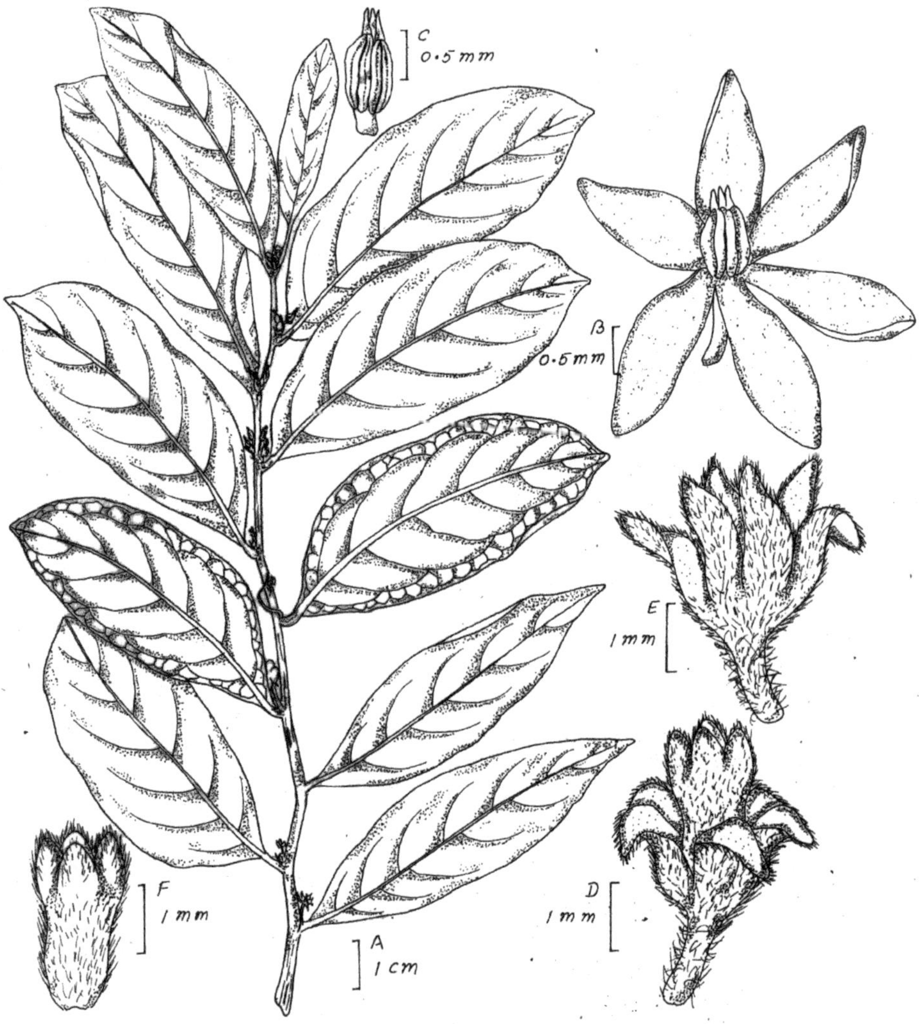

Figure 72. **Glochidion heyneanum** (Wight and Arn.) Wight ex Bedd. (Syn.: *Glochidion velutinum* Wight)

A. Twig, ♂ B. Flower, C. Androecium, ♀ D. Flower, E. Tepals, F. Pistil.

Glochdion lanceolarium (Roxb.) Voigt, Hort. Suburb. Calcutt. 153. 1845; FBI 5: 308. 1887 p.p.; N.P.Balakr. & Chakrab., Fam. Euphorb. India 348. 2007; Chakrab. *et al.*, in N.P.Balakr. *et al.*, Fl. India 23: 425. 2012.

Bushy shrubs or tree, 1.5-2 m high. Leaves elliptic, oblong or oblanceolate, apex apiculate to acuminate, base unequal, subacute; petiole 6-8 mm long. Male flowers numerous, clustered,axillary, yellow; perianth 2-seriate, outer spreading, inner erect, anthers 3-4. Female flowers 1-3, green, outer tepals oblong, inner narrower; ovary depressed or subglobose, styles united into a 6-8-toothed column. Capsule depressed-subglobose; seeds red.

Frequent in forests of Odisha Eastern Ghats. Fl. & Fr.: September – June.

Similipahar (MBJ), *SX* 316, 413, 437 (RRL-B), *MB* 5462, 5510 (RRL-B); Gandhamardhan (BGR), *SX* & *MB* 6200; Bamra, *Bal* 724 (DD); Govindnagar, Bamra, *HFM* 3747 (DD).

INDIA: Uttarakhand, Jharkhand, West Bengal, Sikkim, Assam, Meghalaya, Odisha, Chhattisgarh and Karnataka.

WORLD: Nepal, Bhutan, Bangladesh, Myanmar, Thailand, Cambodia, Laos, Vietnam and China.

Glochidion talakonense Sankara Rao *et al.*, Bangladesh J. Plant Taxon. 23(1): 59. 2016.

Trees, up to 7 m high; branches spreading; branchlets terete, puberulous when young, glabrescent at maturity, greenish-yellow. Leaves oblong to oblong-elliptic, 4.0-12.5 x 2.5-6.2 cm, asymmetric-rounded at base, entire, acuminate at apex, thinly coriaceous, glabrous, green above, glaucous beneath; lateral veins 7-8 pairs, prominent on both sides; petioles 4-5 x c. 2 mm, glabrous; stipules subulate, 1-2 mm long, puberulous. Inflorescences axillary, sessile, fascicled, 8- many-flowered, unisexual or bisexual. Male flowers: up to 18 in each fascicle, c. 2.0 x 1.8 mm, brownish-pink; pedicels filiform, up to 4 mm long; tepals 3+3, free, unequal, spreading and recurved, puberulous outside, glabrous inside; outer ones 1.8 x 1.3 mm, ovate, acute at apex; inner ones 1.8-2.0 x c. 1 mm, elliptic-oblong, rounded at apex; stamens 5; anthers connate into an oblong mass, c. 0.8 x 0.6 mm, connectives c. 0.3 x 0.3 mm; anther thecae linear, longitudinally dehiscent. Female flowers: many in each fascicle, 4 -5 x 4.5-4.7 mm, greenish with purple tinge; pedicels 2-4 mm long; tepals 3+3, free or occasionally shortly connate at base, unequal, tawny puberulous on both surfaces; outer ones 1.8-2.2 x 1.6-1.8 mm, ovate, acute at apex; inner ones 1.4- 2.0 x 0.8-1.2 mm, oblong, rounded or acute at apex; ovary subglobose, c. 2.3 x 2.6 mm, glabrous or puberulous, 6-locular, locules biovulate; styles columnar, inflexed at apex, c. 2.3 x 0.9 mm; apical lobes 6, linear or triangular, tawny-puberulous. Fruits capsular, 6-7 x 8-9 mm, shallowly lobed, slightly depressed, glabrous or puberulous. Seeds 12, glabrous.

Along waterfalls in moist deciduous forests at about 852 m elevation in Talakona forest growing in association with *Allophylus cobbe, Syzygium alternifolium* and *Phoenix loureiroi*. Fl. & Fr.: April - September.

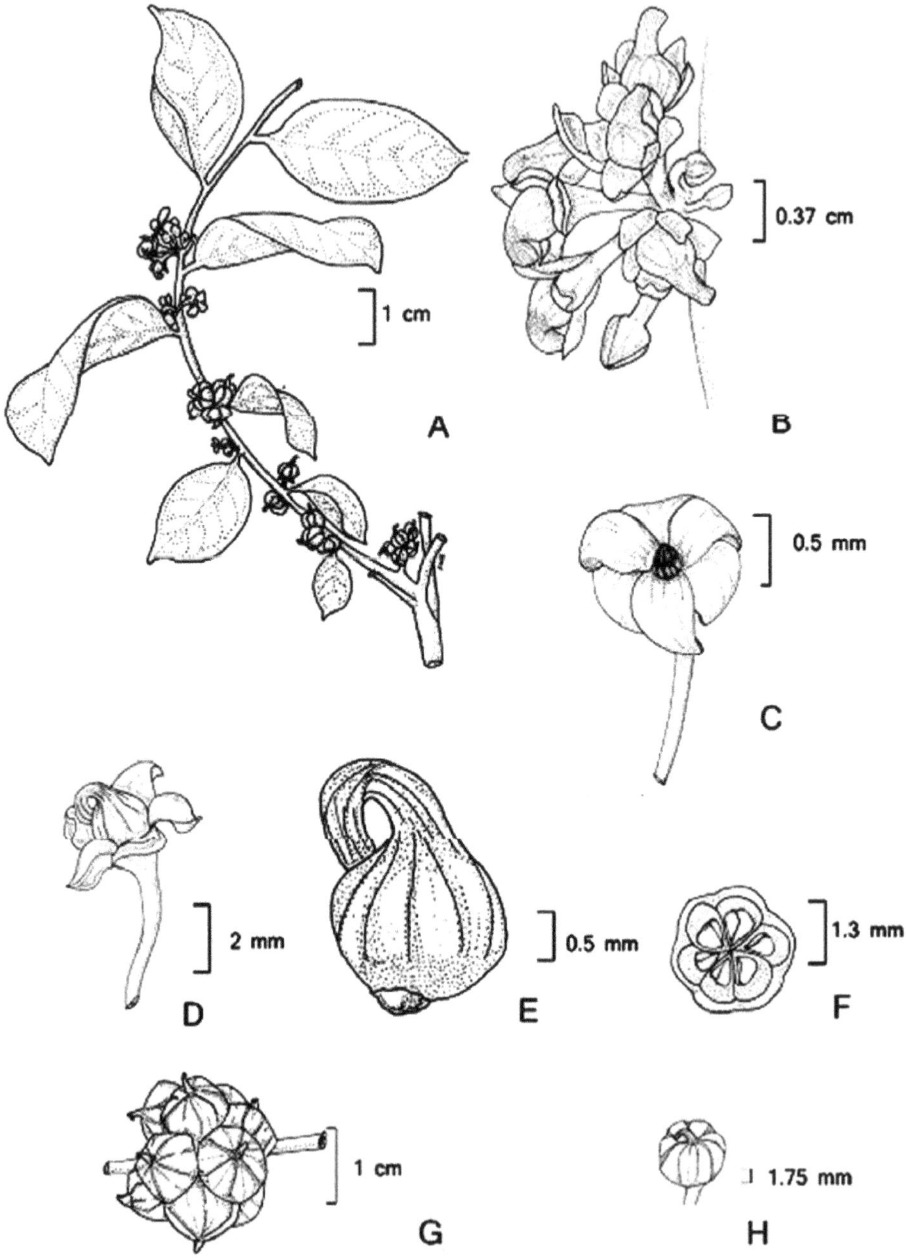

Figure 73. **Glochidion talakonense** Sankara Rao *et al.*
A. Flowering branch; B. Inflorescence; C. Male flower; D. Female flower;
E. Gynoecium; F. T.S. of ovary; G-H. Fruits
(After Sankara Rao *et al.*, Bangladesh J. Plant Taxonomy 23: 59-63. 2016).

Andhra Pradesh, Chittoor Dist.,Talakona hills, 852 m, 16 Apr 2014, *M. Sankara Rao & Party*, 4465 (*Holotype*: CAL; *Isotypes*: BSID); ibid.,9 Sep 2014, *M. Sankara Rao & Party* 5503 (*Paratypes*: BSID).

Etymology: The species is named after the type locality, "Talakona", a famous water fall in Seshachalam Biosphere Reserve in Chittoor district of Andhra Pradesh.

Glochidion tirupathiense Rasingam *et al.*, Taiwania 59: 9. 2014.

Small spreading tree, up to 4 m high; branchlets terete, puberulous when young, turning glabrescent on maturity. Leaves simple, alternate, oblong-elliptic to obovate, 6 12 × 2 4 cm, base cuneate or acute, margin entire, apex acute to apiculate or sometimes rounded, coriaceous, glabrous; midrib raised on both surfaces, lateral nerves 10 pairs, prominent on both sides; tertiary nerves scalariform, faint above, prominent beneath; petioles 3 4 mm long, glabrous; stipules ovate, acuminate, *ca.* 2 mm long, puberulous. Inflorescence axillary, fascicled, 3–6-flowered; both male and female flowers mixed or in separate fascicles. Male flowers: few, green; pedicels filiform, *ca.* 5 mm long, glabrous; tepals 3 + 3, free, subequal, oblong-elliptic, *ca.* 1.5 × 1 mm, glabrous; anthers 3, sessile with filaments connate into an oblong mass, *ca.* 0.8 mm long, connectives produced into an erect, ovate appendages, pistillode absent. Female flowers: many; pedicels 2 5 mm long, glabrous; tepals 6, free, subequal, ovate, obtuse, *ca.* 1.5 × 1 mm, white puberulous, ciliate along margins; ovary urn-shaped, hairy, 4-5-locular, each locule bi-ovulate; style columnar, *ca.* 2 × 1 mm, unlobed, as broad as the ovary from base. Fruits capsular, depressed, *ca.* 1 cm in diam., 4 or 5-locular, deeply lobed; each lobe further bilobulate, coriaceous, glabrous with *ca.* 1 mm long remnant of columnar style at apex; seeds hemispherical, 3.3 × 2.7 mm, with a red arillate coat, shining, glabrous.

Rare along stream banks in moist deciduous forests at about 1000 m altitude growing in association with *Mimusops elengi* L., *Memecylon umbellatum* Burm.f., *Syzygium alternifolium* (Wight) Walp.,*Terminalia pallida* Brandis, *Phoenix loureiroi* Kunth, *Buchanania angustifolia* Roxb. and *Carallia brachiata* (Lour.) Merr. Fl. & Fr.: June November.

Tirumala hills, on the way to Kumaradara Pusupudara Dam (CTR), *LR, M. Sankara Rao & Alok R. Chorghe* 2901 (Holo: CAL; Iso: BSID); Paratype: Tirumala hills, on the way to Kumaradara Pusupudara Dam (CTR), *PVP & M. Sankara Rao* 3205 (BSID).

INDIA: Andhra Pradesh. Endemic.

Glochidion zeylanicum (Gaertn.) A. Juss., Euphorb. Gen. 107. 1824; FBI 5: 310. 1887; Gamble 2: 1306. 1925; N.P.Balakr. & Chakrab., Fam. Euphorb. India 353. 2007; Chakrab. *et al.*, in N.P.Balakr. *et al.*, Fl. India 23: 429. 2012. *Bridelia zeylanica* Gaertn., Fruct. Sem. Pl. 2: 128. t. 109. 1790.

Key to Varieties

1. All parts (except ovary) glabrous .... var. **zeylanicum**
1. All parts pubescent .... var. **tomentosum**

Glochidion zeylanicum (Gaertn.) Juss. var. **tomentosum** (Dalzell) Trimen, Handb. Fl. Ceylon 4: 29. 1898. *G. tomentosum* Dalzell, Hooker's J. Bot. Kew. Gard. Misc. 2: 38. 1850; FBI 5: 309. 1887; Gamble 2: 1307. 1925; Surya Prakash Babu in Pullaiah & Moulali, Fl. Andhra Pradesh 2: 870. 1997.

Small tree, up to 8 m high; monoecious; bark black, rough; branchlets tomentose. Leaves elliptic-oblong, 8-14 × 3.5-6 cm, base obliquely rounded, margin entire, apex acute-acuminate; pubescent on nerves above, finely grey tomentose beneath; secondary nerves ca. 7 pairs; petiole to 3 mm. Flowers pedicellate in axillary fascicles. Capsule depressed globose, 7 mm across, conical, 10-12-lobed, pubescent, beaked.

Occasional, in the deciduous forests at 800-1200 m. Fl.: January-February. Fr.: April-July.

Galikonda (VSKP), *SSR* 15929; Maredumilli (EG), *GVS* 24273, 24695 (MH), *MM* 100814 (BSID).

INDIA: Almost through out the country.

WORLD: Sri Lanka, Bhutan, Bangladesh, Myanmar, Thailand, China and Taiwan.

Glochidion zeylanicum (Gaertn.) Juss. var. **zeylanicum**

Tree, up to 8 m high; bark grey, thin, corky; branchlets zig-zag, coppery, glabrous. Leaves coriaceous, ovate-lanceolate or oblong, 13-17.5 × 5-7.5 cm, glabrous, glaucous above, paler beneath, base unequal, truncate-rounded, margin entire, apex sub-acuminate; secondary nerves 6-10 pairs; petiole 5 mm, thick. Flowers greenish, in axillary subumbellate clusters. Male flowers: pedicels 5-10 mm long; sepals 3 + 2-3, oblong, ovate, suborbicular or obovate, anther 3-8. Female flowers: sessile to pedicellate; pedicels 1-8 mm long; sepals 3 + 2-3, triangular, deltoid, oblong, ovate or suborbiculaar, occasionally subglobose; ovary subglobose, depressed, 4-7-locular, styles free, connivent to erect or united and columnar, conical to clavate. Capsule globose, 0.6 × 0.8 cm, obscurely lobed, glabrous, often beaked; seeds shining, wedge-shaped.

Common, along the streams and in moist places of semi-evergreen forests. Fl. & Fr.: Throughout the year. Vern.: Tel.: *Pageri, Itepulla, Neeru mamidi*; Tam.: *Kumbalaa, Kokkamani maram*; Ori.: *Kalchia, Berlu.*

Saddicudumadugu (ATP), *PSPB* 9173; Gundlabrahmeswaram (KNL), *SSR* 15995; Tirumala (CTR), *SSR* 15269, 15284; Japali (CTR), *KS* 7862 (MH); Talakona (CTR), *SKW* 5789 (BLAT); Malleswara kona, Venkatagiri (NLR), *ASR* 6108 (BSID); Peddarutla (PKM), *VBH* 84940; Dummakonda (EG), *GVS* 68631 (MH); Nulakamaddi (EG), *GVS* 24542 (MH); Sapparla gadda (VSKP), *GVS* 42781 (MH); Gandhamardhan (BGR), *SX* & *MB* 6175, 6402, 6429 (RRL-B); Barogharo (KPT), *VNS* 5985 (MH).

INDIA: Assam, Tripura, Odisha, Chhattisgarh, Andhra Pradesh, Maharashtra, Goa, Karnataka, Kerala, Tamil Nadu and Andaman and Nicobar Islands.

WORLD: Sri Lanka, Bangladesh, Myanmar, Thailand, Malesia to N. Australia.

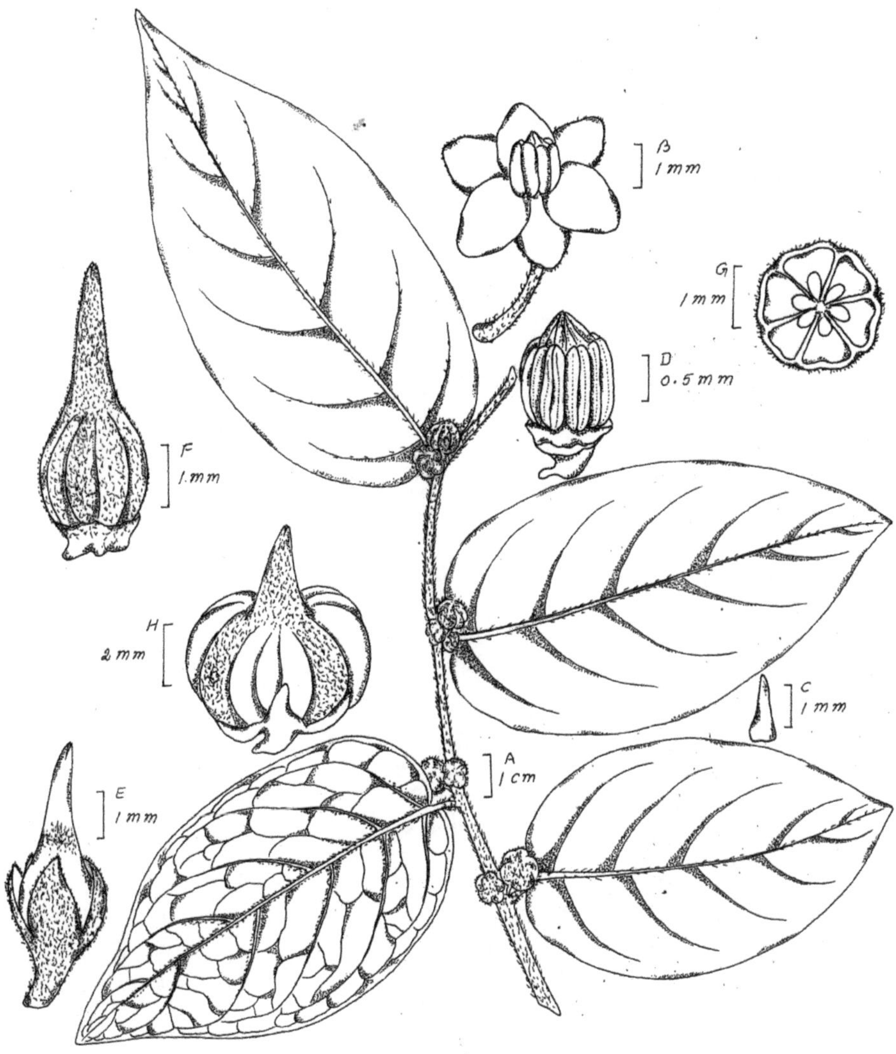

Figure 74. **Glochidion zeylanicum** (Gaertn.) Juss. var. **tomentosum** (Dalzell) Trimen
A. Twig, ♂ B. Flower, C. Bract, D. Androecium, ♀ E. Flower, F. Pistil, G. Ovary T.s.

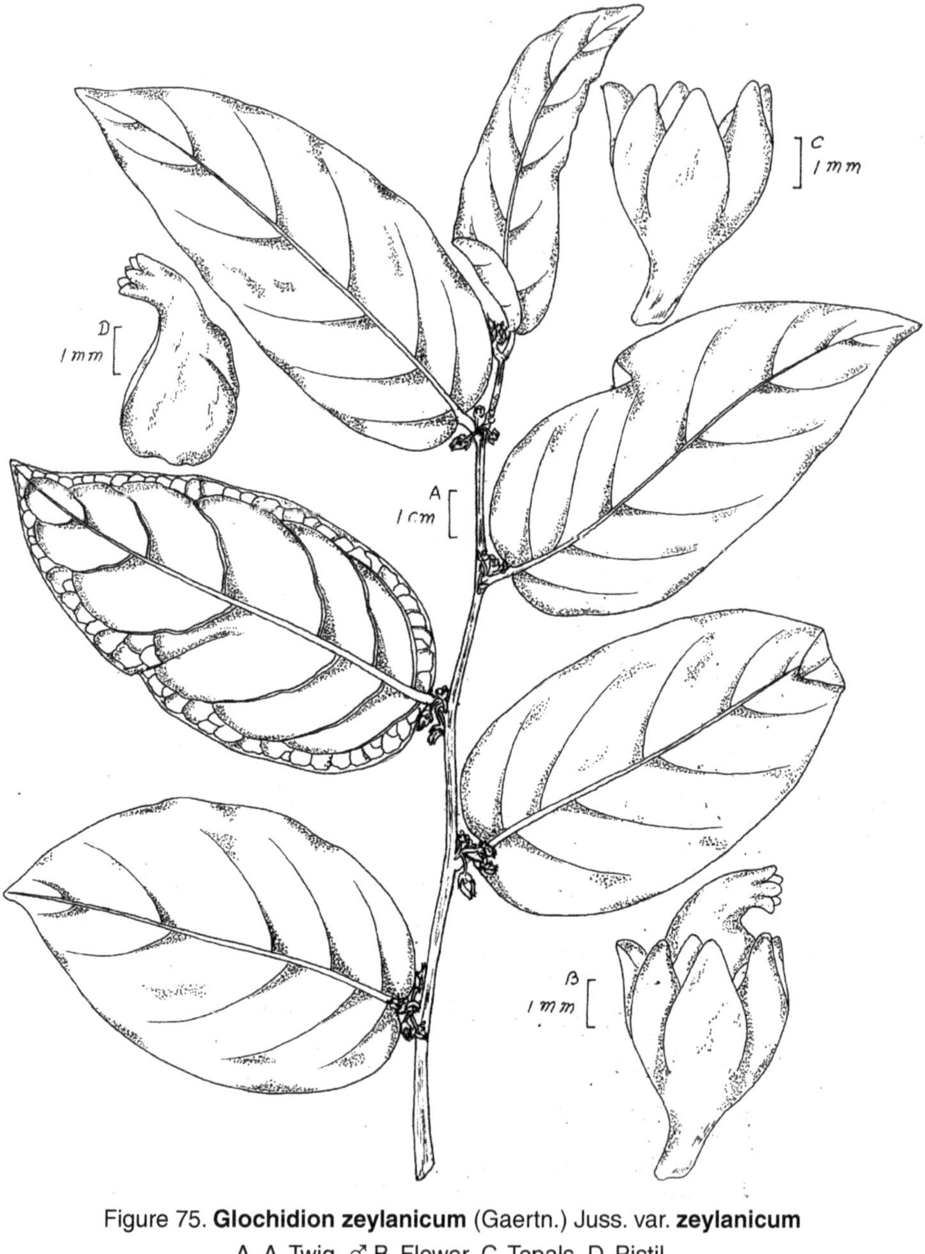

Figure 75. **Glochidion zeylanicum** (Gaertn.) Juss. var. **zeylanicum**
A. A. Twig, ♂ B. Flower, C. Tepals, D. Pistil.

HOMONOIA Loureiro

1. Leaves densely lepidote-scaly beneath; spikes 5-10 cm long .... **H. riparia**
1. Leaves not densely lepidote-scaly beneath; spikes not more than 2 cm long:
    2. Leaves sparsely lepidote-scaly beneath; male and female spikes 1-2 cm long .... **H. retusa**
    2. Leaves moderately lepidote-scaly beneath; male spikes up to 1 cm lomg, female spikes 3-5 cm long............... **H. intermedia**

Homonoia intermedia Haines, Bot. Bihar Odisha 2: 111. 1921; N.P.Balakr. & Chakrab., Fam. Euphorb. India ?137. 2007; Susila Rani & N.P.Balakr. in N.P. Balakr. *et al.*, Fl. India 23: 123. 2012.

Shrub, 2-3 m high. Leaves oblong to obovate, 5-11 × 1-1.6 cm, apex rounded-truncate, margin coarsely serrate, base attenuate, petiole 2-5 mm long; stipules subulate. Flowers in axillary, few-flowered spikes, male spikes 1 cm long, female spikes 3-5 cm long; bracts ovate-lancceolate; bracteoles 2, lanceolate; Male flowers clustered; sepals 3, ovate-lanceolate; stamens numerous. Female flowers sessile, calyx segments 5 or 6, ovate-lanceolate, ovary globose, appressed pubescent, styles 3, connate. Fruits globose with 2 or 3-valved cocci, dark brown.

In Odisha Eastern Ghats along river and stream banks. Fl. & Fr.: March – September.

Likapara, Angul, near Mahanadi, *Haines* 2505 (DD).

INDIA: Bihar, Odisha, Madhya Pradesh, Maharashtra, Goa, Karnataka, Kerala, Tamil Nadu. Endemic.

Homonoia retusa (J. Graham ex Wight) Müll.Arg., Linnaea 34: 200. 1865; FBI 5: 456. 1887; Gamble 2: 1333. 1925; N.P.Balakr. & Chakrab., Fam. Euphorb. India 137. 2007; Susila Rani & N.P.Balakr. in N.P. Balakr. *et al.*, Fl. India 23: 125. 2012. *Adelia retusa* J. Graham ex Wight, Icon. Pl. Ind Orient. 5: t 1869. 1852.

Shrub, up to 2 m high, dioecious; branchlets stout, woody. Leaves 1.5-3.5 × 1-2.5 cm, obovate or obovate – oblanceolate, base acute – crenate, margin serrate – dentate, apex obtuse – retuse, coriaceous, sparsely glandular below; petiole to 2 mm. Flowers in short spikes, sessile; bracts ovate, acute, pubescent. Male flowers 3 mm across; tepals 3, 3 × 2.5 mm, broadly elliptic, concave; stamens many, anthers 0.4 × 0.3 mm. Female flowers 2 mm across; tepals 6, 1.8 × 0.7 mm, linear lanceolate, acute; ovary 1.5 mm, globose, styles 3, simple, papillose. Capsule 3.5 mm, globose; seeds rounded on the back, slightly angular on the inner face, smooth, yellow.

Rare in the river beds. Fl. & Fr.: January – April.

Koruturu (EG), *DCSR* 66 (CAL), *DN* 84285, 84286; Godavari river (EG), *JSG* 15857 (DD).

INDIA: Odisha, Madhya Pradesh, Andhra Pradesh, Maharashtra, Karnataka, Kerala and Tamil Nadu. Endemic.

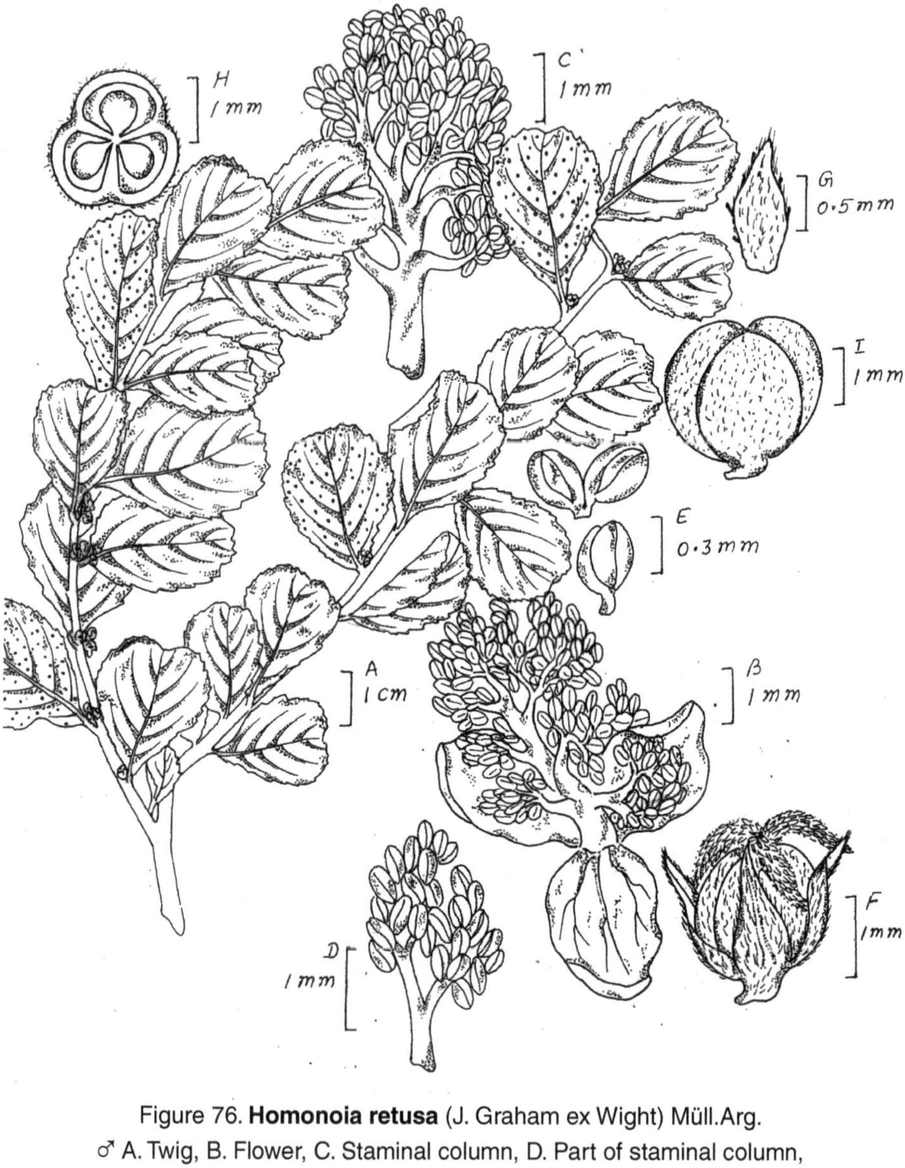

Figure 76. **Homonoia retusa** (J. Graham ex Wight) Müll.Arg.
♂ A. Twig, B. Flower, C. Staminal column, D. Part of staminal column,
E. Anthers, ♀ F. Flower.

Homonoia riparia Lour., Fl. Cochinch 2: 637. 1790; FBI 5: 455. 1887; Gamble 2: 1333. 1925; N.P.Balakr. & Chakrab., Fam. Euphorb. India 137. 2007; Susila Rani & N.P.Balakr. in N.P. Balakr. *et al.*, Fl. India 23: 125. 2012. *Adelia neriifolia* Heyne ex Roth, Nov. Pl. Sp. 375. 1821.

Shrub, up to 3 m high, dioecious; branchlets pubescent. Leaves 13-18 × 1-2 cm, linear lanceolate, base acute – rounded, margin entire, apex acute, orbicular, scaleous and pubescent on veins below; petiole to 9 mm; stipules linear – lanceolate, acute. Flowers in axillary spikes, sessile, peduncle pubescent; bracts ovate, acuminate, pubescent. Male flowers: 4 mm across; tepals 3, 3 × 2 mm, free, broadly elliptic, concave, acute, puberulous without; stamens many; anthers 0.2 mm across. Female flowers: 3 mm across; tepals 5, 1.5 × 0.6 mm, free, ovate, acute – subacuminate, puberulous without; ovary 1 mm, globose, puberulous, styles 3, simple, papillose. Capsule 4 mm, puberulous; seeds rounded on the back, slightly angular on the inner face, smooth, yellow.

Common in the rocky crevices of streams and along riverbeds mixed with rocky boulders. Fl. & Fr.: November-January. Vern.: Tel.: *Taniki chettu*; Tam.: *Aatthu Inchi;* Ori.: *Pani begonia, Pani jhovali, Godamalla.*

Gundlabrahmeswaram (KNL), *PSPB* 9130; Chalama (KNL), *PSPB* 11188; Nandikomma (KNL), *JSG* 10933 (DD); Udaroda falls (CDP), *JSG* 20973 (DD); Talakona (CTR), *SKW* 5783 (BLAT); Venkatagiri, M. Kona (NLR), *ASR* 6495 (BSID); Rampachodavaram (EG), *VNS* 212 (CAL), *SKW* 1835 (BLAT); Pilligedda (VSKP), *GVS* 47399 (CAL); Singupuram (SKLM), *GVS* 62328 (MH).

INDIA: Almost through out India.

WORLD: Sri Lanka, Myanmar, China, Cambodia, Laos, Vietnam, Thailand, Malesia.

JATROPHA Linnaeus

1. Stipules well-developed, divided into narrow lobes or often modified into decurrent series of stalked glands:
    2. Lobes of leaves entire, glandular hairy along margins; flowers reddish or purplish red .... **J. gossypiifolia**
    2. Lobes of leaves serrate, each tooth ending in gland-tipped bristle; flowers green with pinkish tinge;
        3. Leaves shortly palmately lobed, up to or above the middle; stipules short, with only a few gland-tipped segments; petals in female flowers connate to one-third the length; staminodes present in female flowers, almost stamen-like but much smaller and sterile .... **J. tanjorensis**
        3. Leaves deeply palmately lobed up to below the middle; stipules prominent, deeply divided into several gland-tipped segments, petals free or almost so to the base in female flowers; staminodes absent in female flower .... **J. glandulifera**

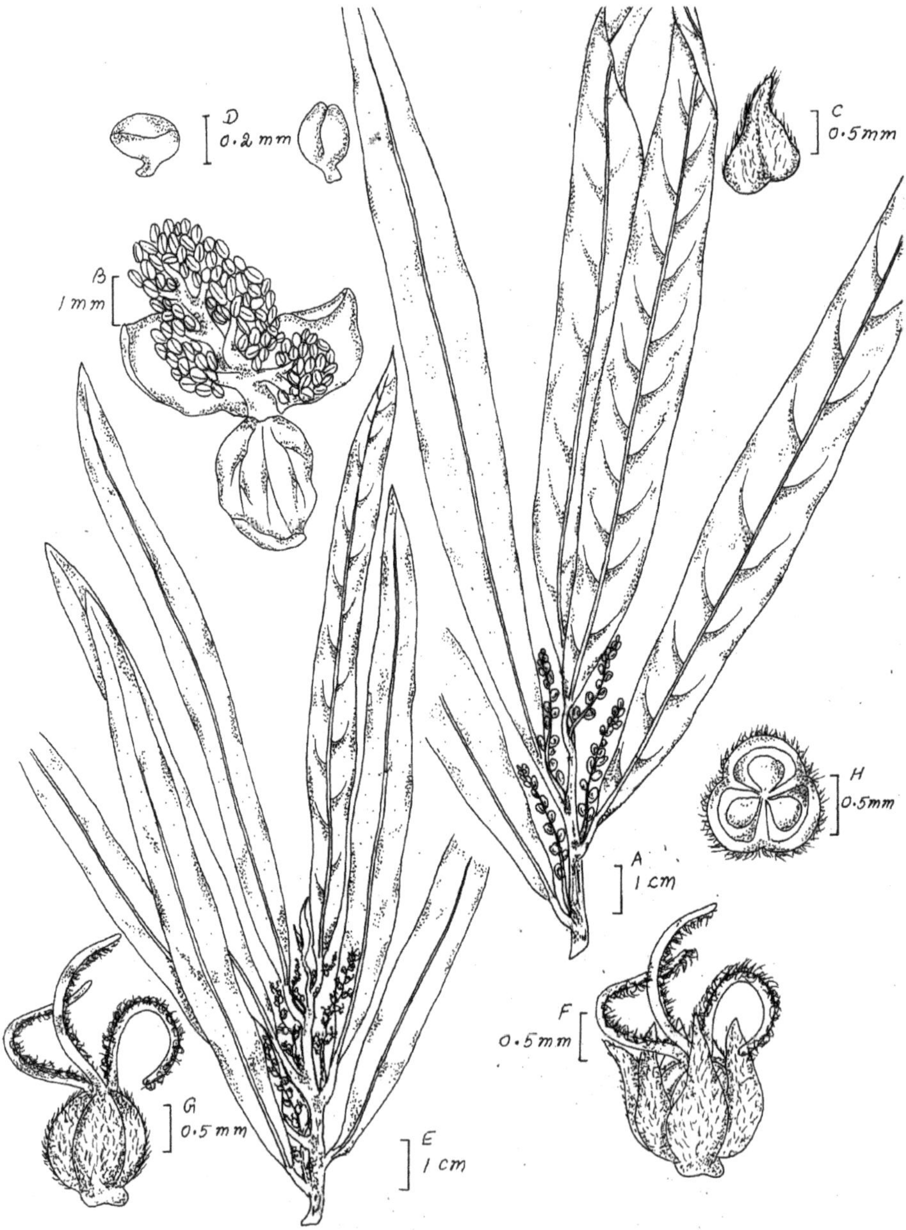

Figure 77. **Homonoia riparia** Lour.
♂ A. Twig, B. Flower, C. Bract, D. Anthers, ♀ E. Twig, F. Flower, G. Ovary, H. Ovary T.s.

1. Stipules absent or minute and subulate:
    4. Leaves peltate, 3-7-lobed .... **J. villosa**
    4. Leaves not peltate, 3-7-angular or lobed:
        5. Small trees or large shrubs; stem base not tuberous; leaves not glaucous beneath; petals united for half the length .... **J. curcas**
        5. Unbranched stunted shrubs; stem base tuberous; leaves glaucous beneath; petals free or connate at base only .... **J. heynei**

Jatropha curcas L., Sp. Pl. 1006. 1753; FBI 5: 383. 1887; Gamble 2: 1340. 1925; N.P.Balakr. & Chakrab., Fam. Euphorb. India 223. 2007; N.P.Balakr. in N.P. Balakr. *et al.*, Fl. India 23: 250. 2012.

Large branched shrubs or small trees, up to 5 m high; monoecious; bark greenish-white, smooth, peeling off in thick flakes; wood white, very soft; branchlets glabrous. Leaves orbicular-cordate, 5-12 × 3.5-13 cm, glabrous, shallowly 3 to 5-lobed, lobes triangular, base cordate, margin entire, apex acuminate; petiole to 15 cm. Flowers greenish-white in axillary cymes. Male flowers: sepals ovate or elliptic, 3-4 × 1-2.5 mm; petals 5-10, united up to the middle, oblong or obovate, yellowish green; disc columnar, deeply 5-lobed; stamens 8-10, outer filaments free, inner 5 connate at base. Female flowers: sepals longer than in male flowers; petals absent or same as in male flowers, but free and smaller; ovary 3-locular, styles short and connate at base. Capsule globose, 3 cm across, rugose.

Common, semi-wild in forests and villages; and also grown as hedge plant along roadsides and out skirts of villages. Fl.: April-June. Fr.: December-January. Vern.: Tel.: *Adavi amudam, Adavi-nepalamu, Pedda-nepalamu, Kondamudam,Nepalam;* Tam.: *Kattukkottai, Kaatamanakku;* Ori.: *Bamprijada, Jahajin, Dhuma jara, Jode, Dumia.*

Bukkapatnam (ATP), *NY* 1036; Devipatnam (EG), *SSR* 15254; G. Madugula (VSKP), *SSR* 15958; Sirivaaka, near Koruturu (WG), *DN* 84248 (BSID); Maredumilli (EG), *MM* 100839 (BSID).

INDIA: Rajasthan, Gujarat, Madhya Pradesh, Uttar Pradesh, Bihar, West Bengal, Sikkim, Odisha, Andhra Pradesh, Maharashtra, Goa, Karnataka, Kerala, Tamil Nadu and Andaman & Nicobar Islands.

WORLD: Native of Brazil, introduced and sometimes in cultivation throughout the tropics.

Jatropha glandulifera Roxb., Fl. Ind. 3: 688. 1832; FBI 5: 382. 1887; Gamble 2: 1340. 1925; N.P.Balakr. & Chakrab., Fam. Euphorb. India 224. 2007; N.P.Balakr. in N.P.Balakr. *et al.*, Fl. India 23: 251. 2012.

Shrub, up to 2.5 m high, monoecious; branchlets stout, short. Leaves 7-10 × 8-11 cm, palmately 3 to 5-lobed, posterior lobes smaller than anterior lobe, lobes obovate, base cordate, margin serrate, serratures gland tipped, apex acuminate; petiole to 5.5 cm, with glandular hairs; stipules divided into many glandular headed, filiform

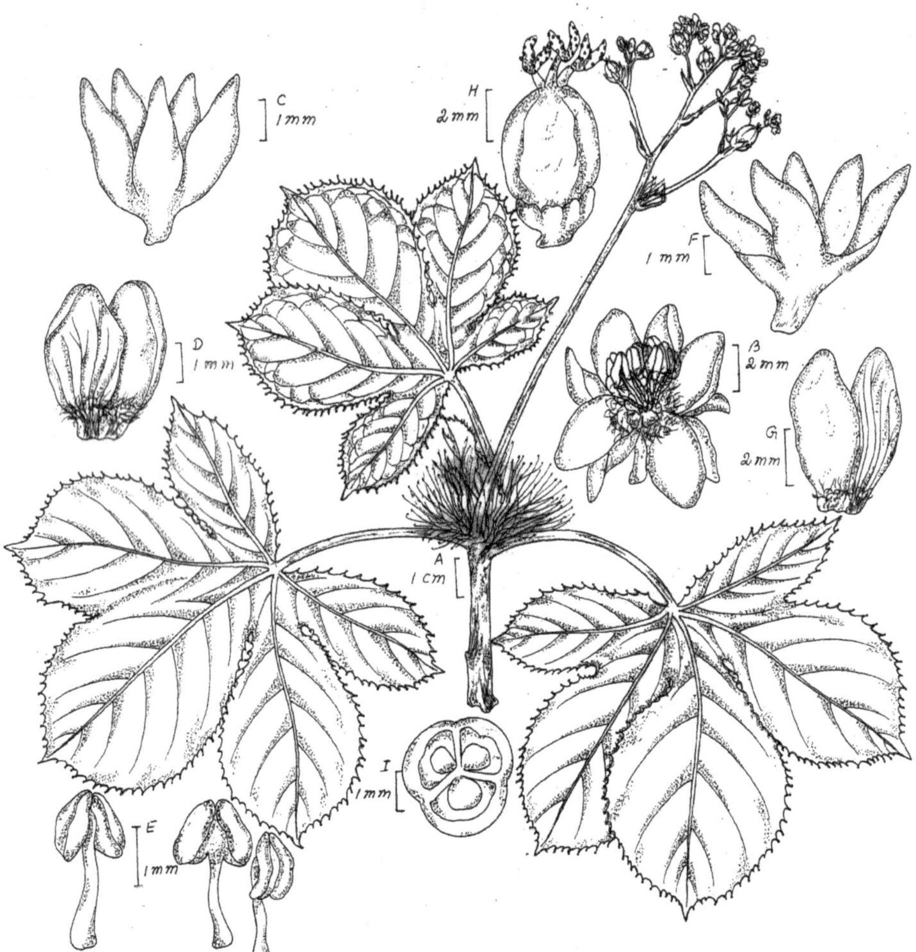

Figure 78. **Jatropha glandulifera** Roxb.
A. Twig, ♂ B. Flower, C. Outer tepals, D. Inner tepals (part), E. Stamens,
♀ F. Outer tepals, G. Inner tepals (part), H. Pistil, I. Ovary T.s.

segments. Flowers in axillary cymes; bracts linear – lanceolate, acute, glandular hairy. Male flowers: outer (sepals) 3.5 × 1.5 mm, slightly connate at base, lanceolate, obtuse; inner (petals) 4 × 2 mm, slightly connate at base, obovate, rounded, villous at base within; stamens 8, biseriate, 5 + 3; filaments 1.5 mm; anthers 1.1 mm across; disc of glands 5, free. Female flowers 5.5 mm across; tepals in two whorls, 5 + 5; outer (sepals) 2.5 × 1.2 mm, slightly connate at base, lanceolate, acute; inner (petals) 6 × 2.8 mm, slightly connate at base, obovate, rounded, villous at base within; ovary 4.5-3.5 mm; styles 3, each 2-fid, tuberculate; disc of glands 5, free. Capsule 2 cm, subglobse.

A weed of waste places especially in black cotton soils and along roadsides. Fl. & Fr.: August - January. Vern.: Tel.: *Dundagaku*; Tam.: *Vellaikkaattukottai, Aadalai.*

Somayajulapalli (KNL), *RVR* 1879; Mabbuchintalapalle (CDP), *KRKS* 36907; Kondavedu (GNT), *SKW* 3735, 3736 (BLAT).

INDIA: Andhra Pradesh, Karnataka, Kerala and Tamil Nadu.

WORLD: Sri Lanka.

Jatropha gossypiifolia L., Sp. Pl. 1006. 1753 (*"gossypifolia"*)

Key to Varieties

1. Young leaves dark purple or bronze red, gradually becoming reddish green on maturity, glaucous, sparingly hairy at the divergence of primary veins; all parts including stalks of viscid glands purplish; styles connate at base; capsule smooth or sparingly hirsutulous .... var. **elegans**

1. Young leaves purplish green or greenish as they mature; densely hairy all over; all parts including stalks of viscid glands greenish; styles not forming a column, free; capsule densely hirsutulous .... var. **gossypiifolia**

Jatropha gossypiifolia L. var. **elegans** (Pohl) Müll.Arg. in DC., Prodr. 15 (2): 1087. 1866; N.P.Balakr. & Chakrab., Fam. Euphorb. India 224. 2007; N.P.Balakr. in N.P.Balakr. *et al.*, Fl. India 23: 252. 2012. *Adenoropium elegans* Pohl, Pl. Bras. Icon. Descr. 1: 15. 1827; FBI 5: 383. 1887; Gamble 2: 1340. 1925.

Shrub, to 3 m high, monoecious; branchlets dark purple, glandular hairy. Leaves 4-10 × 4-11 cm, glaucous, cordiform, deeply 3 to 5-lobed, lobes obovate, base cordate, margin entire, glandular hairy, apex acute; petiole to 7 cm, with many fasicled and branched gland tipped hairs; stipules lanceolate, glandular; young leaves including petioles and glands dark purple, gradually becoming green. Flowers in axillary cymes; pedicel to 5 mm; bracts lanceolate. Male flowers 6.5 mm across; tepals in two whorls, 5 + 5; outer (sepals) 3.5 × 1.1 mm, connate at base, lanceolate, acuminate, glandular hairy; inner (petals) 5 × 3 mm, free, campanulate, ovate, obtuse; stamens 8, biseriate, 5 + 3, filaments connate at base, anthers 0.9 mm across. Female flowers 7.5 mm across; tepals in two whorls, 5 + 5, outer (sepals) 6.5 × 2 mm, connate at base, lanceolate, acuminate, glandular hairy; inner (petals) 5.5 × 3.5 mm, free, campanulate, ovate, obtuse; ovary 2 mm, hairy, styles 3, stylar column short, each 2-fid; disc of glands 5, free. Capsule 2 cm, subglobose, sparingly hirsutulous; seeds oblong.

Common in waste lands along road sides. Fl. & Fr.: Throughout the year.

Kalasamudram (ATP), *EC* 2883; Lower Ahobilam (KNL), *RVR* 2755; Atmakur (KNL), *PSPB* 9139; Mabbuchintalapalli cross (CDP), *KRKS* 36942; Lower Srisailam (KNL), *JLE* 16845 (CAL); Kota (EG), *VNS* 527(CAL); Dolphin's nose (VSKP), *TAR* 7175 (CAL), *SKW* 2372, 2374 (BLAT); Salur (VZN), *NPBK* 891.

INDIA: Andhra Pradesh. Introduced.

WORLD: Native of Brazil.

Jatropha gossypiifolia L. var. **gossypiifolia**. N.P.Balakr. in N.P.Balakr. *et al.*, Fl. India 23: 252. 2012. *Adenropium gossypiifolium* (L.) Pohl, Pl. Bras. Icon. Descr. 1: 16. 1827. *Jatropha staphysagrifolia* Mill., Gard. Dict. ed. 8. 1768. *J. gossypiifolia* L. var. *staphysagriaefolia* (Mill.) Müll.Arg. in DC., Prodr. 15(2): 1087. 1866; FBI 5: 383. 1887; Gamble 2: 1340. 1925.

Shrub, to 3 m high, monoecious; branchlets green. Leaves densely puberulous, cordiform, deeply 3 to 5-lobed, base cordate, margin entire, glandular hairy, apex acute; petiole with many fascicled and branched, gland tipped hairs; young leaves all parts including petioles and glands green. Flowers in axillary cymes; ovary densely hirsutulous; styles 3, independent. Capsule densely hirsutulous; seeds oblong.

Common in waste lands along road sides, near the tanks. Fl. & Fr.: Throughout the year. Vern.: Tel.: *Seema nepalamu*; Tam.: *Aatalai;* Ori.: *Johajim, Baigoba, Baigawa, Nonkakalo.*

Mudurlanka (EG), *MM* 102568 (BSID).

INDIA: Gujarat, Rajasthan, Uttar Pradesh, Bihar, West Bengal, Assam, Odisha, Madhya Pradesh, Maharashtra, Andhra Pradesh, Karnataka, Kerala, Tamil Nadu and Andaman Islands.

WORLD: Native of New World, introduced and often naturalized in many tropical countries.

Jatropha heynei N.P.Balakr., Bull. Bot. Surv. Ind. 3: 40. 1962; N.P.Balakr. & Chakrab., Fam. Euphorb. India 225. 2007; N.P.Balakr. in N.P.Balakr. *et al.*, Fl. India 23: 253. 2012. *J. heterophylla* Heyne ex Hook. f., Fl. Brit. India 5: 382. 1887, non Steud. 1840; Gamble 2: 1339. 1925.

Shrub, up to 2 m high, monoecious; branchlets terete. Leaves 7-11 × 3.5-6.5 cm, oblanceolate, base cuneate, shortly decurrent, margin entire, apex acute; petiole to 3.5 cm, stipules linear – lanceolate. Flowers in terminal cymes; bracts lanceolate. Male flowers 4 mm across; tepals in two whorls, 5 + 5; outer (sepals) 3.3 × 1.1 mm, connate at base, lanceolate at base, oblong, obtuse; stamens 8-10; filaments mostly connate, to 2 mm; anthers 1.6 mm across. Female flowers 6 mm across; tepals in two whorls, 5 + 5; ovary subglose, glabrous; styles 3, connate at base, each 2-fid; disc of glands 5, free. Capsule subglobose; seeds ovoid, carunculate.

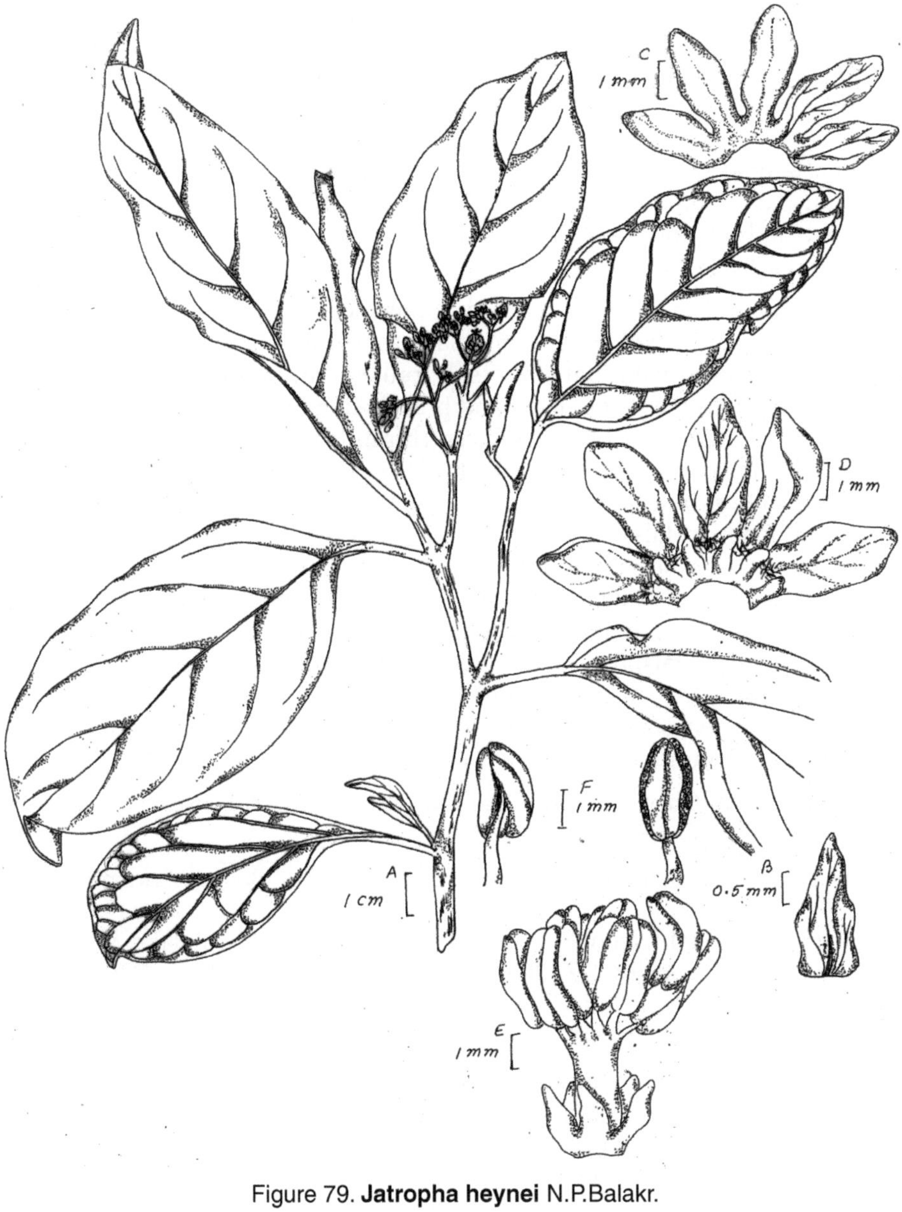

Figure 79. **Jatropha heynei** N.P.Balakr.

A. Twig, ♂ B. Bract, C. Outer tepals split open, D. Inner tepals split open, E. Androecium, F. Anthers.

Occasional in dry deciduous forests. Fl. & Fr.: Throughout the year. Vern.: Tel.: *Karuamudam dumpa, Adavi donda.*

Diguvametta (KNL), *CECF* 4655 (CAL); Lanjabanda (KNL), *RVR* 1589; Malakondapenta (PKM), *RKM* 0797 (CAL), *VBH* 83984 (BSID); Titlagarh, Bolangir, *HFM* 3454 (DD).

INDIA: Madhya Pradesh, Andhra Pradesh, Karnataka and Tamil Nadu. Endemic.

Jatropha tanjorensis J.L.Ellis & Saroja, J. Bombay Nat. Hist. Soc. 58: 834. 1962; Babu *et al.*, J. Bombay Nat. Hist. Soc. 91: 163. 1994; N.P. Balakr. & Chakrab., Fam. Euphorb. India 227. 2007; Susila Rani & N.P. Balakr. in N.P. Balakr. *et al.*, Fl. India 23: 254. 2012.

Shrub, up to 1 m high, polygamous, branchlets striate. Leaves 3-8 x 3.5-7 cm, cordiform. Shortly 5-lobed, ovate-triangular, base cordate, margin serrate, serratures gland-tipped, apex acuminate, sparsely pubescent; petiole to 6 cm, sparsely pubescent; stipules many, fascicled and branched, gland tipped hairs. Flowers in terminal cymes; pedicel to 3 mm, pubescent; bracts lanceolate, glandular hairy. Male flowers 4-5 mm across, tepals in two whorls, 5+5, outer (sepals) 4-4.5 x 1.8-2 mm, slightly connate at base, unequal, lanceolate, acuminate, hairy without, glandular at margins; inner (petals) 2.5 x 1.5 mm, slightly connate at base or free, obovate, obtuse, villous near the base within; stamens 8, free, filaments 0.5 mm, anthers 0.9 mm across, berded. Bisexual flowers 9 mm across, tepals in two whorls, 5+5, outer (sepals) 7-9 x 2.5-3 mm, connate at base, unequal, lanceolatee, acuminate, glandular and hairy; inner (petals) 5 x 2.5 mm, free, obovate, obtuse, villous near the base within, stamens 8, free, filaments to 1.1 mm, anthers 1 mm across, bearded; ovary 2 x 1.5 mm, subglobose, styles 3, connte at base, each 2–fid; disc of glands 5, free. Capsule 3 x 2.5 mm, subglobose.

Occasional in Chengalpattu, Dharmapuri, South Arcot districts (Bahadur *et al.*, 2013). Fl. & Fr.: October.

INDIA: Kerala, Tamil Nadu, Andhra Pradesh, West Bengal. Endemic.

Jatropha villosa Wight, Icon. Pl. Ind. Orient 4(1): 1. 1848; N.P.Balakr. & Chakrab., Fam. Euphorb. India 227. 2007; N.P.Balakr. *et al.*, Fl. India 23: 256. 2012. *J.wightiana* Mull.Arg. in DC., Prodr. 15 (2):1080. 1866; Hook.f., Fl. Brit. India 5: 383. 1887; Gamble, 2: 1340. 1925.

Undershrub, up to 2 m high, monoecious; branchlets striate, pubescent. Leaves 4-5.8 x 4-7 cm, cordiform, base cordate, margin entire, apex acute, pubescent along nerves; petiole to 18 cm; stipules deciduous. Flowers in terminal cymes; peduncle pubescent. Male flowers 5 mm across; tepals in two whorls, 5+5, outer (sepals) 4 x 2 mm, lanceolate, subacute, thinly pubescent; inner (petals) 3.5 x 1.6 mm, obovate, obtuse-subacute, villous at base within; stamens 5-8, filaments to 1.5 mm, anthers 1.5 x 4 mm; disc of glands 5, free. Female flowers 6 mm across; tepals as in the male flower; ovary 2 mm, globose, thinly pubescent; styles 3, connate at base, each 2-fid. Capsule globose, thinly pubescent.

Rather common about Mettur dam (Matthew, 1991). Fl. & Fr. October – December.

INDIA: Tamil Nadu, Andhra Pradesh, Karnataka. Endemic.

LASIOCOCCA Hook.f.

Lasiococca comberi Haines, Bull. Misc. Inform. Kew 1920: 70. 1920; N.P. Balakr. & Chakrab., Fam. Euphorb. India 138. 2007; Susila Rani & N.P. Balakr. in N.P. Balakr. *et al.*, Fl. India 23: 126. 2012. *Homonoia comberi* (Haines) Merr. Lingnan Sci. J. 19: 188. 1980; Surya Prakash Babu in Pullaiah & Moulali, Fl. Andhra Pradesh 2: 871. 1997.

Evergreen tree, up to 8 m high; dioecious, often buttressed; bark pale or white; wood brown and white; branchlets puberulous, greyish. Leaves coriaceous, slightly panduriform or obovate, 7-17 × 2.3-5 cm, glabrous on both surfaces, glandular especially on the veins of upper surface, dark green above, paler beneath, base cordiform, margin slightly undulate, apex abruptly-acuminate, primary lateral nerves 6-12 pairs rather strong; petiole 4 to 8 mm, tomentose. Male flowers short, pedicellate in axillary racemes; bracts caducous; sepals 3, ovate, densely hirtellous; stamens highly branched. Female flowers solitary, long peduncled in axils of terminal leaves; calyx lobes 5 or 7, unequal, elliptic-lanceolate, persistent in fruit; ovary ovoid-conical, styles 3, filiform, connate at base, free above, 5-6 mm long. Capsule 3-lobed, 1.2 cm dia., depressed, covered with bristles or conical tubercles; seed subglobose, brown, smooth.

Occasional along rocky ravines, above 700 m. Fl.: June-August; Fr.: September.

Kambakkam hills (CTR), *PSPB* 11856; Minumuluru (VSKP), *GVS* 30007 (MH); Paderu (VSKP), *SKW* 2324-26 (BLAT); Similipahar-*SX, MB* & *Prabhakar Rao* 4580 (RRL-B).

INDIA: West Bengal, Odisha and Andhra Pradesh.

WORLD: Thailand, China.

MACARANGA Du Petit-Thouars

1. Male inflorescence axis zig zag; bracts linear-spathulate, with a glandular appendage; stamens 3-8, basally connate; leaf margin clearly toothed .... **M. indica**
1. Male inflorescence axis not zig-zag; bracts foliaceous, orbicular or transversely oblong, without a glandular appendage; usually stamens 3, free; leaf margin entire or toothed:
    2. Leaves fulvous tomentose, margin denticulate .... **M. denticulata**
    2. Leaves glabrous, margin entire .... **M. peltata**

Macaranga denticulata (Blume) Müll.Arg. in DC. Prodr. 15: 1000. 1866; FBI 5: 446. 1887. *Mappa denticulata* Blume, Bijdr. 625. 1826.

Figure 80. **Lasiococca comberi** Haine [Syn.: *Homonoia comberi* (Haines) Merr.]
♂ A. Twig, B. Perianth, C. Pistil.

Small tree; bark smooth, light-coloured; tender stem tips and tender leves fulvous tomentose. Leaves ovate or deltoid-ovate, peltate, 10-30 × 8.5-22 cm, often coarsely shortly sinuate-denticulate, sharply acuminate or obtuse, mature glabrous, often whitish, closly covered with minute glands beneath, with 3 strong principal nerves and ca. 6 other basal ones, secondary nerves 7-12, strong, base usually rounded or somewhat retuse; petiole 6-12 cm long; stipules lanceolate, caduous. Male panicles 7.5-10 cm long, pyramidal; bracts small, suborbicular or broadly ovate, 1.5-1.8 cm long with broad sessile base. Female panicles densely scar; flowers usually paired in the lower bracts with stout pedicels. Tepals broadly ovate, 5-7.5 mm. Ovary 2-celled, closely covered with yellow glands; style short and stout with 2-curved stout subulate, terminal stigmas. Capsule 6.2 mm diam., didymous.

Similipahar, occasional in valleys. Fl.: April-June. Fr.: July.

Similipahar (MBJ), *HS* 439 (RRL-B), *HS, MB* & *Prabhakar Rao* 4590 (RRL-B).

INDIA: Sikkim Himalaya, 900-1500 m; Assam; Khasia hills

WORLD: Bangladesh, Myanmar, Sumatra, Java.

Macaranga indica Wight, Icon. Pl. Ind. Orient. t. 1883, 1852 & t. 1949. f. 2. 1853; FBI 5: 446. 1887; Gamble 2: 1326. 1925; N.P.Balakr. & Chakrab., Fam. Euphorb. India 141. 2007; Susila Rani & N.P.Balakr. in N.P.Balakr. *et al.*, Fl. India 23: 132. 2012.

Evergreen tree, up to 20 m high; bark grey, smooth; wood greyish-red; branchlets stout, glaucous; young shoots tawny pubescent. Leaves broadly ovate or sub-orbicular, peltate, 12-17 × 11-18 cm, glabrous above, tawny-pubescent and glandular beneath, base 3-ribbed with 6 basal nerves, margin toothed, apex acuminate, secondary nerves 6-10 pairs, prominent, running nearly to margin and giving short branching ending in a tooth; tertiary nerves scalariform, conspicuous; petiole 6-12 cm; stipules ovate, ciliate, caducous. Male flowers: in dense clusters in bract axils along 6-14 cm long subpyramidal panicles; bracts linear-spathulate; calyx 3-5-lobed, lobes ovate; stamens 3-8. Female panicles pyramidally branched, 8 cm long, hairy, flowers 1-3 at the node; 2.5 mm across; calyx 4-lobed or 4-toothed; ovary 1-loculed, globose, glandular-pubescent, style lateral, subulate, 4 - 5 mm long. Capsule globose, 7 mm across, densely pubescent when young.

Rare, in ravines of semievergreen forest at 1000-1400 m. Fl.: September-October, Fr.: October-November.

Sunkarimetta (VSKP), *GVS* 44354 (CAL).

INDIA: Uttar Pradesh, Bihar, West Bengal, Sikkim, Arunachal Pradesh, Assam, Meghalaya, Odisha, Andhra Pradesh, Karnataka, Kerala, Tamil Nadu and Andaman and Nicobar Islands.

WORLD: Sri Lanka, Bhutan, Bangladesh, Myanmar, China, Laos, Vietnam, Thailand, Malaysia and Sumatra.

Macaranga peltata (Roxb.) Müll.Arg. in DC., Prodr. 15: 1010. 1866; Gamble 2: 1326. 1925; N.P. Balakr. & Chakrab., Fam. Euphorb. India 142. 2007; Susila Rani & N.P. Balakr. in N.P. Balakr. *et al.*, Fl. India 23: 137. 2012. *Osyris peltata* Roxb., Fl. Ind. 3:755. 1932. *Macaranga roxburghii* Wight, Icon. Pl. Ind. Orient. t. 5:23. 1852 and t. 1949 f.4.1853; FBI 5: 448. 1887.

Small or moderate-sized, resinous tree; branchlets glaucous; bark pale, with lenticels; tender parts tomentose. Leaves large, peltate, orbicular-ovate, 10-25 × 8-19 cm, margin usually entire or minutely toothed, apex acuminate, more or less hairy on the nerves beneath and glandular, somewhat rusty above or the mature ones almost glabrous, basal nerves 3 strong principal and 6 others, secondary nerves 7-10, ending usually in hirsute glands; petiole usually 7.5-20 cm; stipules ovate to oblong or ovate-lanceolate. Panicles tomentose, mostly from the axils of fallen leaves, male strict, 7.5-10 cm long; bracts large, tomentose, foliaceous at the base of the branches, orbicular or transversely oblong, denticulate, concealing the flower-clusters, 3.7-7.5 mm broad according to the position; stamens 2-5, usually 3, anthers mostly 4-valved. Female flowers in nearly simple racemes or branched at the base, 3 cm long; bracts large, rusty tomentose, deciduous; pedicels 6.2-15 mm long; tepals very minute, rounded; ovary densely covered with glands, glabrous or only slightly puberulous; stigma lateral, sessile, broadly peltate, papillose. Capsule globose, 6-7.5 mm diam., often grooved, with deciduous yellow glands; seed 1, globose, black.

Common on high hills and along the moist valleys. Fl.: March, Fr.: April- June. Vern.: Tel.: *Boddi, Kondajaphera, Konda tamara*; Tam.: *Vattakanni*.

Thumburathirtham (CTR), *SSR* 16167; Papavinasanam (CTR), *BR, SKB* & *BSS* 30414; Tiger camp, Maredumilli (EG), *MM* 101267 (BSID); Kolli hills, way to Seekuparai (NMK), *SKS* 1031 (SGH).

INDIA: Sikkim, West Bengal, Bihar, Odisha, Madhya Pradesh, Andhra Pradesh, Maharashtra, Goa, Karnataka, Kerla, Tamil Nadu and Andaman and Nicobar Islands.

WORLD: Si Lanka and Myanmar.

MALLOTUS Loureiro

1. Leaves penninerved .... **M. resinosus**
1. Leaves palmately 3-9-nerved at base:
    2. Leaves alternate, red glandular beneath; capsule red glandular:
        3. Leaves narrowly oblong-ovate to lanceolate, mostly twice as long as broad, with red granules beneath; stamens 18-32; styles 3; fruits 3-4-loculed, densely clothed with crimson-red resinous powdery glands .... **M. philippensis**
        3. Leaves broadly ovate-deltoid, mostly as long as broad, yellow tomentose and glandular beneath; stamens 40-80; styles 2; fruits 2-loculed, densely fulvous tomentose, not with red resinous powdery glands:

4. Fruits smooth, not echinate or tubercled .... **M. repandus**
4. Fruits echinate or tubercled ....**M. tetracoccus**

2. Leaves opposite and alternate, yellow glandular beneath; capsule stellate pubescent...**M. rhamnifolius**

Mallotus phillippensis (Lam.) Müll.Arg., Linnaea 34: 1196, 1865; FBI 5: 44. 1887; Gamble 2: 1322. 1925; Beddome, Fl. Sylv. S. India t. 289.1872 (*'philippinensis'*); N.P. Balakr. & Chakrab., Fam. Euphorb. India 150. 2007; Susila Rani & N.P. Balakr. in N.P. Balakr. *et al.,* Fl. India 23: 159. 2012. *Croton philippensis* Lam., Encycl. 2: 206. 1786.

Evergreen tree, up to 12 m high; bark grey, nearly smooth, exfoliating in small woody plates; wood grey to light red, smooth, fairly straight-grained, medium fine and even-textured; branchlets striate, densely grey-tomentose. Leaves narrowly oblong-ovate to lancecolate, 4-19 × 2-10 cm, glabrous above, pubescent and numerous red orbicular glands beneath, base cuneate, margin entire, apex acuminate, strongly 3-nerved, secondary nerves 4-7 pairs above basal ones; petiole to 5.5 cm, with 2 small sessile glands one on each side of the summit. Flowers pale yellow in axillary or terminal spikes. Male flowers: bracts triangular, ca 1 mm long; sepals 3 or 4, ovate to elliptic-lanceolate; stamens 18-32, filaments free. Female flowers: bracts triangular; sepals 3-5, connate at base, ovate-lanceolate; ovary 2-3 mm across, densely crimson-red glandular; styles 3, free above, 2-5 mm long. Capsule 3-lobed, red glandular, loculicidally 3-valved, 1 cm across, smooth; seeds globose, black, 4 mm across with a moderately thick brittle testa.

Common, on hills of deciduous forests and open scrub jungles above 700 m. Fl.: October-December. Fr.: February-March. Vern.: Tel.: *Kumkuma chettu, Sindhuramu, Vasantha gandhamu, Chandirmu, Hanumanthuni bottu, Naga-rakta;* Tam.: *Kunnancholai, Thirisalakkai maram, Korangumanjanattai, Kundaanjaalamaram; Kabila;* Ori.: *Kunkumo, Kapilogundi, Sundari gudi.*

Gundlabrahmeswaram (KNL), *PSPB* 9129; Balapalli (CDP), *BR* & *SKB* 30770; Tirumala (CTR), *SSR* 13774, 15267; Rampa (EG), *PSPB* 11836; Velugodu (KNL), *SKW* 4865 (BLAT); Mamandur (CTR), *SKW* 5827 (BLAT); Durgam – Udayagiri (NLR), *PV* 110236 (BSID); Rollapenta (PKM), *VBH* 83962 (BSID); way to Roygudem (WG), *DN* 85508 (BSID); Kuntada RF, Maredumilli (EG), *MM* 101229 (BSID); way to Yeguvaguda (SKLM), *NRR* 79500 (BSID); Sirumalai hills (DGL), *SKS* 116 (SGH).

INDIA: Through out the country in Tropical areas.

WORLD: Sri Lanka, Nepal, Bhutan, Bangladesh, Myanmar, China, Indo-China, Thailand and Malesia to N. Australia.

Mallotus repandus (Willd.) Müll.Arg., Linnaea 34:197. 1865; FBI 5:442. 1887; Gamble 2: 1321. 1925; N.P. Balakr. & Chakrab., Fam. Euphorb. India 150. 2007; Susila Rani & N.P.Balakr. in N.P. Balakr. *et al.,* Fl. India 23: 161. 2012. *Croton repandus* Willd., Ges. Naturf. Freunde Berlin Neue Schriften 4: 206. 1803.

Large shrub, sometimes scandent or subscandent with thorns on the trunk; branchlets stellate-pubescent or tomentose. Leaves broadly ovate-deltoid, 5-10

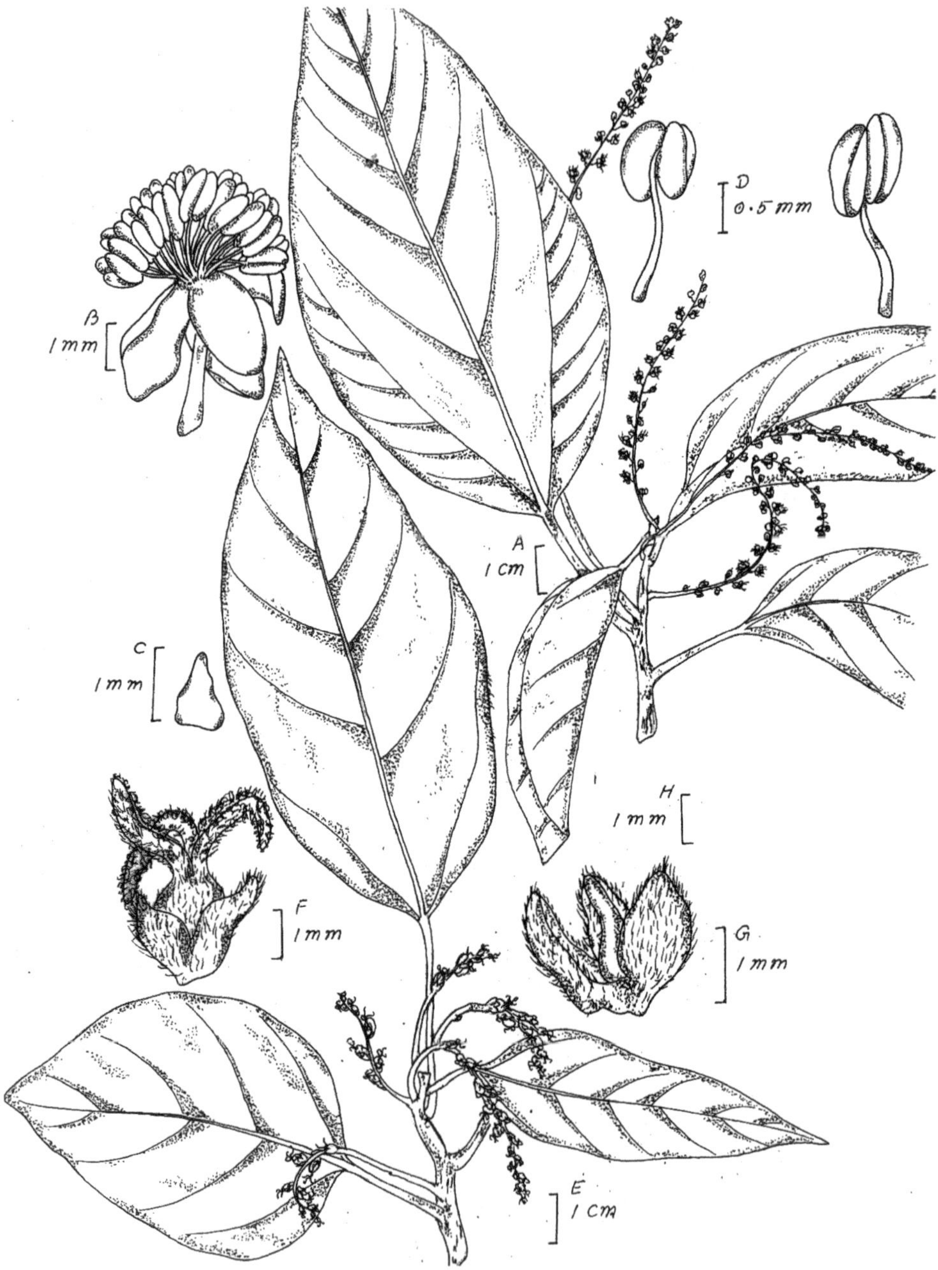

Figure 81. **Mallotus phillippensis** (Lam.) Müll.Arg.
♂ A. Twig, B. Flower, C. Bract, D. Stamens, ♀ E. Twig, F. Flower, G. Tepals, H. Pistil.

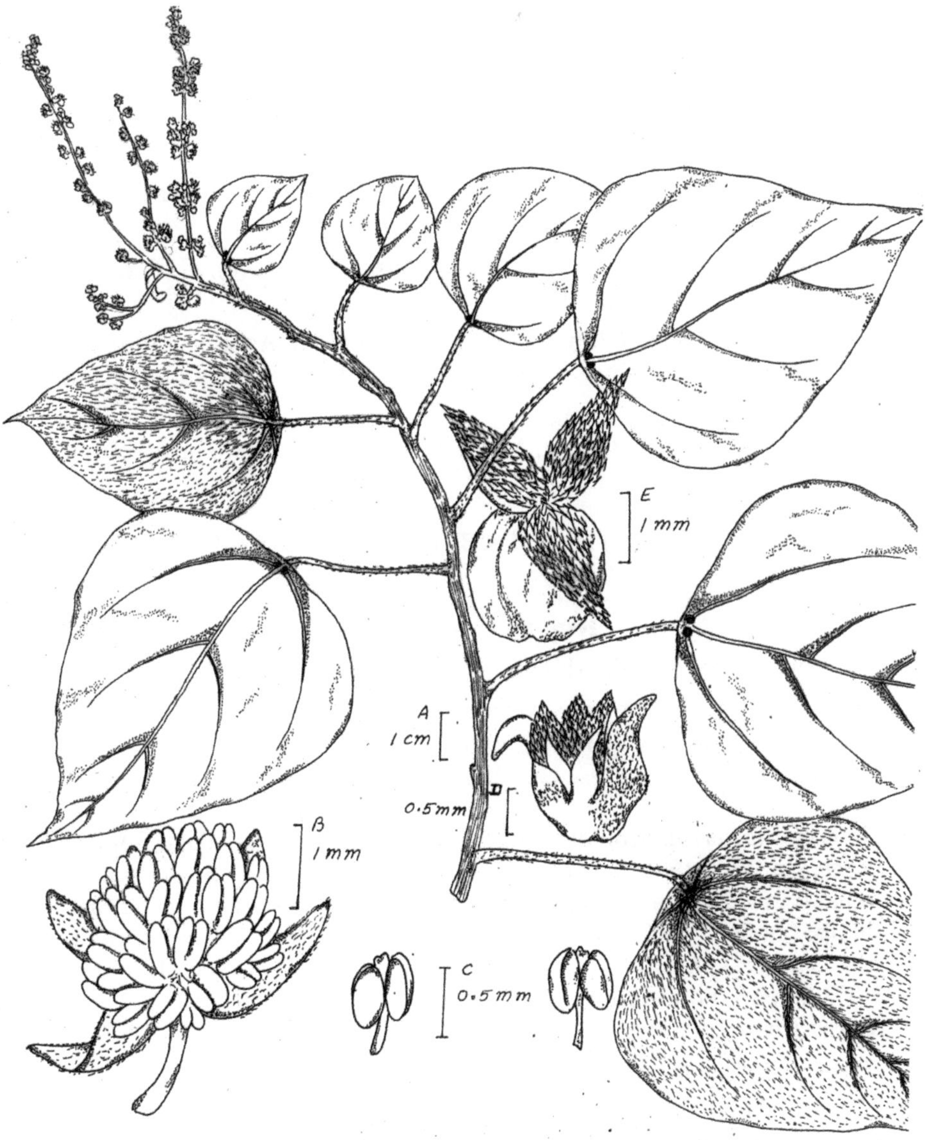

Figure 82. **Mallotus repandus** (Willd.) Müll.Arg.
A. Twig, ♂ B. Flower, C. Stamens, ♀ D. Flower, E. Pistil.

× 4.3-8.7 cm, base cordate, subtruncate, retuse, rarely obtuse, margin entire or somewhat sinuate-toothed, apex acute or acuminate, glabrescent above, densely stellate-pubescent and closely covered with small glands beneath, secondary nerves 2-3 above the 3-nerved; petiole 2.5-6 cm long; stipules minute, tomentose. Inflorescence densely pubescent. Male flowers in terminal panicled racemes; tepals 3-5, yellow, suborbicular or lanceolate, 3 mm long, concave, tomentose, glandular within; anthers with 2 oblong cells, slightly produced beyond the connective. Female flowers green, in simple racemes; tepals linear, caducous; ovary tomentose, 2-lobed and celled; stigma plumose, sessile. Capsule 2-valved, 1.2 cm diam., densely fulvous-tomentose, glandular; seeds subglobose, black.

Occasional in hedges and also in scrub jungles. Fl.: November-February, Fr.: March-April. Vern.: Tel.: *Kanda veltu.*

Antikonda (VSKP), *KCJ* 17247, 17248; Punyagiri (VSKP), *GVS* 23326 (MH); Gadgala fall, Ganjam-*SX* 1200 (RRL-B).

INDIA: Uttar Pradesh, Bihar, West Bengal, Sikkim, Assam, Tripura, Odisha, Andhra Pradesh, Maharashtra, Karnataka, Kerala, Tamil Nadu and Andaman and Nicobar Islands.

WORLD: Sri Lanka, Bhutan, Myanmar, S. China, Thailand, Malesia to N. Australia and New Caledonia.

Mallotus resinosus (Blanco) Merr., Sp. Blancoan. 222. 1918 & Enum. Philip. Fl. Pl. 2: 436. 1923

Key to the varieties

1. Fruits densely yellow glandular with few and scattered spines .... var. **stenanthus**
1. Fruits sparsely yellow glandular with 2-3 linear row of spines .... var. **muricatrus**

Mallotus resinosus (Blanco) Merr. var. **muricatus** (Wight) N.P.Balakr. & Chakrab., Rheedea 1: 39. 1991 & Fam. Euphorb. India 150. 2007; Susila Rani & N.P. Balakr. in N.P. Balakr. *et al.*, Fl. India 23: 162. 2012. *Adelia resinosa* Blanco, Fl. Fillip. (2): 562. 1845. *Claoxylon muricatus* Wight, Icon. Pl. Ind. Orient. 5: 24. t. 1886. 1852. *Mallotus muricatus* (Wight) Beddome, Fl. Sylv. S. India t. 208. 1873, in part, non Müll.Arg. 1865; FBI 5: 436. 1887; Gamble 2: 1322. 1925. *M. intermedius* (Baillon) N.P. Balakr., Bull. Bot. Surv. India 10: 245. 1968.

Trees or shrubs, 4-10 m high; branchlets puberulous. Leaves opposite, rhombic-lanceolate or oblanceolate; up to 12.5 × 3.2 cm, base narrowly cuneate or obtuse-oblique, margin crenate, entire with age, apex acuminate to cuspidate; secondary nerves 8-10 pairs; petiole to 1 cm. Flowers in spikes. Male flowers: bracts ovate or triangular; sepals 3 or 4, ovate; stamens 18-40. Female flowers: bracts ovate-orbicular; sepals 3-6, connate at base, linear-lanceolate or oblanceolate; ovary globose, 3-loculed, covered with conical tubercles, styles 3, 3-5 mm long. Capsule yellow-glandular, with spines; seeds ovoid, angular along faces, shiny dark brown.

Rare, Horsely hills (Fischer, 1925), Tirumala (Thammanna *et al.*, 1976), Kollimalais and Pacchamalais. Fl. & Fr.: October. Vern.: Tam.: *Solai karupichchi, Kaattu pulasai, Karuvalichchi.*

Papanasanam falls (CTR), *KS* 6952 (BSID).

INDIA: Andhra Pradesh, Tamil Nadu and Kerala. Endemic.

Mallotus resinosus (Blanco) Merr. var. **stenanthus** (Wight) N.P.Balakr. & Chakrab., Rheedea 1: 39. 1991 & Fam. Euphorb. India 150. 2007. *Mallotus stenanthus* Müll. Arg. in Linnaea 34: 191. 1865; FBI 5: 437. 1887; Gamble 2: 1322.1925 .

Small trees to 4 m tall. Leaves opposite, rhomboid to obovate, 2-16 x 2-6 cm, drying green; petiole to 3 cm long. Racemes to 8 cm long, terminal, leaf-opposed. Capsule 0.5 cm wide, yellow-glandular, spines sparse, conical; seeds 3.

Rarely found in evergreen forest patches in Sirumalai hills and Alagar hills of Tamil Nadu. Fl. & Fr.: June – August.

Alagar hills (MDU), *SKS* 602 (SGH).

INDIA: Tamil Nadu, Maharashtra, Karnataka and Kerala (Endemic).

Mallotus rhamnifolius Müll.Arg., Linnaea 34: 196. 1865; FBI 5: 440. 1887; Gamble 2: 1322. 1925; N.P. Balakr. & Chakrab., Fam. Euphorb. India 151. 2007; Susila Rani & N.P. Balakr. in N.P. Balakr. *et al.*, Fl. India 23: 163. 2012.

Small tree; branchlets fulvous tomentose. Leaves alternate and opposite, oblong or lanceolate, 7.5-11.8 × 4.3-6.8 cm, glabrous above, pubescent beneath; base rounded or acute, margin entire, apex acuminate with mucro, 3-ribbed, secondary nerves 4-7 pairs above basal-ones, tertiary nerves parallel to the secondary veins, prominent beneath; petiole to 4.5 cm. Spikes erect, axillary with brown tomentum. Male flowers: in clusters; pedicels 1-1.5 mm long; bracts linear-lanceolate; sepals 3 or 4, ovate, 2-3 mm long; stamens 45-50; filaments 2-3 mm long. Female flowers: solitary, bracts ovate to linear-lanceolate, 2-3 mm long; calyx sub-spathaceous, 5 or 6-lobed, lobes linear-lanceolate; ovary 3-loculed, styles 3, 2.2-3 mm long. Capsule globose, 3-valved, finely stellate-pubescent; seeds globose to ovoid, shiny red.

Occasional, in hills of moist forests, especially near the streams above 700 m. Fl.: January-February, Fr.: March. Vern.: Tel.: *Konda-kunkuma.*

Nityapoojakona (CDP), *SKB* & *BSS* 30338; Saddicudumadugu (CTR), *PSPB* 9176; Kambakkam hills (CTR), *PSPB* 9180; Sirivaka (EG), *CAB* 5014 (MH); Galikonda (VSKP), *GVS* 21729 (MH); Anantagiri (VSKP), *GVS* 21729 (MH).

INDIA: Andhra Pradesh, Karnataka, Kerala and Tamil Nadu.

WORLD: Sri Lanka.

Mallotus tetracoccus (Roxb.) Kurz, J. Asiat. Soc. Bengal, Pt. 2, Nat. Hist. 42: 245. 1873; Croizat, J. Arnold Arbor. 21: 503. 1940; Airy Shaw, Kew Bull. 26: 298. 1972; Saldanha & Nicolson, Fl. Hassan 345. 1976. *M. ferrugineus* (Roxb.) Müll.-Arg. Linnaea 34: 188. 1865; Susila Rani & N.P. Balakr. in N.P. Balakr. *et al.*,

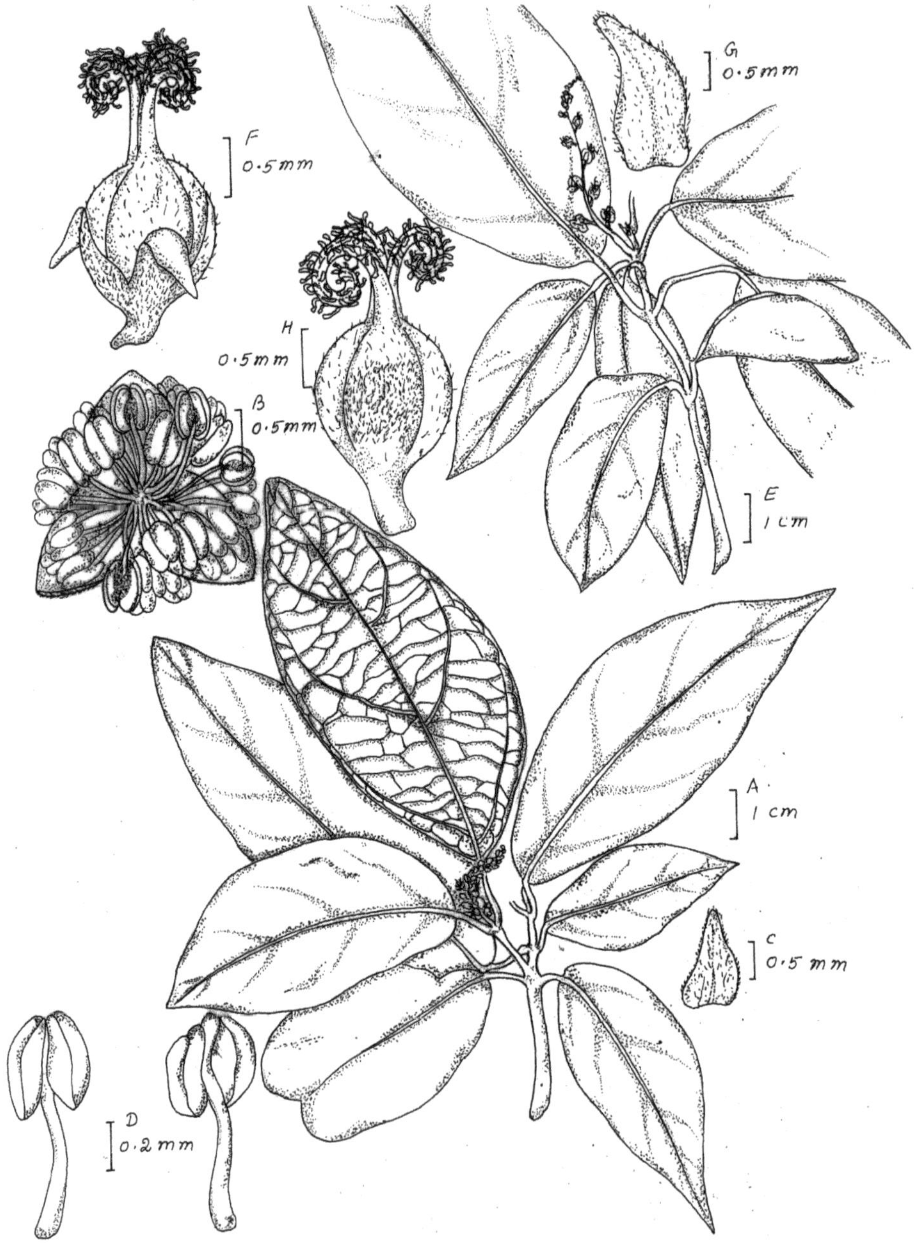

Figure 83. **Mallotus rhamnifolius** Müll.Arg.

♂ A. Twig, B. Flower, C. Bract, D. Stamens, ♀ E. Twig, F. Flower, G. Bract, H. Pistil.

Fl. India 23: 151. 2012. *M. albus* sensu Muell.-Arg. Linnaea 34: 188. 1865 non *Rottlera alba* Roxb. ex Jack. 1820.

Tree, up to 20 m tall; branchlets tomentose. Leaves alternate, broadly ovate to deltoid, 7-20 × 8-18 cm, fulvous-tomentose below, blade entire or lobed when young, base truncate, peltate, apex acuminate, lateral nerves ca. 6 pairs, lower pair longer; petiole ca. 18 cm. Panicles ca. 20 cm; bracts broadly ovate. ♂: 7 mm across. Tepals 4, ovate, 2 mm. Stamens on a fleshy receptacle; filaments to 3 mm; anther cells unequal, 0.7 mm. ♀: 5 mm across, echinate with soft, woolly spines, white-tomentose; seeds angular, black.

Kollimalais and Servarayans above 1200 m, in sholas. Abundant at Yercaud inside plantations, for shade. Fl. & Fr.: June-Aug. Vern.: Tam.: *Mullu polavu, Kumalaa.*

INDIA: Himalayas.

WORLD: Sri Lanka, Himalaya, S.W. China.

MANIHOT P. Miller

1. Shrubs 1.5-2.7 m high, rarely up to 4 m high; leaves epeltate or slightly peltate for less than 5 mm from margin; lobes narrowly oblanceolate, elliptic; calyx 7-12 mm long, hairy inside; fruits 6-winged .... **M. esculenta**
1. Trees, 5-10 m high; leaves peltate for atleast 4-5 mm from margin; lobes broadly obovate to elliptic; calyx 15-18 mm long, glabrous inside; fruits not winged .... **M. glaziovii**

Manihot esculenta Crantz, Inst. Rei. Herb. 1:167. 1766; N.P. Balakr. & Chakrab., Fam. Euphorb. India 230. 2007; Chakrab. & N.P.Balakr. in N.P. Balakr. *et al.*, Fl. India 23: 261. 2012. *M. utilissima* Pohl, Pl. Bras. Ic. 1:32. t.24. 1827; FBI 5: 239. 188; Gamble 2: 1346. 1925.

Shrub, 1.5-2.7 m, with large tuberous roots; stems very nodose. Leaves 3-7-palmatipartite; petiole to 12 cm long; segments oblanceolate, 7-12 × 2.5-4 cm, base cuneate, margin entire or sparsely denate, apex acute to acuminate. Flowers in terminal and pseudoaxillary, laxly paniculate or branched cymes, bracts caducous. Male flowers: pedicel 4-7 mm long; calyx 7-12 mm long, campanulate, 5-lobed up to the middle; disc central; stamens 10, free. Female flowers: pedicel 1-2 cm long; calyx as in male, up to 12 mm long; disc pulveniform; ovary with 6 longitudinal ridges, stigma warty. Fruits globose or ellipsoid, muricate-rugulose.

Frequently cultivated. Fl.: July - December; Fr.: September - March. Vern.: Tel.: *Karra pendalamu.* Ori.: *Simuli konda, Rosokandojado;* Tam.: *Kappakizhangu, Maravalli kizhangu*. English: Cassava, Tapioca.

INDIA: Cultivated in Uttar Pradesh, Bihar, West Bengal, Sikkim, Assam, Odisha, Madhya Pradesh, Andhra Pradesh, Maharashtra, Karnataka, Kerala, Tamil Nadu and Andaman & Nicobar Islands.

WORLD: Native of Brazil now widely cultivatd in tropical regions of Old World.

Note: Variegated type with yellow-green ornamental foliage is often cultivated in the gardens.

Manihot glaziovii Müll.Arg. in C. Martius, Fl. Bras. 11(2): 446. 1874; Gamble 2: 1346. 1925; N.P.Balakr. & Chakrab., Fam. Euphorb. India 230. 2007; Chakrab. & N.P. Balakr. in N.P. Balakr. *et al.*, Fl. India 23: 262. 2012.

Tree, up to 10 m high; latex milky; bark glossy, pale-brown, peeling. Leaf-lobes 3 to 6, elliptic to obovate, 8-18 × 4.5-8 cm, glabrous, dark green above, glaucous beneath, margin entire, apex obtuse with narrow apicule; secondary nerves regular, 12-15 pairs, midvein raised and prominent above; petiole peltate, 5-20 cm. Flowers greenish-yellow in racemose, lax panicles. Male flowers: pedicels ca 7 mm long; perianth 15 - 18 mm long; divided for half way down, greenish without, dark-purple within at base and yellowish towards apex, 4-5-lobed; stamens 10, longer ones alternating with shorter ones, situated in between glands; disc of 10 glands. Female flowers: smaller than male, pedicels 10-12 mm long; perianth as in male; ovary smooth, styles up to 1 mm long, stigma sub-penicillate. Capsule green, globose, 3 cm across, pendulous.

Introduced and cultivated, planted in gardens. Fl. & Fr.: Throughout the year. Vern.: Tel.: *Siyara-rabbaru, Sima-pittati, Badama*.

INDIA: Introduced and naturalized in Andhra Pradesh, Tamil Nadu and Kerala.

WORLD: Native of Tropical Brazil, introduced and naturalized in many tropical Asian countries.

MARGARITARIA Linnaeus f.

Margaritaria indica (Dalzell) Airy Shaw, Kew Bull. 20: 387. 1966; N.P. Balakr. & Chakrab., Fam. Euphorb. India 358. 2007; Chakrab. *et al.*, in N.P. Balakr. *et al.*, Fl. India 23: 436. 2012. *Prosorus indicus* Dalzell, Hooker's J. Bot. Kew. Gard. Misc. 4: 346. 1852; Gamble 2: 1294. *Phyllanthus indicus* (Dalzell) Müll.Arg., Linnaea 32: 52. 1863; FBI 5: 305. 1877.

Evergreen tree, up to 20 m high; bark smooth, whitish; branchlets glabrous, lenticellate. Leaves membranous, elliptic, 7.5-12.5 × 3.5-5 cm, glabrous, glaucous beneath, base rounded- acute, margin entire, apex acute-subacuminate, secondary nerves 8-9 pairs; petiole to 8 mm. Male flowers: very small, green, in fascicles; pedicels 5-12 mm long; sepals broadly oblong, elliptic, suborbicular or obovate; disc flat, adnate, filaments ca 1mm long. Female flowers: larger, in cluster; pedicels 6-20 mm long; sepals ovate to oblong; disc entire; ovary globose, styles 3, 1.5-2 mm long, almost free above; stigmas bifid. Capsule green, smooth, depressed globose, with 3 vertical furrows, 0.8-1.2 cm; seed blue, subtrigonous.

Rare, in the semi-evergreen forests along streams 600-900 m in hills. Fl.: April-June. Fr.: August-October.

Siddulaiah kona (NLR), *SSR* 19675; Mamandur (CTR), *GVS* 1953 (MH); Chitvelu to Rajampeta (NLR), *PV* 110087 (BSID).

INDIA: Odisha, Andhra Pradesh, Maharashtra, Karnataka, Kerala, Tamil Nadu and Andaman Islands.

WORLD: Sri Lanka, China, Taiwan, Vietnam, Thailand, Sumatra to Philippines.

MEINECKIA Baillon

Meineckia parvifolia (Wight) G.L.Webster, Acta Bot. Neerl. 14: 342. 1965; N.P. Balakr. & Chakrab., Fam. Euphorb. India 360. 2007; Chakrab. *et al.*, in N.P. Balakr. *et al.*, Fl. India 23: 439. 2012. *Peltandra parvifolia* Wight, Icon. Pl. Ind. Orient. t. 1892. 1852. *Neopeltandra suberosa* (Müll.Arg.) Gamble Fl. Pres. Madras P2: 1286. 1925. *Phyllanthus thwaitesianus* Müll.Arg., Linnaea 32: 11. 1863; FBI 5: 287. 1887. *P. suberosus* Müll.Arg., Linnaea 32: 10. 1863; FBI 5: 287. 1887.

Herb, up to 40 cm high, monoecious; branchlets zigzag, quadrangular. Leaves 1.5-4.5 × 1-3 cm, ovate, acute-rounded at base, entire, acute; petiole to 6 mm; stipules narrowly triangular. Flowers axillary, solitary, bracteate. Male flowers 2 mm across; pedicel to 1 cm; tepals 5, 1.2 × 1 mm, free, elliptic-ovate, acute; stamens 5, on gynophore; anthers 0.3 mm across; disc of glands 5, connate. Female flowers 3 mm across; pedicel to 4 cm; tepals 5, 1.3 × 1 mm, slightly connate at base, broadly ovate, acute; ovary 1 mm, globose, styles 3, each 2-fid, from base, stigma capitate. Capsule 5 mm, depressed globose; seeds pitted.

Occasional in the hills and in scrub jungles as forest under growth in coastal districts. Fl. & Fr.: January-March.

Bhiravakona (NLR), *PSPB* 9111; Simhachalam (VSKP, *PSPB* 11807, *GVS* 19372, 21507, 22561 (MH); Rapur hills (NLR), *PV* 110031 (BSID); Mellavagu (GNT), *SKW* 3695, 3696 (BLAT); Mantur (EG), *JSG* 4954 (MH); Sirivaka (EG), *JSG* 5248 (MH); Dolphin's nose hill (VSKP), *TAR* 7139 (BSID).

INDIA: Andhra Pradesh, Tamil Nadu, Kerala.

WORLD: Sri Lanka.

MICROCOCCA Bentham

Micrococca mercurialis (L.) Benth. in Hook. Niger. Fl. 503. 1849; Gamble 2: 1328. 1925. *Tragia mercurialis* L., Sp. Pl. 980. 1753. *Claoxylon mercurialis* (L.) Thwaites, Enum. Pl. Zeyl. 271. 1861; FBI 5: 412. 1887; N.P.Balakr. & Chakrab., Fam. Euphorb. India 131. 2007; Susila Rani & N.P.Balakr. in N.P.Balakr. *et al.*, Fl. India 23: 114. 2012.

Herb, up to 80 cm high, monoecious; branchlets sparsely hairy. Leaves 3-5 × 1-2.5 cm, ovate-lanceolate, base acute, margin serrate, apex subacuminate, puberulous; petiole to 3.5 cm; stipules minute. Flowers in long slender interrupted axillary spikes; bracts minute. Male flowers 1.8 mm across; tepals 3, 1.3 × 0.8 mm, slightly connate at base, ovate, acute; stamens 3-5, free, filaments 0.9 mm, anthers 0.3 × 0.25 mm. Female flowers 2.8 mm across; sepals 3, 1.8 × 1.2 mm, slightly connate at base, ovate-lanceolate, acute-acuminate, sparsely pubescent; styles 3, linear, horizonatally spreading, plumose-laciniate; disc of glands linear. Capsule 4 mm, sparsely pubescent; seeds globose, pitted, minutely arillate.

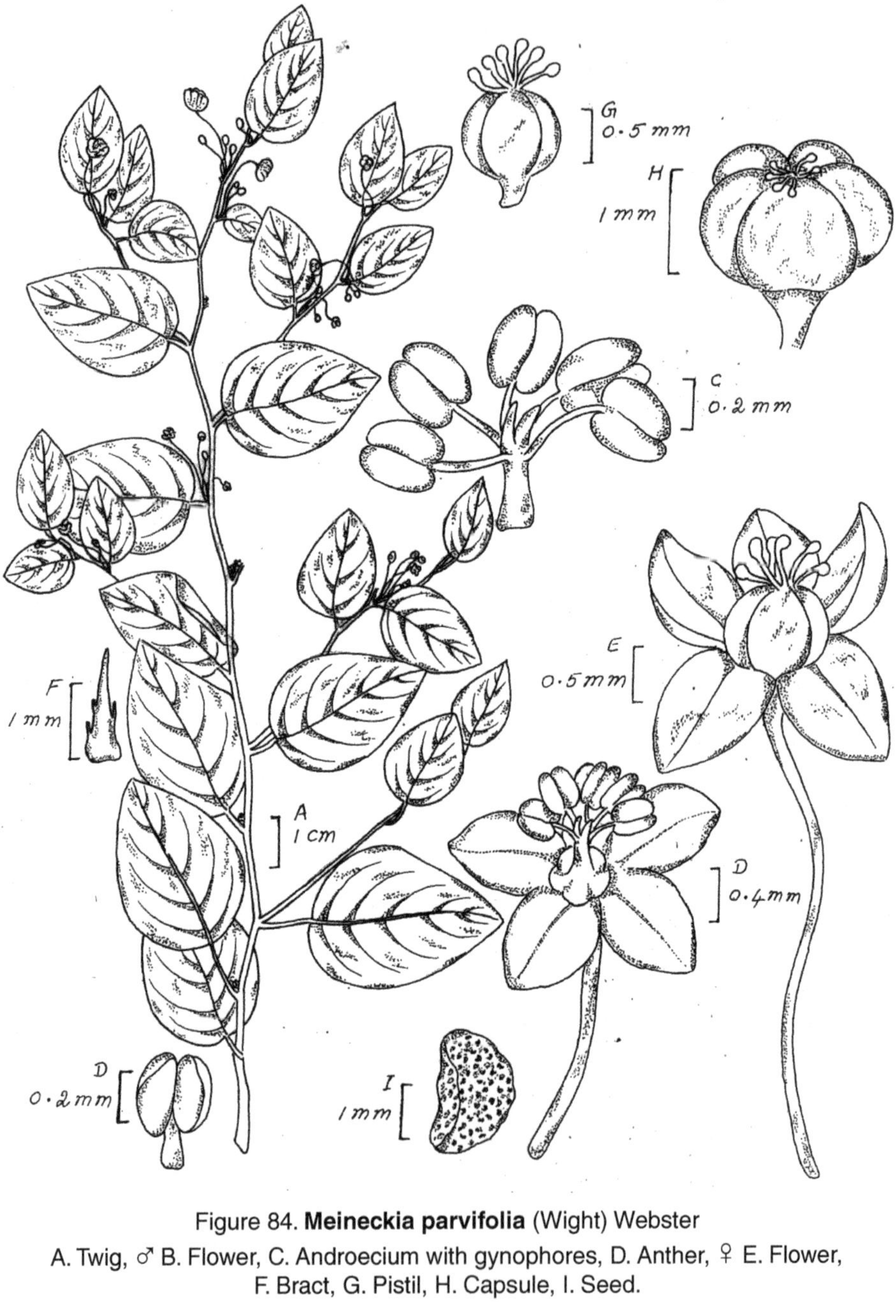

Figure 84. **Meineckia parvifolia** (Wight) Webster

A. Twig, ♂ B. Flower, C. Androecium with gynophores, D. Anther, ♀ E. Flower, F. Bract, G. Pistil, H. Capsule, I. Seed.

Common along roadsides and wastelands. Fl. & Fr.: Throughout the year.

Shivdurg (NLR), *SKW* 6757 (BLAT); Simhachalam (VSKP), *SKW* 4395 (BLAT).

INDIA: Himachal Pradesh, Haryana, Rajasthan, Uttar Pradesh, Bihar, West Bengal, Odisha, Madhya Pradesh, Andhra Pradesh, Maharashtra, Goa, Karnataka, Kerala and Tamil Nadu.

WORLD: Tropical Africa, Arabia, Sri Lanka, Myanmar and Thailand.

Note: Following Prain (1911), Sagun and van Welzen (2002) treat *Micrococca* and *Claoxylon* as independent genera. N.P.Balakrishnan and Chakrabarty (2007) are of the the opinion that these two should be treated as single genus *Claoxyl*on.

MICROSTACHYS Adr. Jussieu

Microstachys chamaelea (L.) Müll.Arg., Linnaea 32: 95. 1863; Esser, Blumea 44: 176. 1999; N.P.Balakr. & Chakrab., Fam. Euphorb. India 293. 2007; Chakrab. & N.P.Balakr. in N.P.Balakr. *et al.*, Fl. India 23: 348. 2012. *Tragia chamaelea* L., Sp. Pl. 1391. 1753. *Sebastiania chamaelea* (L.) Müll.Arg. in DC., Prodr. 15: 1175. 1866; FBI 5: 475. 1888; Gamble 2: 1344. 1925; Surya Prakash Babu in Pullaiah & Moulali, Fl. Andhra Pradesh 2: 886. 1997.

Herb, up to 60 cm high, monoecious; branchlets terete, striate, puberulous. Leaves 2.5-6.5 × 0.3-0.5 cm, linear – lanceolate, cuneate at base, minutely dentate, apiculate; petiole to 0.3 mm, thinly hairy; stipules minute, lanceolate or triangular, ciliate. Flowers in axillary spikes; male flowers many; female flowers solitary at base; bracts lanceolate. Male flowers 1.6 mm across; tepals 5, unequal, ovate – lanceolate, acute, ciliate; stamens 3, free, filaments 0.3 mm, anthers 0.3 mm across. Female flowers 1.9 mm across; tepals 3, ovate or obovate, acute, ciliate; ovary 1.2 mm, echinate, styles 3, simple. Capsule 5 mm, subglobose, 3 cocci, each with 2-vertical rows of spinules; seeds oblong, smooth, yellow.

Common weed along forest paths. Fl. & Fr.: July – February.

Idupulapaya RF (CDP), *KRKS* 39065; Penchalakona (NLR), *PSPB* 9106; Balapalle (CDP), *A.M. Rao* 406 (BSID); Tirumala (CTR), *SKW* 5620 (BLAT); Tirupati (CTR), *KS* 6869; Eethakayalamandapam (CTR), *GVS* 46898; Polavaram (EG), *DCSR* 232 (CAL); Basavapuram (KNL), *JSG* 10948 (DD); Anantagiri (VSKP), *SKW* 21474 (BLAT); Araku (VSKP), *NPBK* 525, 10802 (CAL); Sirumalai hills (DGL), *SKS* 94 (SGH).

INDIA: Uttar Pradesh, West Bengal, Jharkhand, Madhya Pradesh, Chhattisgarh, Odisha, Gujarat, Maharshtra, Goa, Karnataka, Andhra Pradesh, Tamil Nadu, Kerala and Andaman Islands.

WORLD: Africa, Sri Lanka, Myanmar, China to Malesia, Australia and Pacific Islands.

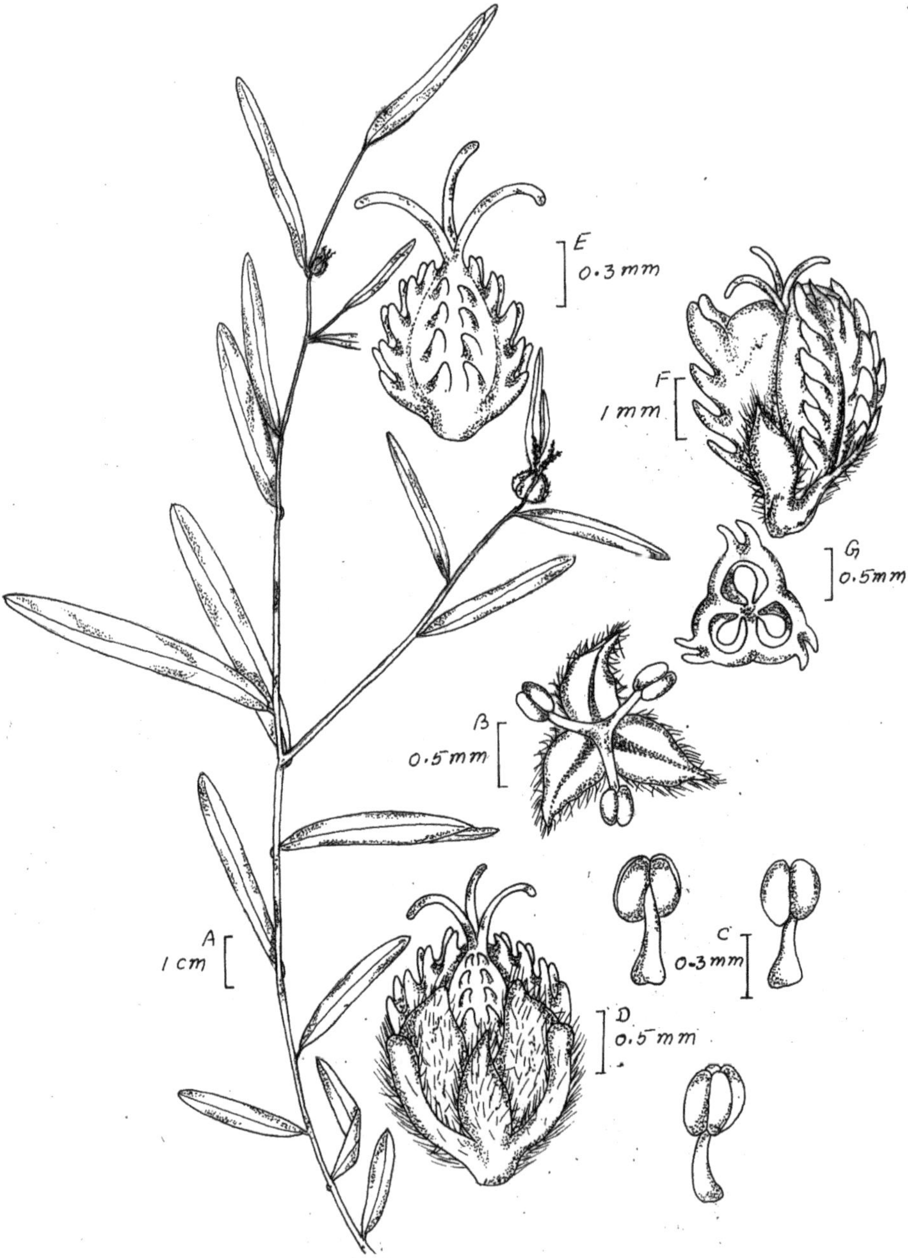

Figure 85. **Microstachys chamaelea** (L.) Müll.Arg.
[Syn.: *Sebastiania chamaelea* (L.) Müll.Arg.]
A. Twig, ♂ B. Flower, C. Stamens, ♀ D. Flower, E. Pistil, F. Capsule, G. Ovary T.s.

PACHYSTYLIDIUM Pax & K. Hoffman

Pachystylidium hirsutum (Blume) Pax & Hoffm. in Engl., Pflanzenr. IV. 147. ix-xi (heft 68): 108. 1919; Panda & P.Das, Rheedea 7: 60. 1997; N.P. Balakr. & Chakrab., Fam. Euphorb. India 178. 2007; Chakrab. & N.P. Balakr. in N.P. Balakr. *et al.*, Fl. India 23: 193. 2012. *Tragia hirsuta* Blume, Bijdr. Fl. Ned. Ind. 630. 1826. *T. gagei* Haines, J. Asiat. Soc. Bengal, n.s. 15(7): 317. 1920 & Bot. Bihar & Odisha, 116. 1921.

Twining subshrubs, ca 3 m high with stinging hairs. Stipules triangular; petiole 0.8-4 cm. Leaves ovate or elliptic, base cordate, margin serrate, apex abruptly acuminate. Inflorescence 1-8 cm, 1-5 female flowers and many male flowers. Male flowers: pedicel 1.5-2.3 mm, calyx lobes ovate. Female flowers: pedicel 1 mm, sepals lanceolate, elliptic or ovate, ovary densely hirsute; styles connate into a cylindric column, free style tips recurved. Capsule drying dark brown; seeds subglobose.

Evergreen forests and scrub jungles, sea level to 600 m altitude in Andhra Pradesh (N.P.Balakrishnan and Chakrabarty, 2007; N.P.Balakrishnan, *loc.cit.*). Fl. & Fr.: March - May.

INDIA: Odisha, Madhya Pradesh and Andhra Pradesh.

WORLD: China, Laos, Vietnam to Java and Philippines.

PEDILANTHUS Poit. *nom. cons.*

Pedilanthus tithymaloides (L.) Poit., Ann. Mus. Natl. Hist. Nat. 19: 390. t.19. 1812; FBI 5:239. 1887; Gamble 2: 1346. 1925; N.P. Balakr. & Chakrab., Fam. Euphorb. India 281. 2007 & in N.P. Balakr. *et al.*, Fl. India 23: 337. 2012. *Euphorbia tithymaloidea* L., Sp. Pl. 453. 1753.

Scandent undershrub, up to 1 m high, monoecious; branchlets zigzag, succulent; latex milky. Leaves 4-11 × 2.5-5.5 cm, brittle, base attenuate, margin entire, apex acute, subsessile. Cyathia terminal; bracts 4, unequal, thin – fleshy, scarlet red, posterior 2 shorter, anterior 2 longer, obtuse, ciliate, caducous; involucre 1.5 cm, obliquely bilateral; glands 4, unequal; appendages 0. Male florets 22-25, ebracteolate; filaments to 1 mm, each 2-fid, above. Female floret: solitary, raised by an erect pedicel, ovary 3-locular, styles stout, cylindrical, ca 7 mm long, stigmas 3, each bifid. Fruits not seen.

Common in plains, in gardens and often in hedges. Fl.: March, Fr.: not seen. Vern.: Tel.: *Kantapala, Seema citramulam*; Tam.: *Kannaadikkalli.*

INDIA: Cultivated in gardens in most states.

WORLD: Native of Tropical America now grown in most tropical countries.

PHYLLANTHUS Linnaeus

1. Fruit a berry or drupe – Trees:
    2. Fruit small, fleshy; calyx lobes 5 .... **P. reticulatus**
    2. Fruit large, of 3-6 hard cocci in a fleshy pericarp:
        3. Sepals 4; stamens 4, the stamina filaments free .... **P. acidus**

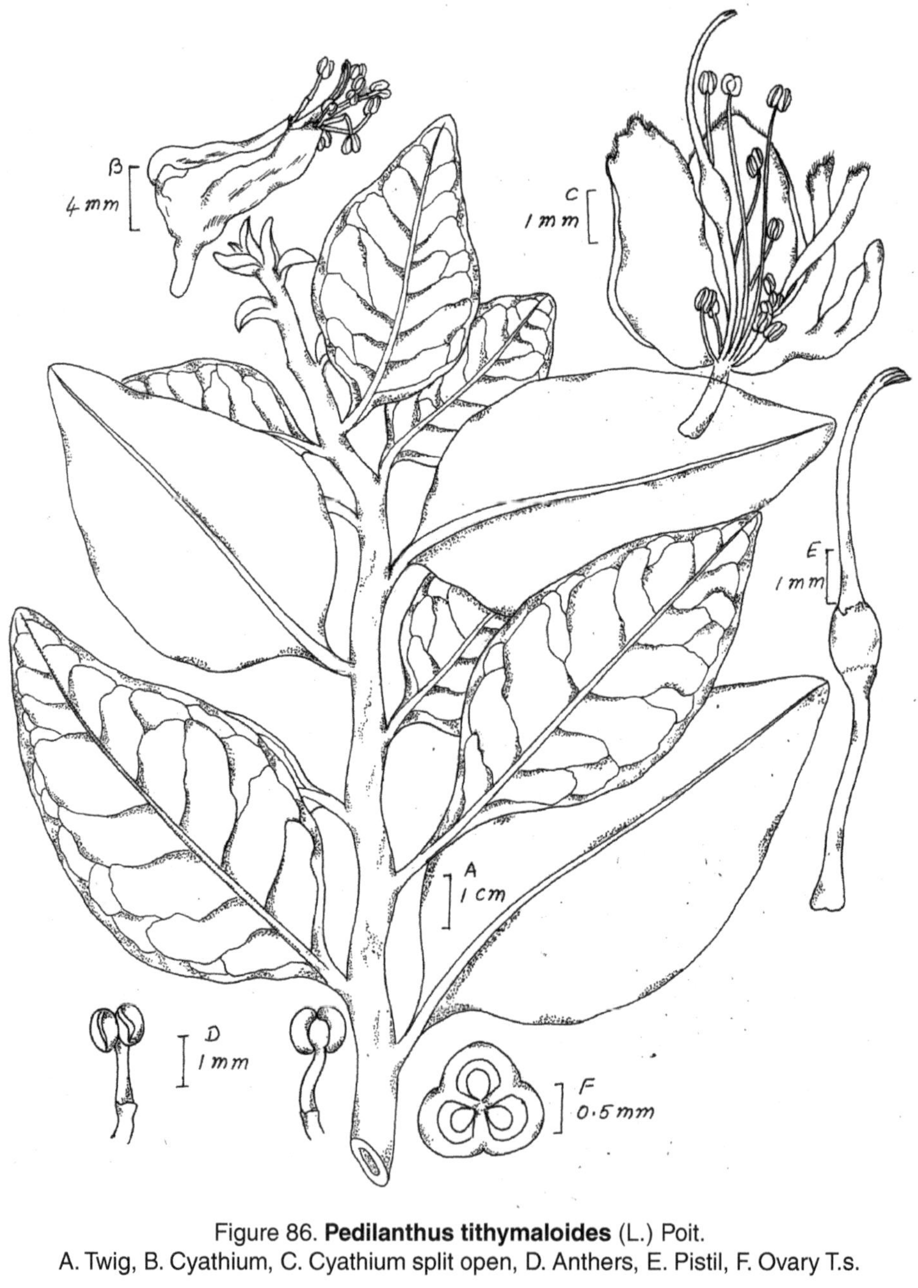

Figure 86. **Pedilanthus tithymaloides** (L.) Poit.
A. Twig, B. Cyathium, C. Cyathium split open, D. Anthers, E. Pistil, F. Ovary T.s.

3. Sepals 5-6, stamens 3, the stamina filaments connate:
   4. Leaves linear-oblong, 1-1.8 × 0.3-0.5 cm, 100 or more on each branchlet; medium sized trees .... **P. emblica**
   4. Leaves elliptic-oblong, 1-2 × 0.3-1.2 cm, ca 50 on each branchlet; small trees .... **P. indo-fischeri**

1. Fruit a dry capsule of 3, 2-valved cocci:
   5. Shrubs or trees:
      6. Stamens 6, free .... **P. pinnatus**
      6. Stamens 3, filaments connate:
         7. Fruit densely reddish hairy, fruting pedicel 4 cm long .... **P. chandrabosei**
         7. Fruit glabrous, fruting pedicel up to 3 mm long
            8. Branchlets 5-15 cm long; leaves oblong; petiole to 8 mm .... **P. polyphyllus**
            8. Branchlets 2.5-7.5 cm long; leaves elliptic-oblong; petiole to 4 mm .... **P. lawii**
   5. Annual herbs, sub-shrubs or undershrubs:
      9. Capsule verrucose:
         10. Leaf margin serrulate; stipules lanceolate; filaments connate; seeds transversely furrowed.................... **P.urinaria**
         10. Leaf margin entire; stipules sagittate-peltate; filaments free; seeds minutely tuberculate .... **P. simplex**
      9. Capsule smooth or other than verrucose:
         11. Tepals 5; disc glands 5:
             12. Stamens 3.. **P. amarus**
             12. Stamens 5.. **P. tenellus**
         11. Tepals 6; disc glands 6:
             13. Stamens 3:
                 14. Proximal leaves of leaf-bearing branches double the size of distal leaves .... **P. airy-shawii**
                 14. All leaves of branchlets more or less of the same size:
                     15. Undershrubs.. **P. narayanaswamii**
                     15. Herbs:
                         16. Prostrate or slightly ascending herb ....**P. rotundifolius**
                         16. Erect herbs:
                             17. Branches winged and hispidulous........**P. scabrifolius**

17. Branches not winged, glabrous:
    18. Stipules peltate, cordate or sagittate at base.. **P. maderaspatensis**
    18. Stipules neither peltate nor cordate or sagittate at base.................... **P. fraternus**

13. Stamens 5:
    19. Male flowers 1-3 .... **P. debilis**
    19. Male flowers 5-6 .... **P. rheedei**

Phyllanthus acidus (L.) Skeels, U.S.D.A. Bur. Pl. Industr. Bull. 148: 7.1909; Chakrab. *et al.*, in N.P. Balakr. *et al.*, Fl. India 23: 445. 2012. *Averrhoa acida* L., Sp. Pl. 428. 1753. *Cicca acida* (L.) Merr., Interpr. Herb. Amb.314. 1917. *C. disticha* L., Mant. 1: 124. 1767. *Phyllanthus distichus* Müll.Arg. in DC., Prodr. 15(2): 413. 1866; FBI 5: 304. 1887.

Small to moderate-sized tree; branches thick, tuberculate and closely marked with the scars of the deciduous, leafy, slender branchlets. Leaves ovate or elliptic to obovate 5-7.5 × 2-3 cm, glabrous, pale-glaucous beneath, base rounded; petiole to 3 mm; stipules minute, toothed. Flowers in slender racemes on leafless branchlets. Male flowers red, minute; pedicels 1-2 mm long; sepals 2 + 2, ovate to suborbicular; disc glands 4, suborbicular; stamens 4, free, 2-3 mm long. Female flowers: green, larger, pedicels 2-4 mm long; sepals 2 + 2, ovate to suborbicular; disc lobulate; ovary subglobose, styles free, 1.5-2 mm long, deeply bifid, reflexed. Fruit yellowish, depressed-globose, 6-8-lobed, 1.6 cm dia., pericarp fleshy, acidic.

Cultivated, in gadens, house compounds *etc.* Fl.: May. Fr.: June-July. Fruit is sour, edile; cooked or pickled or is made into jam or jelly. Vern.: Tel.: *Racha usiri, Chinna usiri;* Tam.: *Aranelli;* Ori.: *Ansokoli, Aura-koli, Bungarada, Narakoli.* English: Star gooseberry.

INDIA: Widely cultivated in West Bengal and all southern states.

WORLD: Native of Brazil, widely grown in Bangladesh, Myanmar, Vietnam, Cambodia, Laos, Thailand, Malaysia, Indonesia and Philippines.

Phyllanthus airy-shawii Jean F. Brunel & J.P.Roux, Nordic J. Bot. 4: 470. 1984; Mitra & Nayar, Taxon 37:470. 1988; N.P.Balakr. & Chakrab., Fam. Euphorb. India 385. 2007; Chakrab. *et al.*, in N.P.Balakr. *et al.*, Fl. India 23: 481. 2012. *P. debilis* 'Herb'. Ham. ex Hook.f., Fl. Brit. India 5: 299. 1887, *p.p.* maj. non Klein ex Willd. 1805; Surya Prakash Babu in Pullaiah & Moulali, Fl. Andhra Pradesh 2: 879. 1997.

Herb, 20-80 cm high; stem subcylindric, smooth, first producing 4-5 basal leaves then triangular cataphylls. Cataphylls slightly tinged with anthocyanin, 1.8-2.8 mm long with stipules 1.9-3.1 mm; branchlets glabrous, strong, 15-30 cm long, hardly compressed. Leaves early caducous when dry, ovate to elliptic, 1.2-3 × 0.8-1.5 cm, apex acute or obtuse, margin entire, base cuneate, yellowish green; petiole 0.6-1.1

mm; stipules 2.2-3.1 mm. Two to four basal axils of the branchlets each with a male multiflorous cymule, the distal axils with a single female flower. Male flower early caducous after anthesis; pedicel 1-1.3 mm long; tepals 0.6-1 × 0.5-0.6 mm, in 2 unequal whorls; disc of 6 papillose glands alternating with the tepals. Filaments fused into column but distally free; anthers transversely dehiscing. Female flowers: pedicels elongating to 3.1-3.8 mm; tepals 6, 1.5-2.8 × 0.9-1.6 mm, in 2 unequal cycles; disc glandular, annular with denticuate margin; ovary smooth, glabrous; styles bifid. Capsule 2.4-2.8 mm diam; seeds brownish, ungula-shaped, with 6-7 longitudinal ribs on the back.

Common weed in cultivated fields, waste places and road sides (Saxena and Brahmam, 1995; Chakrabarty *et al.,* 2012). Fl. & Fr. May-January.

INDIA: Uttar Pradesh, Jharkhand, West Bengal, Sikkim, Assam, Meghalaya, Odisha, Madhya Pradesh and Andhra Pradesh.

WORLD: Bangladesh, Myanmar and Thailand.

Phyllanthus amarus Schumach. & Thonn., Beskr. Guin. Pl. 421 & Kongl. Danske Vidensk. - Selsk. Skr. 4: 195. 1829; N.P. Balakr. & Chakrab., Fam. Euphorb. India 386. 2007; Chakrab. *et al.,* in N.P. Balakr. *et al.,* Fl. India 23: 482. 2012. *P. niruri* auct. non L. 1753; Hook.f., Fl. Brit. India 5: 298. 1887 *p.p.*; Gamble 2: 1290. 1925. *P. nanus* Hook. f., Fl. Brit. India 5: 298. 1887.

Herb, up to 90 cm high, monoecious; branchlets slender, spreading, stem cylindrical, sometimes woody at base. Leaves 4-9 × 2-5 mm, elliptic-obovate or oblong, base rounded – obtuse, margin entire, apex obtuse or apiculate, glaucous below; petiole 0.5 mm, very short; stipules lanceolate, acute, to 1.5 mm. Flowers axillary; male flowers few, 1-3 in upper axils; female flowers many, solitary in the remaining lower axils. Male flowers 1.2 mm across; tepals 5, 0.8 × 0.4 mm, free, ovate, acute; stamens 3, filaments 0.3 mm, connate, anthers 0.3 mm across, basifixed, extrorse, dehiscing transversely, disc of glands 5, minute, flat or saucer-shaped. Female flowers 1.6 mm across; tepals 5, 0.9 × 0.4 mm, free, ovate-oblong, acute – obtuse, with white central margin; ovary 0.5 mm, 6-lobed, styles 3, each 2-fid, at above; disc of 5 glands. Capsule 2.5 mm, depressed globose, scarcely lobed; seeds trigonus, vertically muricate.

Common weed of open waste lands and in cultivated fields. Fl. & Fr.: Throughout the year. Vern.: Tel.: *Nela usiri;* Tam.: *Kizhaa nelli;* Ori.: *Bhui ania, Bhui aola, Badi aonla,Bariamla.*

Anantapuram (ATP), *PSPB* 9182; Kalasamudram (ATP), *KRKS* 33687; Palakonda (CDP), *CS* 7798; Thummalapalle (CDP), *KRKS* 33687; Obulapuram thanda (PKM), *BSS* & *SKB* 32797; Palem (GNT), *VRK* 3266; Nelapattu (NLR), *PV* 110168 (BSID); Kakinada (EG), *SKW* 1941 (BLAT); Araku (VSKP), *SKW* 4679, *HS* 20407 (BLAT); Bhimilipatnam (VSKP), *TAR* 7247 (BSID).

INDIA: Through out the country.

WORLD: Native of tropical America, now a pantropical weed.

Phyllanthus chandrabosei Govaerts & Radcl.-Sm., Kew Bull. 51: 176. 1996; N.P. Balakr. & Chakrab., Fam. Euphorb. India 386. 2007. *Reidia stipulacea* Gamble, Bull. Misc. Inform. Kew 1925: 332.'1925. *Phyllanthus stipulaceus* (Gamble) Kumari & Chnadrab. in A.N. Henry *et al.*, Fl. Tamil Nadu, India, Ser. I. Analysis 2: 239. 1987 (non Bojer, 1837).

Undershrubs, to 5 m tall; branchlets terete, spreading, with persistent stipules. Leaves obliquely ovate or falcate-ovate, 5-10 x 2.5-5 mm, unequal at base, apiculate or shortly acuminate at apex, membranaceous, hispid beneath, lateral nerves 5 or 6 pairs, faint; petiole to 0.5 mm long; stipules lanceolate, stiff, persistent. Inflorescence axillary; male cymules 1-flowered; bisexual cymules with female flowers towards the tips; bracts linear-lanceolate, up to 2 mm long, entire. Male flowers; pedicels to 5 mm long, filifom, glabrous; sepals 2 + 2, broadly ovate to deltoid-suborbicular, entire, pellucid-punctate; disc glands 4, obcordate to squarish, entire; stamens 2, subsessile, horizontal. Female flowers; pedicels ca 10 mm long; sepals 3 + 3, broadly ovate to deltoid-ovate, 2 x 1.8 mm, entire; pellucid-punctate, glabrous; disc glands 6, suborbicular, entire; ovary depressed globose, 0.8 mm across, villous; style 3, 1 mm long, free, bifid in midway. Fruits subglobose, depressed, 4 x 6 mm, densely soft redish hairy.

Forest undergrowth in Kolli hills above 1000 m. Fl. & Fr.: June August.

Kolli hills (NMK), *SKS* 949 (SGH).

INDIA: Tamil Nadu (Endemic).

Phyllanthus debilis Klein ex Willd., Sp. Pl. ed. 4, 4: 582. 1805; Gamble 2: 1290. 1925; N.P. Balakr. & Chakrab., Fam. Euphorb. India 386. 2007; Chakrab. *et al.*, in N.P. Balakr. *et al.*, Fl. India 23: 484. 2012. *P. niruri* auct. non L. 1753: FBI 5: 298. 1887, *p.p.*

Herb, up to 1 m high, monoecious; branchlets terete. Leaves 4-20 × 0.8-7 mm, elliptic-obovate, base rounded-acute, margin entire, apex acute-broadly acute, glaucous below; petioles to 1.5 mm; stipules narrowly triangular, acuminate, scarious. Flowers axillary; male flowers few, 1.3 in the lower axils; female flowers many, solitary in the upper axils. Male flowers 3 mm across; tepals 6, 1 × 0.8 mm, obovate – oblong, subacute, with scarious margin; stamens 3, filaments 0.5 mm, connate, anthers 0.25 mm across; disc of glands 6, stellately lobed. Female flowers 4 mm across; tepals 6, 1.8 × 1 mm, obovate-oblong, acute-subacute, with broad scarious margin; ovary 0.8 mm, globose, 6-lobed; styles 3, each 2-fid, from above the middle; disc of glands 6, stellately lobed. Capsule 4 mm, depressed globose, smooth; seeds trigonus, globose, vertically striate.

Common in shady places in the hill forest and near the foothills. Fl. & Fr.: August -January. Vern.: Tel.: *Gunne sintha koora.*

Rollapenta (KNL), *RVR* 2347; Guvvalacheruvu (CDP), *R.V.Reddy* 7954; Penchalakona (NLR), *PSPB* 9112; Basavasena (KNL), *VBH* 84917; Kodur (CDP), *SKW* 7546; Medarikonda (CTR), *SKW* 6263 (BLAT); way to Tirumala hills (CTR), *LR* & *MSRao* 2534 (BSID); Addatigala (EG), *SKW* 4014 (BLAT); Tadepalli (EG), *NRR* & *DN* 84310 (BSID); Araku (VSKP), *HS* 20489 (BLAT).

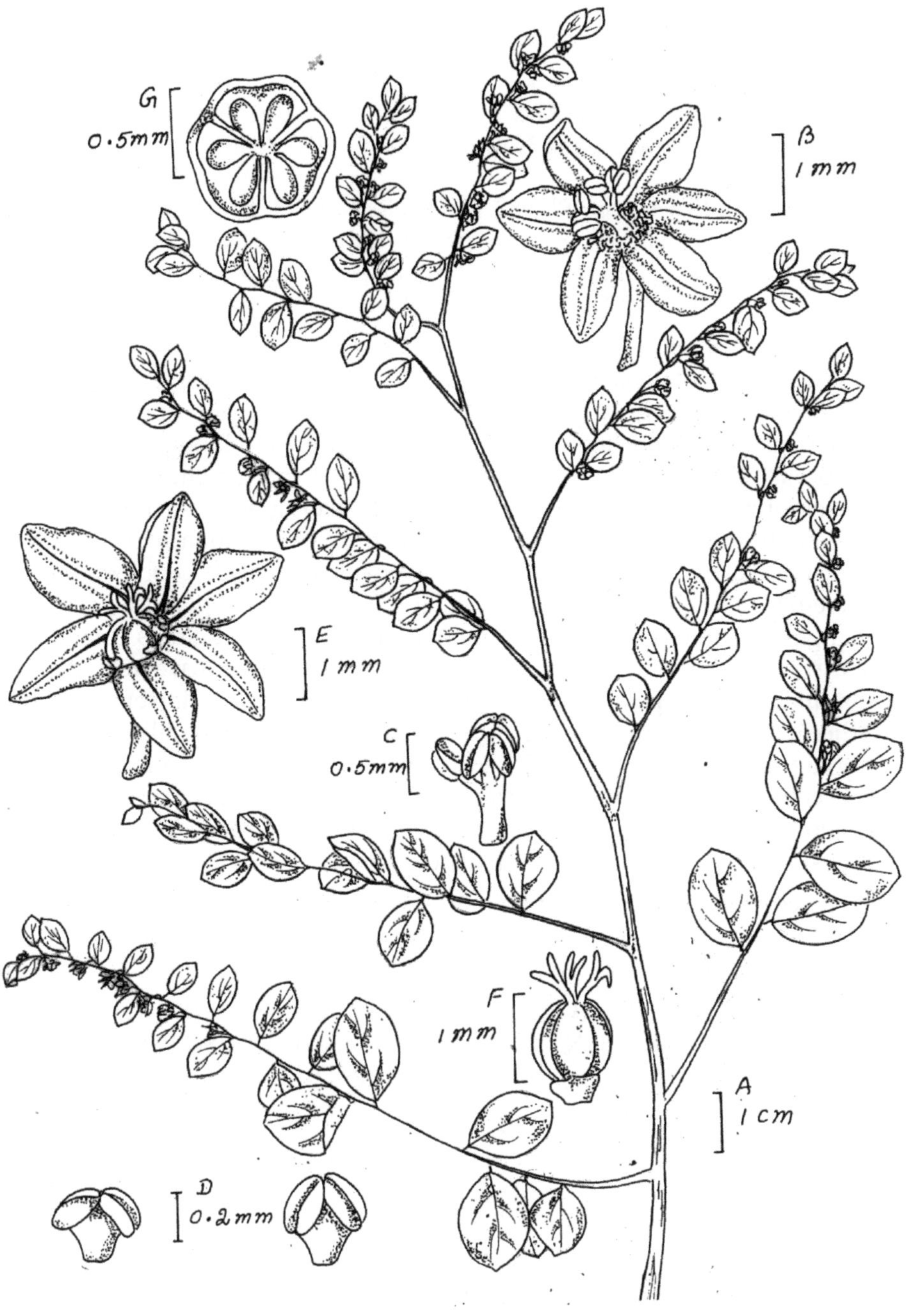

Figure 86. **Phyllanthus debilis** Klein ex Willd.
A. Twig, ♂ B. Flower, C. Androecium, D. Anthers, ♀ E. Flower, F. Pistil, G. Ovary T.s.

INDIA: Rajasthan, West Bengal, Odisha, Madhya Pradesh, Andhra Pradesh, Goa, Karnataka, Kerala, Tamil Nadu and Andaman and Nicobar Islands.

WORLD: Sri Lanka, Myanmar, Thailand, Malaysia, Indonesia, Pacific, West Indies.

Note: *Phyllanthus debilis* described by Hook.f. in Fl. Brit. India is *P. airy-shawii* Brunel & Roux (Mitra and Nair in Taxon 37: 470. 1988).

Phyllanthus emblica L., Sp. Pl. 982. 1753; Hook. f., Fl. Brit. India 5: 289. 1887; N.P.Balakr. & Chakrab., Fam. Euphorb. India 369. 2007; Chakrab. *et al.*, in N.P. Balakr. *et al.*, Fl. India 23: 446. 2012. *Emblica officinalis* Gaertn., Fruct. Sem. Pl. 2: 122. t. 108. f. 2. 1790; Gamble 2: 1295. 1925.

Deciduous, monoecious tree, up to 15 m high; branches spreading; bark greyish-white, smooth, peeling off in flakes; wood red or reddish-brown, more or less shallowly and irregularly interlocked grained, coarse and uneven textured; branchlets glabrous and finely pubescent. Leaves linear-oblong, 1-1.8 × 0.3-0.5 cm, glabrous, light green, base sub-cordate, unequal, margin entire, apex obtuse, apiculate; petiole 1 mm, stipules 1.2 mm, linear. Flowers greenish-yellow in axillary fascicles on the leaf bearing branchlets, often on the naked portion below the leaves, trimerous, cyclic. Male flowers: pedicels 1-4.5 mm long; sepals 3 + 3, oblong-obovate; flabellate-obovate to spathulate-oblanceolate, subequal; disc glands 6, clavate or subtriangular; stamens 3, filaments connate into a column, anthers connate into a squarish or globose mass, basifixed, extrorse, dehiscing transversely; disc of 6 glands small, cup-shaped, alternating perianth lobes. Female flowers: subsessile or pedicels up to 0.5 mm long; sepals 3 + 3, oblong-obovate to oblanceolate; disc shortly cupular-annular, deeply 6-lobed; ovary ovoid, styles 3, 2-3 mm long, shortly connate below into a column, bifid to quadrifid above, recurved; disc present. Drupe pale yellow, depressed globose, 1-2 cm in dia., 3-celled, 6-seeded; seeds reniform, 3-gonous, shining, reddish-brown, small, 6 × 7 mm.

Very common, in dry deciduous forests above 600 m. Fl.: March-May. Fr.: October-February. Vern.: Tel.: *Pedda usiri chettu, Triphalamu, Pulla usirika,Usiri;* Tam.: *Nelli;* Ori.: *Anola, Amla.* Eng.: Indian gooseberry.

Balugram (KNL), *TP* & *RVR* 2751; Tirumala (CTR), *SSR* 13782; Chintala forest (PKM), *SSR* 16124; Anupu (GNT), *VRK* 3964; Kadapa (CDP), *JSG* 11092 (DD); Near Dabuguda, Laxmipuraam (VSKP), *NRR* 83792 (BSID).

INDIA: Through out the country.

WORLD: Sir Lanka, Pakistan, Nepal, Bhutan, Bangladesh, Myanmar, Thailand, China, W. Malesia and Lesser Sunda Islands.

Phyllanthus fraternus G.L.Webster, Contrib. Gray Herb. 176: 53. 1955 & J. Arn. Arb. 38: 309, f. 19G and H. 1957; N.P. Balakr. and Chakrab., Fam. Euphorb. India 387. 2007; Chakrab. *et al.*, in N.P. Balakr. *et al.*, Fl. India 23: 485. 2012. *P. niruri* auct. plur. non L. 1753: Hook.f., Fl. Brit. India 5: 298. 1887, *p.p.*

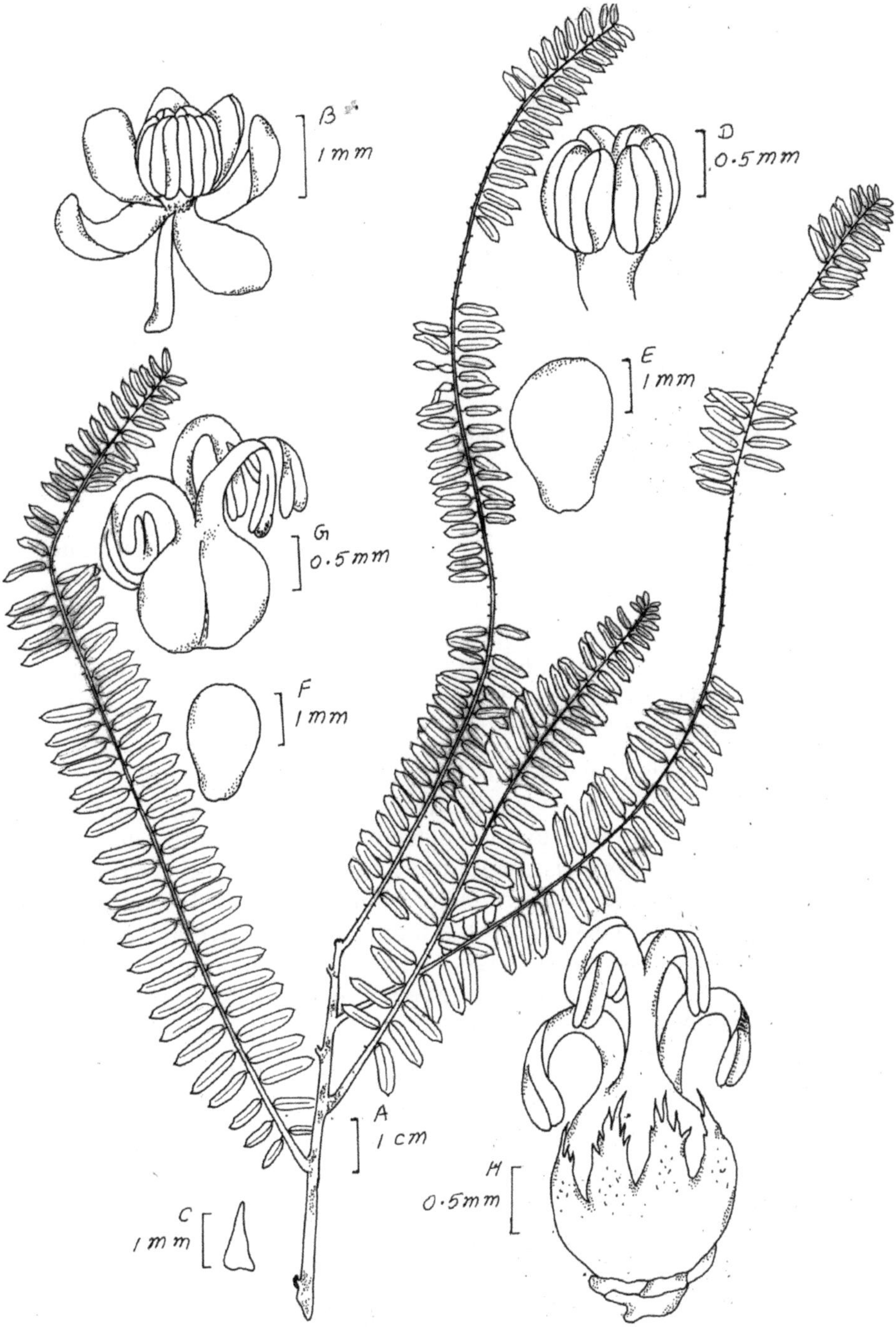

Figure 88. **Phyllanthus emblica** L.

A. Twig, ♂ B. Flower, C. Bract, D. Androecium, ♀ E and F. Tepals, G and H. Pistil.

Slender, erect, annual herb, 30-60 cm high, stem angular, rarely cylindrical. Leaves elliptic-oblong to elliptic-oblanceolate, 5-13 × 1.5-5 mm, apex obtuse or rounded, margin entire, base cuneate or obtuse, glabrous, membranous, lateral nerves 4-7 pairs, indistinct; petioles 0.5 mm long; stipules 1.3 mm, lanceolate. Flowers in cymule. Male flowers 1-3, pedicels 1 mm long; sepals 6, suborbicular-obovate, disc glands 6, lobulate; stamens 3, filaments united into a short column. Female flowers: pedicels 1.5-2 mm long; sepals 6, oblong-oblanceolate, disc thin, flat, irregularly deeply lobed into 6-10 segments, ovary sessile, subglobose, styles minute, free. Fruits trilobite-subglobose, 1.7-2 mm diam.; seeds longitudinally 7-8-ridged on the back, concentrically 7-8-ridged on the sides.

Weed in waste places and cultivated fields in the plains at low altitudes (Saxena and Brahmam, 1995; N.P. Balakrishnan and Chakrabarty, 2007). Fl. & Fr.: August – October.

INDIA: Haryana, Delhi, Uttar Pradesh, Bihar, West Bengal, Tripura, Odisha, Madhya Pradesh, Rajasthan, Gujarat, Maharashtra, Andhra Pradesh, Karnataka and Tamil Nadu.

WORLD: Probably native to Pakistan and Western India; introduced into West Indies and Africa.

Phyllanthus indofischeri Bennet, Indian Forester 109: 221. 1983; N.P. Balakr. & Chakrab., Fam. Euphorb. India 369. 2007; Chakrab. *et al.*, in N.P. Balakr. *et al.*, Fl. India 23: 447. 2012. *Emblica fischeri* Gamble, Bull. Misc. Inf. Kew 1925: 330. 1925 & Fl. Madras 2: 1295. 1925.

Tree, up to 12 m high; monoecious; branchlets puberulous, stout, erect, interior solid, bark mostly smooth. Leaves many on the branchlets, up to ca. 60, oblong, 1-2.8 × 0.3-1.2 cm, glabrous, green above and glaucous below, base truncate, cordate or obtuse, margin entire, apex retuse, secondary nerves obscure; petioles 1-5 mm long; stipules 1mm, linear, acute. Flowers greenish, subsessile in axillary fascicles. Male flowers: many in upper axils of branchlets; tepals 6 in two whorls, free, stamens 3, filaments connate, disc glands large. Female flowers few in lower axils; disc urceolate; styles 3, thin, twice bifid, recurved. Drupe globose, fleshy, 2.5-4 cm in diam., with 3 pyrenes; seeds trigonous, crustaceous.

Occasional in Nallamalais, Seshachalam and Kambakkam hills. Fl.: November, Fr.: January.

Talakona (CTR), *PSPB* 11861, *GVM* 112976 (BSID); Balapalli (CDP), *JLE* 15018; Mahanandi (KNL), *JLE* 25475 (MH).

INDIA: Andhra Pradesh, Karnataka and Tamil Nadu. Endemic.

Phyllanthus lawii Graham ex Müll.Arg. in DC., Prodr. 15: 353. 1866; FBI 5: 290. 1887; Gamble 2: 1289. 1925; Chakrab. *et al.*, in N.P. Balakr. *et al.*, Fl. India 23: 465. 2012. *P. juniperinoides* Müll.Arg., Linnaea 32: 18. 1863 & in DC., Prodr. 15(2): 358. 1866.

Shrub, up to 1.5 m high, monoecious; branchlets terete, striate. Leaves 0.3-0.5 × 0.2-0.3 cm, elliptic – oblong, base rounded or subcordate, margin entire, rounded – apiculate; petiole 0.4 mm; stipules lanceolate – subulate. Flowers axillary; male flowers solitary or few; female flowers solitary. Male flowers: 3 mm across; pedicels filiform, ca 5 mm long; sepals 3+3, linear – oblong, obtuse; stamens 3, filaments connate. Female flowers similar to male flowers; ovary globose, styles 3, short, each 2-fid at above, recurved. Capsule globose, slightly depressed; seeds trigonus, smooth.

Occasional along the river banks. Fl. & Fr.: September – January.

Palakonda (CDP), *JSG* 20929 (DD), 20974 (MH); Koruturu (WG), *DN* 84276 (BSID); Matur (EG), *JSG* 4944 (MH); Velavaram (KSN), *JSG* 18633 (MH).

INDIA: West Bengal, Odisha, Jharkhand, Madhya Pradesh, Andhra Pradesh, Maharashtra, Karnataka and Tamil Nadu. Endemic.

Phyllanthus leschenaultii Müll.Arg., Linnaea 32: 37. 1863 & in DC. Prodr. 15(2): 398. 1866; Chakrab. *et al.*, in N.P. Balakr. *et al.*, Fl. India 23: 486. 2012. *P. missionis* Hook. f., Fl. Brit. India 5: 297. 1887; Gamble 2: 1290. 1925.

Slender, glabrous shrub, 2-2.5 m high. Leaves alternate, very thin, petioled, elliptic-oblong, base acute, apex obtuse. Flowers very small, pedicels much shorter than leaves, nerves 6-8 pairs, distinct, stipules small, ovate-lanceolate. Flowers very small, pedicels much shorter than the leaves; calyx lobes with obscure scarious margins. Male flowers 3-nate; sepals slightly enlarged, but not reflexed in fruit; stamens 3, filaments free; ovary 3-celled, style short, reflexed. Capsule smooth, coriaceous; seeds ribbed.

Occasional in Visakhapanam district. Fl. & Fr.: August – December. Vern.: Tel.: *Pedda amla.*

Near Suthiguda, Laxmipuram (VSKP)), *NRR* & *DN* 84391 (BSID).

INDIA: Sikkim, West Bengal, Arunachal Pradesh, Meghalaya, Mizoaram, Adhra Pradesh, Kerala and Tamil Nadu.

WORLD: Bhutan.

Phyllanthus maderaspatensis L., Sp. Pl. 982. 1753; FBI 5: 292. 1887; Gamble 2: 1289. 1925; N.P. Balakr. & Chakrab., Fam. Euphorb. India 378. 2007; Chakrab. *et al.*, in N.P. Balakr. *et al.*, Fl. India 23: 469. 2012.

Herb, up to 70 cm high, monoecious; branchlets subterete. Leaves 0.2-0.7 × 0.2-0.9 cm, obovate, base cuneate, margin entire, apex retuse – obtuse or rounded – truncate, glabrous, glaucous below; petiole to 3 mm; stipules peltate, lanceolate, acute. Flowers axillary, pedicellate; male flowers few, minute, 1-3 in upper axils; female flowers many, solitary, in remaining lower axils, subsessile; bracts 2, narrowly triangular, acute. Male flowers 1.8 mm across; tepals 6, 0.7 × 1 mm, obovate, obtuse, with scarious margin; stamens 3, filaments 0.3 mm, connate, anthers 0.5 mm acorss; disc of glands 6, minute. Female flowers 2.5 mm across; tepals 6, 1.5 × 0.6-1.2 mm, obovate – broadly ovate, acute, unequal; ovary 0.8 mm, globose, styles 3, small,

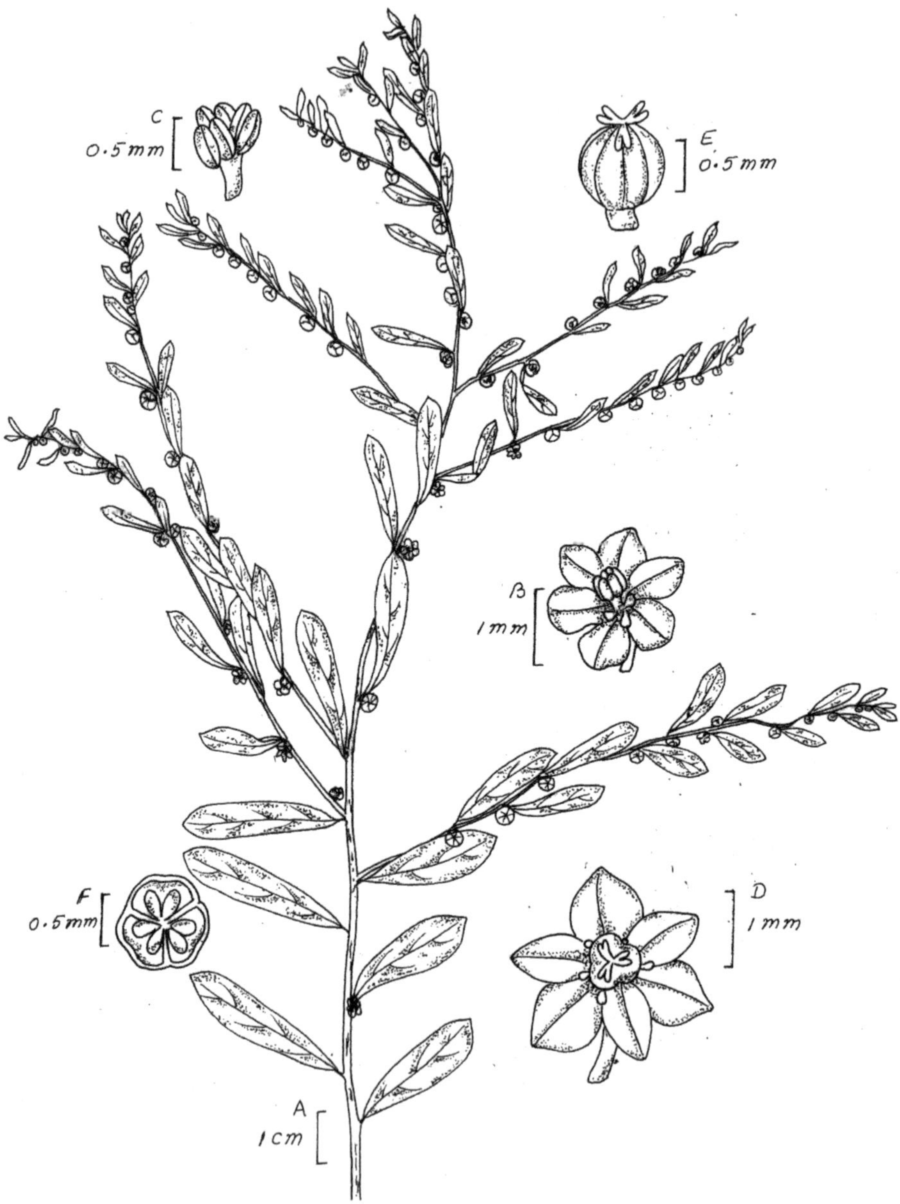

Figure 89. **Phyllanthus maderaspatensis** L.
A. Twig, ♂ B. Flower, C. Androecium, ♀ D. Flower, E. Pistil, F. Ovary T.s.

each 2-fid, at above. Capsule 4 mm, depressed globose, 3-lobed; seeds trigonus, brown, vertically muricate.

Common in wastelands and weed in cultivated fields. Fl. & Fr.: June – April. Vern.: Tel.: *Nalla usirika.*

Akuthotapalli (ATP), *NY* 181; Veldurthi (KNL), *RVR* 1757; Pacherla (KNL), *BSS* & *SKB* 34238; Palakonda (CDP), *CS* 9377; Mabbuchintalaplle (CDP), *TP* & *party* 33671; Macherla (GNT), *VRK* 3279; Simhachalam (VSKP), *PSPB* 11806; Nandyal (KNL), *SKW* 5241 (BLAT); Venkatagiri (NLR), *BSN* 4117 (BSID); Muzumamidivalasa, Maredumilli (EG), *MM* 102526; Addatigala (EG), *SKW* 3009, *HS* 20913 (BLAT).

INDIA: Bihar, Jharkhand, Odisha, Madhya Pradesh, Rajasthan, Maharshtra, Andhra Pradesh, Karnataka, Kerala and Tamil Nadu.

WORLD: Tropitcal Africa to Arabia and eastwards to Bangladesh, Myanmar, S. China, Java and Australia.

Phyllnthus narayanaswamii Gamble, Bull. Misc. Inf. Kew 1925: 329. 1925 & in Fl. Madras 2: 1289. 1925; Rolla S. Rao, J. Bombay Nat. Hist. Soc. 61: 322. 1964; N.P. Balakr. & Chakrab., Fam. Euphorb. India 379. 2007; Chakrab. *et al.,* in N.P. Balakr. *et al.,* Fl. India 23: 470. 2012.

Undershrub, up to 60 cm high, monoecious; branches many, from a stout rootstock, branchlets terete, winged. Leaves 0.6-1.2 × 3-0-6 cm, elliptic, base obliquely cordate, margin entire, thickened, recurved, apex rounded – obtuse, sometimes slightly emarginated or mucronate; petiole short; stipules peltate, lacerate, acuminate, hastate at base. Flowers axillary, solitary; pedicels of male flowers to 5 mm, female flowers to 5.5 mm; bracts linear – lanceolate, to 1 mm. Male flowers 1 mm across; tepals 6, free, 0.5 × 0.45 mm, ovate – obovate, rounded – obtuse; stamens 3, free, anthers 0.3 × 0.5 mm; disc of 6 glands, minute. Female flowers 2.5 mm across; tepals 6, free, 1.3 × 0.9 mm, elliptic oblong, subacute; ovary 1 mm across, verrucose glandular; styles 3, connate at base, each 2-fid, near the base. Capsule 4 mm across, 3-lobed, verrucose glandular; seeds to 2 mm, golden brown, covered with red glands.

Rare on the hills. Fl. & Fr.: July – October.

Tirumala hills (CTR), *LR* & *MSRao* 2533 (BSID); Cherukonda (VSKP), *GVS* 28184 (MH).

INDIA: Andhra Pradesh and Tamil Nadu. Endemic.

Phyllanthus pinnatus (Wight) G.L.Webster, J. Arnold Arbor. 38: 52. 1957. in adnot. Chakrab. *et al.,* in N.P. Balakr. *et al.,* Fl. India 23: 478. 2012. *Chorisandra pinnata* Wight, Icon. Pl. Ind. Orient. t. 1994. 1853; Gamble 2: 1285. 1925.

Large shrub, up to 5 m high, monoecius; branchlets subterete. Leaves variable, membranous, elliptic – sub orbicular or obovate, base cuneate, margin entire, apex acute, glaucous below; petiole to 4 mm, stipules minute. Flowers in axillary clusters; male flowers many; female flowers few; pedicels of male to 5 mm, filiform, female to 2 mm. Male flowers 3 mm across; tepals 6, free, ovate, fimbriate, subacute; stamens 6, free, filaments 2 mm, anthers 0.7 mm across; disc of glands 6, free. Female flowers

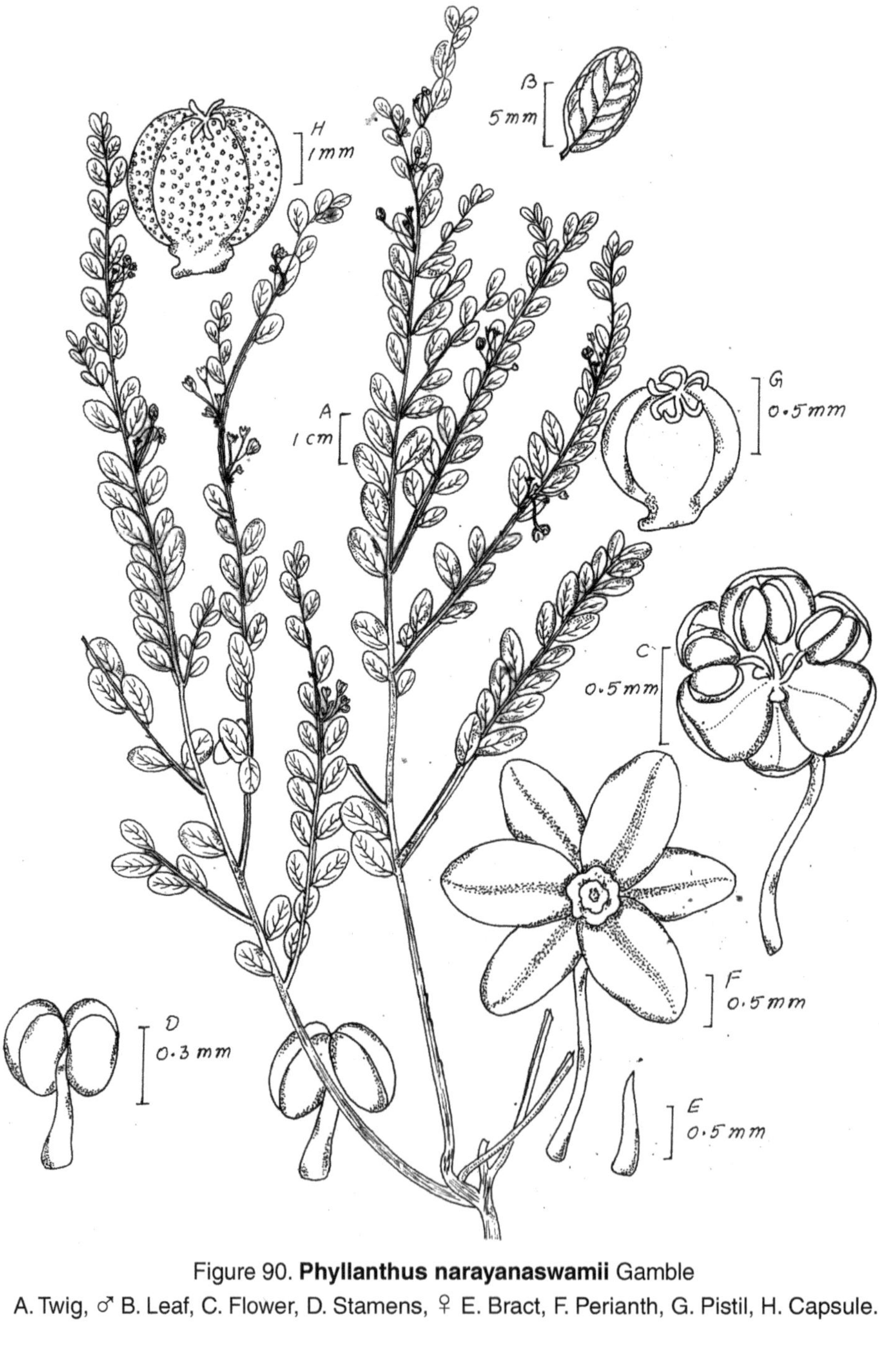

Figure 90. **Phyllanthus narayanaswamii** Gamble
A. Twig, ♂ B. Leaf, C. Flower, D. Stamens, ♀ E. Bract, F. Perianth, G. Pistil, H. Capsule.

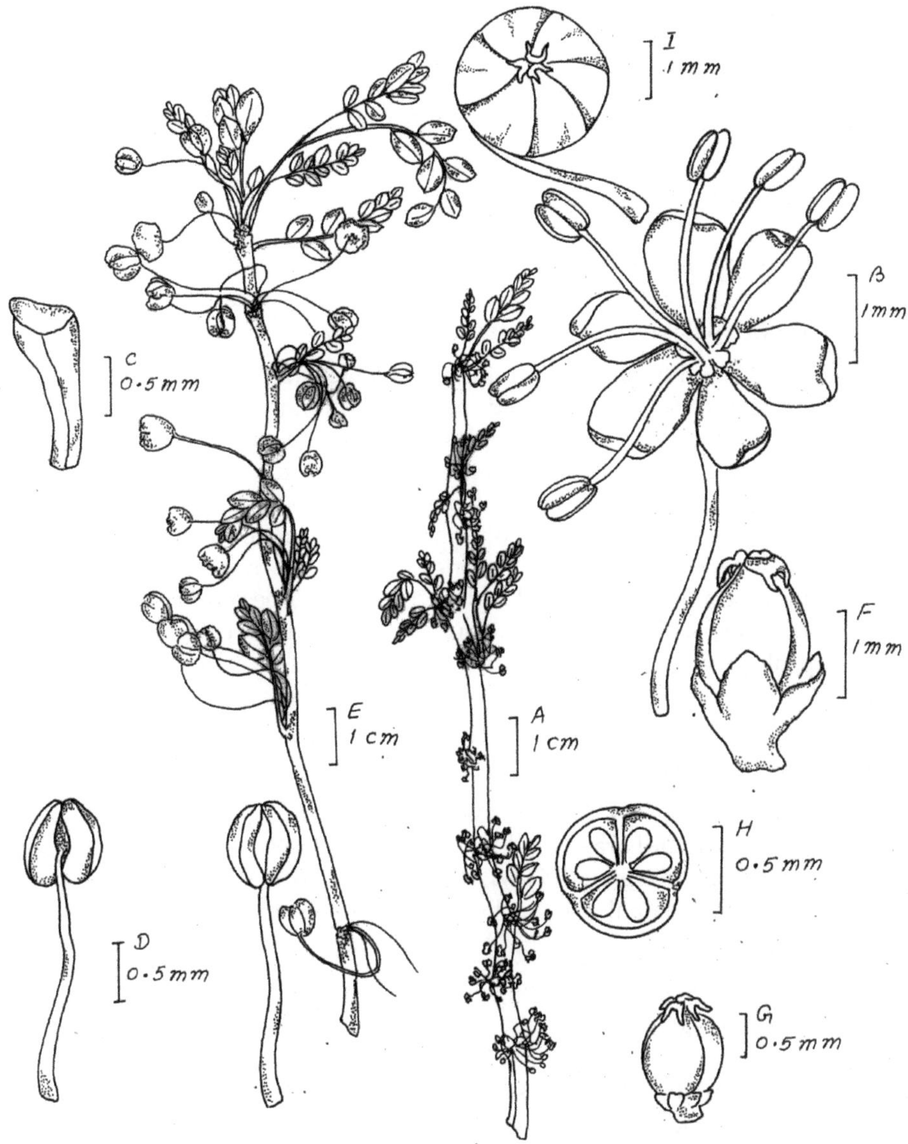

Figure 91. **Phyllanthus pinnatus** (Wight) Webster
♂ A. Twig, B. Flower, C. Bract, D. Stamens, ♀ E. Twig, F. Flower, G. Pistil, H. Ovary T.s., I. Capsule.

2 mm across; tepals 6, free, ovate, fimbriate, subacute; ovary 1 mm, globose, 3-lobed, styles 3, each 2-fid, at above; disc of glands 6, connate. Capsule 2.8 mm, depressed globose; seeds trigonous, smooth

Common in dry deciduous forests and also in open rocky places. Fl. & Fr.: June – September. Vern.: Tel.: *Nalla pooli.*

Madhavaram RF (KNL), *TP* & *RVR* 1326; Palakonda (CDP), *CS* 9325; Seshachalam (CDP), *PSPB* 11845; Mabbuchintalapalle (CDP), *KRKS* 36905; Durgam – Udayagiri (NLR), *PV* 110227; Kondapalli (KSN), *PV* 5315 (MH).

INDIA: Andhra Pradesh and Tamil Nadu.

WORLD: Sri Lanka and East Africa.

Phyllanthus polyphyllus Willd., Sp. Pl. 4: 586. 1805; FBI 5: 290; Gamble 2: 1288.1925; N.P. Balakr. & Chakrab., Fam. Euphorb. India 379. 2007; Chakrab. *et al.*, in N.P. Balakr. *et al.*, Fl. India 23: 472. 2012.

Deciduous small tree, up to 8 m high; branchlets terete. Leaves oblong, 0.5-2 × 0.2-0.5 cm, glabrous, base rounded, margin entire, apex apiculate, secondary nerves 6-8 pairs; petiole to 8 mm, stipules lanceolate. Flowers small, light yellow, monoecious, along the leaf rachis on ventral side, axillary. Male flowers: pedicels 1.5-5 mm long; sepals 3 +3, ovate, orbicular-ovate to oblanceolate; disc glands 6; stamens 3, filamentss connate into a column, anthers connate into an ellipsoid mass. Female flowers few, pedicels 1- 3 mm long; sepals 3 + 3, ovate-obong, obovate or oblanceolate; disc annular, shallowly 6-lobed with the lobes often undulate; ovary subglobose, styles 3, 1-1.5 mm long, connate below into a column, deeply bifid above. Capsule greenish, globose, 3-valved, 4 mm across, 3-lobed; seeds triquetrous, foveolate.

Occasional, in moist deciduous forests and scrub jungles. Fl.: December-February. Fr.: February-March. Vern.: Tam.: *Karunelli.*

Talakona (CTR), *SSR* 13791; Satrasala (GNT), *VRK* 6777; Tammapalli (KNL), *JSG* 10903 (DD); Panchadhara – Paladhara (KNL), *VBH* 84913 (BSID); Chitvel (CDP), *A.M.Rao* 3454 (BSID); Tirupati (CTR), *DRC* 2073, *GVS* 31844 (MH); Durgam (NLR), *PV* 111150 (BSID); Bodalanka (EG), *DN* 85577 (BSID); Kolli hills (NMK), *SKS* 1032 (SGH).

INDIA: Andhra Pradesh, Karnataka, Kerala and Tamil Nadu.

WORLD: Sri Lanka.

Phyllanthus reticulatus Poir. in Lam., Encycl. 5: 298. 1804; FBI 5: 288. 1887; N.P. Balakr. & Chakrab., Fam. Euphorb. India 382. 2007; Chakrab. *et al.*, in N.P. Balakr. *et al.*, Fl. India 23: 479. 2012. *Kirganelia reticulata* (Poir.) Bail., Etud. Gen. Euphorb. 613. 1858; Gamble 2: 1294. 1925.

Large shrub, up to 6 m high, monoecious; branchelets subterete, pubescent. Leaves 0.8-1.2 × 0.4-1 cm, elliptic – oblong, base acute, margin entire, apex acute – rounded; petiole 2 mm, pubescent; stipules lanceolate, acute. Flowers axillary; male flowers many, 3-6 in fascicles in upper axils; female flowers few, solitary, in lower

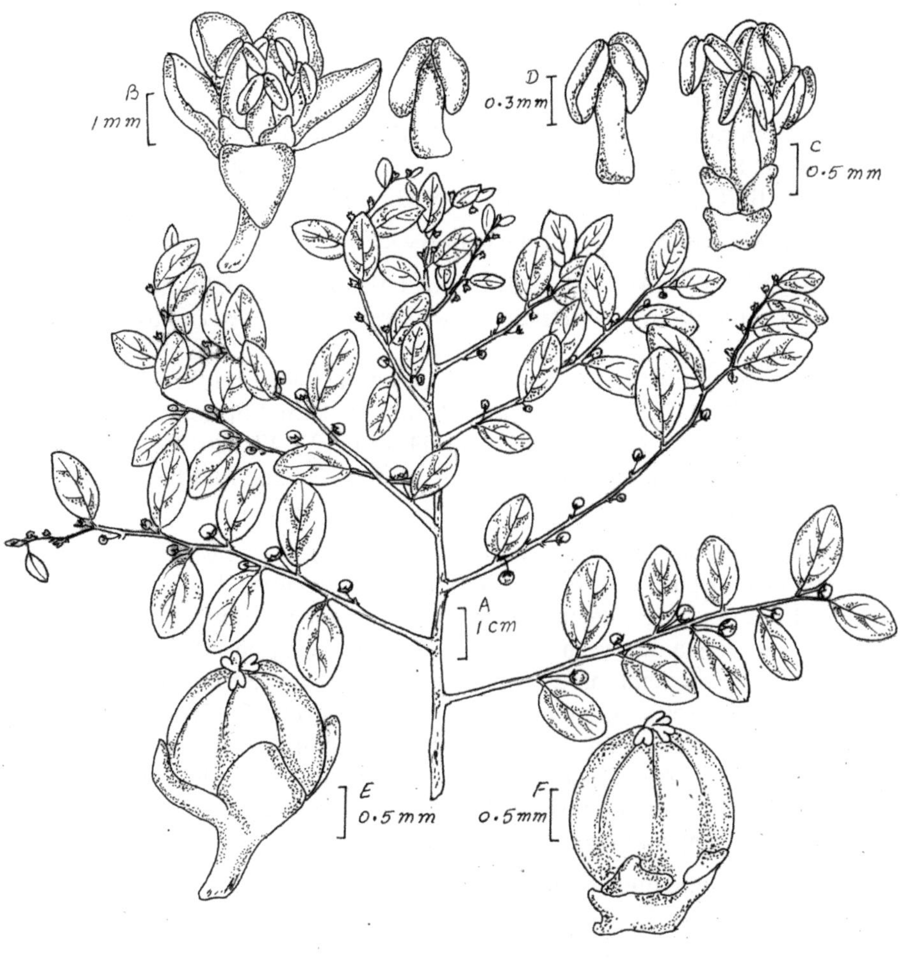

Figure 92. **Phyllanthus reticulatus** Poir.
A. Twig, ♂ B. Flower, C. Androecium, D. Stamens, ♀ E. Flower, F. Pistil.

axils; pedicels to 5 mm, pubescent; bracts small, lanceolate. Male flowers 2.5 mm across; tepals 5 or 6, obovate-oblong, oblong, obtuse, unequal; stamens 5, biseriate, 2 + 3; outer shorter, free, 1.1 mm, inner longer, connate, 1.9 mm; anthers 0.5 mm across; disc of glands 5. Female flowers 2 mm across; tepals 5 or 6, 1.5 × 1 mm, connate at base, obovate- broadly ovate, obtuse, unequal; ovary 1.5 mm, globose, styles 3, minute, each 2-fid; disc of glands 5. Drupe 5 mm, globose, smooth, fleshy, dark pink when ripened; seeds trigonus, crustaceous.

Common in hedges and at the foot hills of forests. Fl. & Fr.: July – March. Vern.: Tel.: *Nalla purugudu*; Tam.: *Karumpoolan, Poolan*.

Kalasamudram (ATP), *NY* 647; Chinnagunitam (KNL), *PSPB* 9148, 9149; Mungalpadu (KNL), *AA* 2037; Thummalapalle (CDP), *KRKS* & *KRS* 33675; Anupu (GNT), *VRK* 3600; Palakonda (CDP), *EKK* 79301 (DD); Sivadurg (NLR), *SKW* 6789 (BLAT); Sitarampuram (NLR), *PV* 111186 (BSID); Kondur (KSN), *SKW* 3376 (BLAT); Sirivaka, near Koruturu (WG), *DN* 84242 (BSID); Tiger camp, Maredumilli (EG), *MM* 101259 (BSID); Kolli hills (NMK), *SKS* 932 (SHG).

INDIA: Almost through out the country.

WORLD: Tropical W. Africa, Sri Lanka, Pakistan, Nepal, Bhutan, Bangladesh, Myanmar, Thailand, China, Vietnam, Laos, Cambodia, Malesia up to NE Australia.

Phyllanthus rheedei Wight, Icon. Pl. Ind. Orient. t. 1895. f.l. 1852; FBI 5: 293. 1887; Gamble 2: 1289. 1925; N.P. Balakr. & Chakrab., Fam. Euphorb. India 379. 2007; Chakrab. *et al.*, in N.P. Balakr. *et al.*, Fl. India 23: 474. 2012. *P. kozhikodianus* Sivar. & Manilal, J. Indian Bot. Soc. 56: 165. 1977; C.S. Reddy *et al.*, Suppl. Fl. Andhra Pradesh 107. 2008.

Undershrub, up to 50 cm high, monoecious; branchlets terete, angular, pubescent. Leaves 1-2.3 × 0.4-0.9 cm, elliptic – elliptic oblong, base rounded – acute, margin entire, apex apiculate, thinly pubescent; petiole to 4 mm, pubescent; stipules lanceolate. Flowers axillary; male flowers few, 5-6 in upper axils; female flowers many, solitary, in the remaining lower axils. Male flowers 2.5 mm across; tepals 6, 2 × 1.1 mm, obovate – oblong, subacute; stamens 5, biseriate, 2 + 3; outer shorter, free, inner longer, connate, anthers 0.5 m across; disc of glands 6, free. Female flowers 4 mm across; tepals 6, 2.4 × 1.4 and 2.2 × 1 mm, unequal, obovate – broadly oblong, subacute; ovary 0.5 mm, globose, styles 3, each 2-fid from above the middle; disc of glands 6, free. Capsule 4-5 mm, globose; seeds striate.

Occasional in the hills. Fl. & Fr.: October – December.

Kekati RF (ATP), *TP* & *NY* 780; Batrepalli (ATP), *KRKS* 39633; Ramannapenta (KNL), *JLE* 22201 (MH); Balepalli (CDP), *JLE* 14948 (MH); Durgam (NLR), *PV* 111154 (BSID); Agsipalli (KSN), *PV* 5223 (MH); G. Kondur (KSN), *PV* 6082 (MH); Rampachodavaram (EG), *MM* 105074 (BSID); Alagar hills (MDU), *SKS* 88 (SGH).

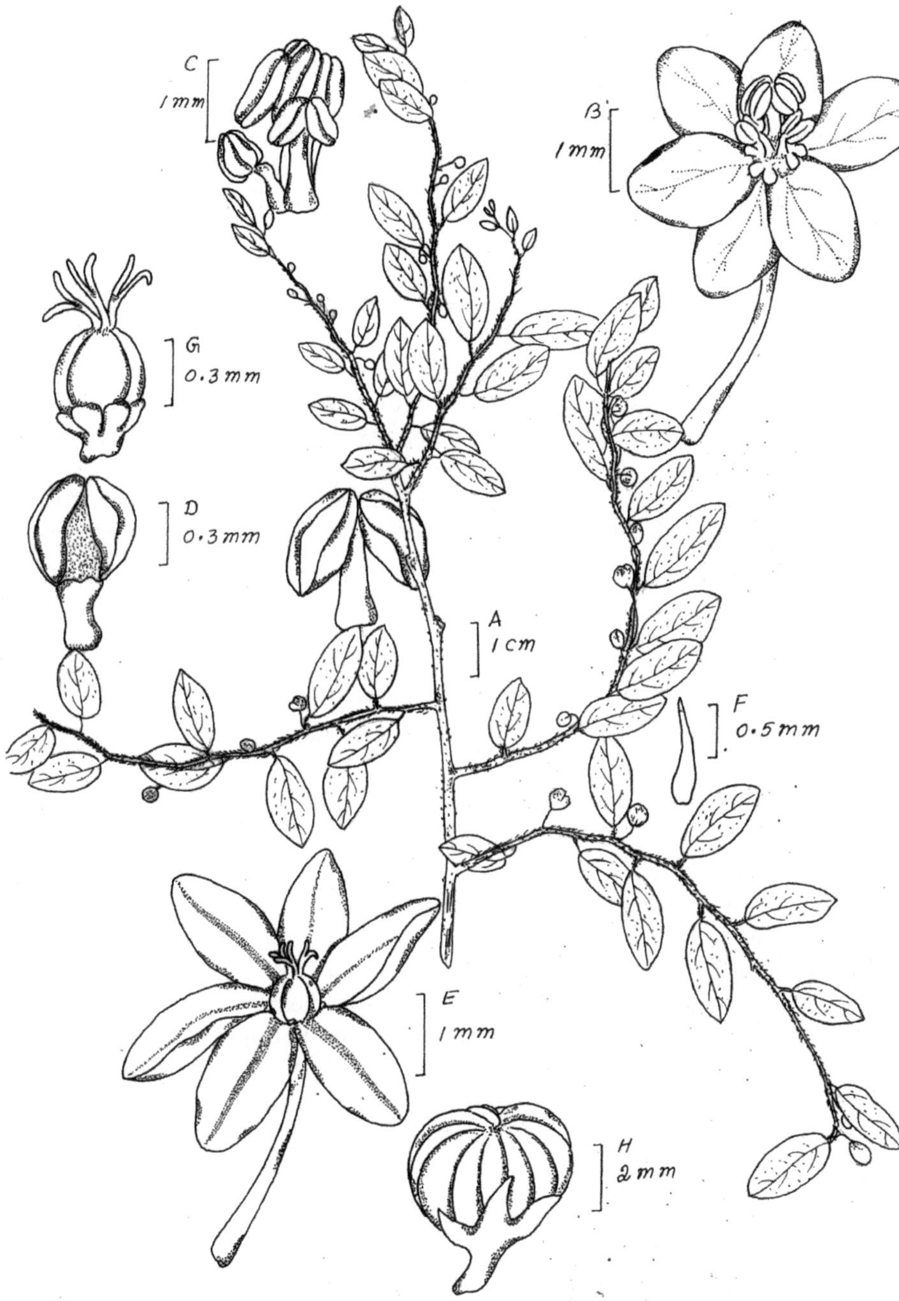

Figure 93. **Phyllanthus rheedei** Wight

A. Twig, ♂ B. Flower, C. Androecium, D. Anthers, ♀ E. Flower, F. Bract, G. Pistil, H. Capsule.

INDIA: Arunachal Pradesh, Meghalaya, Assam, West Bengal, Sikkim, Rajasthan, Maharashtra, Andhra Pradesh, Karnataka, Kerala, Tamil Nadu and Andaman & Nicobar Islands.

WORLD: Sri Lanka, Myanmar, Indonesia, Pacific zislands and West Indies.

Phyllanthus rotundifolius Klein ex Willd., Sp. Pl. 4: 584. 1805; FBI 5: 299. 1887; Gamble 2: 1290. 1925; N.P. Balakr. & Chakrab., Fam. Euphorb. India 389. 2007; Chakrab. *et al.*, in N.P. Balakr. *et al.*, Fl. India 23: 489. 2012.

Prostrate or slightly ascending herb; branches trailing from a stout root stock. Leaves 3-6 × 2-5 mm, ovate – orbicular, margin entire, apex obtuse – acute or rounded; petiole 0.5 mm; stipules lanceolate, acute. Flowers axillary, solitary; male flowers few in the upper axils; female flowers many in the remaining axils; pedicels of female longer than male, to 3 mm; bracts 0.5 mm, lanceolate. Male flowers 1.5 mm across; tepals 6, 1 × 0.6 mm, ovate – oblong, obtuse; stamens 3; filaments connate at base; anthers 0.3 mm across; disc of glands 6, free. Female flowers 2.5 mm across; tepals 6, 1.5 × 1mm, oblong – broadly ovate, obtuse, with scarious margins; ovary 0.6 mm, globose; styles 3, each 2-fid, at above, recurved; disc annular, cushion shaped. Capsule 2.5 mm, smooth; seeds with vertically arranged minute tubercles.

Occasional in wastelands mostly in sandy areas near the coasts. Fl. & Fr.: August-January.

Pulivendula Ghat (ATP), *KRKS* 39624; Satyavedu (CTR), *MCB* 45218 (MH).

INDIA: Odisha, Andhra Pradesh, Karnataka, Kerala and Tamil Nadu.

WORLD: Sri Lanka, Pakistan to Arabia and tropical Africa.

Phyllanthus scabrifolius Hook.f., Fl.Brit. India 5: 299. 11887; Chakrab. *et al.*, in N.P. Balakr. *et al.*, Fl. India 23: 474. 2012; Lalithamba, J. Indian Bot. Soc. 93: 284. 2014.

Annual, erect, herbs, up to 75 cm high; main stem and lateral branches winged, hispidulous. Leaves subsessile, elliptic, oblong-elliptic or obovate to oblanceolate, 4-20 × 2-10 mm, base cuneate or rounded, margin entire to denticulate, apex apiculate to acuminate; petioles 1-2 mm long; stipules linear-lanceolate or triangular-lanceolate. Flowers unisexual, female flowers solitary in upper axils, male flowers 1-3-flowered cymules at lower axils. Male flowers: pedicels ca 1 mm long; sepals 3+3, lanceolate, thickened along midrib region and broadly membranous along margin, glabrous; disc saucer-shaped, 6-lobed; stamens 3, filaments connate below, free and spreading above. Female flowers: pedicels ca 2 mm long; sepals 3 + 3, lanceolate, oblanceolate or oblong, styles 3, free, erect when young, spreading later. Fruits depressed-globose; seeds trigonous.

Dry gravelly soils in hill slopes in Nellore district (Lalithamba, 2014). Evergreeen forest floors in Andhra Pradesh (Chakrabarty *et al., loc cit.*). Fl. & Fr.: August - Septmber.

INDIA: Madhya Pradesh, Maharashtra, Andhra Pradesh, Karnataka, Kerala and Tamil Nadu. Endemic and Rare.

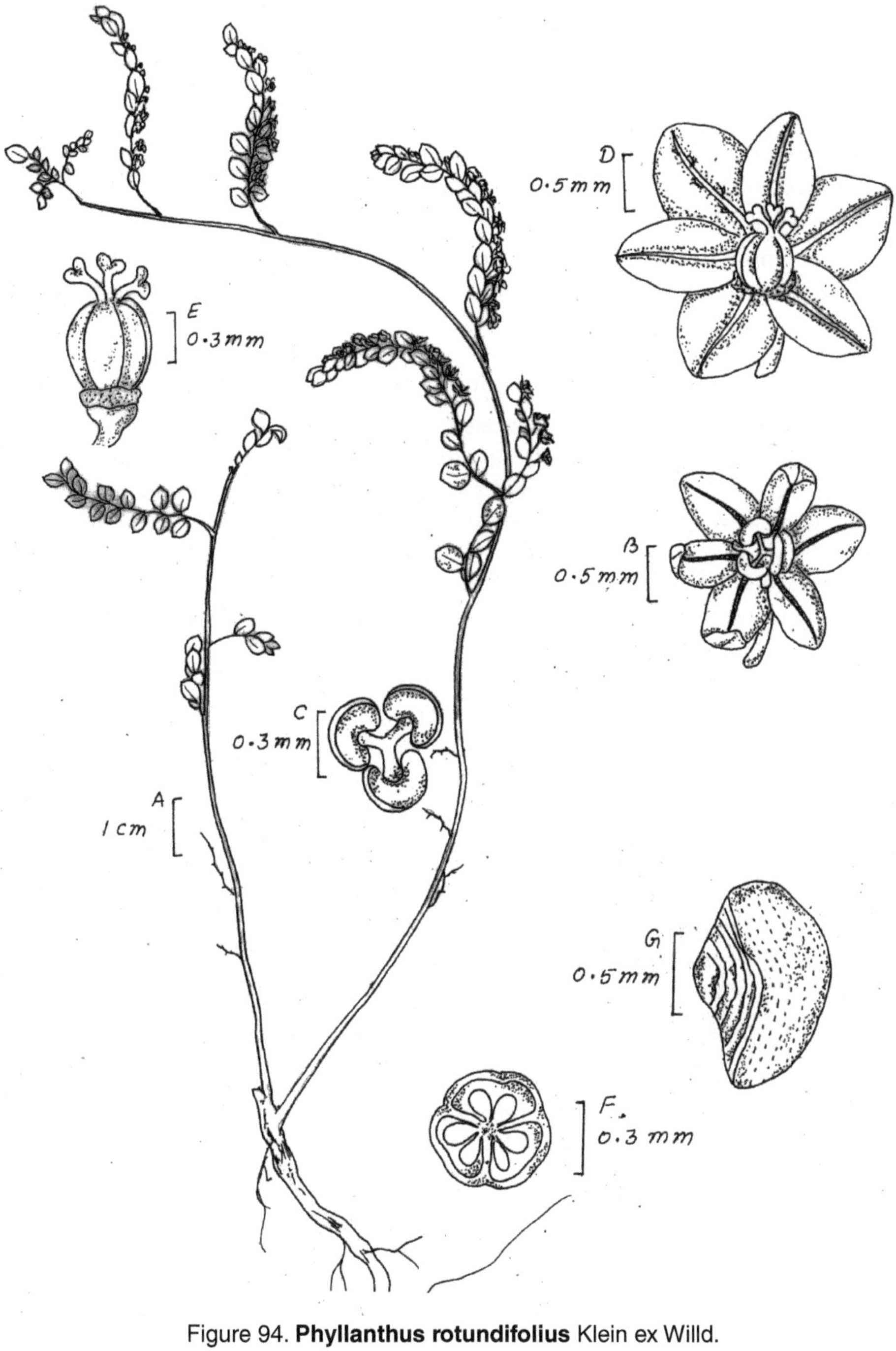

Figure 94. **Phyllanthus rotundifolius** Klein ex Willd.
A. Twig, ♂ B. Flower, C. Androecium, ♀ D. Flower, E. Pistil, F. Ovary T.s., G. Seed.

Phyllanthus simplex Retz., Observ. Bot. 5: 29. 1788; FBI 5: 295. 1887; Gamble 2: 1289. 1925; N.P. Balakr. & Chakrab., Fam. Euphorb. India 380. 2007; Chakrab. *et al.*, in N.P. Balakr. *et al.*, Fl. India 23: 476. 2012. *P. virgatus* auct. non G. Forst.; Surya Prakash Babu in Pullaiah & Moulali, Fl. Andhra Pradesh 2: 883. 1997.

Key to Varieties

1. Leaves mostly thinly coriaceous; male pedicels 2-5 mm long; male sepals 1.2-1.5 mm long; fruiting pedicels 7-12 mm long; seeds ca 2 mm long.......... var. **gardnerianus**

1. Leaves mostly chartaceous; male pedicels and sepals ca 0.5 mm long; fruiting pedicels 2-6 mm long; seeds ca 1.5 mm long.......... var. **simplex**

Phyllanthus simplex Retz. var. **simplex**

Herb, up to 50 cm high, monoecious; branchlets winged. Leaves 0.6-2 × 2-0.4 cm, linear – oblong or lanceolate, entire, acute – apiculate, subsessile; stipules sagittate – peltate, acuminate, scarious. Flowers solitary in axils; male flowers few, in lower axils; female flowers many, in upper axils; pedicels of male to 3 mm, female to 11 mm. Male flowers 1.1 mm across; tepals 6, 0.8 × 0.3 mm, oblong, obtuse; stamens 3, free, filaments 0.35 mm, anthers 0.35 mm across; disc of glands 6, free. Female flowers 2 mm across; tepals 6, 1 × 0.4 mm, oblong, obtuse, mucronate; ovary 0.5 mm, tuberculate, styles 3, horizontally spreading, each 2-fid, from above the middle; disc annular, crenulate. Capsule 3 mm, globose, 3-lobed, minutely tuberculate; seeds trigonus, minutely tuberculate.

Common in moist localities and along the streams. Fl. & Fr.: August – May. Vern.: Tel.: *Uchchi usirika.*

Bukkapatnam (ATP), *NY* 617, 1047; Owk (KNL), *RVR* 1469; Palakonda (CDP), *CS* 7633; Kottala (CDP), *TP* & *SSR* 36424; Bhiravakona (NLR), *PSPB* 9110; Atchempeta (GNT), *VRK* 5869; Kodur (CDP), *SKW* 7552 (BLAT); Papavinasanam (CTR), *JHFB* 117058 (BSID); Udayagiri (NLR), *SKW* 6691 (BLAT); Devaraapally, Maredumilli (EG), *MM* 105066 (BSID); Minumuluru (VSKP), *SKW* 2781 (BLAT); Elwinpet (SKLM), *SKW* 7057 (BLAT).

INDIA: Almost through out the country.

WORLD: Sri Lanka to SE Asia, S. China, Indo-China and Malesia.

Note: Mueller (1866) merged *P. virgatus* G.Forst. (1786) with *P. simplex* Retz. (1789), but incorrectly used the later name for the species and reduced *P. virgatus* as variety *virgatus.* Webster (in Pacific Science 40: 94. 1986) indicates that the populations in Asian region including India and Sri Lanka are *P. simplex* Retz. and separable from *P. virgatus* which appear to be restricted to Pacific Islands. *P. simplex* has strongly papillate ovaries and entire female disc, whereas *P. virgatus* has smooth ovaries and irregularly dissected female disc (Chakrab. *et al.*, 2012).

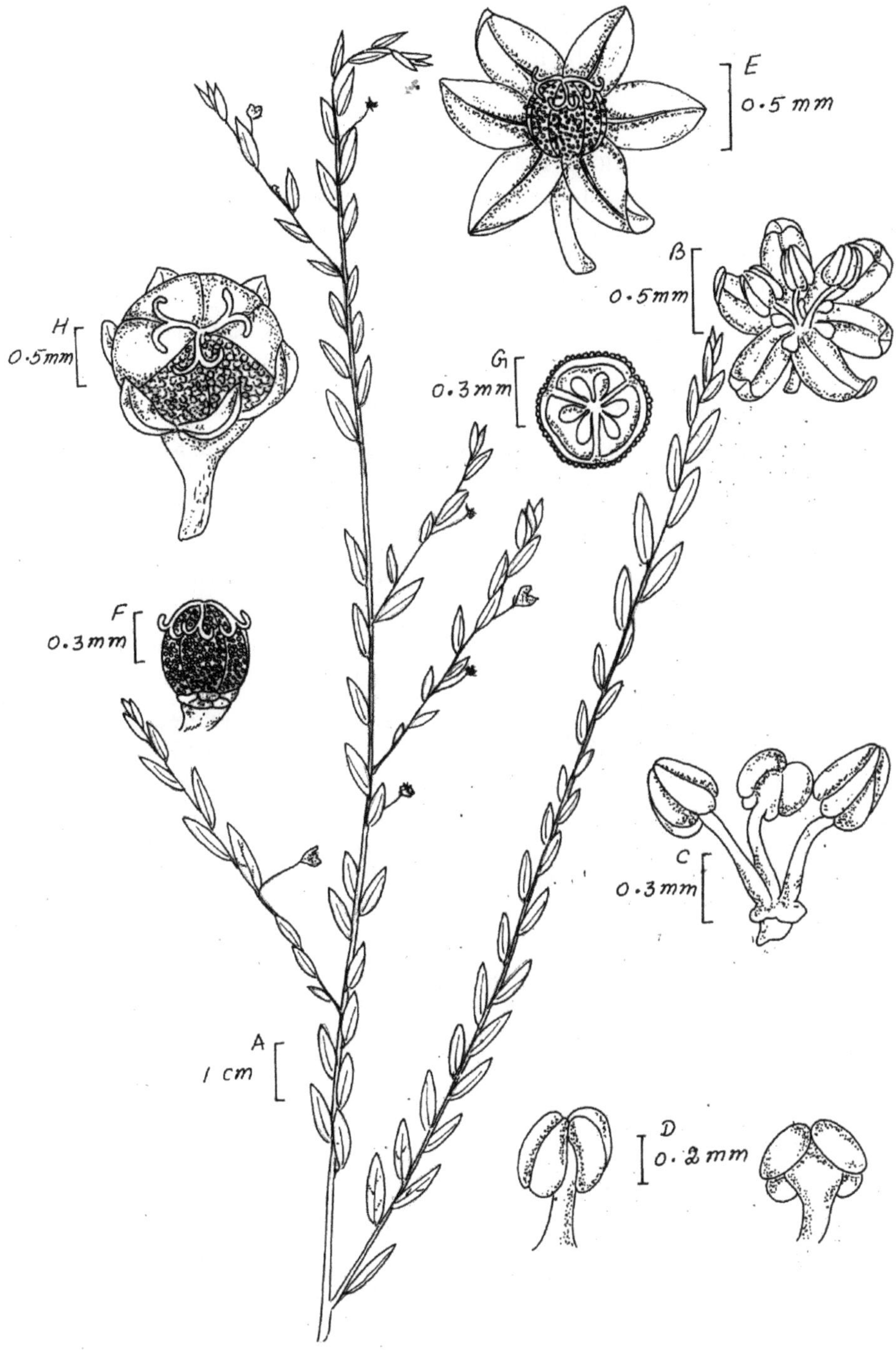

Figure 95. **Phyllanthus simplex** Retz. (Syn.: *P. virgatus* Forst. f.)
A. Twig, ♂ B. Flower, C. Androecium, D. Anthers, ♀ E. Flower, F. Pistil, G. Ovary T.s., H. Capsule.

Phyllanthus simplex Retz. var. **gardnerianus** (Wight) Müll.Arg., Linnaea 32: 32. 1863; FBI 5: 295. 1887; N.P. Balakr. & Chakrab., Fam. Euphorb. India 380. 2007. *Macraea gardneriana* Wight, Icon. Pl. Ind. Orient. t. 1902. f. 3. 1852. *Phyllanthus gardnerianus* (Wight) Baill., Etud. Gen. Euphorb. 628. 1858; Gamble 2: 1289. 1925.

Undershrub, monoecious; branches long, slender. Leaves elliptic – oblong, slightly cordate at base, revolute, obtuse, subsessile. Flowers axillary; male flowers in fascicles, with short pedicels; female flowers solitary, with long pedicels, filiform; ovary smooth, globose. Capsule depressed globose, minutely verrucose; seeds minutely tubercled.

Common in the crevices of shaded hard rocks in Srikakulam district (Rolla S. Rao and Hara Sreeramulu, 1986). Fl. & Fr.: June-November.

Kuruturu (WG), *DN* 84289 (MH, BSID).

INDIA: Odisha, Andhra Pradesh, Maharashtra, Karnataka, Kerala and Tamil Nadu.

WORLD: Sri Lanka.

Phyllanthus tenellus Roxb., Fl. India 3: 668. 1832; Webster, Brittonia 22: 63. 1970; Mitra, Bull. Bot. Surv. India 27: 154. 1985; C.S. Reddy & V.S.Raju, J. Econ. Taxon. Bot. 26: 197. 2002; N.P.Balakr. & Chakrab., Fam. Euphorb. India 382. 2007; Chakrab. *et al.*, in N.P.Balakr. *et al.*, Fl. India 23: 480. 2012.

Annual herb, up to 60 cm high; main stem smooth, subterete with spirally arranged cataphylls. Lateral branchlets *ca* 5-15 cm long, with *ca* 10-25 leaves; leaf blades membranous, elliptic to obovate, 6-25 × 4-11 mm, pale below, base acute to rounded, apex acute or obtuse, stipules lanceolate, acuminate. Monoecious, cymules axillary on lateral branchlets; proximal cymules with 2 or 3 male flowers and 1 or 2 female flowers, distal cymules with 1-2 female flowers, male flowers with pedicels 0.5-0.7 mm long; disc segments 5; stamens 5, filaments free, distally bent, anthers dehiscing horizontally. Female flowers with smooth flexuous, capillary pedicels becoming 3-8 mm long, basally dilated into a pulvinus; calyx lobes 5, ovate, acute or obtuse, whitish except for the unbranched midrib, disc platelliform, ovary smooth, styles free, spreading, bifid ½ to their length. Capsules greenish, 1.7-1.9 mm broad; seeds pale brown, densely papillose.

Weed of gardens in Chittoor district (C.S. Reddy and V.S. Raju, 2002). Fl. & Fr.: Throughout the year.

Tirupati (CTR), *VSR* 3659 (KU).

INDIA: Maharashtra, Andhra Pradesh, Tamil Nadu and Kerala.

WORLD: Sri Lanka, Mascarene Islands, Japan and New World.

Phyllanthus urinaria L., Sp. Pl. 982. 1753; FBI 5: 293. 1887; Gamble 2: 1289. 1925; N.P. Balakr. & Chakrab., Fam. Euphorb. India 389. 2007; Chakrab. *et al.*, in N.P. Balakr. *et al.*, Fl. India 23: 490. 2012 var. **urinaria**. *Phyllanthus nozeranii*

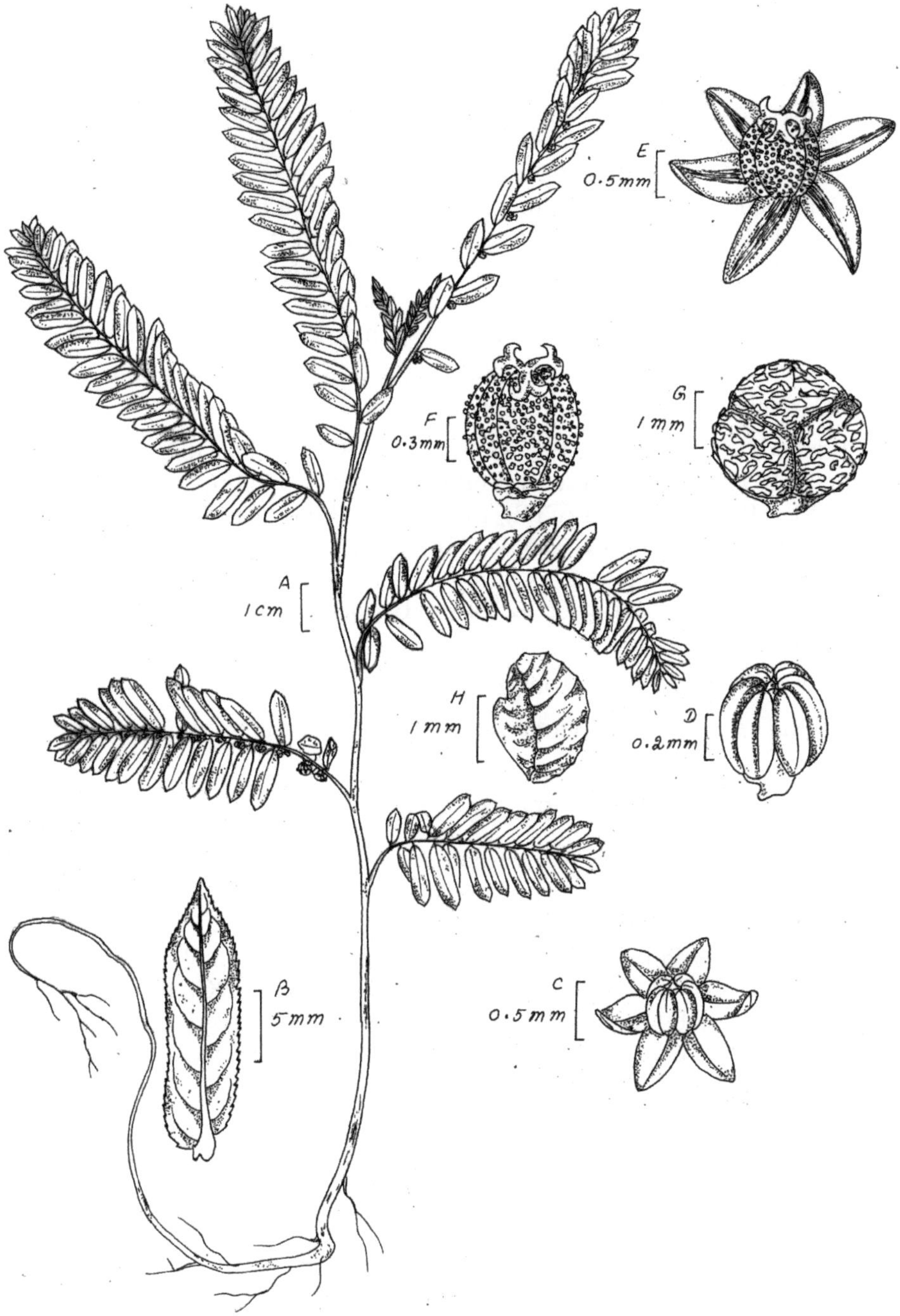

Figure 96. **Phyllanthus urinaria** L.
A. Twig, B. Leaf, ♂ C. Flower, D. Androecium, ♀ E. Flower, F. Pistil, G. Capsule, H. Seed.

Rossignol & Haicour, Amer. J. Bot. 74: 1858. 1987 (publ. 1988); V.S.Raju *et al.*, Bangladesh J. Plant Taxon. 6: 60. 2011.

Herb, to 70 cm high, monoecious, slightly branched; branchlets flattened or slightly winged, thinly puberulous. Leaves 0.5-1.5 × 0.2-0.4 cm, oblong – linear oblong, base obliquely rounded, margin serrulate, apex apiculate; petiole very short; stipules lanceolate, acuminate. Flowers minute, solitary in axils; male flowers few in upper axils; female flowers many in remaining lower axils, subessile; bracts 1 mm, linear lanceolate, acute, scarious. Male flowers 1.4 mm across; tepals 6, 0.8 × 0.35 mm, ovate – oblong, subacute; stamens 3, connate, subsessile; anthers 0.5 mm across; disc of glands 6. Female flowers 2 mm across; tepals 6, 1 × 0.5 mm, oblong, subacute; ovary 0.8 mm, globose, verrucose, styles 3, each 2-fid, adpressed to ovary. Capsule 3 mm, globose, verrucose; seeds trigonus, transversely furrowed.

Occasional in shady, moist localities in the forests as a forest undergrowth. Fl. & Fr.: December – March. Vern.: Tel.: *Erra usirika.*

Talakona (CTR), *PSPB* 9199; Rampa (EG), *PSPB* 1135, 11841; Bukkapatnam (ATP), *JSG* 20929 (DD); Pecheruvu (KNL), *VBH* 83954 (BSID); Gunjana river (CDP), *JLE* 14981 (MH); Satyavedu (CTR), *MCB* 45259 (MH); Diguvametta (PKM), *JLE* 22163 (MH); Addatigala (EG), *SKW* 4031, 4032 (BLAT); Anantagiri (VSKP), *GVS* 19484; Srungavarapukota (VZN), *NPBK* 11018 (MH); Vazeedu (KMM), *V.S. Raju* 1616 (KUH); Lukke (KNL), *JLE* 42224 (MH); Forest near S. Kota (VSKP), *NPBK* 11018 (MH).

INDIA: Through out the country.

WORLD: Native of southern Asia, now widespread in tropical and subtropical regions.

Key to Varieties

1. Seeds without foveoles .... **P. urinaria** var. **hookeri**
1. Seeds with foveole .... **P. urinaria** var. **urinaria**

Phyllanthus urinaria L. var. **hookeri** (Muell.-Arg.) Hook. f., Fl. Brit. India 5:294. 1887. *P. hookeri* Mull.-Arg., Linnaea 32: 19. 1863. *et* in DC., Prodr. 15(2): 366. 1866; Rossignol *et al.*, Amer. J. Bot. 74: 1862.1988. *P. leprocarpus* Wight, Icon. Pl. Ind. Or. 5: 25, t. 1895, f.4. 1852. *Diasperus hookeri* (Mull.-Arg.) Kuntze, Revis. Gen. Pl. 2: 599.(1891.

Perennial erect herbs, up to 60 cm high. Main stem woody, hard, smooth; cataphylls arranged spirally; internodes relatively short, angular. Leaves narrowly oblong, distichous, hispidulous along the margins; leaf base slightly oblique, 1.1-1.4 × 0.3-0.4 cm; petiole 1 mm long; stipules acuminate. Phyllanthoid branchlets plagiotropic up to 15 cm long, with 47-50 leaves. Plants monoecious; 22-25 proximal nodes with axillary, solitary, pistillate flowers while the distal nodes 30-37 with solitary or at the most 2 staminate flowers. Staminate flowers: pedicels 0.5 mm long, tepals 6, white with red strip along the midvein, elliptic-oblong, entire, obtuse; filaments form a column (less than 0.5 mm); anthers free, divergent, sub-globose, yellow, dithecous, extrorse and dehiscing by longitudinal slits; disk glands discrete, alternating the tepals, granular; pollen grains 5- colporate. Pistillate flowers: pedicels 0.5 mm long, tepals 6, white, linear-oblong, apex rounded; glabrous below; upper rim of disk finely crenulate. Fruits capsular, depressed globose, exocarp mamillate-rugulose, with fleshy scales. Seeds 6, brown, trigonous, 1.3- 1.5 × 1.5 mm; ornamentation radiating from the rounded hilum with 5-7 ridges on lateral faces and 13-14 distinct transverse ridges on dorsal side; non-foveolate.

Fl. & Fr.: Throughout the year.

Satyavedu to Madras (CTR), *MCB* 45259 (MH); on the side of Gunjana river (CDP), *JLE* 14981 (MH); Kotha Isukapalli (EG), *V.S. Raju* 1615 (KUH);

INDIA: Andhra Pradesh, Kerala, Tamil Nadu, Karnataka.

WORLD: India to Philippines.

Phyllanthus wightianus Müll.Arg., Linnaea 32: 47. 1863; FBI 5: 303. 1887. *Reidia floribunda* Wight, Icon. Pl. Ind. Orient. 5(2): 28. t. 1903. 1852; Gamble 2: 1293. 1925; Rani in K.M.Matthew, Fl. Tamil Nadu Carnatic 3(2): 1470. 1983. *Phyllanthus cinereus* Müll.Arg., Linnaea 32: 48. 1863; FBI 5: 303. 1887; Chakrab. *et al.*, in N.P. Balakr. *et al.*, Fl. India 23: 490. 2012.

Subshrubs, to 1 m high, branches in close spirals. Leaves alternate, distichous, elliptic to oblong, 1.5-3 x 1-1.5 cm, dark green above, glaucous below, base obtuse or rounded, apex apiculate or cuspidate. Flowers reddish, solitary, axillary; tepals 4 in male flowers, 6 in female flowers; stamens 2, opposite tepals, exserted,filaments connate around pistillode; pistillode conical; ovary 3-lobed, glochidiate, 3-locular, styles 3, bifurcate, stigma obtuse; disc glands 4 in male, 6 in female. Capsule globoseof 3, bivalved cocci; seeds 6, trigonous, appressed hairy.

Hills, 1000 m, on the floor and border of forests in Tamil Nadu Eastern Ghats (Rani *loc. cit.*). Fl. & Fr.: April – Decmber.

INDIA:Tamil Nadu, Karnataka

WORLD: Sri Lanka.

PLUKENETIA Linnaeus

Plukenetia corniculata Sm., Nov. Act. Upsal. 6.4. 1799; FBI 5: 464. 1887; N.P. Balakr. & Chakrab., Fam. Euphorb. India 179. 2007; N.P. Balakr. in N.P. Balakr. *et al.*, Fl. India 23: 194. 2012. *Pterococcus corniculatus* (Sm.) Pax & Hoffm. in Engler, Pflanzenr. 68: 22. 1919; Airy Shaw, Kew Bull. 26: 327. 1972 & Kew Bull. Add. Ser. 4: 187. 1975; Surya Prakash Babu in Pullaiah & Moulali, Fl. Andhra Pradesh 2: 884. 1997.

Twining shrub, monoecious; stems obscurely ribbed. Leaves alternate, membranous, ovate – ovate oblong, base cordate and 2 glandular, margin dentate, apex caudate, puberulous; petioles 2-6 cm long, with 2 glands on either side of basal nerves. Flowers in axillary or leaf-opposed racemes, male flowers many, crowded at top, female flowers few, at base; bracts 3-fid, lanceolate; bracteoles 2, lanceolate. Male flowers pedicelss 1.5- 2.5 mm long; sepals 4-5, ovate; stamens 12-30, sessile on a cushion-like receptacle. Female flower pedicels ca 2.5 mm long; sepals 4, lanceolate; ovary hirsute. Capsule 4-winged.

Occasional in East Godavari district. Fl. & Fr.: July – August. Vern.: Tel.: *Kodigandham.*

Boddalanka (EG), *DN* 85554 (MH).

INDIA:Arunachal Pradesh, Assam, Sikkim, West Bengal and Andhra Pradesh.

WORLD: Myanmar, Thailand, Malesia to Celebes and the Philippines.

PUTRANJIVA Wallich

Putranjiva roxburghii Wall., Tent. Fl. Nepal. 61. 1826; FBI 5. 1336. 1887; Gamble 2:1311.1925; N.P. Balakr. & Chakrab., Fam. Euphorb. India 335. 2007; Chakrab. & N.P. Balakr. in N.P. Balakr. *et al.*, Fl. India 23: 405. 2012. *Drypetes roxburghii* (Wall.) Hurursawa, J. Fac. Sci. Univ. Tokyo Sect. 3.Bot. 6: 337. 1954; Suryprakash Babu in Pullaiah & Moulali, Fl. Andhra Pradesh 2: 854. 1997.

Moderate-sized, dioecious evergreen tree; bark dark grey with horizontal lenticels; wood greyish-white, close-grained; branchlets puberulous, lenticellate. Leaves coriaceous, elliptic-oblong, 5-12.5 × 1-8 cm, glabrous, base obtuse, some what oblique, margin distantly serrulate, apex shortly acuminate, secondary nerves ca. 11 pairs, not prominent; petiole to 1 cm. Flowers greenish-yellow, dense, sessile, in axillary clusters around branchlets. Male flowers: pedicels 1.5-2 mm long; sepals 3-5, suborbicular; stamens 2-4, 1.5-2 mm long. Female flowers: pedicels 6-15 mm long; sepals 5 or 6, ovate to elliptic; ovary ovoid, styles 2 or 3, 3-7 mm long, often connate below; stigmas flabellate. Drupe yellowish, globose or ellipsoid, 2 × 1.5 cm, pubescent; seed 1, crustaceous.

Ocasional, usually grows on alluvial soil along the rivers or in swamps or evergreen forests at 600-1000m; sometimes cultivated in gardens for hedges and on roadsides. Fl.: March-April, Fr.: October-December. Vern.: Tel.: *Kudrajuvi, Kaduru, Putrajivika, Veku, Mahaputta-jivi.*

Mamandur-Thumburatheertham (CTR), *SSR* 16172; Salivendrakona (CDP), *JSG* 16785 (DD); Durgam – Udayagiri (NLR), *PV* 110235 (BSID).

INDIA: Through out India except Andaman and Nicobar Islands.

WORLD: Sri Lanka, Bangladesh, Myanmar, Thailand, Indo-China, Java, Moluccas, Lesser Sunda Islands and New Guinea.

RICINUS Linnaeus

Ricinus communis L., Sp. Pl. 1007. 1753; FBI 5: 457. 1887; Gamble 2: 1335. 1925; Purseglove, Trop. Crops, Dicots 180. 1977; N.P. Balakr. and Chakrab., Fam. Euphorb. India 154. 2007; Susila Rani & N.P. Balakr. in N.P.Balakr. *et al.*, Fl. India 23: 166. 2012.

Monoecious, evergreen shrub, up to 4 m high. Leaves alternate, palmatifid, 6-10-lobed, each 1-nerved with many lateral nerves, peltate; lobes lanceolate, 10-18 × 3-7 cm, thinly pubescent below, margin serrate, apex acuminate; petiole to 20 cm. Paniculate racemes terminal, to 10-20 cm. Male flowers below, 1.5 cm across,

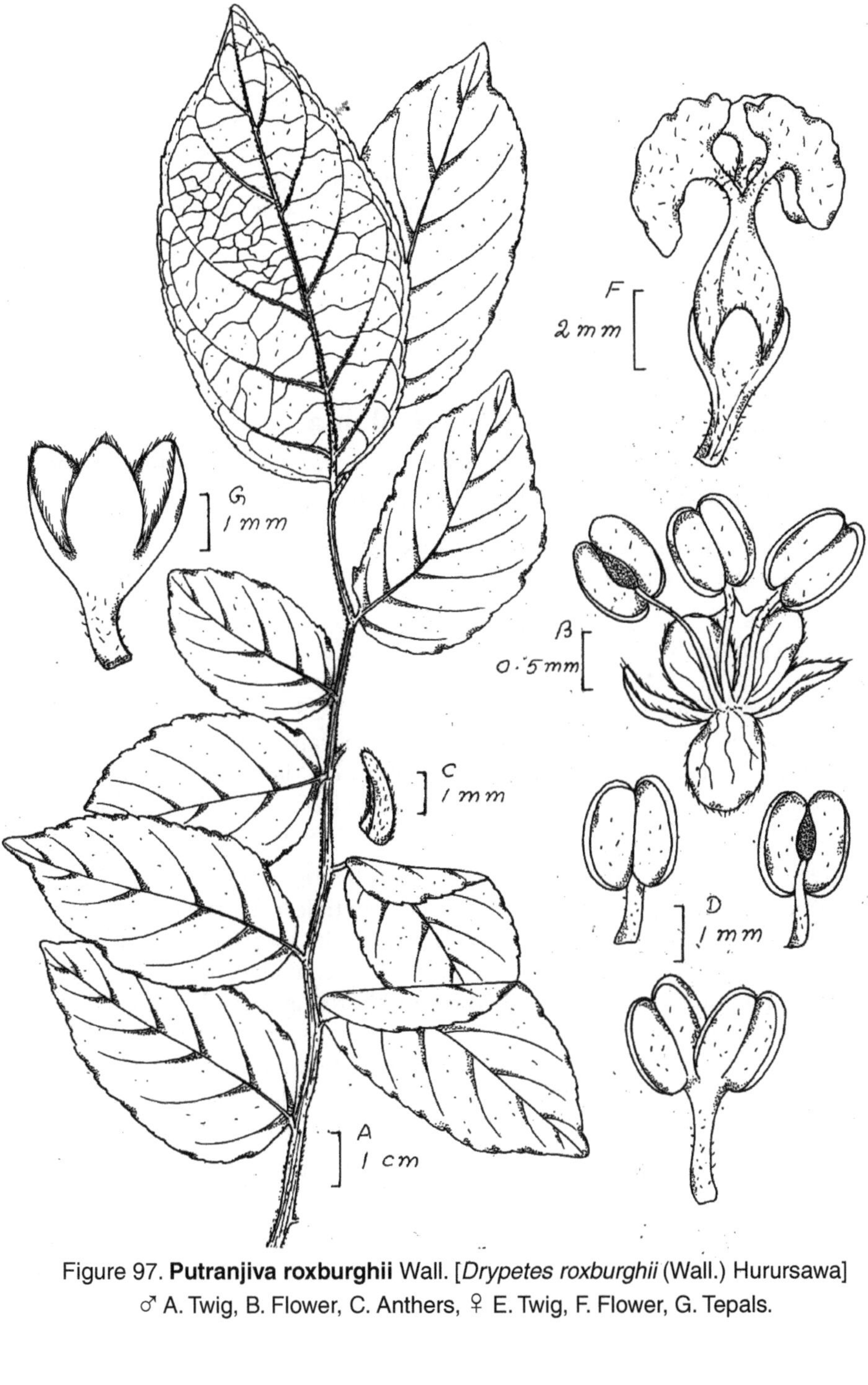

Figure 97. **Putranjiva roxburghii** Wall. [*Drypetes roxburghii* (Wall.) Hurursawa]
♂ A. Twig, B. Flower, C. Anthers, ♀ E. Twig, F. Flower, G. Tepals.

female ones above, 6 mm across. Male flowers: perianth cupular, splitting into 3-5 lobes, lanceolate, valvate, 4 mm, margin inrolled, acuminate; stamens numerous; filaments connate and repeatedly branched; anther cells divergent, 4 mm. Female flowers: tepals 5, sub-equal, lanceolate, valvate, 5 mm, acute, ovary globose, echinate, 3 mm, 3-locular, ovules 3, pendulous, styles 3, stout, 4 mm, papillose, stigmatiferous, disc 0. Capsule 3-lobed, 2 cm across, prickly; seeds oblong; testa smooth, marbled, shiny, carunculate.

Plains from the coast, to 1400 m. Planted for its oil seeds. Fl. & Fr.: Through the year. Vern.: Tel.: *Amudamu*; Tam.: *Aamanakku*. English: Castor.

Somulavaripalli (ATP), *KRKS* 37819; Venkatagiri (NLR), *ASR* 254 (BSID); Maddiveedu, near Tadepalli (EG), *DN* 85536 (BSID).

INDIA: Cultivated for its seeds through out the country.

WORLD: Pantropical. Native of N. Africa.

SAUROPUS Blume

1. Segments of male calyx unlobed, entire .... **S. quadrangularis**
1. Segments of male calyx emarginated to deeply bilobulate:
    2. Petioles up to 0.6 mm long; male calyx 1 – 1.5 mm in diam .... **S. bacciformis**
    2. Petioles 1-1.5(-2.5) mm long; male calyx 2-3.5 mm in diam .... **S. compressus**

Sauropus bacciformis (L.) Airy Shaw, Kew Bull. 35: 685. 1980; N.P.Balakr. & Chakrab., Fam. Euphorb. India 393. 2007; Chakrab. *et al.*, in N.P. Balakr. *et al.*, Fl. India 23: 494. 2012. *Phyllanthus bacciformis* L., Mant. Pl. 294. 1771. *Agyneia bacciformis* (L.) Juss., Euphorb. Gen. 24. 1824; FBI 5: 285. 1887; Gamble 2: 1284. 1925.

Herb, to 40 cm high, monoecious; branchlets terete, striate. Leaves 0.7-2 × 0.2-1 cm, elliptic – obovate, acute at base, entire, acute; petiole to 2 mm; stipules narrowly triangular, acuminate. Male flowers in axillary short clusters; bracts lanceolate, acute; flowers 2 mm across; tepals 6, 1 × 0.8 mm, slightly connate at base, ovate, obtuse, hooded at apex; stamens 3-6, connate on gynophores, anthers 0.5 mm across; disc of glands 6. Female flowers axillary, solitary; flowers 4.5 mm across; tepals 6, 1.5 × 1.2 mm, slightly connate at base, ovate, acute, gland – dotted at apex; ovary 1 mm, truncate, styles 3, sunken, each 2-fid, spreading; disc of glands 0. Capsule 4 × 3 mm, subglobose; seeds curved.

Occasional along the bunds of fallow lands and grass lands towards coast. Fl. & Fr.: July – February.

Kalasamudram (ATP), *KRKS* 39531; Musalreddipalle (CDP), *BSS* & *SKB* 30824; Nagalapuram (CTR), *PSPB* 9184; on the way to Balapalle (CDP), *JLE* 14908 (MH); Ipur (NLR), *SKW* 6665, 6666 (BLAT).

INDIA: West Bengal, Andhra Pradesh, Karnataka, Kerala and Tamil Nadu.

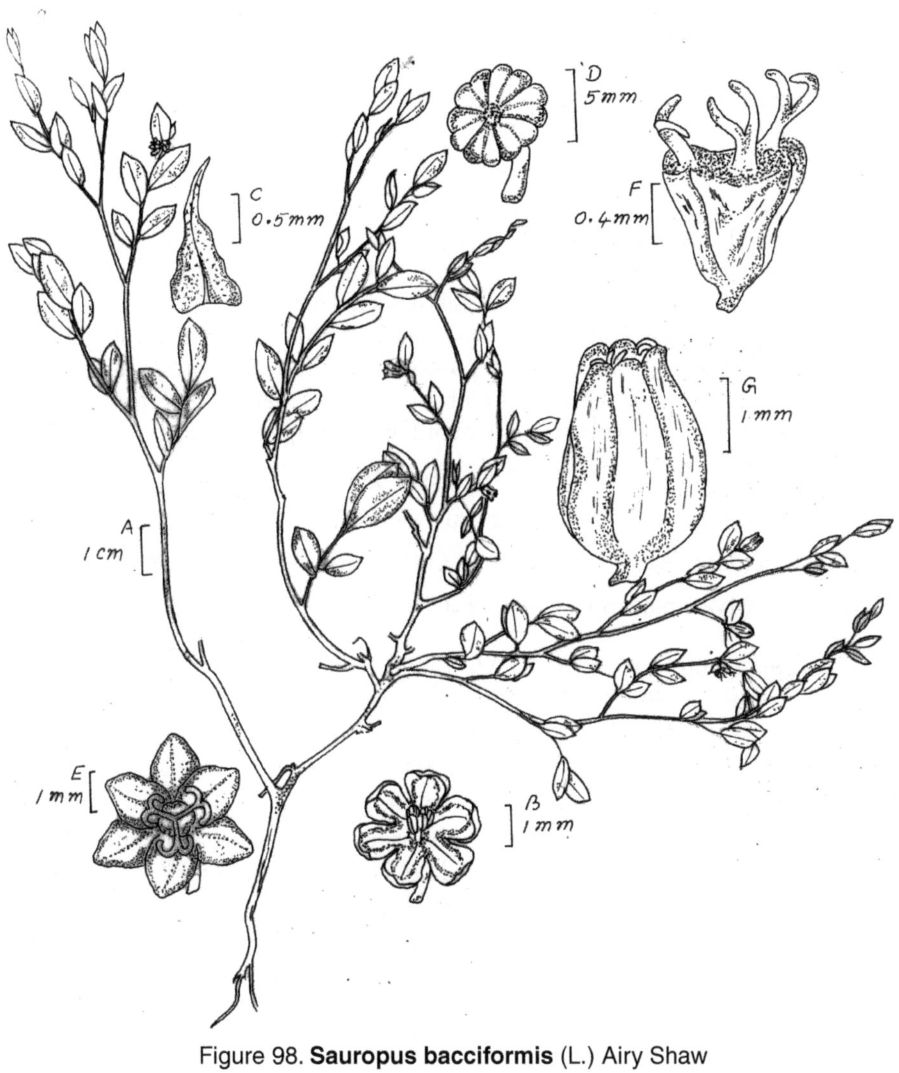

Figure 98. **Sauropus bacciformis** (L.) Airy Shaw
A. Twig, ♂ B. Flower, C. Bract, D. Androecium, ♀ E. Flower, F. Pistil, G. Capsule (premature).

WORLD: Mauritius, Sri Lanka, Myanmar, Thailand, S. China, Malay Peninsula, Indonesia, Philippines and Lesser Sunda Islands.

Sauropus compressus Müll.Arg. in DC., Prodr. 15(2): 243. 1866; FBI 5: 336. 1887. var. **puberulus** (Kurz) Chakrab. & M. Gangop., J. Econ. Taxon. Bot. 20: 528. 1996; Chakrab. *et al.*, in N.P.Balakr. *et al.*, Fl. India 23: 495. 2012. *S. quadrangularis* (Willd.) Müll.Arg. var. *puberulus* Kurz., For. Fl. Brit. Burma 2: 350. 1877; Hook.f., Fl. Brit. India 5: 335. 1887; Surya Prakash Babu in Pullaiah & Moulali, Fl. Andhra Pradesh 2: 885. 1997 (as var. *pubescens*). *S. pubescens* Hook.f., Fl. Brit. India 5: 335. 1887; Gamble 2: 1303. 1925.

Undershrub, to 1 m high, monoecious, tomentose when young; branchlets quadrangular, compressed, often subulate, pubescent. Leaves 1-4 × 0.7-2 cm, elliptic or broadly thinly above, densely below; petiole very short. Flowers in axillary cymes or solitary, minute; pedicels to 3 mm, filiform, pubescent; bracts lanceolate, pubescent. Male flowers 2 mm across; tepals 6, 1 × 0.8 mm, connate at base, triangular or obovate, emarginated, pubescent; stamens 3, connate at base, anthers 0.35 mm across. Female flowers 4 mm across; tepals 6, 3 × 1.8 mm, free, orbicular – obovate, rounded, pubescent; ovary 1.2 mm, globose, styles 3, each-2-fid, from above the middle, recurved. Drupe sub-baccate, ovoid, with truncate top.

Rare in the hills. Fl & Fr.: August – December.

Jeedipalem (EG), *GVS* 24518; Paderu (VSKP), *GVS* 28092 (MH).

INDIA: Sikkim, West Bengal, Bihar, Odisha and Andhra Pradesh.

WORLD: Myanmar, Indo-China and Thailand.

Sauropus quadrangularis (Willd.) Müll.Arg., Linnaea 32: 73. 1863; FBI 5: 335. 1887; Gamble 2: 1303. 1925; N.P.Balakr. & Chakrab., Fam. Euphorb. India 395. 2007; Chakrab. *et al.*, in N.P. Balakr. *et al.*, Fl. India 23: 497. 2012. *Phyllanthus quadrangularis* Willd., Sp. Pl. 4: 585. 1805.

Shrub, to 1 m high, monoecious; branchlets quadrangular. Leaves 1-2 × 0.8-1.4 cm, elliptic – oblong, base acute – obtuse, margin entire, apex apiculate; petiole to 2 mm; stipules minute, lanceolate – subulate. Flowers in axillary cymes; male 2-3, female solitary; pedicels to 6 mm, filiform. Male flowers 3 mm across; tepals 6, 1 × 1.8 mm, connate at base, triangular, rounded – emarginated; stamens 3, connate at base, anthers 3.5 mm across. Female flowers 4 mm across; tepals 6, 2.2 × 1.8 mm, free, obovate, rounded; ovary 1.5 mm, globose, styles 3, each 2-fid, recurved. Drupe 7 mm, depressed globose.

Occasional in the foothills. Fl. & Fr.: September – November.

Batrepalli (ATP), *KRKS* 39595; Penchalakona (NLR), *PSPB* 9107; Galikonda (VSKP), *GVS* 19593 (MH); Karakakonda (VSKP), *JSG* 1707; Krishnapuram (VSKP), *JSG* 1944 (MH).

INDIA: Uttar Pradesh, Jharkhand, Madhya Pradesh, Andhra Pradesh, Maharashtra, Goa, Karnataka and Kerala.

WORLD: Sri Lanka.

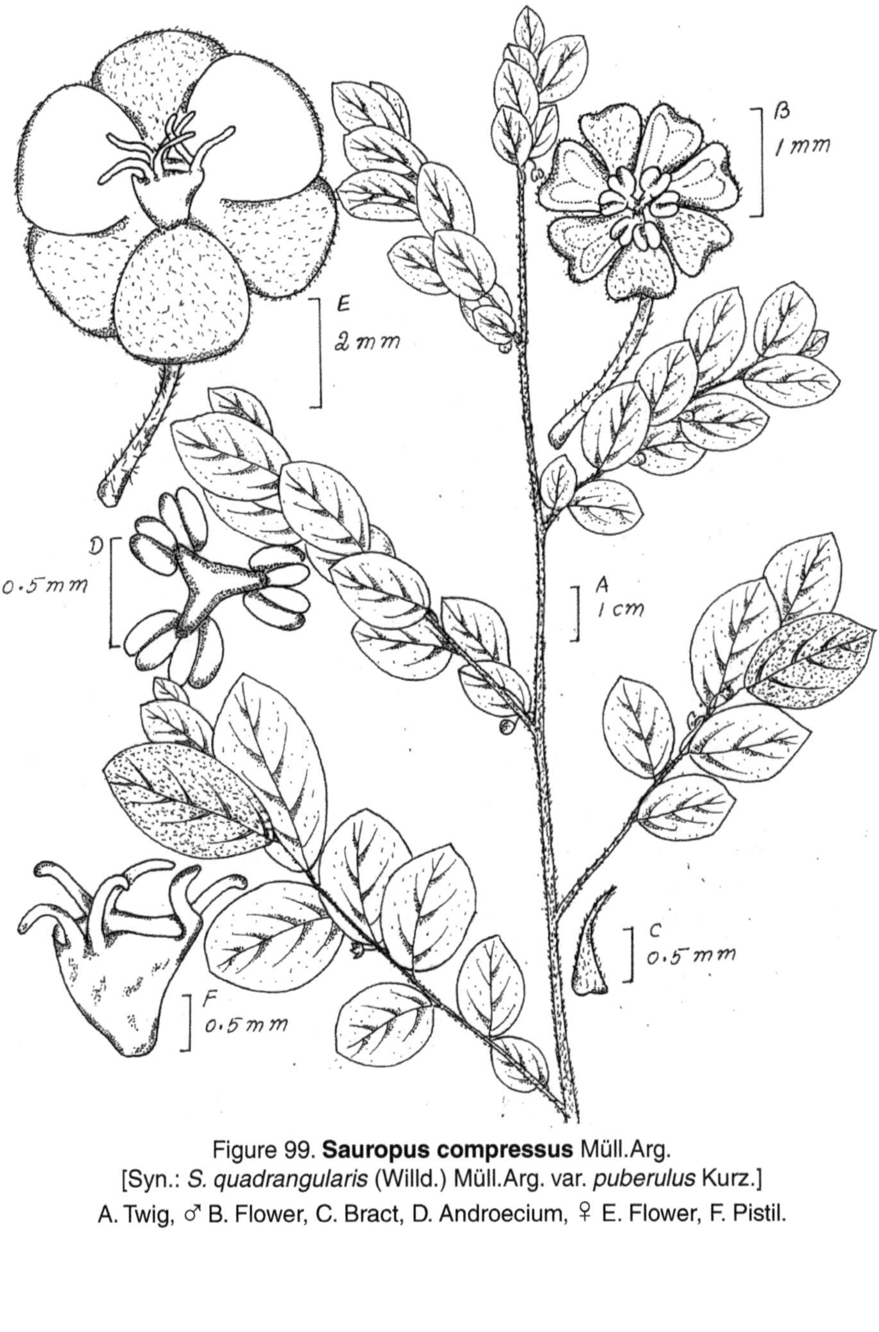

Figure 99. **Sauropus compressus** Müll.Arg.
[Syn.: *S. quadrangularis* (Willd.) Müll.Arg. var. *puberulus* Kurz.]
A. Twig, ♂ B. Flower, C. Bract, D. Androecium, ♀ E. Flower, F. Pistil.

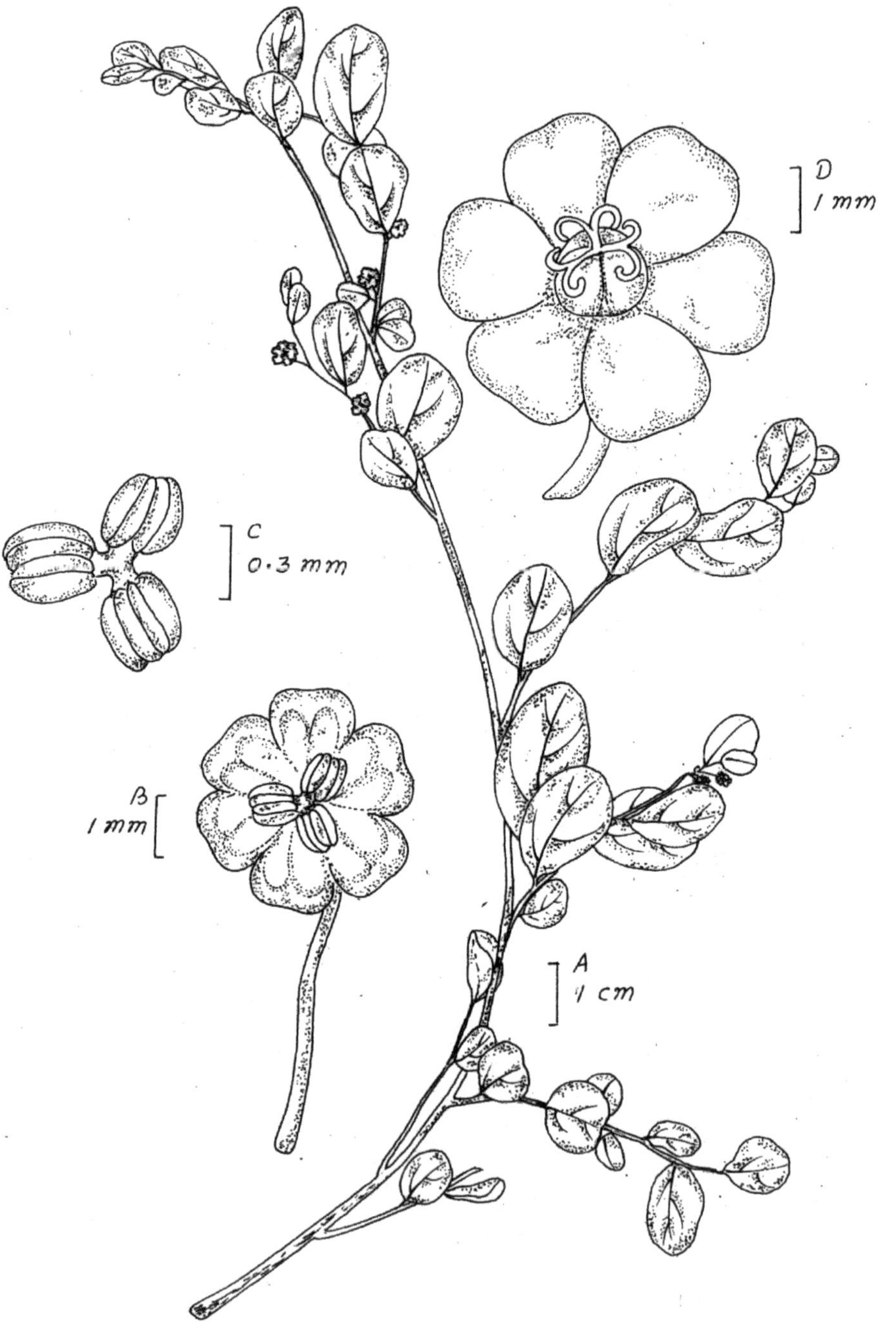

Figure 100. **Sauropus quadrangularis** (Willd.) Müll.Arg.
A. Twig, ♂ B. Flower, C. Androecium, ♀ D. Flower.

SUREGADA Roxburgh ex Rottler

1. Flowers yellow, rather large, many in contracted cymes; leaves lanceolate or elliptic-oblong; capsule rough, obscurely lobed .... **S. multiflora**
1. Flowers greenish, small, few in fascicles; leaves obovate-oblanceolate or elliptic-oblong; capsule smooth, deeply lobed .... **S. angustifolia**

Suregada angustifolia (Bail. ex Müll.Arg.) Airy Shaw, Kew Bull. 23: 128. 1969; N.P. Balakr. & Chakrab., Fam. Euphorb. India 219. 2007; N.P. Balakr. in N.P.Balakr. *et al.*, Fl. India 23: 247. 2012. *Gelonium angustifolium* Bail. ex Müll.Arg. in DC., Prodr. 15:1128. 1866. *G. lanceolatum auct. non* Willd. 1806; FBI 5: 459. 1887; Gamble 2: 1343. 1925.

Evergreen, dioecious tree, up to 10 m high; bark smooth, greyish; wood flesh coloured; branchlets glabrous, striate. Leaves coriaceous, obovate-oblanceolate or elliptic-oblong, 3.5-12 × 1.5-5.7 cm, glabrous, shining and pale; base acute, margin entire, apex subacute or obtuse, secondary nerves 6-8 pairs; petiole to 4 mm. Flowers green, small. Male flowers ca. 10, axillary in sub-umbellate fascicles; pedicels ca 5 mm long; calyx lobes 5 or 6, concave, 2-3 mm long, scaly; stamens 15-25, mixed with rugose glands. Female flowers solitary or 2-4 together; calyx-lobes 5 or 6, ca 3 mm acrosss, ovate; disc cupular; ovary globose, styles 3. Capsule 3-lobed, 0.8-1 cm in dia., smooth, keeled on the lobes, yellowish; seeds globose, crustaceous, hollow-pitted.

Occasional in dry deciduous forests, chiefly along streams. Fl.: January-March. Fr.: May-June. Vern.: Tel.: *Ellachi, Vella dabba, Surgada, Karugugilam.*

Talakona (CTR) *SSR* 13800, 13259; Penchalakona (NLR), *PSPB* 9113; Chintala (PKM), *SSR* 16133; Donubai RF (SKLM), *SSR* 15586; Balapalle (CDP), *JLE* 15713 (MH); Thumburathirtham (CTR), *JHFB* & *GVM* 11745 (MH, BSID); Diguvametta (PKM), *SKW* 5339, 5342 (BLAT).

INDIA: Andhra Pradesh, Tamil Nadu and Kerala.

WORLD: Sri Lanka.

Suregada multiflora (Juss.) Bail., Etud. Gen. Euphorb. 396.1858; N.P. Balakr. & Chakrab., Fam. Euphorb. India 220. 2007; N.P. Balakr. in N.P. Balakr. *et al.*, Fl. India 23: 247. 2012. *Gelonium multiflorum* Juss., Euphorb. Gen. Tent. 3: 5. to f. 31A. 1824; FBI 5: 459. 1887. *G. bifarum* Roxb. ex Willd., Sp. Pl. 4: 831. 1805 *p.p.*; FBI 5: 459. 1887.

Tree, up to 10 m high; bark nearly smooth; branchlets with stipular scars. Leaves lanceolate or elliptic-oblong, 10-20 × 3.2-5.7 cm, glabrous, base acute or cuneate, margin entire to slightly serrate, apex acuminate, secondary nerves 10 pairs; petiole 5 mm. Flowers yellow, rather large in contracted cymes or clusters. Male flowers: pedicels 5 - 10 mm long; sepals 5, suborbicular; stamens many; glands many between filaments. Female flowers: pedicels 5-10 mm long; sepals as in male but larger, disc annular; ovary ovoid, styles 2 or 3, 1 mm long, bifid. Capsule globose or subglobose, orange-brown, tardily dehiscent.

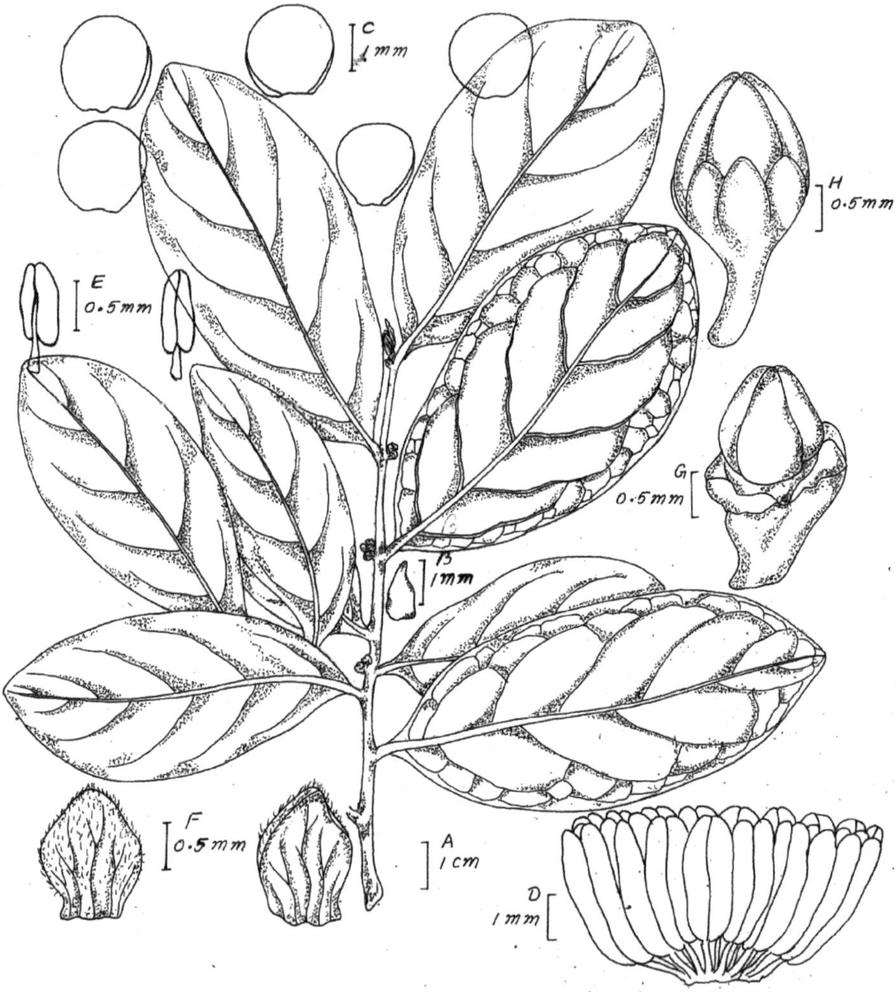

Figure 101. **Suregada angustifolia** (Bail. ex Müll.Arg.) Airy Shaw
A. Twig, ♂ B. Bract, C. Tepals, D. Androecium, E. Stamens, ♀ F. Tepals lobes, G. Pistil.

Occasional, in semievergreen forests. Fl.: March-April. Fr.: April-June. Vern.: Tel.: *Surugaata, Pindemaredu.*

Mamandur RF (CTR), *JHFB* 116948 (MH); Girijan coffee plantation (SKLM), *GVS* 62452 (MH).

INDIA:West Bengal, Sikkim, Assam, Andhra Pradesh and Andaman and Nicobar Islands.

WORLD: Bangladesh, Myanmar, China, Indo-China, Thailand and Malaysia.

SYNADENIUM Boissier

Synadenium grantii Hook. f. in Curtis Bot. Mag. t. 5633. 1867; N.P.Balakr. Chakrab., Fam. Euphorb. India 281. 2007 & in N.P. Balakr. *et al.*, Fl. India 23: 338. 2012.

Succulent shrub, up to 5 m high. Leaves 10-16 × 4-6 cm, broadly lanceolate cuneate at base, decurrent, entire, obtuse – subacute, alternate; petiole to 1 cm. Cyathia in axillary or subterminal cymes; rim cushion like, recurved, papillose; involucre 5.5 mm across, broadly cupular; bracts 5, concave, fimbriate. Male florets 25-30, in 5 groups; stalk 2.5 mm; bracteolate, laciniate; anther 0.6 × 0.8 mm. Female floret laterally pendulous; ovary 2 mm, 3-lobed; styles 3, each 2-fid, at above. Capsule 3-lobed, thinly pubescent; seeds oblong.

Occasional as a hedge plant. Fl. & Fr.: November – January.

Horseley hills (CTR), *PSPB* 9198.

INDIA: Cultivated as fence through out India.

WORLD: Native of Tropical Africa, now introduced into many Asian countries.

Note: In The Plant list it is treted as *Euphorbia umbellata* (Pax) Bruyns.

TRAGIA Linnaeus

1. Twining hispid herb; leaves simple; styles circinate-revolute:
    2. Leaves rounded, obtuse, acute to cuneat at base:
        3. Leaves 1.5-5 times longer than broad, 3.5-6.5 × 1-3.5 cm, serrate to serrulate or dentate to denticulate along margins, caudate at apex, prominently trinerved at base with the basal lateral nerves extending 30-80 per cent way up to the lamina; secondary venation predominantly semicraspedromous or often craspedromous; lateral nerves 3-7 per side; fruiting calyx lobes larger (1.5-3 mm long) .... **T. involucrata**
        3. Leaves 6-12 times longer than broad, 4-20 × 0.4-2.6 cm, irregularly and distantly serrulate to entire along margins, attenuate or attenuate-caudate at apex, penninerved or very weakly trinerved at base with the basal lateral nerves extending 10-30 per cent way up

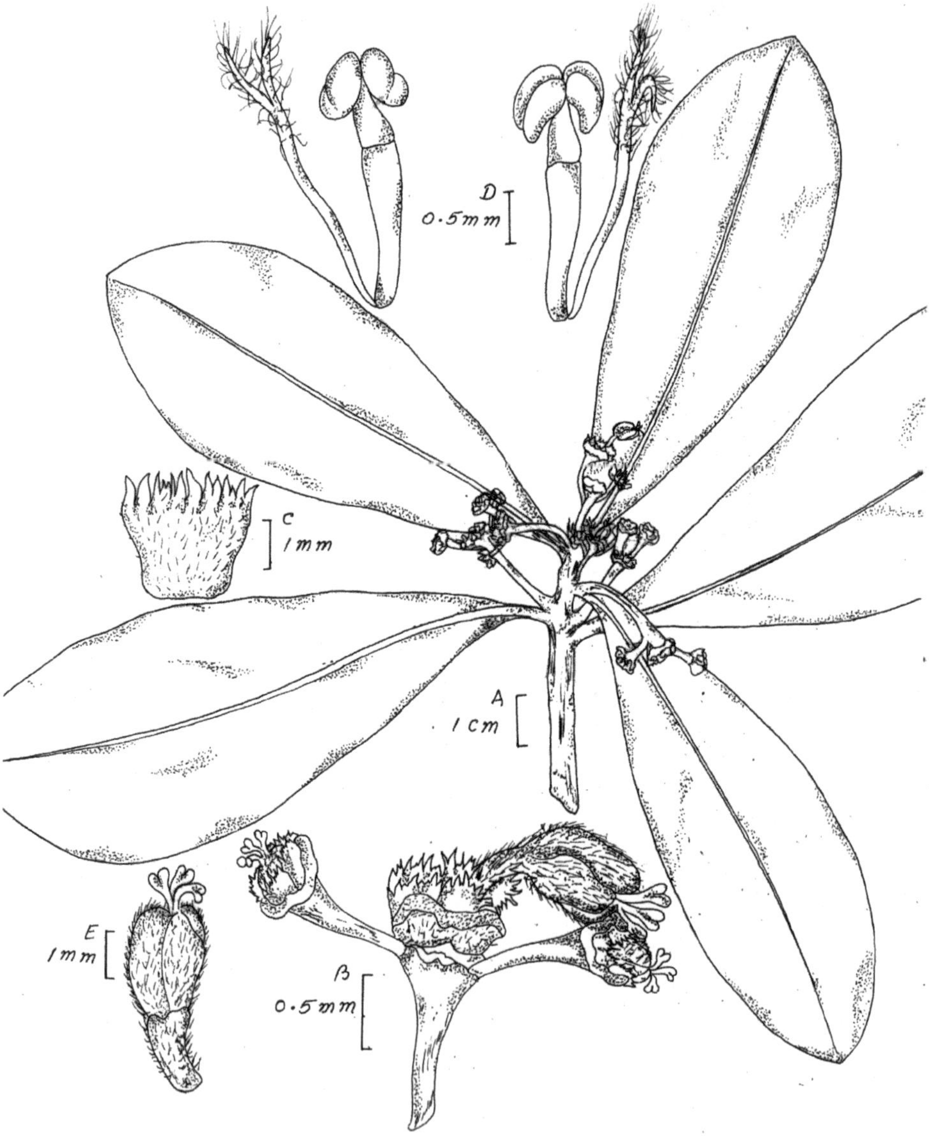

Figure 102. **Synadenium grantii** Hook.f.
A. Twig, B. Inflorescence, C. Involucre bract, D. Stamens with bracteoles, E. Pistil.

the lamina; secondary venation mostly brochidromous; lateral nerves 7-15 per side above the basal region, fruiting calyx lobes very small (0.3-2 mm long) .... **T. sanjappae**

2. Leaves cordate or narrowly at base .... **T. bicolor**

1. Hispid herb or under shrub; leaves simple or palmately 3-partite; styles entire .... **T. plukenetii**

Tragia bicolor Miq., Linnaea 26: 222. 1853; FBI 5: 465. 1888; Gamble 2: 1332. 1925; Saxena & Brahmam, Fl. Orissa 2: 1687. 1995.N.P. Balakr. in N.P. Balakr. *et al.*, Fl. India 23: 195. 2012.

Slender, fulvous-tomentose climbing herb. Leaves cordate-ovate, orbicular-ovate to ovate-lanceolate, 4-16 x 1.5-6.5 cm, base cordate to subcordatee, margin serrate to serrulate, apex caudate, densely fulvous-tomentose. Flowers in racemes. Male flowers: pedicels ca 1 mm, sepals 3, broadly ovate to suborbicular, stamens 3. Female flowers: sessile to subsesile; sepals 6, enlarge and stellateely spreading in fruit, ovary depressed, densely hirsute, styles 2.5 mm long, connate below into a column, trifid at apex. Fruit deeply 3-lobed, brown when dry, fulvous hirsute.

Mahendragiri hills (Saxena and Brahmam, 1995). Fl. & Fr.: Throughout the year.

Mahendragiri (GJM), *VNS* 5600 (MH).

INDIA: Karnataka, Tamil Nadu, Odisha. Endemic.

Tragia involucrata L., Sp. Pl. 980. 1753; FBI 5: 465. 1888; Gamble 2: 1332. 1925; N.P. Balakr. & Chakrab., Fam. Euphorb. India. 181. 2007; N.P. Balakr. in N.P. Balakr. *et al.*, Fl. India 23: 196. 2012.

Twining hispid berb, with stinging hairs, monoecious; branchlets striate. Leaves 3-6.5 × 1-3.5 cm, linear – lanceolate or oblong lanceolate, rounded or cuneate at base, serrate, acuminate, sparsely hispid; peiole to 1.5 cm, hispidate; stipules ovate, acuminate. Flowers in axillary and terminal leaf-opposed racemes, subsessile; male flowers many, minute, from above the middle; female flowers few, near the base; bracts 3 mm, lanceolate, acute. Male flowers 2.5 mm across; tepals 3-4, 1.4 × 1 mm, slightly connate at base, broadly ovate or orbicular, concave, subacute; stamens 3-4, anthers 0.45 mm across; pistillode small, 3-fid. Female flowers 5.8 mm across; tepals 6, 5 × 2 mm, free, accrescent, pinnatifid, lanceolate, laciniate, hispid without; ovary 2 mm, globose, hispid, styles 3, connate at base, circinately revolute. Capsule 5 × 9 mm, hispid; seeds globose, smooth.

Frequent along forest outskrits and in scrub jungles. Fl. & Fr.: December – March. Vern.: Tel.: *Dulagondi*; Tam.: *Senthotti.*

Jambugumpala (ATP), *NY* 574; Palakonda (CDP), *CS* 7717; Chandragiri (CTR), *PSPB* 9126; Saddikudumadugu (CTR), *PSPB* 9174; Galikonda (VSKP), *PSPB* 11821; Rapur hills (NLR), *PV* 110030 (MH); Upputeru – Mandapadu (KSN), *PV* 116547 (MH); Gokavaram (EG), *SKW* 1780 (BLAT); Kothuru (VSKP), *NPBK* (CAL); Sunkarimetta (VSKP), *HS* 20512 (BLAT); Salur (VZN), *NPBK* 956 (CAL).

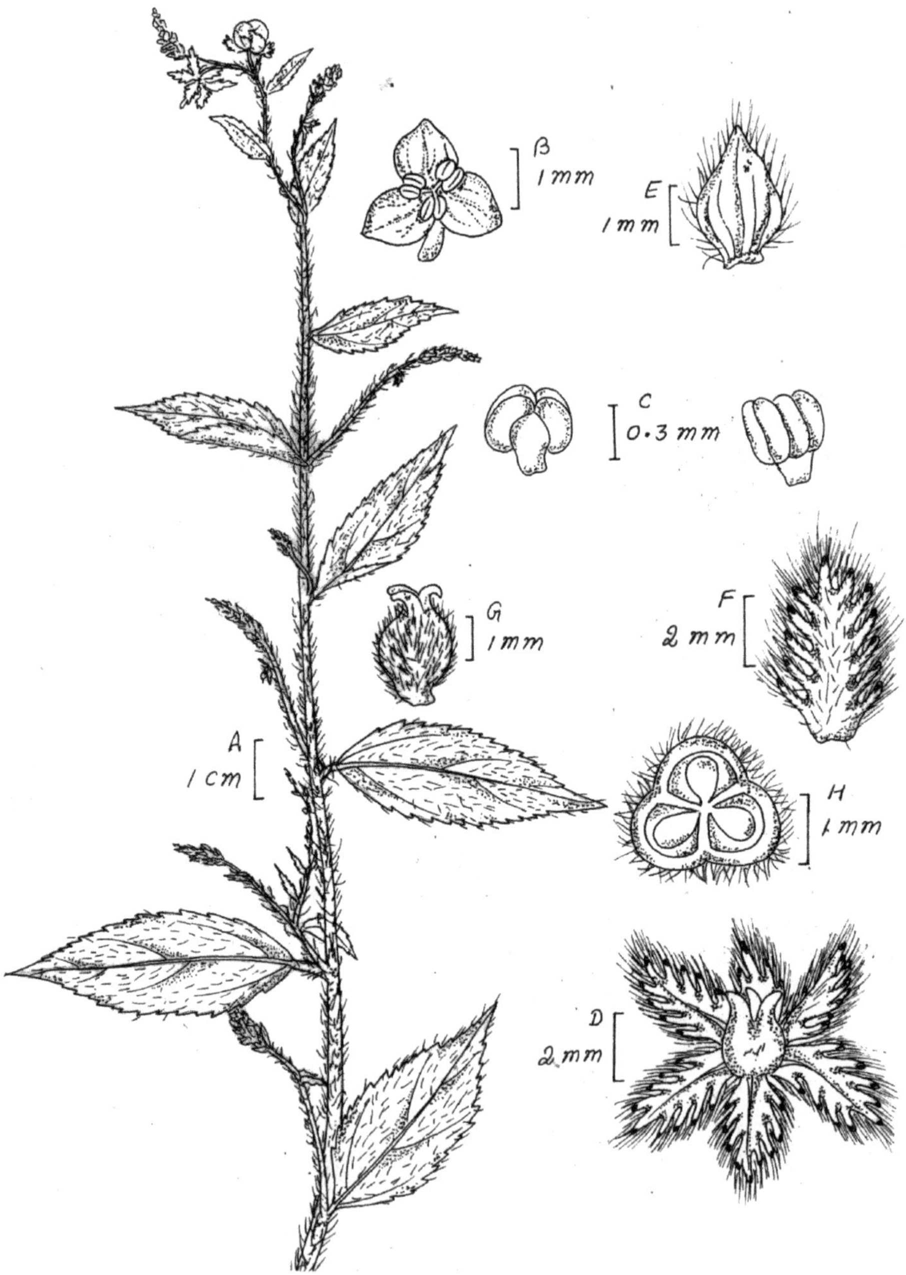

Figure 103. **Tragia involucrata** L.

A. Twig, ♂ B. Flower, C. Anthers, ♀ D. Flower, E. Bract, E. Involucre lobe, G. Pistil, H. Ovary T.s.

INDIA:Uttar Pradesh, West Bengal, Tripura, Odisha, Jharkhand, Madhya Pradesh, Maharashtra, Andhra Pradesh, Karnataka, Kerala and Tamil Nadu.

WORLD: Sri Lanka and Bangladesh.

Tragia plukenetii Radcl.- Sm., Kew Bull. 37: 688. 1983; N.P. Balakr. & Chakrab., Fam. Euphorb. India. 182. 2007; N.P. Balakr. in N.P. Balakr. *et al.*, Fl. India 23: 199. 2012. *T. cannabina* L. f., Suppl. Pl. 415. 1781, *nom. illeg.;* Gamble 2; 1333. 1925. *T. involucrata* L. var. *cannabina* (L.f.) Müll.Arg. in DC., Prodr. 15: 944. 1866; FBI 5: 465. 1888.

Herb or undershrub, to 1 m high, with sparsely hispid stinging hairs, monoecious; branchlets striate. Leaves 1.5-7 × 1.5-4 cm, lanceolate or plamately 3-lobed, middle longer than laterals, base truncate – cordate, margin serrate – crenate dentate, apex acuminate, sparsely hispid; petiole 1.5 cm, hispidate; stipules ovate, acuminate. Flowers in axillary racemes; male flowers many, at above; female flowers few at base; bracts 1.4 mm, lanceolate. Male flowers 2 mm across; tepals 3, 1.5 × 1 mm, free or slightly connate at base, broadly ovate – broadly elliptic, concave, subacute; stamens 3, filaments 0.3 mm, anthers 0.4 mm across. Female flowers 6 mm across; tepals 4-6, 4.2 × 2.4 mm, free, ovate, pinnatifid, densely hispid without and near the margin within; ovary 2.5 mm, subglobose, hispid, styles 3, slightly connate at base, not revolute. Capsule 4 × 7 mm, hispid; seeds globose, smooth.

Common in hedges and in forest outskirts. Fl. & Fr.: December – March. Vern.: Tel.: *Pedda dulagondi*.

Tadipatri (ATP), *TP* 404; Mamalapadu (KNL), *AA* 1969; Rammallakota (KNL), *RVR* 1834; Mabbuchintalapalli cross (CDP), *KRKS* 36940; Tirumala (CTR), *PSPB* 9128; Balasamudram (CDP), *JSG* 15799 (MH); Tapeswaram (EG), *Bourne* 3248 (CAL); Mellavagu (GNT), *SKW* 3641 (BLAT).

INDIA: Rajasthan, Gujarat, Madhya Pradesh, West Bengal, Andhra Pradesh, Maharashtra, Goa, Karnataka and Tamil Nadu.

WORLD: Sri Lanka, Africa.

Note: Linnaeus f. when uniting Linnaeus' *Croton hastatus* and *C. urens* (Sp. Pl. ed. 1: 1005. 1753) and transferring them to *Tragia,* failed to choose between one of these epithets. Hence, his name *Tragia cannabina* becomes superfluous and illegitimate. As the epithets *hastatus* and *urens* have since been preoccuipied A. Radcliffe-Smith has named it after Leonard Plukenet (1642-1706).

Tragia sanjappae Chakrab. & N.P.Balakr. in N.P.Balakr. & Chakrab., Fam. Euphorb. India. 184. 2007; N.P.Balakr. in N.P.Balakr. *et al.*, Fl. India 23: 201. 2012. *Tragia involucrata* L., Sp. Pl. 980. 1753. var. *angustifolia* Hook. f., Fl. Brit. India 5: 465. 1888; Gamble 2: 1332. 1925; Surya Prakash Babu in Pullaiah & Moulali, Fl. Andhra Pradesh 2: 888. 1997.

Climbing or twining hispid herb, with stinging hairs, monoecious; branchlets striate. Leaves linear lanceolate – linear elliptic, base cuneate – rounded, margin irregularly and distantly serrulate to entire, acuminate, sparsely hirsute at above,

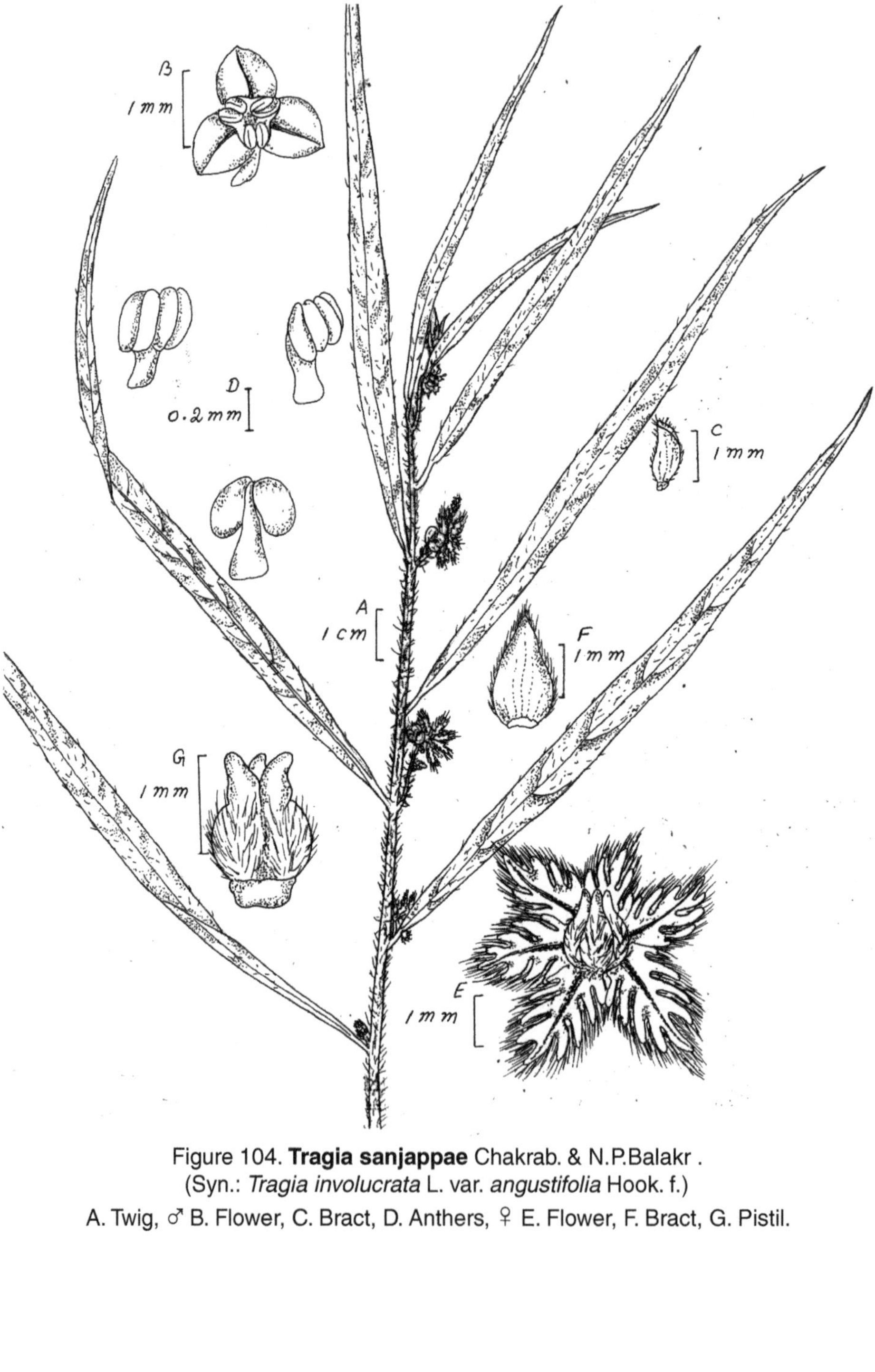

Figure 104. **Tragia sanjappae** Chakrab. & N.P.Balakr .
(Syn.: *Tragia involucrata* L. var. *angustifolia* Hook. f.)
A. Twig, ♂ B. Flower, C. Bract, D. Anthers, ♀ E. Flower, F. Bract, G. Pistil.

densely below; petiole to 4 mm; stipules 3 mm, lanceolate. Flowers in axillary and terminal leaf-opposed racemes; male flowers many, minute, from above to middle; female flowers few, near the base; bracts 1.9 mm, spathulate, acute, hispid without. Male flowers 1.7 mm across; tepals 3, 1 × 0.8 mm, free, orbicular, subacute; stamens 3, subsessile, connate at base, anthers 0.35 mm across. Female flowers 4.5 mm across; tepals 6, 2.3 × 1.3 mm, free, accrescent, pinnatifid, ovate-elliptic, laciniate, hispid without; ovary 1.5 mm, globose, hispid, styles 3, connate at base. Capsule 5 mm, hispid; seeds globose, smooth.

Rare in the interior forests in Chittoor. Fl. & Fr.: December– March. Vern.: Tel.: *Teega dulagondi*.

Talakona (CTR), *PSPB* 9121; Galikonda (VSKP), *PSPB* 11803; Boddalanka (EG), *DN* 85558 (MH); Satraldhari (CTR), *JHFB* 117415 (MH).

INDIA: Andhra Pradesh, Tamil Nadu and Kerala. Endemic.

TREVIA Linnaeus

Note: The original spelling used by Linnaeus (Sp.Pl. ed. 1, 2: 1193. 1753) is *'Trevia'*, but later he inconsistenty used both *'Trevia'* and *'Trewia'*. All subsequent authors have adopted the spelling *'Trewia"*. However, Art. 133.4 of the present code (2006) gives protection to the original spelling given in Species Plantarum, ed. 1 (1753) and therefore *'Trevia'* cannot be altered.

Trevia nudiflora L., Sp. Pl. 1193. 1753; Hook.f., Fl. Brit. India 5: 423. 188; Gamble 2: 1319. 1925; var. **nudiflora;** N.P. Balakr. and Chakrab., Fam. Euphorb. India. 155. 2007; Susila Rani & N.P.Balakr. in N.P. Balakr. *et al.,* Fl. India 23: 168. 2012. *T. polycarpa* Benth. in Gen. Pl. 3: 318. 1880; FBI 5: 424. 188.

Deciduous, small tree; bark light grey, flaking-off in thin plates; wood white or pale brownish grey, straight-grained, medium, fine and even-textured; young branches fulvous-pubescent. Leaves broadly ovate, 10-17 × 6-11 cm, glabrous, base truncate or cordate, margin entire-irregularly crenate, apex acuminate, 3-ribbed, secondary nerves 5-pairs above basal ones; petiole to 12 cm. Male flowers: in pendulous, many-flowered racemes; pedicels 3-5 mm long; bracts ovate-lanceolate, stamens 20-25, filaments free. Female flowers: axillary, solitary or in 5-15 cm long stout racemes; pedicels 5-9 mm long; sepals 3-5, broadly ovate, densely woolly tomentose outside, caducous; ovary 3-5-loculed; styles 3-5, shortly connate at base, 1.5-3 cm long. Drupe oblong, woody; seeds 3-5, globse to ovoid, smooth, dark brown.

Rare, along streams and moist hilly regions of Visakhapatnam and Srikakulam districts, also cultivated for ornament. Fl.: February Fr.: March-May. Vern.: Tel.: *Botku, Eru ponaku*; Tam.: *Aatharasu*. Eng.: False white teak.

Sirivaka near Koruturu (WG), *DN* 84250 (BSID); Narsipatnam (VSKP), *GVS* 30078 (MH); Regadi (SKLM), *GVS* 62480 (MH).

INDIA: Uttar Pradesh, Bihar, West Bengal, Sikkim, Assam, Tripura, Meghalaya, Odisha, Madhya Pradesh, Andhra Pradesh, Maharashtra, Karnataka, Kerala and Tamil Nadu.

WORLD: Sri Lanka, Nepal, Bhutan, Bangladesh, Myanmar, S. China, Indo-China, Thailand to Malesia.

TRIADICA Loureiro

Triadica cochinchinensis Lour., Fl. Cochinch. ed. 1, 2: 610. 1790 & ed.2, 2: 749. 1793; Chakrab. & N.P. Balakr. in N.P. Balakr. *et al.*, Fl. India 23: 350. 2012. *Stillingia discolor* Champ. ex Benth., Hook. J. Bot. Kew Gard. Misc. 6: 1. 1854. *Sapium discolor* (Champ. ex Benth.) Müll.Arg., Linnaea 32. 121. 1863; FBI 5: 469. 1888. *S. eugeniifolium* Buch.-Ham. (in Wall. Cat. No. 7970. 1847) (as *eugeniafolium*) ex Hook. f., Fl. Brit. India 5: 470. 1888; Surya Prakash Babu in Pullaiah & Moulali, Fl. Andhra Pradesh 2: 884. 1997.

Tree, monoecius, up to 15 m high; crown spreading; bark greyish, smooth. Leaves coriaceous, elliptic-lanceolate, 3-11.5 × 2.3-4.6 cm, red when very young, glabrous, base acute, margin entire, apex narrowly acute to very shortly acuminate, secondary nerves 12-15 pairs, irregular, 3-nerved from base, not prominent; petiole to 6 cm, red, 2-glands at top. Flowers greenish in terminal solitary spikes, male ones above, female ones below. Male flowers: pedicels 2-3 mm long, calyx ca 1 mm long; stamens 0.7-1 mm long. Female flowers: pedicels 2-4 mm long; calyx 1-2 mm long; style 1-3 mm long. Capsule subglobose, to 6 mm long; seeds globose, white or brownish when dry.

Rare, in semi-evergreen forests of Visakhapatnam district at above 800 m. Fl.: February-April, Fr.: May-June.

Anantagiri (VSKP), *SSR* 19603 and *GVS* 19537 (MH); Madalammagudi (VSKP), *GVS* 47349 (MH); Padalammagudi (VSKP), *GVS* 30046 (MH).

INDIA: Kerala, Karnataka, Tamil Nadu, Andhra Pradesh and Himalaya.

WORLD: Myanmar, Thailand, Indo-China, China, Malay Peninsula, Indonesia, Philippines and Celebes.

URTICACEAE

1. Plants with stinging hairs:
    2. Flowers in interrupted racemes, cymes or spikes; ovary oblique:
        3. Shrubs or small trees .... **Dendrocnide**
        3. Herbs .... **Laportea**
    2. Flowers in heads; ovary straight .... **Girardinia**
1. Plants without stinging hairs:
    4. Lithophytes or epiphytes .... **Procris**
    4. Terrestrials:
        5. Stigmas penicillate (except **Maoutia**):
            6. Leaves opposite .... **Pilea**
            6. Leaves alternate

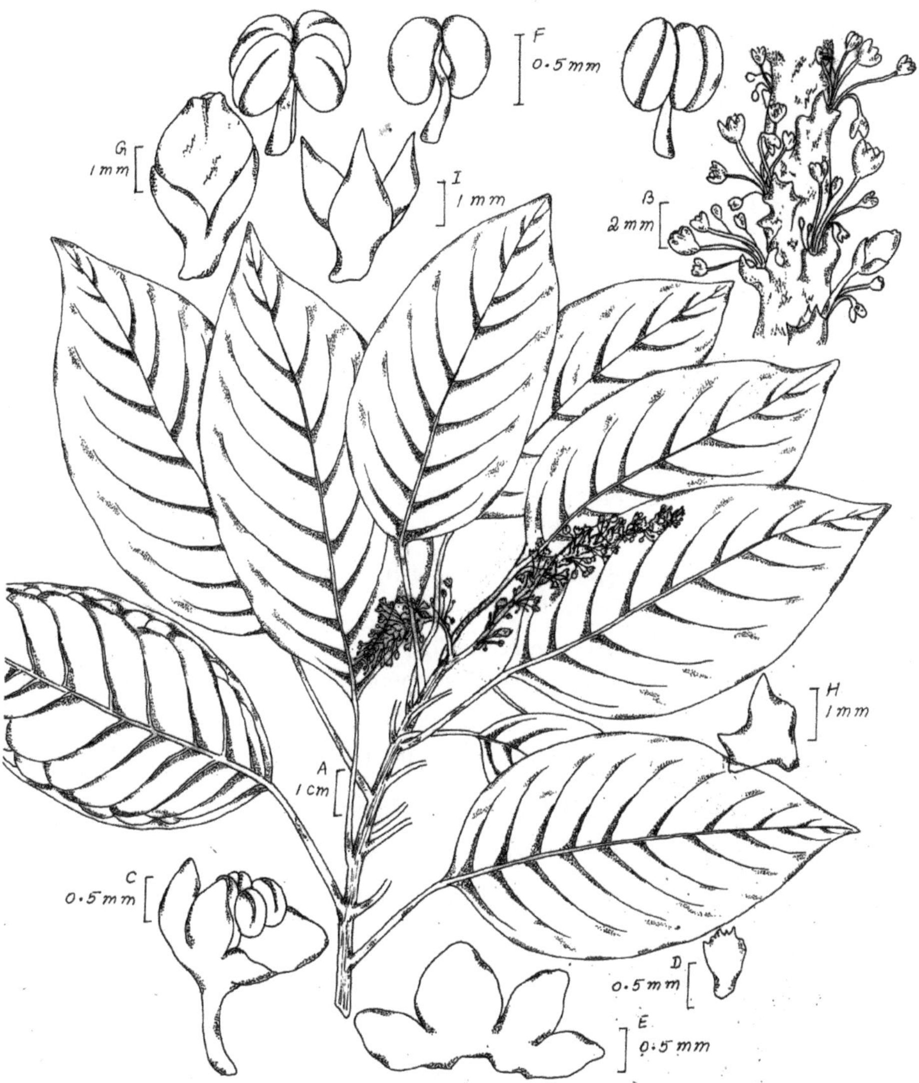

Figure 105. **Triadica cochinchinensis** Lour.
(Syn.: *Sapium eugenifolium* Buch.-Ham.)
A. Twig, B. Inflorescence (part), ♂ C. Flower, D. Bracteole, E. Tepals,
F. Stamens, ♀ G. Flower, H. Bract, I. Tepals.

7. Stipules simple .... **Elatostemma**
7. Stipules 2-fid .... **Debregeasia**

5. Stigmas filiform (not penicillate):
   8. Leaves entire along margins; female perianth often beaked; stigma jointed, deciduous:
      9. Clusters many-flowered, male flowers with 4 or 5, rarely with 3 stamens:
         10. Male buds truncate; perianth with annular, often villous, crown formed from transverse crests of lobes; leaf blade 3-veined from base............ **Gonostegia**
         10. Male flowers without this combination of characters; leaf blade with at least 2 major lateral veins arising clearly above base:
            11. Leaves opposite.. **Pouzolzia**
            11. Leaves alternate:
               12. Stigma peltate with long ciliate hairs along margin; achene adnate to fleshy perianth disc .... **Oreocnide**
               12. Stigma penicillate; achene without fleshy receptacle.. **Maoutia**
      9. Clusters few-flowered or flowers single along spike, male flowers with only 2 stamens .... **Neodistemon**
   8. Leaves serrate, serrulate or crenulate along margins; female perianth not beaked; stigma persistent .... **Boehmeria**

BOEHMERIA Jacquin

1. Leaves alternate; flowers in single axillary cluster .... **B. glomerulifera**
1. Leaves opposite; flowers in numerous clusters arranged on axillary spikes .... **B. macrophylla**

Boehmeria glomerulifera Miq., Syst. Verz. 101. 1854. *B. malabarica* Wedd., Arch. Mus. Hist. Nat. 8: 350. 1855-56; FBI 5: 575. 1888; Fischer 3: 1387. 1928.

Erect shrubs; stems and branches terete, sparsely lenticellate, pubescent. Leaves alternate, alternately smaller, up to 12 × 7 cm, membranous, ovate, base rounded or cordate, margin serrate, apex acute or acuminate, rugose and scabrid with bulbous-based hairs above, pubescent beneath, 3-nerved at base; petioles up to 5.8 cm long, obscurely canaliculate, pubescent, stipules lanceolate, caducous. Flower clusters sessile, dioecious, brownish-green. Male flowers: perianth tubular, silky-hairy without, 4-toothed; ovary minute, ovoid, covered by perianth.

Evergreen plants found in the area with considerable rainfall in Visakhapatnam district (Subba Rao and Kumari, 2008). Fl. & Fr.: August.

INDIA: Odisha, Karnataka, Kerala, Andhra Pradesh and Tamil Nadu.

WORLD: South China, Pakistan, Sri Lanka, Bhutan, Bangladesh Myanmar, Thailand, Laos, Vietnam, Malaysia, Indonesia, Philippines, New Guinea.

Boehmeria macrophylla Hornem., Hort. Bot. Hafn. 2: 890. 1815; Fris & Marais, Kew Bull. 37: 164. 1982. *B. platyphylla* D. Don, Prodr. Fl. Nepal 60. 1825; FBI 5: 578. 1888; Fischer 3: 1387. 1928; Pullaiah & Moulali, Fl. Andhra Pradesh 2: 907 *B. scabrella* Gaud. in Bot. Freyc. Voy. 499. 1846; FBI 5: 578. 1888.

Shrubs or small trees, branches strigose. Leaves 7-23 × 3.2-20 cm, suborbicular to ovate, opposite, rarely alternate, base rounded of cordate, margin acutely serrate with large teeth, apex acuminate or caudate, more or less hairy above, punctiform and lacunose beneath; petiole to 14 cm long. Flowers in clusters on long axillary spike; bracts ovate-lanceolate, strigose without. Male flowers: perianth lobes 4, concave, ovate, up to 1.5 mm long; stamens 4. Female flowers: usually on solitary spikes; perianth up to 1.2 mm long, tubular, contracted above the ovary, 4-toothed at apex, strigose without; style simple, filiform, penicillate. Achenes ovoid, shining, closely invested by the persistent perianth.

Common in moist or shady localities, especially in the valleys in the hills in north coastal districts. Fl. & Fr.: September- March.

Bison Hill (EG), *CAB* 5177 (MH); Near Tiger camp, Maredumilli (EG), *MM* 101286 (MH); Maredumilli towards Chinturu (EG), *GVS* 68521 (MH); Chintapalli (VSKP), *GVS* 28224 (MH); Galikonda (VSKP), *GVS* 32866 (MH); Near Dabuguda, Lakshmipuraam (VSKP), *NRR* & *DN* 84214 (MH).

INDIA: Throughout India.

WORLD: Tropical Asia and Africa.

DEBREGEASIA Gaudichaud Beaupré

Debregeasia longifolia (Burm. f.) Wedd. in DC., Prodr. 16: 235. 1869. *Urtica longifolia* Burm. f., Fl. Ind. 197. 1768. *Debregeasia velutina* Gaudich., Voy Bonite Bot. t. 90. 1844-46; FBI 5: 590. 1888; Fischer 3: 1389. 1928.

Tall shrubs or small trees, erect; branched, branches rough with warts and the scars of fallen leaves; bark greyish brown; female parts hispid-pubescent. Leaves 6-10 × 2-3.5 cm, lanceolate or oblong-lanceolate, simple, alternate, apex acuminate, margin serrulate, base acute to rounded, surface white or ashy – tomentose, and softer beneath, pubescent above, petiole 0.8-2.8 cm long. Flowers subsessile, in dense, globose, stalked, paniculate clusters, from the axils along the branchlets. Male flowers: 5 mm across, peduncles up to 1 mm long, bracts triangular, pubescent without, tepals 4, ovate, 2 mm, basaly connate, stamens 4, filaments 4 mm, pistillode ellipsoid, 1 mm. Female flowers: 1 mm across, subsessile, perianth tube up to 1 mm long, with a minute unequally 4-toothed mouth, ovary obovoid, stigma sessile, penicillate. Fruit yellow when ripe.

Rare in moist deciduous to evergreen forests. Fl.: November-January, Fr.: December- March. Vern.: Tel.: *Kerangi;* Tam.: *Kaatunochi.*

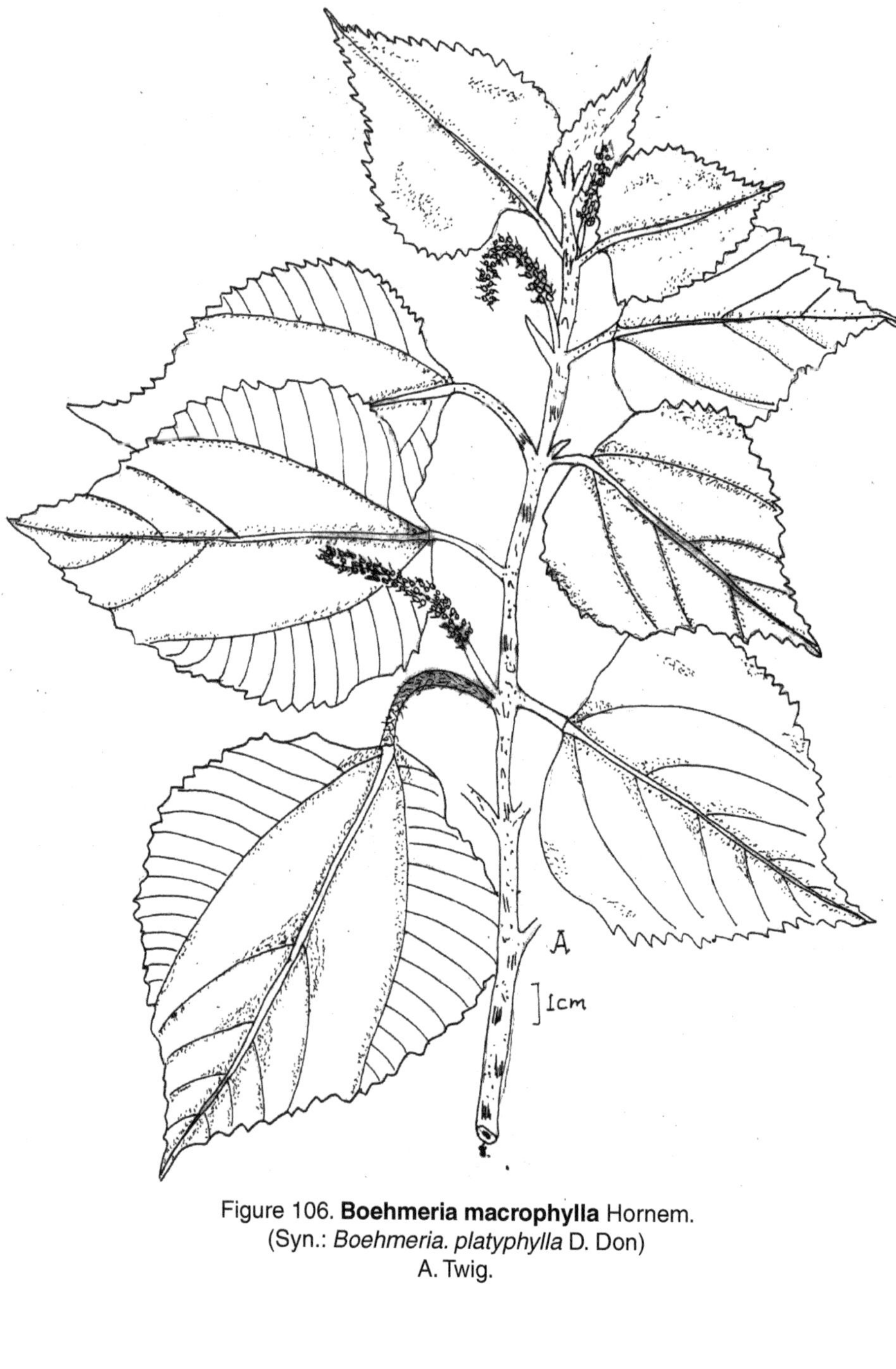

Figure 106. **Boehmeria macrophylla** Hornem.
(Syn.: *Boehmeria. platyphylla* D. Don)
A. Twig.

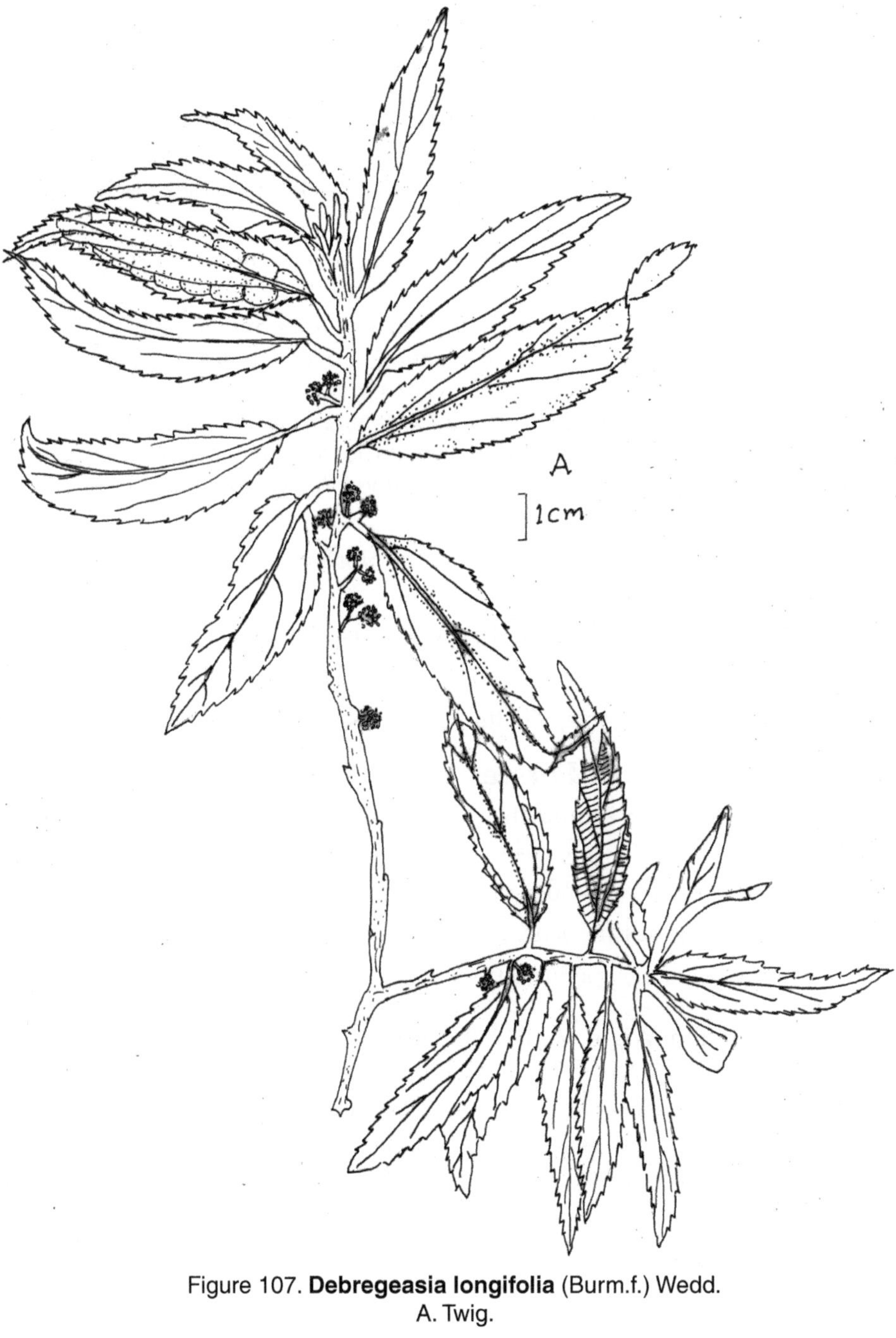

Figure 107. **Debregeasia longifolia** (Burm.f.) Wedd.
A. Twig.

Araku forest (VSKP), *PSPB* & *KJ* 12560; Kerangi, Sunkarimetta (VSKP), *GVS* 22651 (MH); Way to Rajampeta (NLR), *PV* 109975 (MH).

INDIA: Assam, Sikkim, Odisha, South India.

WORLD: Sri Lanka, Myanmar, Indo-China, W. China, Malesia.

DENDROCNIDE Miquel

Dendrocnide sinuata (Blume) Chew, Gard. Bull. Singapore 21: 206. 1965. *Laportea crenulata* Gaudich., Voy. Bonite Bot. 498. 1866; FBI 5: 550. 1888; Fischer 3: 1373. 1928.

Short shrubs or small trees; bark white, smooth, very thin, branchlets covered with minute stinging hairs. Leaves 12-25 × 5-12 cm, lanceolate to broadly elliptic, alternate, base narrowed, rarely rounded, margin entire or crenulate in the upper half, apex shortly acuminate, glabrous, shining above, petioles up to 7.5 cm long, sometimes with many stinging hairs. Flowers monoecious, in axillary cymes. Male flowers: perianth lobes 4, subequal, up to 0.25 cm long, concave, cucullate, pubescent on both sides; stamens 4, inflexed in bud. Female flowers: perianth subcampanulate, lobes 4, unequal, up to 2 mm long, tapering, villous. Achene oblique, broadly ovate, crowned with style.

Occasional in moist deciduous forests of Northern Eastern Ghats. Fl. & Fr.: August – December.

Rampa hills (EG), *JSG* 15940 (MH); Padalammagudi (VSKP), *GVS* 42585 (MH).

INDIA: Throughout.

WORLD: China, Malaysia, Sikkim, Sri Lanka, Thailand and Myanmar.

ELATOSTEMA J.R. Forster & J.G.A. Forster *nom. cons.*

1. Large herbs; leaf apex acuminate, leaves with large conspicuous cystoliths .... **E. lineolatum**
1. Dwarf herbs; leaf apex rounded or sub-acute; cystoliths minute, inconspicuous:
    2. Male flowers sessile; achene ellipsoid, non-ribbed.... **E. cuneatum**
    2. Male flowers long peduncled; achene slender, ribbed .. **E. monandrum**

Elatostema cuneatum Wight, Icon. Pl. Ind. Orient. 6: 35. t. 2094. f. 3. 1853 (*cuneata*); FBI 5: 568. 1888; Fischer 3: 1377. 1928.

Annual herbs, more or less pubescent, stem simple or scarcely branched, stem rooting near the base. Leaves sub-sessile, falcate, base cuneate, margin dentate or crenate, apex rounded or sub-acute, the upper surface with numerous cystoliths, 3-nerved from base; stipules ovate. Flowers sessile, axillary, in heads. Achene ellipsoid, yellow.

In damp and shady places often on rocks, walls, steep slopes *etc.* in Northern Eastern Ghats. Fl. & Fr.: August – December.

Gupteswar (KPT), *SX* & *MB* 6744 (RRL-B); Gandhamardhan, *SX* & *MB* 6084,

6262 (RRL-B); Sirsumuli (GJM), *SX* 2919 (RRL-B); Brudhakol (GJM), *MB* 2781 (RRL-B).

INDIA: Sikkim, West Bengal, Bihar, Uttar Pradesh, Odisha,Central India, Peninsular India.

WORLD: Indo-Malesian region.

Elastostema lineolatum Wight, Icon. Pl. Ind. Orient. t. 1984. 1853; FBI 5: 565. 1888; Fischer 3: 1376. 1928.

Herbaceous or suffruticose, pubescent. Leaves alternate, lanceolate, 12-17 × 3-5 cm, base obliquely acute, margin coarsely serrate along the upper half, apex acuminate, cystoliths linear or scattered. Flowers dioecious; receptacles of both sexes sessile, capitate. Achenes ellipsoid, acute.

Hills above 1000 m, inside forests on woods and rocks in Kolli hills. Fl.: October – May.

Kolli hills, Mathikettanshola (NMK), *SKS* 1207 (SGH).

INDIA: Odisha, Karnataka, Maharashtra, Kerala and Tamil Nadu.

WORLD: Sri Lanka, China, SE Asia.

Elatostema monandrum (Wight) H. Hara, Fl. E Himal. 3rd Repr. 21. 1975. *E. surculosum* Wight, Icon. Pl. Ind. Orient. 6: t. 2094. F.4. 1853; FBI 5: 572. 1888; Fischer 3: 1377. 1928.

Small, delicate, erect, succulent herb, 5-10 cm high, stems puberulous. Leaves alternate, with small or very small filiform, linear, lanceolate or oblong, usually deflexed, monophylls, opposite or subopposite the normal leaves; leaves sessile or shortly petioled, distichous, falcately oblanceolate, 1-8.7 x 0.5-3.2 cm, apex acute or rounded, margin coarsely crenate-serrate; stipules minute. Male receptacles long or short-peduncled, the outer bracts usually horned. Female receptacles smaller, sessile. Achenes slender, ribbed.

On moss-covered rocks in damp and shady places in valleys in Odisha Eastern Ghats. Fl. & Fr.: August – October.

Goyalpathar Ghati,below Kapilas (DKL), *HFM* 2068 (DD); Barapahar (SBP), *HFM* 3621 (DD); near Ranpur, Rairakhool (SBP), *HFM* 3990 (DD).

INDIA: Assam, Khasi hills, Sikkim, Odisha, Western Ghats.

WORLD: Sri Lanka.

GIRARDINIA Gaudichaud Beaupré

Girardinia diversifolia (Link) Friis, Kew Bull. 36: 145. 1981. *Urtica diversifolia* Link, Enum. Hort. Berol. Alt. 2: 385. 1822 (non. Blume 1825). *Girardinia leschenaultiana* Decne in Jacq. Voy. Inde 4: 152. 1844; Fisher 2: 1372. 1928. *G. zeylanica* Decne, *loc.cit.*; Fisher 2: 1373. 1928.

Tall, perennial herbs, 1-1.5 m high, closely armed with slender stinging hairs; stems and branches furrowed. Leaves 7.5-12 × 9-12.5 cm, pedately 3-5-lobed, base

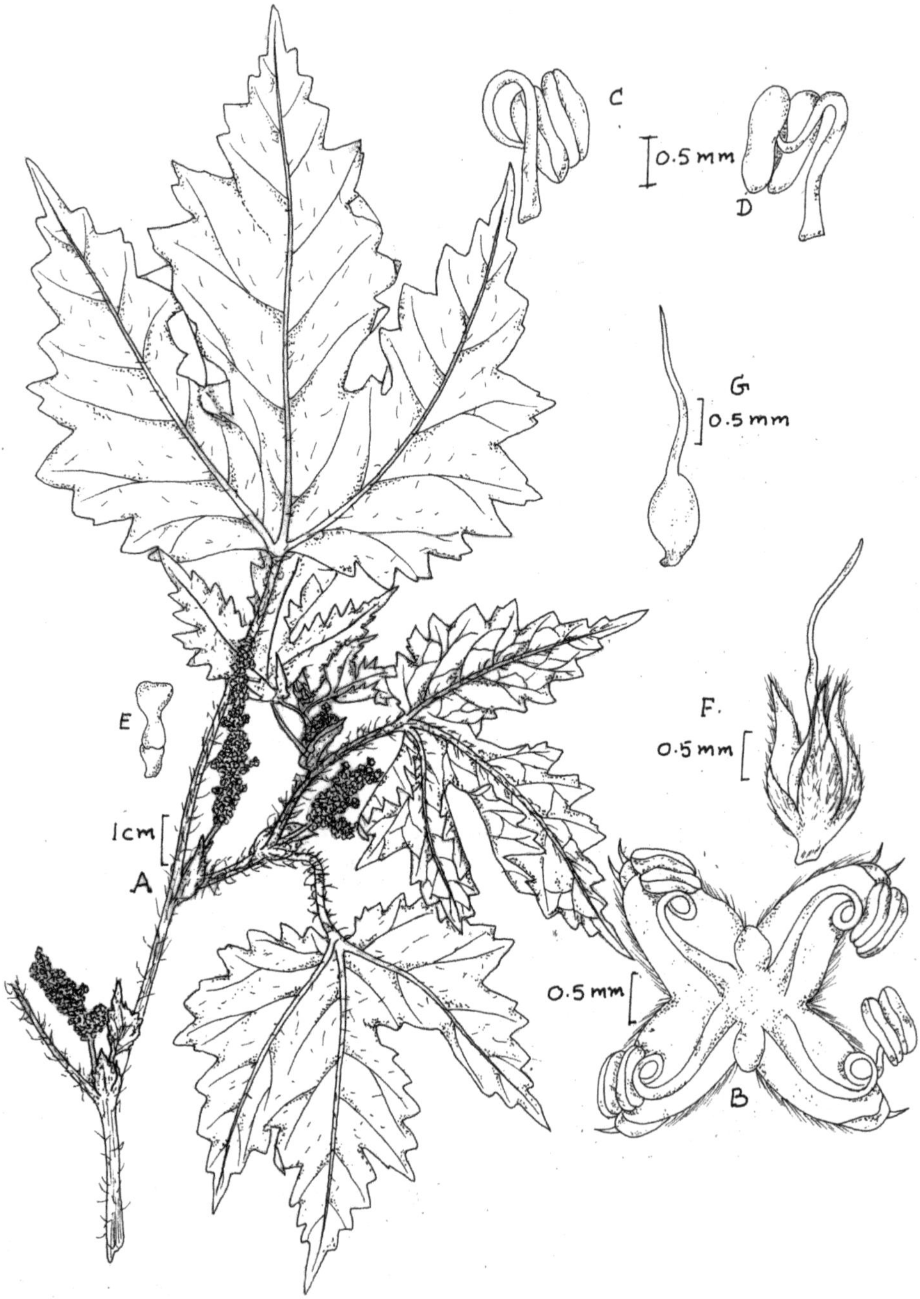

Figure 108. **Girardenia diversifolia** (Link) Friis
A. Twig, B. Male flower, C and D. Anthers, E. Pistillode, F. Female flower, G. Pistil.

truncate to subcordate, alternate, margin coarsely serrate, the serratures triangular, apex acuminate, chartaceous, sparsely scabrous above, densely so along the nerves below, 3-nerved from base; petioles up to 4 cm long, slender; stipules interpetiolar, foliaceous, up to 1 cm long, ovate, cordate. Male flowers on simple heads armed with numerous stinging hairs, 3.5 cm across, tepals 4, ovate, cucullate, stamens 4, 2.8 mm. Female flowers in capitate heads, clustered along the rachis, axillary; tepals connate into a ventricose tube, 2-4-lobed, ovary ovate, 1 × 0.8 mm, flattened, style filiform, 2.5 mm. Achenes compressed, broadly ovate, warted, 4 mm.

Occasional in hills. Fl. & Fr.: February – March. Vern.: Tam.: *Yanaichenthotti.*

Sujanakota (VSKP), *GVS* 33304; Eethakayalamandpam (CTR), *GVS* 46909 (MH).

INDIA: Himalayas up to 2100m, Central India, Deccan Peninsula.

WORLD: Sri Lanka, Myanmar, China, Malesia.

GONOSTEGIA Turc.

Key to Varieties

1. Male flowers 5-merous; achene glabrous .... **G. pentandra** var. **pentandra**
1. Male flowers 4-merous; achene pubescent .... **G. pentandra** var. **wightii**

Gonostegia pentandra (Roxb.) Miq., Ann. Mus. Bot. Lugduno-Batavi 4: 302. 1869 var. **pentandra.** *Urtica pentandra* Roxb., Fl. Ind. 3: 583. 1832. *Pouzolzia pentandra* (Roxb.) Benn. in Bennett *et al.*, Pl. Jav. Rar. 1: 64.t.14. 1838; Hook.f., Fl. Brit. India 5: 583. 1888; Fischer 3: 1383.1928.

Erect, suberect or decumbent herb or undershrub, up to 90 cm or more; stems from a perennial rhizome, often red, usually with raised puberulous lines decurrent from the leaf-bases. Leaves linear or linear-lanceolate, or the lower ovate or ovate-lanceolate 2-7.5 × 0.6-2 cm, lower always opposite, upper and floral bracts alternate and gradually smaller, base rounded or cordate, margin entire, apex acute to subacute, pubescent, hairy or glabrous, 3-nerved, petiole short or 0 in the upper leaves; stipules very small. Stems running out into terminal or axillary spikes due to the leaves becoming smaller and bractform above; spikes up to 40 cm. Flowers very inconspicous in nearly all the axils. Male flowers pedicelled, 5-merous, tepals abruptly inflexed about the middle and angled at the flexure so that the buds are truncate or depressed. Female flowers sessile, most minute in flower, conspicuous in fruit. Perianth at first smooth, tubular, 2-toothed, subsequently ridged on one side and the two tepals becoming alate, finally fruiting perinath with a large, white, concave, cordate or emarginate combined wing, 2.5-3 mm long and a fleshy ridge, enclosing black, shining conical achene, as long as the fleshy ridge.

Along river banks and streams and in wet places. Fl. August-October, Fr. October - December.

INDIA: Tropical Himalya, Assam, Khasi hills, West Bengal, W. Peninsular India.

WORLD: China, Afghanistan, Java.

Gonostegia pentandra (Roxb.) Miq. var. **wightii** (Benn. & Br.) Pullaiah & Karuppusamy *comb. nov. Pouzolzia pentandra* (Roxb.) Bennett var. *wightii* (Benn. and Br.) Friis & Wilmot Dear, *P. wightii* Benn. & Br., Pl. Jav. Rar. 66. 1838; FBI 5: 584. 1888; Fischer 3: 1383. 1928.

Shrubs, up to 70 cm high; tender parts pubescent, gradually tapering into leafy spikes. Leaves 4-12 × 3-5 cm, obovate-lanceolate, alternate, base acute to rounded, margin entire, apex acute, scabrid above, pubescent below: petiole to 1 cm, upper leaves gradually reduced to floral bracts. Flowers in sessile clusters, in the axils of leaves or bracts. Male flowers: 7 mm across, pedicels up to 6 mm long, hairy; tepals 4, free, obovate, acute, margins inflexed; stamens 4, filaments 2.5 mm long, flattened. Female flowers: perianth up to 2 mm long, hairy without, 4-toothed; ovary ovoid, up to 1.2 mm long, style up to 4 mm long, hispid. Achenes prominently winged, pubescent, grooved.

Common in the forests, near streams. Fl. & Fr.: February – April.

Durgam – Venkatagiri (NLR), *PV* 118146 (BSID); Bison hills (EG), *CAB* 5120 (MH); Minumuluru (VSKP), *GVS* 29607 (MH); Gangagudi (VSKP), *GVS* 44383 (MH).

INDIA: Odisha, Tamil Nadu, Karnataka, Kerala and Andhra Pradesh.

WORLD: Sri Lanka.

LAPORTEA Gaudichaud-Beaupré

Laportea interrupta (L.) Chew, Gard. Bull. Singapore 21: 200. 1965 & 25: 145, t. 12. 1969; Kalpana *et al.*, J. Biol. Records 9: 82. 2016. *Urtica interrupta* L., Sp. Pl. 985. 1753; Roxb., Fl. Ind. 3: 585. 1832; Wight, Icon. Pl. Ind. Orient. 2: t. 692. 1843. *Fleurya interrupta* (L.) Gaudich., Voy. Bonite, Bot. 497, t. 8. 1826; Wight, Icon. Pl.Ind. Orient. t. 1975. 1853; Wedd., Monogr. Urtic. 115. 1856; Hook.f., Fl. Brit. India 5: 548. 1888; C.E.C. Fisch., Fl. Madras 3: 1372. 1928.

Annual herb, monoecious, to 80 cm tall. Stems and petioles sparsely with short stinging hairs. Leaves simple, alternate, ovate-cordate, 5-10×3-5 cm, membranous, 3-nerved, apex acuminate, base cuneate-narrowly cordate, margin coarsely serrate, sparsely pubescent with stinging hairs, stipules intrapetiolar, slightly connate at base, apex 2-cleft, deciduous; petiole to 6 cm. Inflorescence axillary, interrupted spike, c. 25 cm long, sparsely pubescent with minute stinging hairs. Male flowers: pedicellate, ca. 1.2 mm; perianth lobes 3-4, connate at base, obovate, puberulous; stamens 3-4; ovary reduced or absent. Female flowers: prianth lobes 4, free, unequal, dorsal tepal ovate, concave, 0.3 mm, 2 lateral lobes large, enclosing the ovary, broadly ovate, 0.5 mm, ventral lobe small, triangular, 0.3 mm. Ovary triangular; stigma filiform, reflexed, 3-fid. Achene obliquely triangular, compressed, with triangular ridge enclosing a warty depression,persistent lateral perianth lobes forming a small basal cup.

Rare near waterfalls and moist shady localities in forests especially in rock crevices. *Pogostemon benghalensis, Rhynchoglossum obliqum, Lindernia* spp., and *Arthraxon* spp.are associated with *Laportea interrupta*. Found in an elevation ranges from 600-800m MSL. Fl. & Fr.: July – December.

Tiger camp, Maredumilli (EG), *MM* 1025 34 (MH); East Godavari district, *DN* 86501 (MH); Mahendragiri (GJM), *SX* & *MB* 3366 (RRL-B); Patna (GJM), *MB* 2789 (RRL-B); Gandamardhan (Balangir), *SX* & *MB* 6261 (RRL-B); Ramgiri- Gupteswar (KPT), *SX* & *MB* 6751 (RRL-B). Saleswaram (MBNR), *B. Sadasivaiah*, 43882; Umamaheswaram Forests (MBNR), *B. Sadasivaiah* & *B. Kalpana*, 1789 (Got. Degree College, Wanaparthy).

INDIA: Karnataka, Kerala, Maharashtra, Odisha, Tamil Nadu and Telangana.

WORLD: Australia, China, Indonesia, Japan, Malaysia, Myanmar, Philippines, Sri Lanka, Thailand, Taiwan and Vietnam.

LECANTHUS Weddell

Lecanthus peduncularis (Wall. ex Royle) Wedd. in DC., Prodr. 16(1): 164. 1869. *Procris peduncularis* Wall. ex Royle, Ill. t. 83. f.2. '1836. *Lecanthus wightii* Wedd., Ann. Sci. Nat. Bot. Ser. 4. 1: 187. 1854; FBI 5: 559. 1888; Fischer 3: 1374.1928.

Small, erect herbs, to 4.5 cm high, stem filiform. Leaves membranous, only 2 to 3 pairs, pairs unequal, ovate, up to 1.2 × 0.8 cm, base rounded, margin subentire, apex acute, sparsely pubescent on both sides, 3-nerved at base; cystoliths scattered; petioles up to 0.5 cm long, slender, stipules scarious. Flowers dioeciuous, crowded on axillary, fleshy receptacles; peduncles up to 0.4 cm long, slender; pedicels up to 0.5 mm long, fleshy; tepals in female flowers 4, unequal, up to 1.2 mm long, hooded; ovary ovoid, style penicillate, caducous. Achenes red, up to 0.8 mm long, ovoid, compressed.

Rare in moist shaded places in Northern Eastern Ghats. Fl. & Fr.: October.

INDIA: Throughout greater parts of India up to 3000 m.

WORLD: Paleotropics.

MAOUTIA Weddell

Maoutia puya (Hook.) Wedd., Ann. Sci. Nat. Bot. ser. 4. 1: 193. 1854; FBI 5: 592. 1888; Saxena & Brahmam, Fl. Orisssa 3: 1740. 1995.

Erect undershrub or shrub, branchlets slender, scabridly hairy; bark dark grey with vertical lines or brown lenticels. Leaves alternate, elliptic or elliptic oblong, 10-20x 5-7 cm, base rounded or subacute, apex acuminate or caudate-acuminate, margin coarsely dentate, crenate or crenate-serrate, strigose above, softly white-felted beneath except the hairy or strigose nerves; petiole 1.2-10 cm long. Flowers in small heads which are disposed in axillary dichotomous cymes or panicles. Male flowers subsessil; tepals ovate, 1 mm long. Achenes ovoid, 1.3 mm long, triquetrous, reddish-brown hairy.

Mahendragiri hills in Odisha, 1000-1300 m. Fl. & Fr.: November.

Mahendragiri (GJM), *SX* & *MB* 3969 (RRL-B).

NEODISTEMON Weddell

Neodistemon indicum (Wedd.) Babu & Henry, Taxon 19: 651. 1970. *Distemon indicum* Wedd., Arch. Mus. Paris (Monog. Urt.) 9: 551. t. 20A. 1857; FBI 5: 588. 1888.

Erect slender herb, 60-90 cm high, stem quadrangular, zig-zag. Leaves broadly ovate, to 5-15 × 5- 6.5 cm, base cuneate or rounded, margin coarsely dentate, apex acute-acuminate, hairy above and with whitish cobwebby pubescence, often in patches beneath between the hairy nerves; stipules lanceolate, 1 cm, with a long setaceous tip. Flowers unisexual, in spikes, male and female flowers in the same axils; bracts sub-amplexicaul, ovate-cordate; stamens exserted. Fruit achene, beaked.

Frequent in moist shady localities and crevices of huge boulders along stream banks in Northern Eastern Ghats. Fl.: & Fr.: July – September.

INDIA: Assam, Bihar, Odisha, Andhra Pradesh.

WORLD: Myanmar, Java.

OREOCNIDE Miquel

1. Shrub; leaves ovate, dentate-serrate, hispidly hairy above .... **O. frutescens**
1. Small tree; leaves elliptic-oblong or oblanceolate, entire or obscurely crenate, glabrous above .... **O. integrifolia**

Oreocnide frutescens (Thunb.) Miq., Ann. Mus. Bot. Lugduno-Batavi 3: 131. 1867; Saxena & Brahmam, Fl. Orissa 3: 1742. 1995. *Urtica frutescens* Thunb., Fl. Jap. 70. 1784. *Villebrunea frutescens* (Thunb.) Bl., Ann. Mus. Bot. Lugduno-Batavum 2: 168. 1856; FBI 5: 590. 1888.

Large straggling shrub with hairy branches. Leaves ovate, 7.5-17.5 cm long, base rounded or very obtuse, margin dentate-serrate, apex acuminate or caudate, hispidly hairy above, softly hairy beneath; stipules subulate,, 1 cm long. Flowers in axillary and from old leaf scars, dense cymose clusters. Achenes somewhat compressed, minutely pitted, adnate to perianth.

In damp valleys in Orissa Eastern Ghats. Fl. & Fr.: December – May.

Karlapat (KHD), *HFM* 1193 (DD).

Oreocnide integrifolia (Gaud.) Miq., Ann. Mus. Bot. Lugdunoo-Batavi 4: 306. 1869; Saxena & Brahmam, Fl. Orrisa 3: 1742. 1995.

Small, evergreen tree; bark smooth, branches pubsescent. Leaves lanceolate, elliptic-oblong, 6.5-40 x 2.2 – 12.5 cm, apex acuminate or caudate, margin entire, somewhat crenulate or obscurely crenate; petiole 0.6-11.2 cm long, usually pubescent; stipules villous. Flowers greenish, in globose clusters, in shortly peduncled, dichotomous, hispid cymes, cauliflorous. Male flowers: Tepals 4, free, concave, stamens 4, pistillode pilose below. Female flowers: Tepals connate into a tube, adnate to pistil, ovary obovoid, stigma discoid, ciliate-fimbriate. Achenes white, embraced by fleshy bracteoles below; seed albuminous.

In ravines in Eastern Ghats of Odisha and Tamil Nadu. Fl. & Fr.: December-March.

Bhalupodar valley near Pottangi (KPT), *HFM* 3860 (DD).

PILEA Lindley *nom. cons.*

1. Dioecious subshrubs; leaves more than 5 cm long, 3-nerved from base, margin serrate, apex acuminate:
   2. Male tepals horned; achenes ovoid with persistent perianth at base ....**P. melastomoides**
   2. Male tepals not horned; achenes ellipsoid without persistent perianth .... **P. scripta**
1. Monoecious herbs; leaves less than 1 cm long, 1-nerved, margin entire, apex subacute or obtuse..**P. microphylla**

Pilea melastomoides (Poir.) Wedd., Ann. Sci. Nat. Bot. (Ser. 4) 1: 186. 1854. *Urtica melastomoides* Poir. in Lam., Encycl. (Suppl. 4) 223. 1816. *Pilea trinervia* Wight, Icon Pl. Ind. Orient. t. 1973. 1853; FBI 5: 557. 1888; Fischer 3: 1379. 1928.

Succulent dioecious susbshrubs, up to 2.5 m high, with swollen internodes. Leaves 12-16 × 4-6.5 cm, elliptic-lanceolate or broadly ovate, opposite in pairs, base acute or obtuse, margin sharply serrate, apex acuminate or finely caudate, thin chartaceous, with numerous minute scattered linear raphides which look like appressed hairs when the leaf is dry; 3-ribbed from the base, lateral ribs prolonged right into the apex. Flowers in short axillary, much branched pedunculate cymes. Male flowers: bracts up to 1 mm long, scarious; tepals 4, up to 1.5 mm long; stamens 4. Female flowers: 4 mm across, tepals 4, unequal, outermost the longest, up to 1.2 mm long, linear, smaller ones deltoid, obtuse; stamens 4, 4 mm long. Female flowers: 1.5 mm, tepals 3-4, unequal, imbricate, staminodes 3, ovary ovoid, 1 mm. Achenes ovoid, biconvex, granular, smooth when young, with persistent perianth at base.

Not common, usually found near stream banks in forests in North coastal districts. Fl. & Fr.: October – December.

Forest near Sunkarimetta (VSKP), *NPBK* 10913 (MH); Galikonda (VSKP), *GVS* 21656 (MH).

INDIA: Deccan Peninsula.

WORLD: Sri Lanka, Malay Islands.

Pilea microphylla (L.) Liebm., Mexic. Meldeagt. Pl. V, 2: 302. 1851; FBI 5: 551. 1888; Fischer 3: 1379. 1928. *Parietaria microphylla* L., Syst. Nat. ed. 10. 1308. 1759. *Pilea muscosa* Lindli., Coll. Bot. t. 4. 1821; Hook.f., Fl. Brit. India 5: 551. 1888. *Urtica microphylla* Sw., Fl. Ind. Occid. 1: 305. 1797.

Prostrate herb, to 20 cm long, monoecious. Leaves decussate, elliptic-ovate in unequal pairs, to 4 × 2 mm, subsucculent, 1-nerved, base acute, margin entire, apex subacute or obtuse; stipules connate, intrapetiolar, raphides transverse; petiole to 1 mm. Flowers in stalked umbellate clusters. Male: 1.5 mm across, tepals 4, connate,

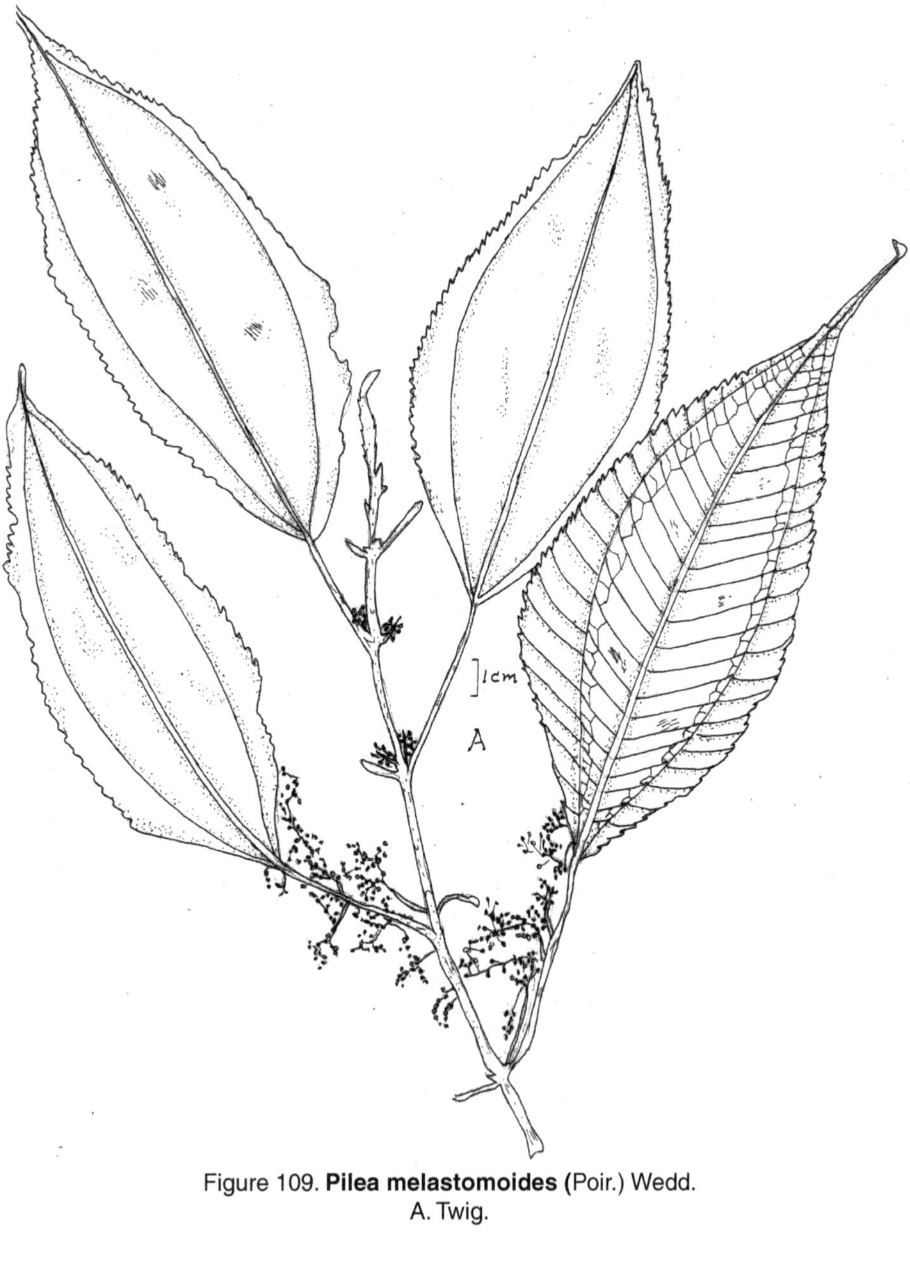

Figure 109. **Pilea melastomoides (**Poir.) Wedd.
A. Twig.

minutely 2-4-toothed, valvate, stamens 4, filaments 0.7 mm, anthers 0.3 mm. Female: 0.4 mm across, ovary ellipsoid, 0.5 mm. Achenes ellipsoid, 1 mm, 1-locular, ovules erect. Stigma shortly penicillate, subsessile, albumen scanty, cotyledons broad.

Common in moist places, on walls, slopes or in flower pots. Fl. & Fr.: July – January.

INDIA: Naturalized in many parts of India.

WORLD: Native of South America.

Pilea scripta (Buch.-Ham. ex D.Don) Wedd., Ann. Sci. Nat. Bot. ser.4.1: 187. 1854; FBI 5: 556. 1888; Saxena & Brahmam, Fl. Orissa 3: 1745. 1995.

Erect, much branched, glabrous herb, 30-90 cm high, stems succulent. Leaves sometimes slightly unequal in the pairs or with unequal petioles, narrowly elliptic, 7.5-15 cm long, apex finely acuminate, margin serrate; petioles slender, 1.8-3.7 cm long; stipules 2.5 mm. Male tepals not horned. Achenes minute, bordered, with strong, intramarginnal ridge, strongly granulate.

In moist shady ravines in Odisha Eastern Ghats. Fl. & Fr.: September – October.

INDIA: Himalaya, up to 2100 m.

WORLD: China, Bhutan, Myanmar, Nepal.

POUZOLZIA Gaudichaud-Beaupré

1. Leaves opposite:
   2. Male flowers sessile; achenes ovoid .... **P. bennettiana**
   2. Male flowers pedicelled; achenes ellipsoid .... **P. cymosa**
1. Leaves alternate:
   3. Lateral ribs of leaves not forked; male flowers 4-merous; achenes winged .... **P. zeylanica**
   3. Lateral ribs of leaves forked; male flowers 5-merous; achenes not winged .... **P. auriculata**

Pouzolzia auriculata Wight, Icon Pl. Ind. Orient. t. 1980. f. 2. 1853 and t. 2099. f. 37. 1853; FBI 5: 582. 1888; Fischer 3: 1383. 1928.

Erect scabrid herbs to under shrubs, up to 1 m high, stems woody, young branches angled, pubescent. Leaves 2-7 × 1-2 cm, elliptic-lanceolate, alternate, base acute to rounded, margin entire; apex acuminate, appressed hairy; petiole up to 5 cm long, pubescent; stipules membranous, ovate-lanceolate. Flowers in axillary, sessile clusters, greenish. Male flowers: 3 mm across, tepals 5, up to 2 mm long, ovate-oblong, concave, pubescent without, filaments up to 2 mm long, glabrous, flattened. Female flowers: perianth up to 2.2 mm long, tepals connate, minutely toothed, pubescent without; ovary up to 2 mm long, ovoid, white, shining, style up to 3 mm long, hairy. Achenes ovoid, reddish brown, shining.

Occasional along the streams in forests. Fl. & Fr.: January – November.

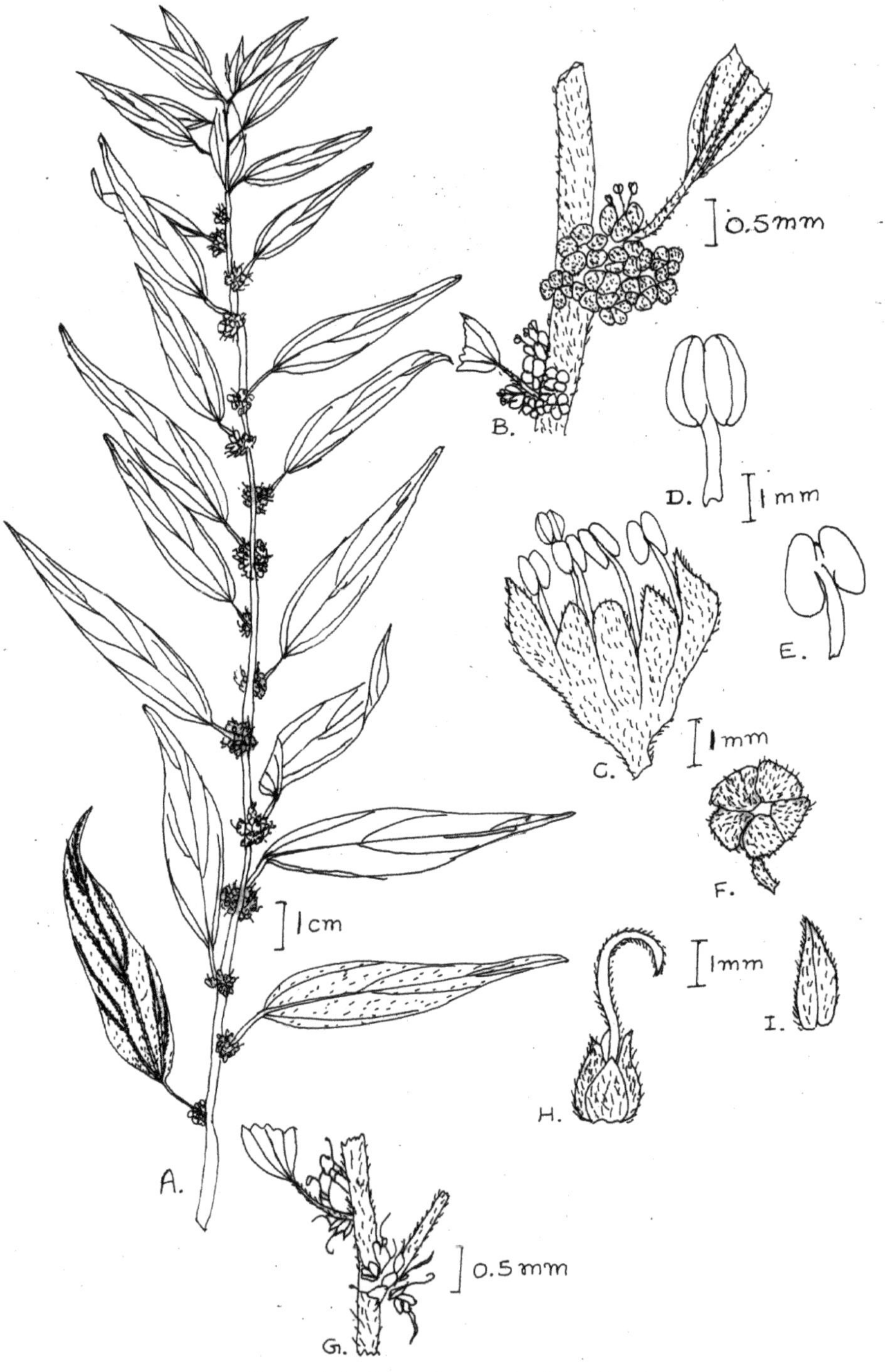

Figure 110. **Pouzolzia auriculata** Wight
A. Twig, B. Male flower branch, C. Male flower, D and E. Anthers, F. Male flower bud, G. Female flower branch, H. Female flower, I. Pistil.

Lankamalai RF (CDP), *SSR* & *KJ* 14655; Ishtakameswari temple (PKM), *BSS* & *SKB* 31005; Galikonda (VSKP), *KH* & *BR* 5514; Gollapuram (KSN), *PVP* 5363; Patalaganga, Srisailam (KNL), *JLE* 16833 (MH); Balapalli (CDP), *JLE* 14997 (MH); Ramakrishnathirtham (CTR), *GVM* 117474 (BSID); Durgam – Udayagiri (NLR), *PV* 110218 (BSID); Nagarjunakonda (GNT), *KT* 9862 (CAL); Near Araku (VSKP), *NPBK* 10814 (MH); Chintapalli (VSKP), *GVS* 28227 (MH).

INDIA: Central India, Bihar, Deccan Peninsula.

WORLD: Sri Lanka.

Pouzolzia bennettiana Wight, Icon. Pl. Ind. Orient. t. 1978. 1853; FBI 5: 585. 1888; Fischer 3: 1384. 1928. var. **bennettiana**

Suffrutescent herbs, stem erect, sparingly branched, 50-100 cm high, tender parts pubescent. Leaves opposite or ternate, 5.5-11 × 1-2 cm, elliptic, lanceolate, opposite, base rounded or somewhat oblique, margin entire, apex acuminate, thinly pubescent on both surfacts; petiole of lower leaves up to 4 mm long, canaliculated, upper leaves subsessile; stipules membranous, deltoid. Flowers in dense, sessile, axillary cymose clusters, greenish white in colour. Male flowers: Tepals 5, up to 3 mm long, ovate-lanceolate; stamens 5, filaments up to 2 mm long, anthers up to 1.5 mm long. Female flowers: 1 mm across, tepals connate, pubescent, ovary obovoid, style linear, 1.5 mm, hairy. Achene with 2 membranous, reticulately veined wings.

Abundant in marshy places in forests. Fl. & Fr.: Throughout the year.

Talakona water falls (CTR), *KJ* 12547; Bodalanka (EG), *DN* 85597 (BSID); Sunkarimetta (VSKP), *GVS* 21658 (MH).

INDIA: Peninsula.

WORLD: Sri Lanka.

Key to Varieties

1. Stem erect; leaves opposite .... var. **bennettiana**
1. Stem prostrate or decumbent; leaves opposite to ternate:
    2. Leaves pubescent above, tomentose below; male flowers 5-merous .... var. **quadrialata**
    2. Leaves pubescent without; male flowers 4-merous .... var. **gardneri**

Pouzolzia bennettiana Wight var. **gardneri** (Wight) Hook.f., Fl. Brit. India 5: 586. 1988; Fischer 3: 1384. 1928.

Shrubby, prostrate or decumbent, sparsely hairy; branches, leaves above and ribs below dark brown when dry. Leaves ternate or opposite. Male tepals more or less hispid above the flexure.

Mahendragiri hill, 1500m.

Mahendragiri hill (MBJ), *CECF* 102 (CAL).

INDIA: Peninsular India.

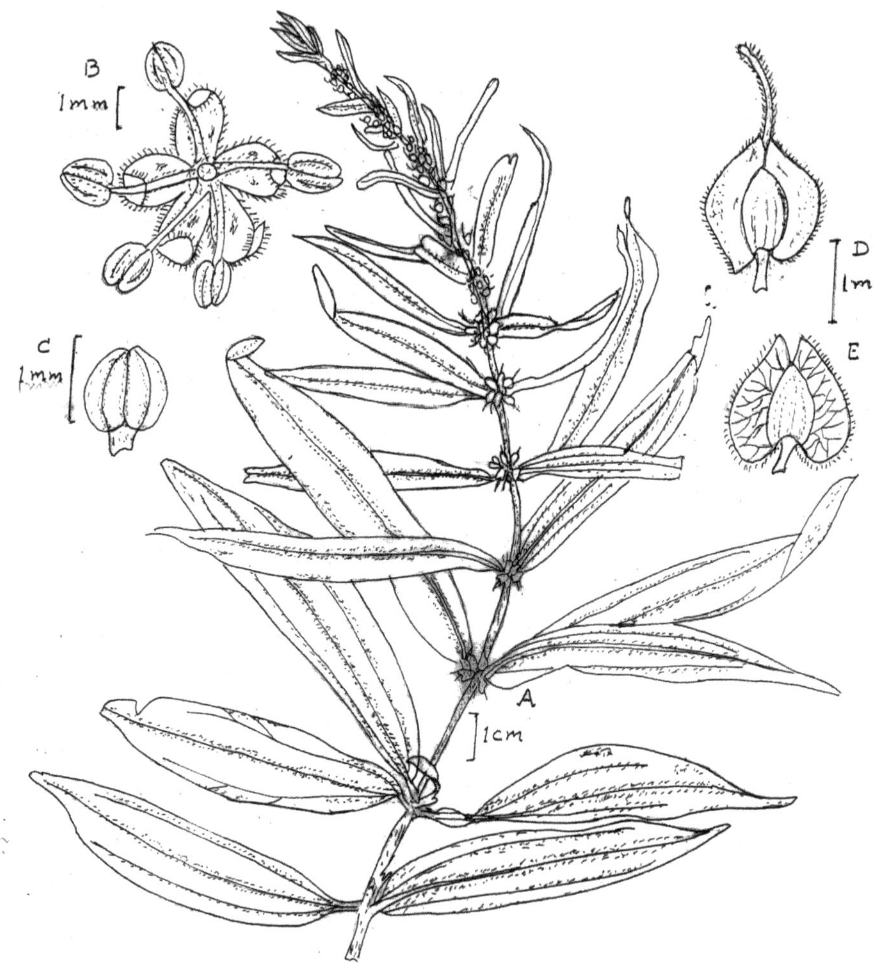

Figure 111. **Pouzolzia bennettiana** Wight
A. Twig, B. Male flower, C. Anthers, D. Female flower, E. Ovary.

Pouzolzia bennettiana Wight var. **quadrialata** (Wight) Hook.f., Fl. Brit. India5: 586. 1888; Fischer 3: 1384. 1928; Saxena & Brahmam, Fl. Orissa 3: 1748. 1995.

Herbs, stem prostrate or decumbent, woody below, tomentose above. Leaves softly pubescent above, tomentose below, up to 7.5 cm long. Flowers in sessile, axillary clusters; perianth of male 5-merous, lobes plicate and abruptly inflexed at about the middle buds truncate, male sepals tomentose or villous above the flexure; stamens 5; pistillode clavate or oblong; ovary included; ovule erect; stigma filiform, jointed to the top of the ovary. Fruit winged or not; seeds with a membranous testa.

In damp, shady valleys, near streams in Odisha Eastern Ghats. Fl.: November – December.

Kalahandi (KHD), *HFM* 1233 (DD); Sirimanda parbat, Pottangi (KPT), *HFM* 3852 (DD).

Pouzolzia cymosa Wight, Icon. Pl. Ind. Orient. 6: 42.1853; *P. auriculata* Wight var. *cymosa* (Wight) Hook. f., Fl. Brit. India 5: 582. 1888.

Subshrubs to 2 m tall; tender parts pubescent. Leaves opposite, elliptic-ovate, 3-5 x 1.5-2 cm, chartaceous, mid-nerve and lateral nerves forked, making for 9 parallel ribs, appressed-pubescent above, except along the impressed nerves, pubescent below, densely so along the nerves, base rounded to sub cordate, apex acuminate; petiole to 0.5 cm. Male flowers shortly stalked, in short panicles, 4 mm across; buds rounded; tepals 5, free, lanceolate, 2 mm, pubescent without, acute; stamens 5; filaments 2 mm; anthers 1mm. Female flowers fewer, sub sessile, 1 mm across; tepals connate, 2 mm, sparsely scabrous without, acute; ovary ellipsoid, 2 mm. Achenes ellipsoid.

Hills above 1200 m in moist places in central Tamil Nadu (Matthew, 1991). Fl. & Fr.: Round the year.

INDIA: Kerala and Tamil Nadu (Endemic).

Pouzolzia zeylanica (L.) Benn. & R. Br., Pl. Jav. Rar. 66. 1838. *Parietaria zeylanica* L., Sp. Pl. 1052. 1753. *P. indica* L., Mant. Pl. 128. 1767. *Pouzolzia indica* (L.) Goud., Voy. Bonitc, Bot. 503. 1826; FBI 5: 581. 1888; Fischer 3: 1382. 1928.

Perennial rooted, weak stemmed herbs or shrubby, to 50 cm high, very variable in size and habit, stems grooved, hirsute. Leaves 2-5 × 1.5-2.5 cm, very variable, gradually reducing in size, elliptic-ovate, alternate, base rounded, margin entire, apex acute, pilose below, less so above; petiole up to 1.3 cm long, slender, pubescent; stipules up to 4 mm long, lanceolate, strigose without. Flowers white in subsessile clusters. Male flowers: shortly pedicellate; tepals 4, up to 1.5 mm long, ovate-oblong, apiculate, strigose without; stamens 4; pistillode clavate. Female flowers: Perianth up to 1.2 mm long; tepals connate, white strigose without; ovary up to 1 mm long, ovoid, styles up to 5 mm long, pubescent. Achenes 1.3 × 1 mm, broadly ovoid, acute, smooth and shining, white with a red coloured tip.

Common as a weed. Fl. & Fr.: January – April.

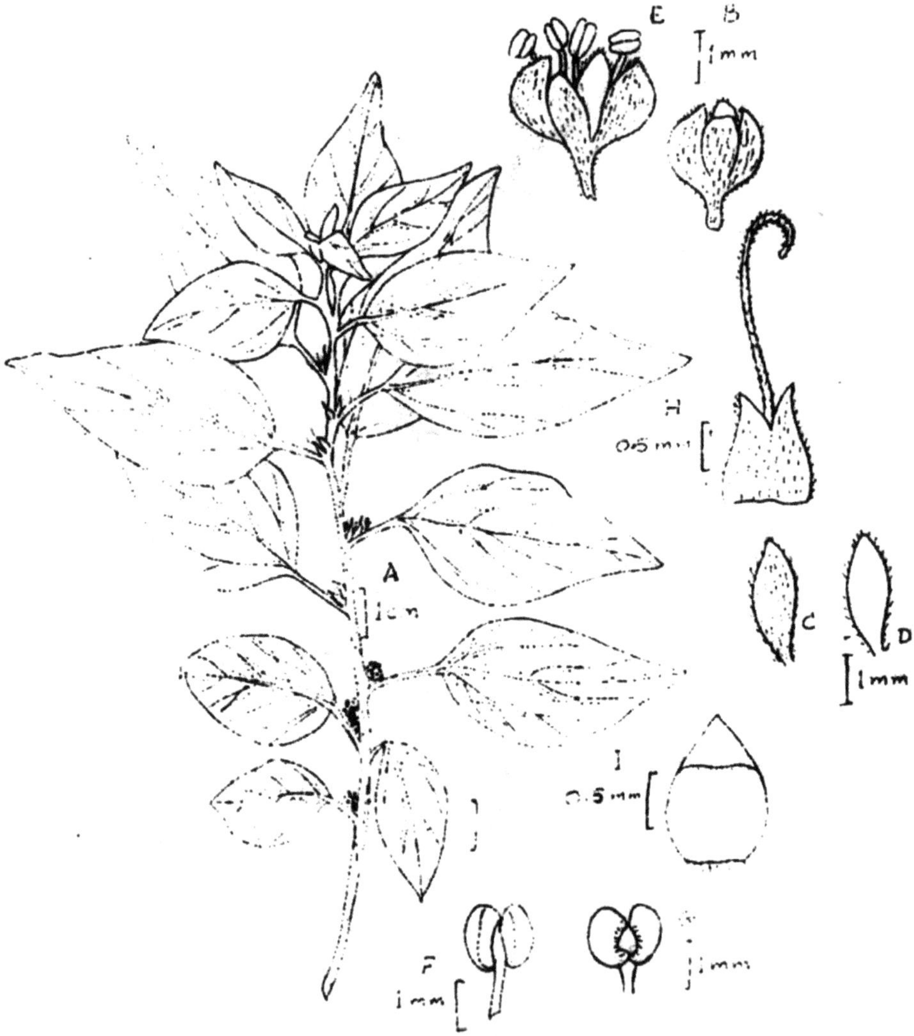

Figure 112. **Pouzolzia zeylanica** (L.) Bennett
A. Twig, B. Flower bud, C and D. Bracteole dorsal and ventral view, E. Male flower, F and G. Anthers, H. Pistil, I. Achene.

Saddikudu madugu, near Nagalapuram (CTR), *PSPB* 9175; Nallamalais (KNL), *JLE* 32623 (MH); Chelama (KNL), *JLE* 18075 (MH); Kallamanaidu peta, Satyavedu (CTR), *MCB* 5627 (MH); Udayagiri (NLR), *BSN* 4709 (VV); Kondapalli (KSN), *PV* 45263 (MH); Nulakamaddi (EG), *GVS* 24532 (MH); Paderu (VSKP), *GVS* 2811 (MH); Kolli hills (NMK), *SKS* 935 (SGH).

INDIA: Throughout India.

WORLD: Sri Lanka, Myanmar, east to China, S Japan, Malesia.

PROCRIS Commerson ex Jussieu

Procris crenata C. B. Rob., Philipp. J. Sci. 5: 507. 1911. *P. wightiana* Wall. ex Wedd., Monog. Utric. 336. 1856. *nom. illegit.*; Fischer 3: 1378. 1928.

Lithophytic undershrubs growing on wet rocks; stems terete, glabrous. Leaves alternate, up to 19 × 3.5 cm, thinly chartaceous, lanceolate, base cuneate and oblique, margin entire or remotely serrate in the upper half, apex acute or acuminate, glabrous, penninerved; petioles up to 7 mm long. Flowers in the axils of leaves as well as in the axils of fallen leaves. Female flowers: on fleshy pedunculate clusters, peduncles up to 3.5 mm long; tepals 5, pale green, cucullate, minute, connate below; ovary ovoid, up to 1 mm long, stigmas sessile. Achenes brown, up to 1.2 mm long, ovoid, somewhat rugose.

Rare in Visakhapatnam district (Subba Rao and Kumari, 2008) and Kolli hills. Fl. & Fr.: October.

Kolli hills, Kulivalavu shola (NMK), *SKS* 437 (SGH).

INDIA: Kerala and Tamil Nadu.

WORLD: Africa, Sri Lanka, SE Asia to Philippines.

ULMACEAE

1. Leaves penninerved; stamens twice the number of periath lobes; plant deciduous:
   2. Fruit drupe .... **Aphananthe**
   2. Fruit samara .... **Holoptelia**
1. Leaves 3-nerved from base; stamens as many as perianth lobes; fruit a drupe; plants evergreen:
   2. Plants dioecious; leaf scabrous above, glaucous beneath.................. **Trema**
   2. Plants with polygamous flowers; leaves glabrous on both surfaces, green beneath .... **Celtis**

APHANANTHE Planchon

Aphananthe cuspidata (Blume) Planch., in DC. Prodr. 17: 209. 1873; Matthew, Fl. Central Tamil Nadu 477. 1991. *Cyclostemon cuspidatus* Blume, Bijdr. 599. 1826.

Evergreen tree, 10-25 m high. Leaves elliptic to broadly lanceolate, 14-20 x 5-9 cm; stipules sheathing, deciduous, leaving circular scars. Flowers in cymes, clustered around young branches, unisexual, sessile; tepals 4-5, stamens 4-5, pistillode discoid, woolly; style arms 2, linear, brown-tomentose. Drupe ripening yellow, pendulous, obovoid, epicarp stony; seed ovoid-globose.

In Kolli hills, 1100 m, in sholas. Fl. & Fr.: February- March.

INDIA: West Bengal, Odisha, Maharashtra, Karnataka, Kerala, Andhra Pradesh and Tamil Nadu.

WORLD: Sri Lanka, SE Asia.

CELTIS Linnaeus

1. Flowers 4-merous; leaves thin, base oblique, margin serrate at the upper half .... **C. tetrandra**
1. Flowers 5-merous; leaves thick, base rounded or acute, margin entire throughout:
    2. Flowers few in racemes; style arms entire, pistillode absent; twigs reddish-brown; leaves dark brown when dry .... **C. timorensis**
    2. Flowers numerous in panicles; style arms 2-fid, pistillode present; twigs grey-brown; leaves green as fresh leaf when dry .... **C. philippensis**

Celtis philippensis Blanco, Fl. Filip. 197. 1837. C. *wightii* Planch., Ann. Sci. Nat. Bot. (Ser.3) 10: 307. 1848; FBI 5: 483; Fischer 3: 1349. 1925. *C. philippensis* Blanco var. *wightii* (Planchon) Soep. in Steenis, Fl Males. 1.8: 62. 1977.

Evergreen tree, up to 12 m high; bark brownish, smooth; branchlets appressed pubescent when young. Leaves thick coriaceous, ovate or elliptic-oblong, glabrous, 7.5-13 × 3.2-5.2 cm, base acute or cuneate, margin entire, apex acuminate, 3-ribbed, lateral veins conspicuous, running throughout the leaf, reticulately veined with transverse parallel nervules; petiole to 2 cm. Flowers yellow, pedicellate in axillary puberulous, 1 cm long, cymose panicles, bisexual flowers above, males below. Drupe ellipsoid, 1 cm across, red, obtuse.

Common, in dry forest of southern Eastern Ghats. Fl.: August-September. Fr.: October-November. Vern.: Tel.: *Koti bira, Chipura, Kaki-mushti*; Tam.: *Elumbiruthi, Pinari, Kodalimuriki, Vellai Thovarai, Kalluviri.*

Tirumala foot path (CTR), *SSR* 13762; Tirumala hills (CTR), *KJ* & *SSR* 12577; Siddulaiah kona (NLR), *SSR* 19570; Nallamalais (KNL), *JLE* 23753 (MH); Dornal (KNL), *JSG* 18695 (DD); Hatakeswaram (KNL), *VBH* 84910 (BSID).

INDIA: Kerala, Karnataka, Tamil Nadu, Andhra Pradesh.

WORLD: Tropical Africa, E and SE Asia, Malesia to NE Australia, Solomon Islands.

Celtis tetrandra Roxb., Fl. Ind. 2: 63. 1832; FBI 5: 482. 1888; Fischer 3: 1348. 1925. *C. serotina* Planch., Ann. Sci. Nat. Bot. (Ser.3) 10: 301. 1848.

Evergreen tree, up to 15 m high; bark greyish-black, smooth; wood greyish-white with brown dots and streaks; young branches pubescent. Leaves coriaceous, obliquely ovate-elliptic, 3.5-10 × 1.4-5 cm, glabrous and green above, greyish puberulous beneath, base obliquely acute, margin serrate above the middle, apex acuminate, 3-ribbed, lateral nerves 2 pairs; petiole to 6 mm. Male flowers 2-5 borne on leafless new shoots, female flowers 1-2 in leaf axils. Drupe ellipsoid, ridged, 0.8 cm across, 1-2 on slender stalks, ripening yellowish orange.

Occasional in moist deciduous forests of Visakhapatnam at 1000-1500. Fl.: Feb. Fr.: April-May. Vern.: Tel.: *Jabjaabal;* Ori.: *Chitimiti;* Tam.: *Kuviya.*

Penchalakona (NLR), *PV* 110142 (BSID); Minumuluru (VSKP), *SSR* 15943; Minumuluru towards Paderu (VSKP), *GVS* 30001 (MH); Mahendragiri (GJM), *SX* & *MB* 2216 (RRL-B). Similipahar (MBJ), *SX* & *MB* 5605 (RRL-B).

INDIA: Lower outer Himalays, Sikkim, Khasia hills, Bihar, Odisha, Deccan Peninsula.

WORLD: Bangladesh, Myanmar, Indo-China, Malesia, Thailand.

Celtis timorensis Span., Linnaea 15: 343. 1841. *C. cinnamomea* Lindli ex Planch, Ann. Sci. Nat. Bot. (Ser.3) 10: 303. 1848; FBI 5: 482. 1888; Fischer 3: 1349. 1925.

Evergreen tree, up to 10 m high; bark pale grey, smooth; wood dull white; young branches pubescent. Leaves membranous, ovate-elliptic, 5.5-11 × 2.2-5.2 cm, glabrous, blackish when dry, base obtuse or rounded, margin entire, apex acute-acuminate, 3-ribbed, basal lateral nerves nearly reaching the apex; petiole to 1.4 cm. Flowers small, whitish, pedicellate on 2 cm long, axillary, dichotomous racemes; tepals 5, ovate-lanceolate, concave, hairy without, ciliate; stamens 5; disc woolly; ovary ovoid, styles 2, simple, villous. Drupe ovoid, globose, 0.6 cm across, orange when ripe.

In dry and semi-evergreen forests. Fl.: Mar.-April, Fr.: May. Ver.: Tel.: *Koti-bira, Chipuru, Kaki-mushti.*

Chintala (PKM), *KJ* 12554, *SSR* 16113; Penchalakona (NLR), *PV* 110142 (BSID); Nagarjuna Konda (GNT), *KT* 9871 (CAL); Galikonda (VSKP), *CBDM* 48761 (MH).

INDIA: Himalayas in Lower hills, Assam, Uttar Pradesh, W.Peninsula.

WORLD: Bangladesh, Sri Lanka, Myanar, Indo-China, Malesia, Thailand.

HOLOPTELEA Planchon ('*Holoptelaea*')

Holoptelia integrifolia (Roxb.) Planch., Ann. Sci. Nat. Bot. (Ser.3) 10: 266. 1848; FBI 5: 481. 1888; Fischer 3: 1348. 1925. *Ulmus integrifolia* Roxb., Pl. Coromandel t. 78. 1798 & Fl. Ind. 2: 68. 1832.

Deciduous tree, up to 16 m high; bark whitish-grey, pustular, smooth, exfoliating in quadrilateral scales; wood light yellowish-grey, interlocked grained, medium and quite even-textured; branchlets terete, glabrous, lenticellate. Leaves chartaceous, elliptic-ovate, 6-14 × 3-8 cm, puberulous along nerves above when young, glabrous later, base rounded, margin entire, apex abruptly acuminate, secondary nerves 5-6 pairs; petiole to 1.3 cm. Flowers yellow in 2 cm long, axillary racemes or in fascicles at the scars of fallen leaves. Samara orbicular, ca. 3 cm in dia., flattened, yellow, wings reticulately-veined.

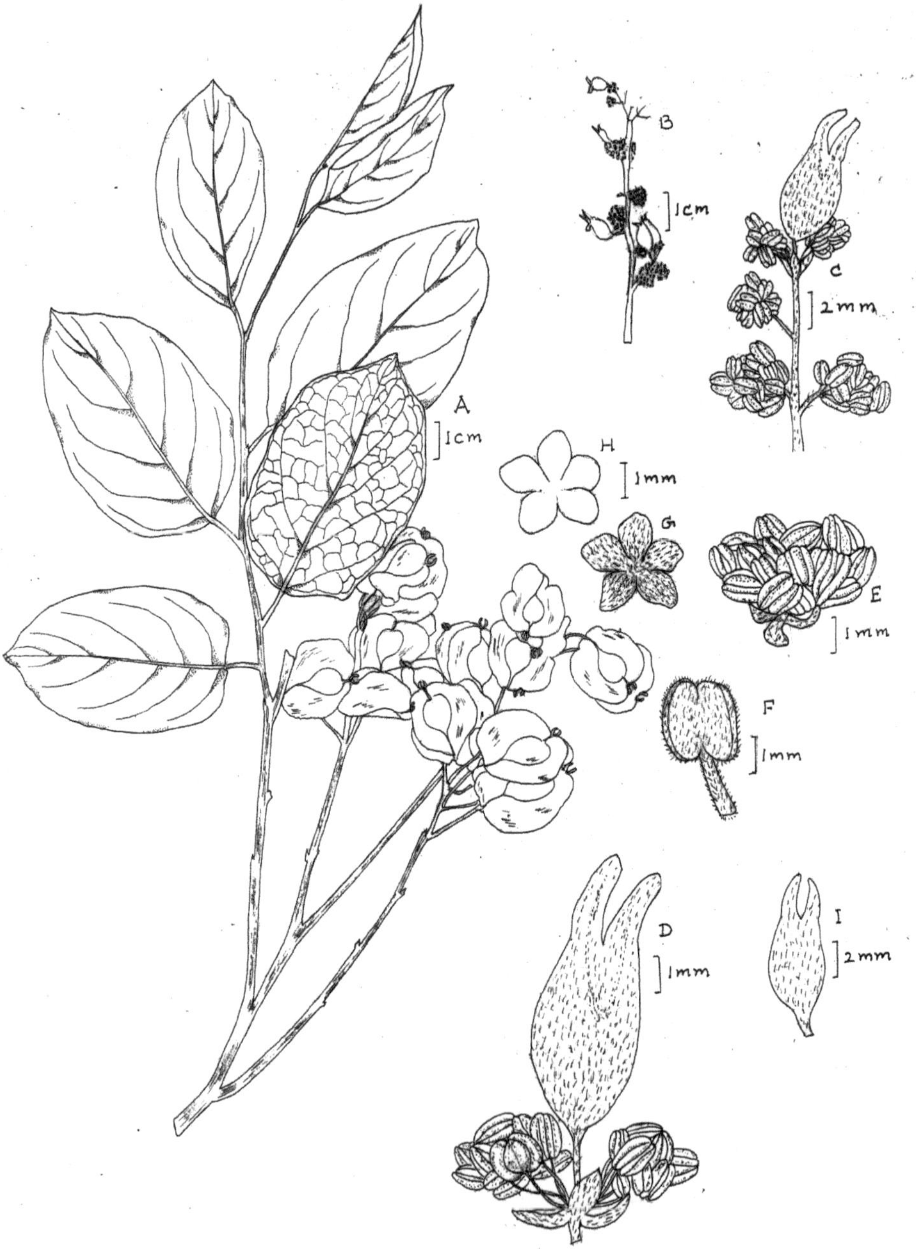

Figure 113. **Holoptelia integrifolia** (Roxb.) Planch.
A. Twig, B. Branchlet (part) with flowers, C and D. Bisexual flowers, E. Male flowers, F. Anthers, G and H. Perianth, I. Ovary.

Common, in deciduous forests and in villages along roadsides to 1200 m. Fl.: January-February, Fr.: April - May. Vern.: Tel.: *Nauli, Pulari, Pedda nevili, Tapasi, Tappachi, Nevili, Namlinara;* Ori.: *Dauranja, Dharanj, Jirappa, Charla;* Tam.: *Aavimaram.*

Bukkapatnam (ATP), *TP* & *NY* 2101; Gundumala RF (ATP), *NY* 1240; Rollapenta (KNL), *SSR* 15511; Srisailam (KNL), *KJ* 12572, *SSR* 15520; Balarajupalli (CDP), *SSR* 16156; Thummala bailu (PKM), *SSR* 16118; Yerramalais (KNL), *JSG* 10946 (DD); Adava (GJM), *SX* & *MB* 3240 (RRL-B).

INDIA: Sub-Himalayan region, Uttar Pradesh, Rajasthan, Bihar, Madhya Pradesh, Peninsula.

WORLD: Sri Lanka, Myanmar, Indo-China.

TREMA Loureiro

Trema orientalis (L.) Blume, Mus. Bot. 2: 62. 1856; FBI 5: 484. 1888; Fischer 3: 1350. 1925. *Celtis orientalis* L., Sp. Pl. 1044. 1753.

Evergreen, dioecious tree, up to 12 m high; trunk straight with spreading branches; bark greyish brown, smooth, with numerous lenticels; wood light reddish-grey; branchlets terete, sparsely pubescent. Leaves chartaceous, ovate or ovate-lanceolate, 7-18 × 2-7 cm, somewhat rough above, whitish beneath, clothed with soft tomentum, base unequal, rounded or sub-cordate, margin serrulate, apex acuminate, secondary nerves 3-5 pairs, basal pairs reaching to the half the length of the leaf; petiole to 1.5 cm, pubescent. Flowers greenish in 2 cm long, axillary, pubescent cymes; the male cymes rather more compact than the female and usually on different trees. Drupe ovoid, 0.3 cm long, black when ripe, glabrous.

Common, in dry and evergreen forests, along perennial streams. Fl. & Fr.: Throughout the year. Vern.: Tel.: *Morali, Boggu chettu, Priyalu, Aava, Konda jonna, Ada nelli;* Tam.: *Ambarathai.* Eng.: Charcoal tree.

Way to Upper Ahobilam (KNL), *TP* & *RVR* 2705, *TP* & *SSR* 15205; Peccheruvu (KNL), *RVR* 2183; Talakona (CTR), *KJ* & *SSR* 12552; Mamandur (CTR), *SKB* & *BR* 34323 (BSID); Pamuleru (EG), *TP* & *KJ* 12523; Tirumala (CTR), *KS* 6930 (MH).

INDIA: More or less throughout India.

WORLD: Tropical Africa, SE Asia through Malesia to Queensland, Melanesia, Micronesia, Polynesia.

CANNABACEAE

CANNABIS Linnaeus

Cannabis sativa L., Sp. Pl. 1027. 1753; FBI 5: 487. 1888; Fischer 3: 1350. 1928.

Large, erect, aromatic, resinous herb or sub-shrub; branchlets angled, sparsely pubescent. Leaves 3.5-6.5 × 0.5-0.8 cm, palmately compound, leaflets sub-sessile, 3-5-fid above, lateral ones smaller, lanceolate, alternate above, opposite below, base cuneate, margin serrate, apex acuminate, chartaceous, thinly scabrid above, appressed pubescent, with sessile glands below. Flowers dioecious, axillary, males

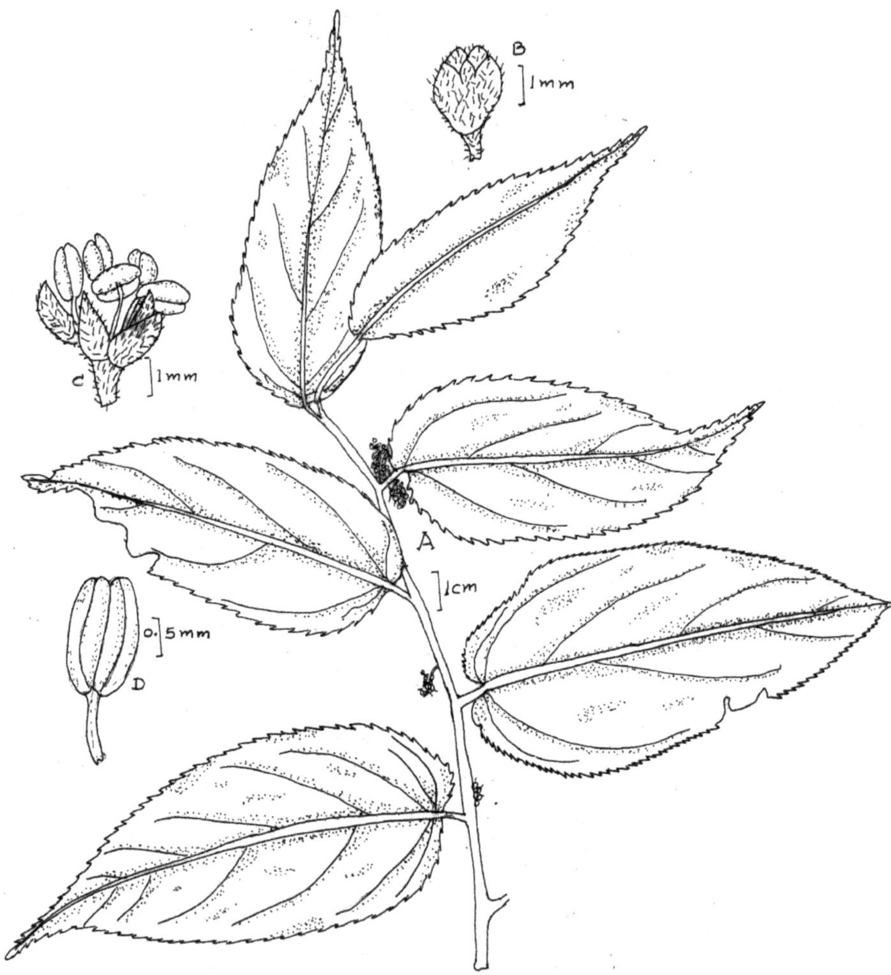

Figure 114. **Trema orientalis** (L.) Blume
♂ A. Twig, B. Flower bud, C. Flower, D and E. Anthers.

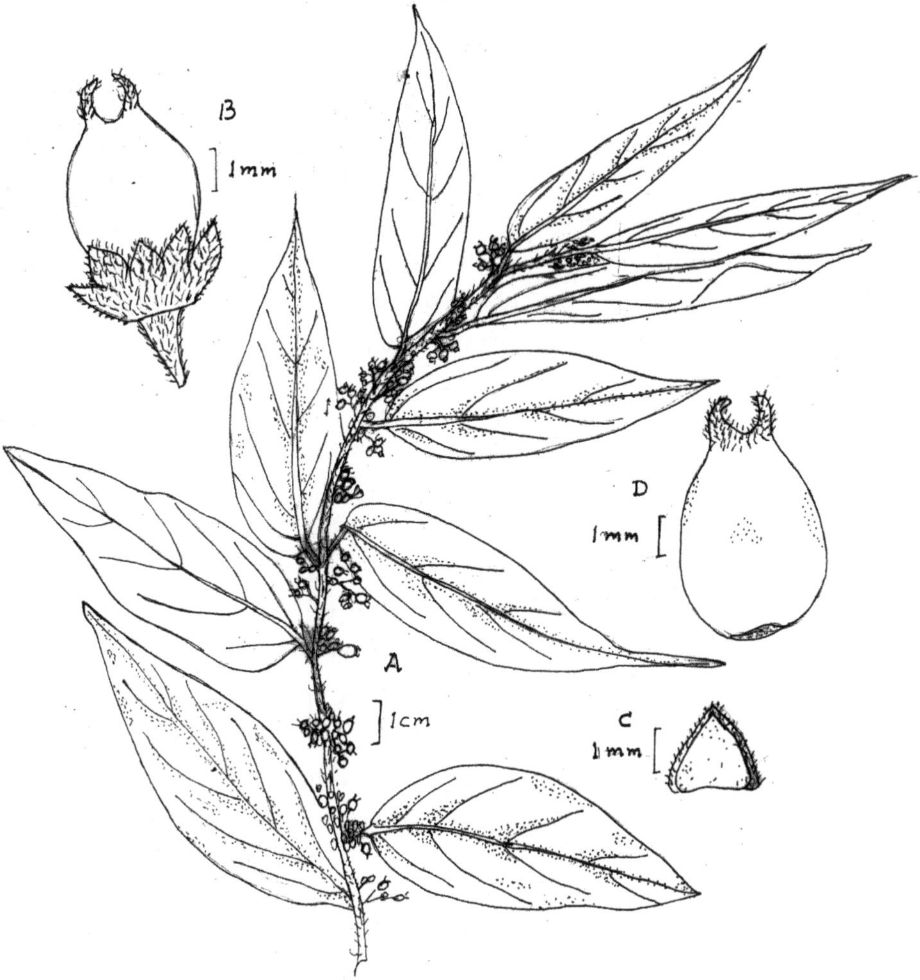

Figure 115. **Trema orientalis** (L.) Blume
♀ A. Twig, B. Flower, C. Perianth, D. Ovary.

cymose or panicled, female in small cymes or heads, with large bracts. Fruit an achene.

Occasionally cultivated clandestinely. Fl. & Fr.: January – April. Vern.: Tel.: *Ganjai-chettu, Ganja-chettu, Bangi-aku;* Tam.: *Kanja.*

INDIA:Throughout the greater part of India, up to 1800 m on the Himalaya.

WORLD: Native of Central Asia, cultivated in many other countries.

MORACEAE

1. Plants thorny:
    2. Stragglers; leaf margin entire; ♂ inflorescence stalked .... **Plecospermum**
    2. Erect shrub; leaf margin serrate; ♂ inflorescence subsessile .... **Streblus (taxoides)**
1. Plants not thorny:
    3. Flowers in a hollow receptacle; flowers hidden .... **Ficus**
    3. Flowers not in a hollow receptacle; flowers patent:
        4. Plants herbaceous; receptacles flattened .... **Dorstenia**
        4. Plants shrubs or trees; receptacles globose or oblong:
            5. Plants with a milky sap:
                6. Stamens monandrous .... **Artocarpus**
                6. Stamens 4:
                    7. Leaf margin serrate .... **Streblus (asper)**
                    7. Leaf margin entire .... **Maclura**
            5. Plants with watery sap .... **Morus**

ARTOCARPUS J.R. Forster & J.G.A. Forster *nom. cons.*

1. Fruit tubercled; leaves glabrous, apex obtuse to shortly acuminate; stipules 5-8 cm long, spathaceous; trees evergreen:
    2. Lateral nerves of leaves 5-7 paris, glabrous beneath; fruits very large, 30-90 cm across .... **A. heterophyllus**
    2. Lateral nerves of leaves 7-11 paris, ferrugineous rusty hairy beneath; fruits 10-30 cm across .... **A. hirsutus**
1. Fruit smooth, 3-8 cm long; leaves pubescent beneath, apex cuspidate; stipules small, not spathaceous; trees deciduous .... **A. lakoocha**

Artocarpus heterophyllus Lam., Encycl. 3: 209. 1789. *A. integrifolius auct. non* L.f. 1781; FBI 5: 541. 1888; Fischer 3: 1369.1928.

Large, evergreen tree; crown dense, round; bark black, mottled green, smooth; wood bright yellow, darkening on exposure; branchlets glabrous. Leaves coriaceous, 10-17 × 5.6-10.1 cm, dark green and shining, base rounded, margin entire, apex obtuse with short acumen, secondary nerves ca. 7 pairs, prominent, 2 cm apart, curved upwards; petiole to 3 cm. Flowers greenish, crowded on globose, oblong or cylindrical, solitary, axillary and cauliflorus receptacles. Fruit oblong-globose, greenish-yellow, large, clothed with fleshy and enlarged perianth, carpels spiny; seeds reniform, greyish, and oily.

Cultivated, in hills of high elevations. It is grown by tribals and farmers for its fruits. It also grows wild. Fl.: November-December. Fr.: July-August. Vern.: Tel.: *Panasa;* Ori.: *Panasa;* Eng.: Jack fruit tree.

Maredumilli - Rampachodavaram (EG), *TP* & *KJ* 12517; Geddapalli (WG), *DN* 85513 (BSID); Anantagiri (VSKP), *GVS* 19528 (MH).

INDIA: Cultivated throughout the warmer parts of India.

WORLD: Widely cultivated throughout the tropics.

Artocarpus hirsutus Lam., Encycl. 3: 210. 1789; FBI 5: 541. 1888; Gamble, 3: 1369 (957). 1928.

Evergreen trees, to 50 m high; bark 10-15 mm thick, surface dull grey-brown, smooth, lenticellate, exudation milky white, sticky, branchlets hirsute. Leaves simple, alternate; stipules to 4 cm long, lateral, densely tawny strigose; petiole 10-30 mm long, stout, hirsute; lamina 13-25 x 7.5-15 cm, broadly ovate, obovate or elliptic, base acute, obtuse or round, apex subacute or very shortly acuminate, margin entire, undulate, coriaceous, glabrous above, hirsute-pubescent beneath; lateral nerves 6-12 pairs, pinnate, prominent; intercostae scalariform, faint. Flowers unisexual, minute, yellowish-green; male in axillary, pendulous, narrowly cylindric spikes up to 15 cm long; tepals 2, united below; stamen 1; anther exserted, ovate, bracteoles chaffy; female flowers in axillary ovoid spikes; perianth tubular, confluent below with the receptacle; ovary superior, straight, ovule pendulous; style exserted; stigma undivided. Fruit a sorosis, 6-7.5 cm across, globose or ovoid, echinate, yellow when ripe, the spines cylindric, straight, hispid, perforate at the apex for filiform style; seeds 16-18 mm long, ovoid, white.

It is a common in ravines of Silambar valley in Alagar hills. Fl. & Fr.: December – June. Vern.: Tam.: *Ayinipala, Kaatupala, Kurangupala, Peipala.*

Alagar hills, Silambar valley (MDU), *SKS* 127 (SGH).

INDIA: Kerala and Tamil Nadu (Endemic).

Artrocarpus lakoocha Roxb., Fl. Ind. 3: 524. 1832; FBI 5: 543. 1888; Fischer 3: 1369. 1928. *A. lacucha* Buch.-Harm. ex D.Don, Prodr. Fl. Nepal 333. 1825.

Deciduous tree, up to 15 m high; crown dense; bark grey, flaky in small, round wood scales; wood red; interlocked grained, very coarse and quite even textured; branchlets tomentose with milky juice. Stipules small, lanceolate, grey pubescent. Leaves coriaceous, oblong or elliptic, 10-30 × 5-15 cm, glabrous and shining above, pubescent beneath, base rounded or truncate, margin entire-serrate, apex cuspidate, main lateral nerves 13 pairs reaching the margin, conspicuous beneath; petiole to 2 cm. Flowers monoecious, globose, axillary; male receptacles orange-yellow, spongy, ovoid to 2.5 cm, sessile, deciduous, sepals 2-3, puberulous, stamen 1; female receptacles irregularly subglobose, pedunculate, smooth. Fruit globose, 6 cm in dia., velvety, yellow; seeds oblong, few.

Occasional, in hills of evergreen forests (1300 m) along the banks of streams and of moist ravines. Commonly cultivated for its edible fruits. Fl.:December,Fr.: May and October-November. Vern.: Tel.: *Nakkarenu, Kamma regu, Lakuchamu*; Ori.: *Jeola, Jautha*. Eng.: Monkey-Jack.

Minumuluru (VSKP), *SSR* 15946; Foot of Galikonda hills (VSKP), *GVS* 19620 (MH); Similipahar (MBJ), *SX* & *MB* 5666 (RRL-B).

INDIA: Kerala, Karnataka, Tamil Nadu and Andhra Pradesh.

WORLD: Myanmar, Sril Lanka, Nepal, Malaysia, Laos, Cambodia, Bhutan and Bangladesh.

DORSTENIA Linnaeus

Dorstenia indica Wall. ex Wight, Icon. Pl. Ind. Orient. t. 1964. 1853; FBI 5: 494. 1888; Fisher 3: 1370. 1928.

Monoecious, slender herb, up to 30 cm high. Leaves alternate, elliptic ovate to oblanceolate, 5-10 × 2.5-3.5 cm, chartaceous, lateral nerves 6-8 pairs, base gradually acute, margin minutely toothed, apex gradually acute; petiole to 4 cm. Inflorescence discoid, 5-angled, 1.5-2.5 cm across, marginal lobes 8-12, to 1 cm long, with several short lobes. Flowers minute, embedded in fleshy receptacle, exposed on a flat receptacle, male more than female; tepals 2, basally connate, 0.5 mm, boat-shaped, puberulous without, acute; stamen(s) 1 or 2, inflexed in bud, filaments 0.6 mm, anthers 0.2 mm. Female: tepals obscurely 2-lobed, connate with the receptacle, ovary stalked, globose or ellipsoid, 1.2 mm across, style lateral, 1.5 mm, unequally 2-fid. Achenes ovoid, minutely verrucose, aggregated into a flat syconium.

Hills, above 1300 m, in deep shade on rotting wood, soil humus, crevices of rocks, *etc.* Fl.: February-April, Fr.: August.

Kolli hills, Kulivalvau shola (NMK), *SKS* 437 (SGH).

INDIA: Maharashtra, Kerala, Karnataka and Tamil Nadu.

WORLD: Sri Lanka,

FICUS Linnaeus

1. Receptacles in dense clusters or fascicles on tubercles or on leafless racemes of panicled branches from the trunk or larger branches:
    2. Receptacles in fascicles on long peduncles:
        3. Evergreen, low spreading trees; leaves shortly pubescent beneath; receptacle purplish, 5-6.2 cm diam .... **F. auriculata**
        3. Deciduous, large trees; leavess glabrous; receptacle yellowish green, pink to red with white streaks and spots....................... **F. variegata**

2. Receptacles dense on leafless branches on short peduncles:
 4. Leaves glabrous on both surfaces, margin entire or subundulate:
  5. Leaves elliptic-oblong, apex acute or obtuse; petiole to 4 cm long; receptacle reddish when ripe, white pilose .... **F. racemosa**
  5. Leaves ovate-rotund, broadly elliptic or ovate-oblong, apex shortly and abruptly acuminate; petiole 5.5-12.5 cm long; receptacles yellowish red, glabrous .... **F. geniculata**
 4. Leaves scabrid-hispid above, pubescent or tomentose beneath, margin toothed or serrate:
  6. Leaves alternate, base semicordate; receptacled branches very long .... **F. semicordata**
  6. Leaves opposite, base rounded or sub-cordate; receptacled branches short .... **F. hispida**

1. Receptacles solitary, paired or 2-6 together in the axils of leaves:
 7. Leaves with very many fine lateral nerves—cultivated .... **F. elastica**
 7 Leaves not as above:
  8. Receptacle distinctly peduncled:
   9. Variable shrub; leaves very variable, linear to ovate to pinnatifid .... **F. heterophylla**
   9. Distinct characteristic trees or climbers; leaves not variable:
    10. Leaves lobed—cultivated .... **F. carica**
    10. Leaves entire or not lobed:
     11. Leaves scabrous above:
      12. Trees .... **F. exasperata**
      12. Climbing shrubs:
       13. Embracing or basketing the trunks of host plants, old walls and buildings; leaves asymmetric, gibbose, lamina variable .... **F. tinctoria**
       13. Not basketing the host; leaves symmetric, not gibbose, lamina ovate or broadly ovate to elliptic or orbicular .... **F. hederacea**
     11. Leaves smooth above:
      14. Nerves not bullate in appearance beneath the leaves; basal bracts persistemt:
       15. Waxy glands two, in the axils of basal lateral nerves beneath of leaves, aerial roots usually

absent; receptacles oblong or obovoid; tepals of male flower united; young foliage greenish to pale beneath .... **F. anamalayana**

15. Waxy gland one, at the base of midrib beneath of leaves; aerial roots present, numerous; receptacles globose to sub-gobose, turbinate; tepals of male flower free; young foliage dark red or pink beneath .... **F. dalhousiae**

14. Lateral nerves prominently raised and tertiary nerves depressed beneath the leaves, and bullate in appearance; basal bracts caducous or absent... **F. nervosa**

8. Receptacles sessile or subsessile:

16. Apices of leaves caudate:

17. Caudate-acumen short, leaf base narrowed...................... **F. talbotii**

17. Caudate-acumen long, leaf base broad:

18. Leaf base cordate.. **F. arnottiana**

18. Leaf base truncate or subcordate:

19. Bark greyish-black; apical caudate-acumen one-sixth as long as blade; receptacles black when ripe .... **F. rumphi**

19. Bark greenish; apical caudate acumen one-third as long as blade; receptacles dark purple when ripe .... **F. religiosa**

16. Apices of leaves obtuse, blunt or acute-shortly bluntly acuminate:

20. Leaves tomentose or pubescent beneath even at maturity:

21. Midrib usually with an elongated gland at base; receptacle scarlet red, puberulous............. **F. benghalensis**

21. Midrib eglandular; receptacles green tomentose.. **F. mollis**

20. Leaves glabrous:

22. Lateral nerves above 20 pairs, inter secondaries as prominent as secondaries:

23. Branches drooping; aerial roots numerous; leaves shining, base indistinctly 3-veined at base .... **F.benjamina**

23. Branches not drooping; aerial roots few; leaves not shining, base distinctly 3-veined at base .... **F. microcarpa**

22. Lateral nerves less than 20 pairs, inter secondaries not prominent as secondaries:

24. Receptacles susbsessile, very shortly peduncled; aerial roots few:

25. Leaves obovate-oblong, or oblanceolate.......... **F. concinna**

25. Leaves elliptic, oblong, oblong-ovate, apex abruptly acuminate with blunt tip............... **F. virens**

24. Raceptacles sessile; aerial roots absent:

26. Receptacle solitary .... **F. drupacea**

26. Receptacles 2 or more:

27. Raceptacles 2-6 together on tubercles from leafless axils... **F. tsjahela**

27. Raceptacles in pairs in leafy axils .... **F. amplissima**

Ficus amplissima J.E. Sm. in Rees, Cyclop. 14: Ficus no. 68. 1810; Sudhakar *et al.*, Figs Eastern Ghats 26. 2017. *F. tsiela* Roxb. ex Buch.-Ham., Trans. Linn. Soc. London. 15: 149. 1827; King in Hook.f., Fl. Brit. India 5: 515.1888; Fischer 3: 1362. 1928.

Spreading tree, up to 15 m high; no aerial roots, trunk cylindric; bark greenish grey, smooth. Leaves coriaceous, ovate-elliptic, 7-12 × 5-6 cm, base rounded, margin entire, apex short bluntly acuminate, lateral nerves 10-13 pairs; petiole to 7 cm. Receptacles monoecious, 1 cm dia., sessile in axillary pairs on leafless branchlets, depressed-globose or pyriform, purple when ripe; basal bracts 3; orifice circular, closed by 3 flat apical scales. Achenes ellipsoid or ovoid, surface verrucate.

Common, in deciduous and evergreen forests up to 1000 m. Fr.: January-March. Vern.: Tel.: *Pedda juvi, Pitta juvvi;* Ori.: *Jori;* Tam.: *Ichi, Kal-ichchi, Sittal.*

Kalasamudram (ATP), *EC* 2947; Gani RF (KNL), *TP* & *RVR* 1798; Bairluri (KNL), *SSR* 15517; Mabbuchintalapalli (CDP), *TP* & *SSR* 33660; Macherla (GNT), *VRK* 3943; Punganur RF (CTR), *GVM* 111984 (BSID); Chitvelu to Rajampeta (NLR), *PV* 109961 (BSID); Dornal (PKM), *JSG* 18688 (MH); Balapalli (CDP), *JLE* 15750 (MH); Malakondapet (PKM), *RKM* 6804 (CAL); Duggeru (SKLM), *GVS* 62519 (MH); Balisoi (GJM), *JSG* 13843 (MH); Harishankar (Bolangir), *SX* & *MB* 5997 (RRL-B).

INDIA: Throughout India except Northeast and West Bengal

WORLD: Maldives, Sri Lanka.

Ficus anamalayana Sudhakar & G.V.S.Murthy, Rheedea 25: 1. 2015; Sudhakar *et al.*, Figs Eastern Ghats 26. 2017. *F. dalhousiae* sensu K.M. Matthew, Mat. Fl. Tamilnadu Carnatic: 344. 1981, non Miq. 1867. F. *albipila* sensu N. Rani, J. Bombay Nat. Hist. Soc. 82: 235. 1985, non (Miq.) King 1888; Chithra in A.N. Henry *et al.*, Fl. Tamil Nadu 2: 251. 1987; K.M. Matthew, Fl. Tamilnadu Carnatic 4: t. 593. 1988, Fl. Central Tamil Nadu 482. 1991, J. Econ. Taxon. Bot., 20: 12. 1996.

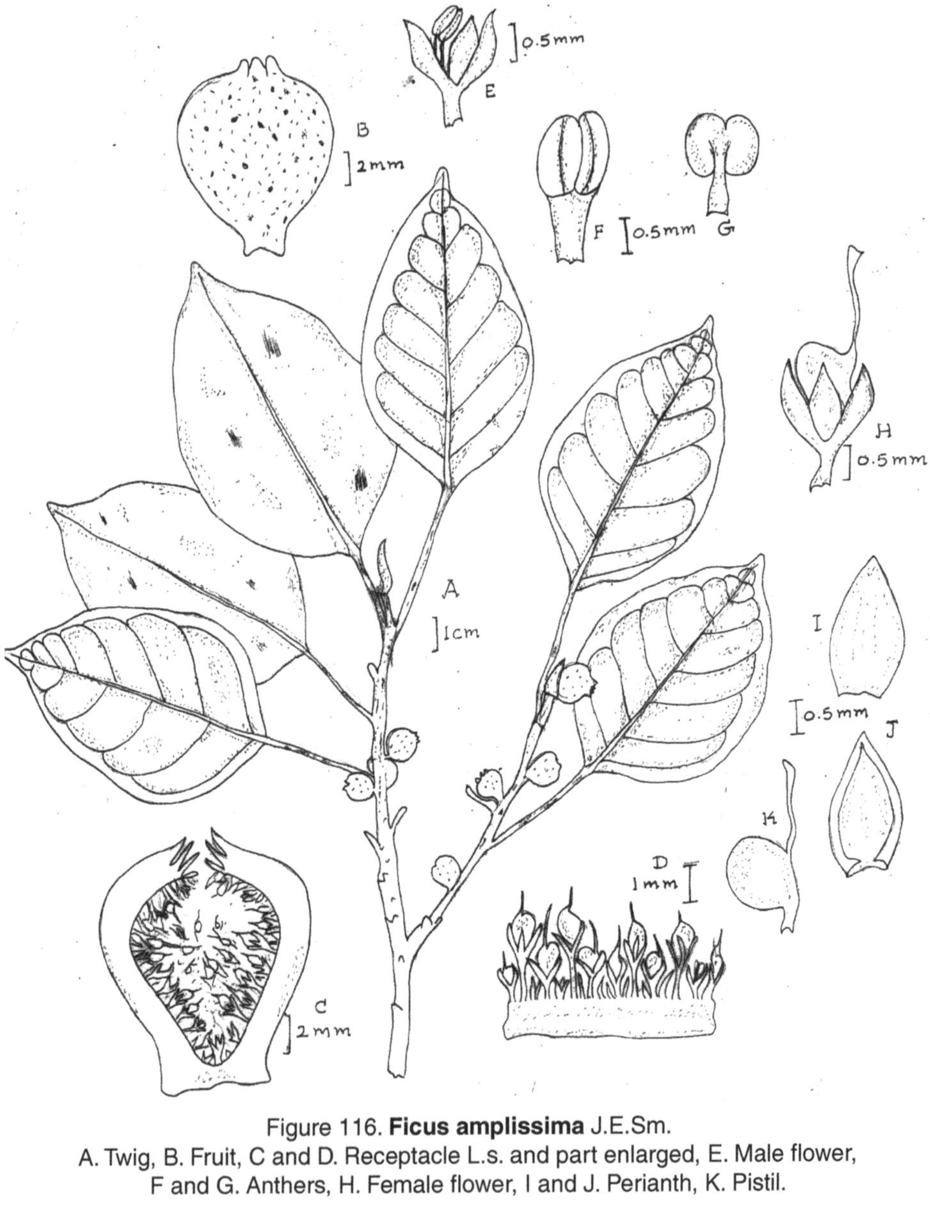

Figure 116. **Ficus amplissima** J.E.Sm.
A. Twig, B. Fruit, C and D. Receptacle L.s. and part enlarged, E. Male flower, F and G. Anthers, H. Female flower, I and J. Perianth, K. Pistil.

Tree, up to 15 m tall, without buttresses; aerial roots usually absent; branchlets spreading as a dense crown, puberulous or pubescent. Leaves simple, alternate, symmetric; lamina ovate or ovate-oblong, 20–32 x 10–18 cm, cordate to rounded at base, entire or faintly undulate along margins, acuminate to subacute at apex with ca 1 cm long acumen, subcoriaceous, sparsely white puberulous on main nerves above, white puberulous beneath; basal nerves 5–9, midrib flat above, prominent beneath; lateral nerves 10–12 pairs, looped at margin, flat above, raised beneath; tertiary nerves scalariform; waxy glands two, at axils of basal lateral nerves, cystoliths only on upper surface (hypergenous); petiole 4–11cm long, stout, white puberulous or pubescent, 3–4 mm thick, epidermis persistent. Stipules terminal, in pairs, lanceolate or ovate – lanceolate, 3–5 cm long, subglabrous, caducous. Receptacles axillary or below leaves, paired, pedunculate, monoecious, oblong to obovoid, 1–1.5 cm in diam., pale green when young, green at maturity, white dotted, white puberulous, slightly ribbed when dry; peduncle 0.5–1.2 cm long, slender, 1–2 mm thick, pale green to green, puberulous; basal bracts 3, basally connate forming small cupule, shallowly lobed at apex, 2–2.5 mm long, ciliolate, puberulous; ostiole prominent, 1–1.5 mm in diam., open; ostiolar bracts 3, ovate, up to 1 mm long; internal ostiolar bracts ovate or deltoid, 1.3–1.5 mm long, glabrous; interfloral bracts ovate – lanceolate, 0.8 – 1.2 x 0.5 – 0.6 mm, white, glabrous. Male flowers: very few, near ostiole, pedicellate, 1.5–1.8 mm long, pedicel up to 0.5 mm long; tepals 3, completely united, rarely free, spathulate, 1–1.3 mm long, reddish or whitish, glabrous. Stamen 1, 0.6–0.8 mm long, inserted; filament c. 0.2 mm long or absent; anther ovoid or oblong, c. 0.6 mm long. Gall flowers pedicellate, 2.5–3.5 mm long, pedicel 1.4–1.6 mm long; tepals 3, free or united, ovate, c. 1.6 mm long, glabrous; ovary obovoid, 0.5–1 mm; style lateral or terminal, 0.3–0.5 mm long, glabrous; stigma simple or capitate. Female flowers sessile, 2.3–3.5 mm long; tepals 3, completely united, spathulate or tubular, 1.5–2 mm long, pale or whitish, glabrous; ovary ovoid or globose, c. 1 mm long; style 1–1.2 mm long, glabrous; stigma bifid, smooth. Achenes ovoid or tetragonous, 1–1.5 mm in diam., surface striate or reticulate, rarely verrucate.

In deciduous and semi-evergreen forests, especially on rocks. Fl. & Fr.: June-December.

Kolli hills, Ariadurai aaru (NK), *N. Venugopal* 15844 (RHT); Varakur, Vatchukkal aaru (NK), *N. Venugopal* 15969 (RHT); On the way to Aruvithurai (NK), *P. Perumal* 22491; Arappuleeswarer (NK), *K.M. Matthew* 62336 (RHT); Periyakalrayans, shola of Pattivalavu Nagaluru (SLM), *K.M. Matthew* & *D.I. Arockiasamy* 5455 (RHT); Shevaroys, Yercaud ghat road SLM), *K.M. Matthew* 23478 (RHT); Pacchaimalais, Kottathumaduvu (TRP), *K.M. Matthew* & *V.A. Amalraj* 3911 (RHT); Kannimar Shola, Foot path from Moopanvayal to Yerumaipetty (TRP), *D.I. Arockiasamy* 7037; Kannimar shola, Erumaippetty – Kaikatty (TRP), *K.M. Matthew* 12486; way to Kaikatty (TRP), *C. Manoharan* 22601 (RHT).

INDIA: Tamil Nadu, Kerala (Endemic).

Ficus arnottiana (Miq.) Miq., Ann. Mus. Bot. Ludg.-Bat. 3: 287. 1867; FBI 5: 513. 1888; Fischer 3: 1363. 1928; Sudhakar *et al.*, Figs Eastern Ghats 30. 2017. *Urostigma arnottianum* Miq. in Hook. London J. Bot. 6: 564. 1847.

Tree, up to 7 m high; glabrous, often epiphytic, also creeping over rocks, without aerial roots; bark pale, smooth; wood white. Leaves subcoriaceous, broadly ovate, 5-11 × 4-7 cm, base rounded or cordate, margin entire or undulate, apex shortly caudate-acuminate, 5-7 ribbed; lateral primary nerves 5-7 pairs, flattened above, raised below; petiole 5-12 cm. Receptacles 6 mm across, sessile in pairs or clusters on tubercles in axils of fallen leaves, depressed-globose, reddish when ripe, smooth; orifice plane; bracts 3, ovate, membranous.

Common, throughout Eastern Ghats in moist deciduous and semi-evergreen forests mainly in rocky places. Fr.: August onwards. Vern.: Tel.: *Kalla-ravi, Banda ravi;* Ori.: *Jhinde, Plokhyo;* Tam.: *Kagoli, Kal-arasu, Kallaralu, Kodiarasu, Kodiyaraku.*

Kalasamudram (ATP), *SSR* 13755; Near Sikharam, Srisailam (KNL), *KJ* & *SSR* 12579; Talakona (CTR), *KJ* & *SSR* 12549; Tirumala (CTR), *KS* 7840 (MH); Bhupatipalem (EG), *TP* & *KJ* 12508; Battigudam (KNL), *NRR* & *TRS* 86587 (MH); Kondapalli forest (KSN), *PV* 5744 (MH); Karaka (VSKP), *CAB* 1681 (MH); Bhurlijharan (KHD), *MB* & *Dhal* 7475 (RRL-B).

INDIA: Throughout India except Northeast.

WORLD: Nepal, Sri Lanka.

Ficus auriculata Lour., Fl. Cochinch. 666. 1790; Sudhakar *et al.*, Figs Eastern Ghats 33. 2017. *F. roxburghii* Wall. ex Steud., Nom. Bot. ed. 2, 1: 637. 1840; FBI 5: 534. 1888. *F. macrophylla* Roxb. & Buch.-Ham. ex J.E.Smith, Rees Cyclop 14: Ficus no. 32. 1810 (non Desf. ex Fees. 1807). *F. oligodon* Miq., Ann. Mus. Bot. Lugd. Bat. 3: 234. 297. 1867; Jayamma in Pullaiah & Moulali, Fl. Andhra Pradesh 2: 901. 1997.

Evergreen, low spreading tree; bark greyish brown, smooth; branchlets hollow; young shoots pubescent. Leaves broadly elliptic-ovate, large, 20-40 × 15-28 cm, glabrescent above, shortly-softly pubescent beneath, base cordate, margin repand-toothed or coarsely serrate, apex shortly acuminate, 5-7-ribbed; secondary nerves 5-7 pairs, tertiaries scalariform, prominent; petiole 6-10 cm. Receptacles purplish, pyriform, 5-6.2 cm dia., clustered, pendulous, cauline from tubercles on short, lateral leafless branches and trunk; peduncle 4-7 cm long; pubescent, bracteate at their apex.

Rare, along streams in semi-evergreen forest at 800-1500m. Fl. & Fr.: Throughout the year. Vern.: Tel.: *Raach bodda. Raj dummiri;* Ori.: *Raj dimri.*

Gudem (VSKP), *SSR* 15966; Sapparla (VSKP), *SSR* 19640, *GVS* 30092 (MH); Chintapalli (VSKP), *KJ* & *KSM* 12598; Galikonda (VSKP), *SSR* 15916; Near Dabuguda, Lakshmipuram (VSKP), *NRR* & *DN* 84222 (BSID); Sunkarimetta (VSKP), *GVS* 22644 (MH); Karlapat(KHD), *MB* & *Dhal* 7574 (RRL-B); Rohila (DKL), *MB* & *Dhal* 8347 (RRL-B).

INDIA: Throughout India.

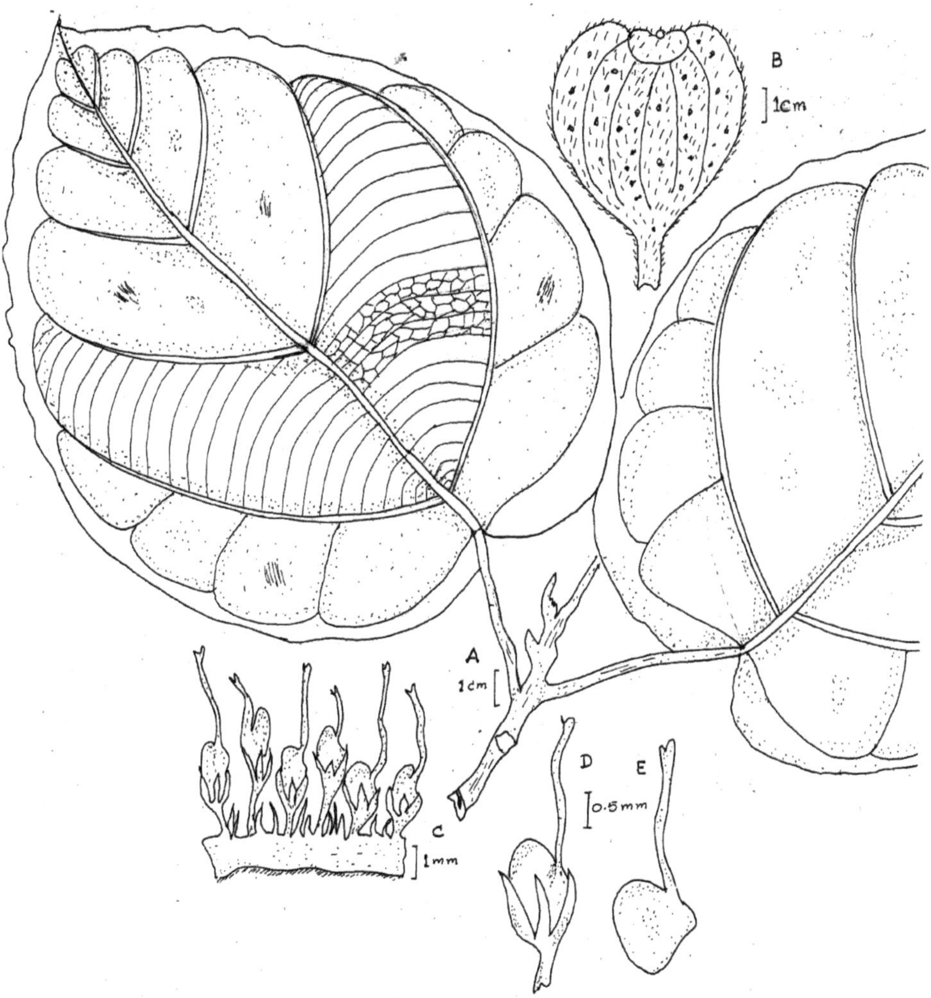

Figure 117. **Ficus auriculata** Lour. (Sym.: *Ficus oligodon* Miq.)
A. Twig, B. Fruit, C. Receptacle L.s. part enlarged, D. Female flower, E. Pistil.

WORLD: Cambodia, China, Japan, Laos, Malesia, Myanmar, Nepal, Pakistan, Thailand and Vietnam.

Ficus benghalensis L., Sp. Pl. 1059. 1753; FBI 5: 499. 1888; Fischers 3: 1361.1928; Sudhakar *et al.*, Figs Eastern Ghats 33. 2017. *F. indica* L., Sp. Pl. 1060. 1753. *Urostigma benghalensis* (L.) Gasp. Nov. Gen. Fic. 7. 1844.

Evergreen tree, up to 25 m high; sending down many aerial roots from the spreading branches and form support, thus extending the growth of the tree indefinitely; bark greyish white, smooth; wood grey or greyish white, more or less irregularly grained, rather coarse and uneven textured; branchlets softly pubescent. Leaves thick coriaceous, ovate or ovate-oblong, 8-16 × 6-11 cm, glabrescent above, puberulous beneath, base subcordate, rounded, or rarely acute, margin entire, apex obtuse, 5-7-ribbed; 4-7 pairs of lateral nerves above the basal nerves; petiole to 6 cm. Receptacle monoecious, globose, 1.7 cm dia, red dotted, sessile in axillary pairs, puberulous, basal bracts 3, obtuse; orifice slightly raised, circular.

Very common, on roadsides, wild in village surroundings and old walls up to 1000 m. It is hardy and drought resistant and withstands mild frost. It is sacred tree of Hindus and most popular shade providing tree. Syconia: September-March. Vern.: Tel.: *Marri, Pedda marri, Rohi, Vatairilsja*; Ori.: *Baro, Boru*; Tam.: *Al, Ala, Alamaram, Arla, Pudavam*. Eng.: Banyan, Indian fig.

Gootybayalu (ATP), *TP* 1243; Bhumaiahgaripalle (CDP), *KRKS* & *KRS* 376021; Tirumala (CTR), *KJ* & *SSR* 12558; V.P. South (GNT), *VRK* 3941; Mahanandi (KNL), *JLE* 42344 (MH); Horsley hiil (CTR), *JHFB* 111931 (BSID); Rapur (NLR), *PV* 113655 (BSID); Bhojeripakala, Maredumilli (EG), *MM* 101313 (BSID); Simhachalam hill (VSKP), *GVS* 19365 (MH); Nunna (KSN), *PVK* 5903 (MH).

INDIA: Throughout India.

WORLD: Afghanistan, Bangladesh, Malesia, Nepal, Pakistan, Iran and Sri Lanka. Introduced and widely cultivated in tropics and elsewhere.

F. benghalensis L. var. **krishnae** (C.DC.) Corner, Gard. Bull. Singapore 21(1): 14. 1965; Chaudhary *et al.*, Taiwania 57: 210. 2012. *F. krishnae* C.DC., Bull. Herb. Boiss., ser. 2, 2: 760. 1902; Sudhakar *et al.*, Figs Eastern Ghats 118. 2017.

Cultivated in gardens and considered as Horticultural variety of *F. benghalensis.*

Ficus benjamina L., Mant. Pl. 129. 1767; King, Ann. Roy. Bot. Gard. (Calcutta) 1. t. 52A. 1887 & in FBI 5: 508. 1888; Fischer 3: 1367. 1928; Sudhakar *et al.*, Figs Eastern Ghats 40. 2017.

An evergreen tree, up to 20 m high; crown spreading, aerial roots numerous, trunk cylindric; bark whitish-grey, smooth; branches drooping, branchlets glabrous. Leaves subcoriaceous, elliptic-obovate, 5.5-8.5 × 2.8-4.8 cm, glabrous and shining, drying grey-green to olivaceous brownish, base rounded or acute, margin entire, apex abruptly acuminate, 3-nerved, lateral nerves 6-14 pairs, secondary nerves almost as prominent, raised on both sides, connected by an intramarginal nerves from base; petiole to 2.2 cm. Receptacles monoecious, obovoid, ellipsoid or

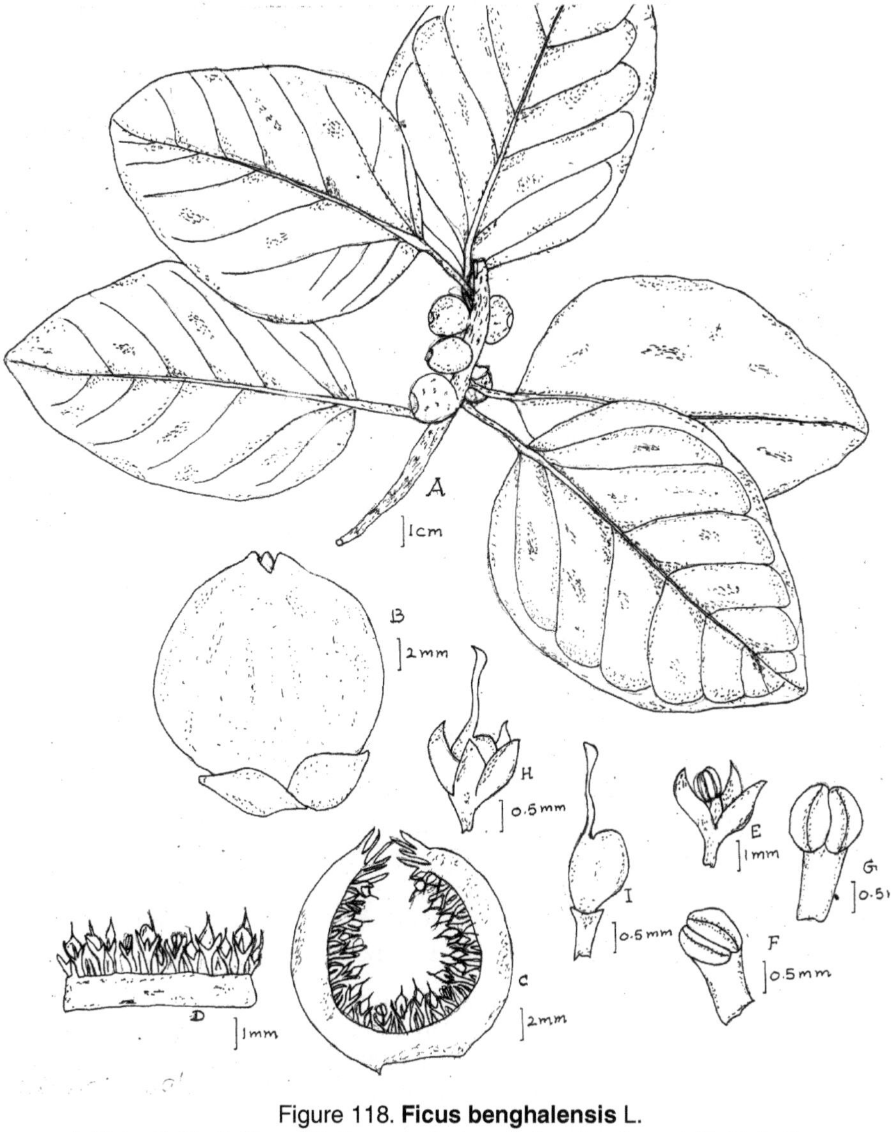

Figure 118. **Ficus benghalensis** L.
A. Twig, B. Fruit, C and D. Receptacle L.s. and part enlarged, E. Male flower, F and G. Anthers, H. Female flower, I. Pistil.

subglobose, sessile or stalked, pyriform, axillary, paired, ripening yellow to orange, basal bracts 2-3, crescentic; orifice slightly raised, circular, closed by 3 small flat apical brackts.

Often planted for shade in avenues and as an ornamental tree, rarely near water canals or moist localities up to 1200 m. Syconia: March- May. Vern.: Tel.: *Konda-juvvi., Putta-juvvi, Konda golugu;*Tam.: *Malai-ichchi, Ponichchi.*

Talakona (CTR), *KJ* & *SSR* 12551; Srisailam (KNL), *KJ* 12573; Anantagiri (VSKP), *KJ* & *SSR* 15933; Kilagada (VSKP), *GVS* 22584; Maredumilli (EG), *GVS* 27297 (MH); Sunki-Ampabali (KPT), *SX* & *MB* 6667 (RRL-B); Koinpur- Mahendragiri (GJM), *SX* & *MB* 3901 (RRL-B).

INDIA: Throughout India.

WORLD: Bangladesh, Bhutan, China, Malesia to the Solomon Islands, Myanmar, Nepal,New Guinea, North Australia, Pakistan, Philippines, Sri Lanka, Thailand and Vietnam.

Ficus carica L., Sp. Pl. 1753; Bailey, Man. Cult. Pl. 339. 1949; Sudhakar *et al.*, Figs Eastern Ghats 108. 2017.

Much branched, small deciduous tree, up to 6 m high, branching from the base, aerial roots absent. Leaves thick, broad-ovate nearly orbicular, 3-5-lobed, 10-30 × 8-30 cm, rough above, pubescent beneath, base cordate, margin entire, apex acuminate, 3-5-nerved; secondary nerves 6-8 pairs above basal pairs; petiole to 6 cm. Receptacles solitary or paired, axillary and subterminal, pear-shaped, variable in size, yellowish- purple when ripe.

Introduced and cultivated; commercially grown for its fruits. Syconia: January-June. Vern.: Tel.: *Anjur, Anjuram;*Tam.: *Sima-atti, Ten-atti.*

INDIA: Cultivated throughout India.

WORLD: Asia, Australia, East Mediterranean, North Africa, South Europe, USA.

Ficus concinna (Miq.) Miq., Ann. Mus. Bot. Lugduno-Batavi 3: 286. 1867; Sudhakar *et al.*, Figs Eastern Ghats 43. 2017. *Urostigma concinna* Miq., London J. Bot. 6: 570. 1847. *Ficus affinis* Wall. ex Kurz. J. Asiat. Soc. Bengal, Pt. 2, Nat. Hist. 42(2): 105. 1873. *F.concinna* Miq. var. *subsessilis* Corner, Gard. Bull. Singapore 17: 376. 1959; C.S. Reddy *et al.*, Suppl. Fl. Andhra Pradesh 108. 2008.

Trees, up to 20 m tall, monoecious, aerial roots few; young parts glabrous or pubescent. Leaves obovate-oblong or oblanceolate, 5-12 × 3-5 cm, base acute or cuneate, margin entire, apex short acuminate to obtuse, subcoriaceous, 3-nerved at base; petioles 2-2.5 cm long, glabrous; stipules ovate-lanceolate. Receptacles axillary, paired, from the scars of fallen leaves, subsessile or with a peduncle up to 4 mm, globose or pyriform, 10-212 mm across, dark bluish purple, smooth. Male flowers sessile; tepals 2 in male flowers, 4 in female flowers; female and gall flowers sessile or shortly pedicellate. Achenes ovoid or spherical.

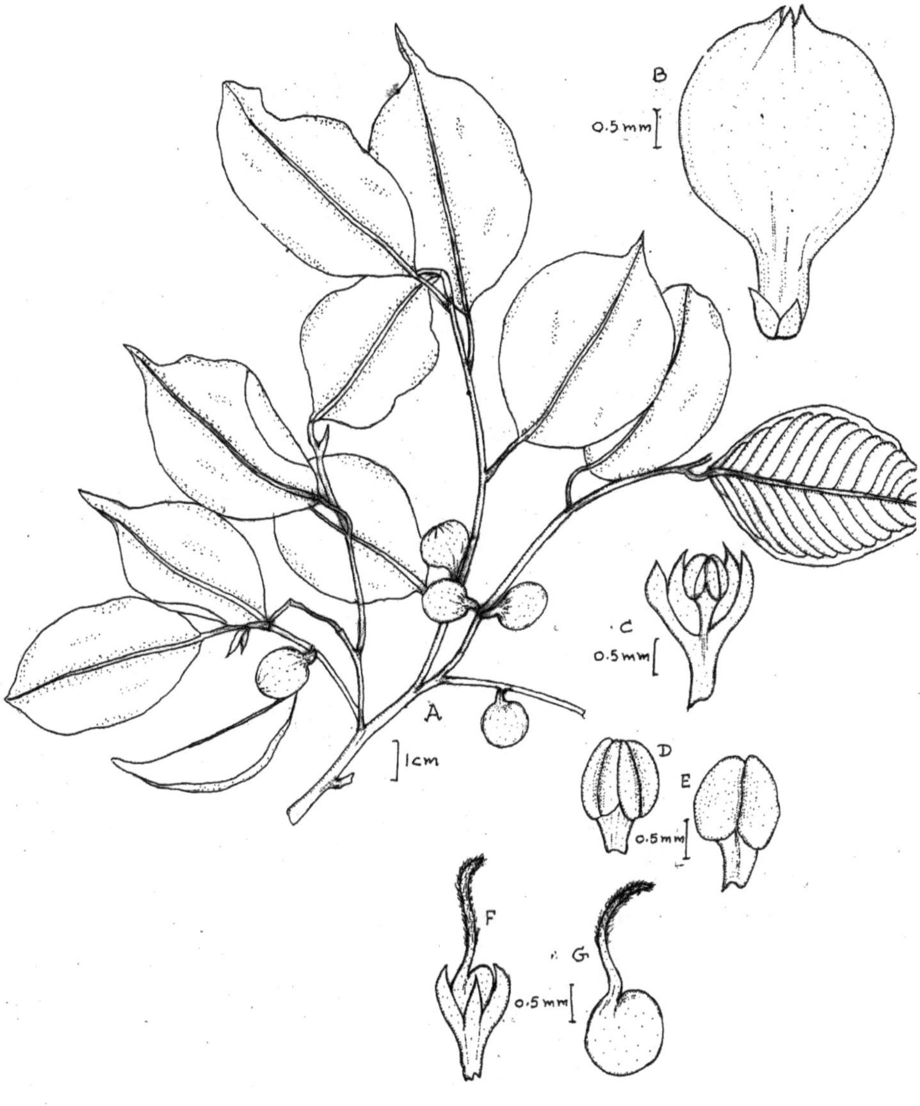

Figure 119. **Ficus benjamina** L.
A. Twig, B. Fruit, C. Male flower, D and E. Anthers, F. Female flower, G. Pistil.

Rare in moist deciduous forests and sholas of Northern Eastern Ghats. Fl. & Fr.: April – June.

INDIA: Andhra Pradesh, Odisha, Bihar, Chhattisgarh, Madhya Pradesh, Jharkhand and Northeast India.

WORLD: Bangladesh, Bhutan, China, Malesia, Myanmar, Thailand and Vietnam.

Ficus dalhousiae (Miq.) Miq., Ann. Mus. Bot. Lugd.-Bat. 3: 285. 1867; FBI 5: 499. 1888; Fischer 3: 1364. 1928; Sudhakar *et al.*, Figs Eastern Ghats 45. 2017. *Urostigma dalhousiae* Miq., Hook., London J. Bot. 6: 571. 1847.

Tree to 15 m high, spreading; young branches softly pubescent. Leaves coriaceous, ovate-elliptic, broadly ovate or ovate-oblong, 14.5-26 × 7.5-15 cm, glabrous above, white pubescent beneath, base subcordate, margin entire, apex obtuse to shortly acuminate, lateral nerves 12-14 pairs, inter-secondary nerves absent, petiole 3-6 cm long. Receptacles yellowish, 0.8 cm across, pubescent, obovoid, pedunculate in pairs.

Rare, in rocky ravines of deciduous forest of Chittoor and Kadapa districts in Andhra Pradesh and Namakkal and Salem districts in Tamil Nadu at 700-1000 m. Endemic to Peninsular India. Fl.: & Fr.: March -September. Vern.: Tam.: *Pei-aal, Kal-al, Kalloali.*

Tirumala-Seshatheertham (CTR), *DRC* 1197 (SVU).

INDIA: South India, Endemic

Ficus drupacea Thunb., Diss. Ficus 6, 11. 1786; Sudhakar *et al.*, Figs Eastern Ghats 48. 2017. var. **pubescens** (Roth) Corner, Gard. Bull. Singapore 17: 381. 1960; Chaudhary *et al.*, Taiwania 57: 211. 2012. *F. mysorensis* Heyne ex Roth var. *pubescens* Roth, Roem & Schult. Syst. Veg. 1: 508. 1817; King, Ann. Roy. Bot. Gard. (Calcutta) 1: t. 14. 1887 & in FBI 5: 500. 1888; Fischer 3: 1361. 1928.

Tree, up to 25 m high; branchlets fulvous-tomentose. Leaves alternate, broadly elliptic-ovate, 10-22 × 6-10 cm, subcoriaceous, 3-nerved from base, sometimes with an additional pair, lateral nerves ca. 14 pairs, flattened above, raised below, intercostals transverse, ± zigzag, prominent on either side, glabrous above and below, tender leaves puberulous below, base broadly acute to cordate, margin entire, apex abruptly acuminate, cusp 1-1.5 cm; petiole to 3 cm, glandular at apex below; stipules broadly lanceolate, 2.5 cm, tomentose. Fig monoecious, axillary, solitary, sessile, ellipsoid-globose, 1.5-2 cm across; fig-wall thick, ripening orange, tomentose without; basal bracts orbicular, concave, 6 mm, brown-pilose without, obtuse; orifice umbonate, closed by 3-4, apical bracts; internal bristles a few, linear-lanceolate, to 4 mm. Tepals 3-4, free, ovate-concave, 1 mm, brown, acute. Male: disperse, pedicellate; stalk to 4 mm, stamen 1, exserted, filament 0.5 mm, anther oblong, parallel, 1 mm. Female: sessile, ovary obovoid, 0.7 mm, brown; style filiform, 2 mm. Achenes smooth.

Hills, above 800 m. Occasional. Syconia: Through out the year. Vern.: Tel.: *Kondamarri chettu*; Tam,: *Kallala maram, Kozhi alamaram, Sonnai aalam.*

Parlakhemudi (GJM), *Corey & Ramamurthy* 20168 (BSIS).

INDIA: Tamil Nadu, Andhra Pradesh, Odisha, Punjab, Rajasthan and Northeast.

WORLD: Bangladesh, Laos, Malesia, Myanmar and Sri Lanka.

Ficus elastica Roxb. ex Hornem., Hort. Bot. Hafn. Suppl. 7: 1819; FBI 5: 508. 1888; Sudhakar *et al.*, Figs Eastern Ghats 115. 2017.

Small to large, evergreen tree with numerous aerial roots; trunk much buttressed and fluted. Leaves thick, coriaceous, oblong or elliptic, 15-25 cm long, glossy green, base rounded or somewhat narrowed, margin entire, apex with sharp point, lateral nerves numerous, fine parallel, petiole to 4 cm; stipules pink, about half as long as leaf lamina. Receptacles yellow, paired, sessile on old leaf axils.

Grown as an ornamental tree in gardens for its glossy ornamental leaves or for shade (Thammanna *et al.*, 1996). Vern.: Tel.: *Rubber chettu, Seema marrichettu;* Tam.: *Seemaiaal, Shimai-alai-maram.* Eng.: Indian Rubber tree.

INDIA: Cultivated throughout India.

WORLD: Bhutan, Indonesia, Malesia, Myanmar, Nepal and North America.

Ficus exasperata Vahl, Enum. Pl. 2: 197 & 402. 1805; Sudhakar *et al.*, Figs Eastern Ghats 52. 2017. *F. asperrima* Roxb., Fl. Ind. 3: 554. 1832; FBI 5: 522. 1888; Fischer 3: 1366. 1928.

Deciduous tree, 8-10 m high; without aerial roots, bark whitish-grey, smooth; wood whitish; branchlets scabrous. Leaves subcoriaceous, oblong-elliptic to obovate, 10.2-14.5 × 5.2-6.8 cm, scabrid with stout stiff white hairs on both surfaces, raised dots conspicuous on upper surface, more or less densely tomentose beneath, drying grey-green, base often inequilateral, acute, margin serrate, toothed, apex blunt to shortly acuminate, 3-ribbed, elongate; lateral nerves 3-5 pairs above basal ones, prominent on the lower surface; petiole to 1.5 cm. Receptacles dioecious, globose, 1.3-1.5 × 0.9-1.3 cm, solitary, axillary, pedunculate, hispid without, yellow when ripe; basal bracts 0, orifice circular, raised.

In mixed dry deciduous forests at 800-1200 m. Syconia: Mar.-May. Vern.: Tel.: *Karaka boddu, Siribodda, Karasaana;* Ori.: *Koritasano;* Tam.: *Irambarattam, Irumbaruttan, Maram-thinni-atthi, Paeperunkai.*

Pamuleru (EG), *TP & KJ* 12521; Borra caves (VSKP), *KJ* 12587; Kusimi RF (SKLM), *SSR* 15904; Seethampet (SKLM), *GVS* 62444 (MH); Atmakur (NLR), *KCJ* 383 (MH); Maredumilli (EG), *GVS* 27264 (MH); Rampa hills (EG), *JSG* 16011 (MH); Dharakonda (VSKP), *GVS* 47410 (MH); Kapilas (DKL), *SX* 2074 (RRL-B); Koinpur (GJM), *SX* 1247 (RRL-B); Mahendragiri (GJM), *SX & MB* 2247 (RRL-B), *JSG* 13933, 14114 (MH).

INDIA: Tamil Nadu, Andhra Pradesh, Odisha.

WORLD: Arabia, East Africa, Sri Lanka.

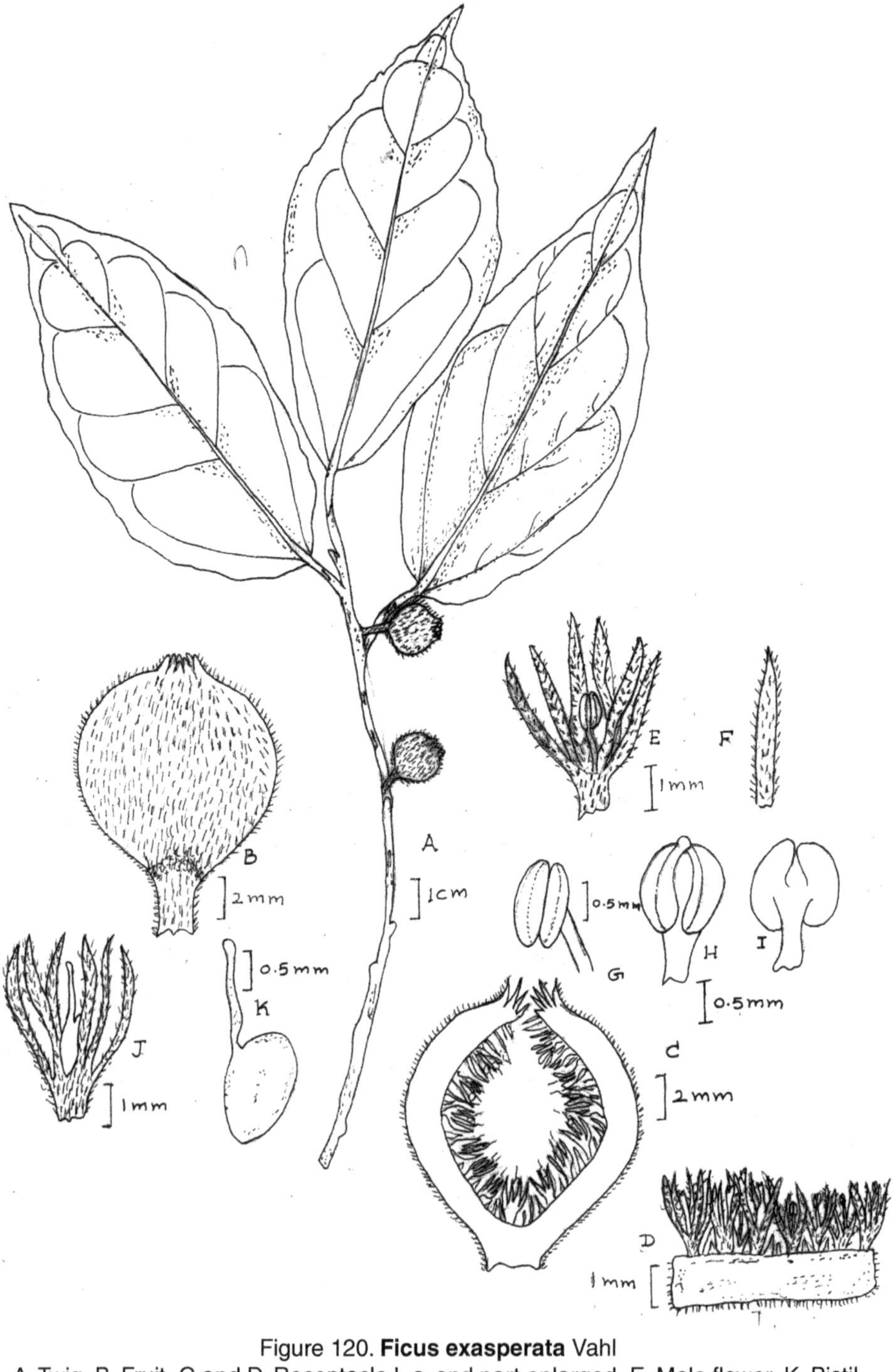

Figure 120. **Ficus exasperata** Vahl
A. Twig, B. Fruit, C and D. Receptacle L.s. and part enlarged, E. Male flower, K. Pistil.

Ficus geniculata Kurz, J. Asiat. Soc. Bengal, pt. 2, Nat. Hist. 42(2): 105. 1873 & For. Fl. Brit. Burma 2: 447. 1877; FBI 5: 516. 1888; Sudhakar *et al.*, Figs Eastern Ghats 45. 2017.

Large tree, up to 15 m high, deciduous to semideciduous, epiphytic in early stage, aerial roots absent; branchlets stout, glabrous, marked with scars of fallen leaves, lenticellate, young branchlets silky, white pubescent; bark brownish-grey. Leaves ovate-rotund, broadly elliptic or ovate-oblong, 9-20 × 4.5-10 cm long, base broadly cuneate to rounded, narrowed, margin entire or subundulate, apex shortly and abruptly acuminate; acumen 1-1.7 cm long, coriaceous, glabrous; basal nerves 3, 2 accessary nerves inconspicuous; lateral nerves 6-12 pairs, nearly horizontal, looped at margin; petiole 5.5-12.5 cm long, glabrous, slightly channelled; stipules terminal, broadly ovate or ovate-lanceolate, 0.6-1.2 cm, pubescent, silky when young, caducous. Receptacles clustered on raised tubercles or woody burs on old twigs and in axils of leaf scars on branchlets, monoecious, sessile or subsessile, depressed of subglobose, 5-8 mm in diam., green in early stage, yellowish red with dots when ripe, glabrous; peduncle 0.5-1 mm long, pubescent; basal bracts 3, free, broadly ovate, obtuse, 2-3 mm long, bifid or lobed, pubescent, persistent, ostiole slightly raised, 2.4-2.6 mm in diam., opened; ostiolar bracts 3, ovate-obtuse, 0.6-1 mm long, pubescent. Internal bristles ovate-rhomboid, 1-1.2 mm long; interfloral bracts lanceolate, 0.7-0.8 mm long. Male flowers: near the ostiole, sessile, 0.8-1 mm long; stamen 1, 0.6-0.7 mm long, anther oblong, filament 0.3 mm long. Female flowers: sessile, 1.5-2 mm long; tepals 3, united, tubular or cup-shaped, 0.5-0.6 mm long; ovary ovoid, 0.6-0.7 mm, style 1-1.2 mm, stigma simple, sometimes lobed. Gall flowers sessile or shortly pedicellate, 1.3-1.6 mm long, pedicel 0.5-0.6 mm long, tepals 3, free, lanceolate, 0.8-1 mm long, style short, 0.4-0.7 mm long, stigma simple. Achenes ovoid, 1-1.2 mm long.

In evergreen, mixed deciduous and deciduous forests in Northern Eastern Ghats. Fl. & Fr.: March – November.

Maredumilli, towards Bodaluru (EG), *J.V.Sudhakar* 126724, 126725 (MH); way to Maredumilli to Kundada, Near Nandanavanam (EG), *J.V.Sudhakar* 126735 (MH); Chintapalli to Gudem way, Chappagadda (VSKP), *J.V.Sudhakar* 126760 (MH).

INDIA: Andhra Pradesh, Odisha, Bihar, Jharkhand, West Bengal and Northeast.

WORLD: Bangladesh, China, Indo-china, Myanmar, Nepal, Thailand.

Ficus hederacea Roxb., Fl. Ind. 3: 103. 538. 1832; Sudhakar *et al.*, Figs Eastern Ghats 57. 2017. *F. scandens* Roxb., Fl. Ind. 3: 65. 1832, non Lam.; FBI 5: 526. 1888. *F. fruticosa* Roxb., Fl. Ind. 3: 533. 1832.

Climbing, scandent shrub, up to 8 m long, aerial roots present. Leaves ovate to elliptic or orbicular, 5.5 – 10 × 2.5-5 cm, coriaceous, subscabrid above, glabrous or slightly hairy on nerves beneath; petiole 0.7-1.5 cm long. Figs solitary or in pairs, axillary, pedunculate, green when young, olive yellow to orange to scarlet red when mature, hispid, hispidulous or pubescent, peduncle 0.7-1.2 cm long. Achenes ovoid ovoid or ellipsoid, 1.5-2 mm in diameter.

Occasional in cool valleys, creeping on rocks, hill sides and tree trunks in Eastern Ghats of Odisha. Syconia: January – July.

Similipahar (MBJ), *SX, MB* & *Prabhakar Rao* 4562 (RRL-B).

INDIA: Odisha, Bihar, Himachal Pradesh, Jharkhand, West Bengal, Madhya Pradesh, Northeast, Uttarakhand.

WORLD: Bangladesh, Bhutan, China, Laos, Myanmar, Nepal, Thailand and Tonkin.

Ficus heterophylla L.f., Suppl. Pl. 442. 1781; FBI 5: 518. 1888; Fischer 3: 1366. 1928; Corner in Dassan. & Fosberg, Rev. Handb. Fl. Ceylon 3: 272. 1981; Sudhakar *et al.*, Figs Eastern Ghats 59. 2017.

Very variable shrub, creeping, sarmentose or sometimes erect; branchlets pithy and more or less hollow, brown hairy, hispid or with dense spreading pubescence. Leaves very variable, linear, linear-lanceolate, ovate, elliptic to broadly ovate-oblong, unlobed or shallowly or deeply pinnatifidly 2-many-lobed, 2.5-12.5 cm long, base obtuse, rounded, subcordate or cordate, margin shallowly serrate or dentate, apex acute or shortly acuminate, more or less hispid or scabrid above, softly pubescent, glabrescent or scabrellous beneath, petiole 0.4-7.5 cm long, usually hispid; stipules oblong-lanceolate, 3.7-7.5 mm. Figs axillary, solitary, young more or less pyriform or ellipsoid and hispid, mature globose or subglobose, 1.8-2.5 cm long and 1.2 cm across, mouth broad; peduncle 5-10 mm long. Gall flower pedicelled, tepals 4, linear, 2.5-3 mm. Male flowers: Tepals 3, shorter than in gall flowers, stamen 1.

In wet and watery places. Figs: January-May. Vern.: Tel.: *Buroni;* Or.: *Bhui dumar, Ganthi sahada.*

Araku forest (VSKP), *KJ* 12546; Donkarayi near Mangapadu (VSKP), *GVS* 29660 (MH); Godavari, *JSG* 15850 (MH); Kanobori (KHD), *MB* & *Dhal* 7425 (RRL-B).

INDIA: Tamil Nadu, Andhra Pradesh, Odisha, West Bengal.

WORLD: Borneo, China, Indo-China, Java, Malesia, Myanmar, Sri Lanka.

Ficus hispida L.f., Suppl. Pl. 442. 1781; FBI 5:522. 1888; Fischer 3:1367. 1928; Sudhakar *et al.*, Figs Eastern Ghats 61. 2017.

Tree, up to 8 m high; without aerial roots, trunk straight with root suckers, laxly branched, bark grey, rough; wood dark grey; branchlets stout, hollow. Leaves membranous, obovate, or ovate-oblong 15-25 × 6-10.5 cm, hispid-scabrid above, tomentose, and glaucous beneath, base rounded or subcordate, margin toothed, apex shortly apiculate, 3-ribbed, lateral nerves 6-10 pairs, petiole to 6 cm. Receptacles obovoid, to 2.5 cm across, yellowish, hispid on leafless branches from trunk; bracts triangular. Achene ovoid.

Common, in hedges, thickets, vacant plots up to 900 m. Syconia: December-March. Vern.: Tel.: *Kakimedi, Beddu marri, Kukka bodda, Brahmamedi, Bomma medi;* Ori.: *Balu-dumari, Baidimiri, Dimiri, Kharsen, Tambol;* Tam.: *Chona-atthi, Ottanali, Peatti, Polla parakam.*

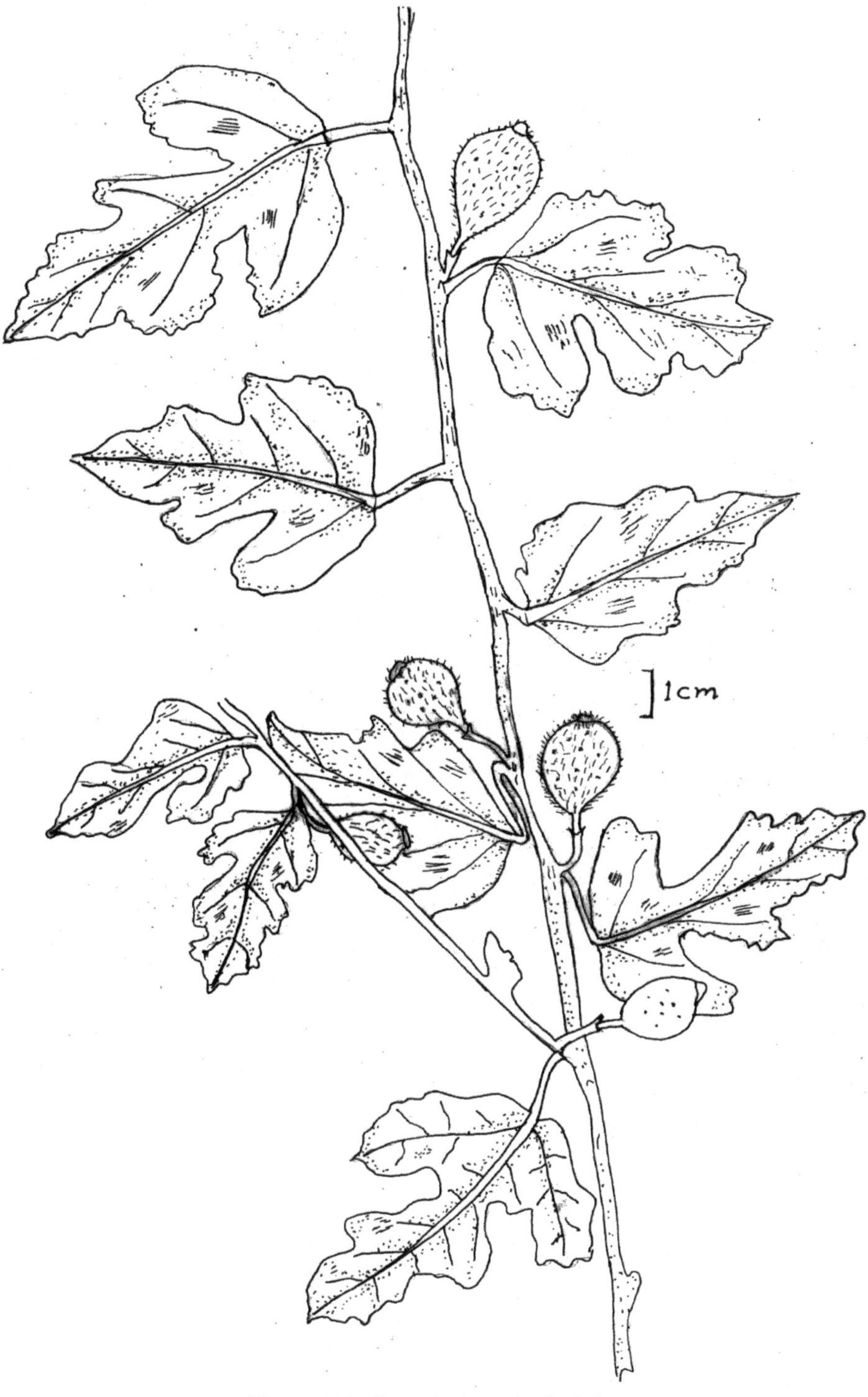

Figure 121. **Ficus heterophylla** L.f.
A. Twig.

Figure 122. **Ficus hispida** L.f.
A. Twig, B. Fruiting twig, C. Fruit, D. Bract, E and F. Receptacle L.s. and part enlarged, G. Male flower, H. Stamen, I. Female flower.

Meredumilli (EG), *TP* & *KJ* 12519; Kalvabugga (KNL), *RVR* 3106; Srisailam, Paladhara – Panchadhara (KNL), *KJ* 12576; Palakonda hills (CDP), *CS* 8274; Tirumala (CTR), *KJ* & *SSR* 12565; Macherla (GNT), *VRK* 3865; Javuku (GNT), *VRK* 3865; Kodur (CDP), *JLE* 14337 (MH); Satyavedu (CTR), *MCB* 45224 (MH); Thumburathirtham (CTR), *JHFB* 117453 (BSID); Kalwai (NLR), *PV* 110200 (BSID); Sirivaka near Koruturu (WG), *DN* 84246 (BSID); Rampachodavaram (EG), *GVS* 24480 (MH).

INDIA: Throughout India.

WORLD: Bhutan, China, Indo-china, Malesia, Myanmar, Nepal, North Australia, Sri Lanka, Thailand, Vietnam.

Ficus microcarpa L.f., Suppl. Pl. 442. 1782; Sudhakar *et al.*, Figs Eastern Ghats 64. 2017. *F. retusa auct. non* L.; King, Ann. Roy. Bot. Gardn. Cal. 1:50 1887 & in FBI 5:511.1888; Fischer 3: 1362. 1928.

Deciduous tree, up to 15 m high; aerial roots few; bark brown, nearly smooth; wood light reddish-grey; branchlets glabrous, aerial roots few. Leaves coriaceous, elliptic or obovate, 3-9.5 × 1.5-5 cm, glabrous, shining, base narrowed or cuneate, margin entire, apex rounded to acute or retuse, 3-nerved; lateral nerves 10-13, closely and reticulately veined; petiole to 1.6 cm. Receptacles monoecious, 8 mm dia., sessile in axillary pairs, rarely solitary, globose, yellowish red when ripe.

Common, often planted in avenues above 500 m. Syconia ripening by October. Vern.: Tel.: *Juvvi, Konda juvvi, Yerra-juvvi, Konda pillara, Nandireka, Billa juvi, Hama vudaga*; Ori.: *Jida, Mal jodi;* Tam.: *Kal-atthi, Kal-ichchi, Pon-ichchi.*

Kalasamudram (ATP), *NY* 1193; Ankireddipalli (KNL), *TP* 962; Rangapuram RF (KNL), *RVR* 1487; Upper Ahobilam (KNL), *SSR* 13792, *SSR* & *KJ* 15206; Palakonda hills (CDP), *CS* 9333; Tirumala (CTR), *SSR* & *KJ* 13740; Horsley hills (CTR), *SSR* 13792; way toThumbura theertham (CTR), *SSR* 1678; S.R.Puram (NLR), *BR* & *BSS* 30743; Araku forest (VSKP), *KJ* 12583; Peddarutla (PKM), *VBH* 84938 (BSID); Bison hill (EG), *CAB* 5160 (MH); Ramavaram (EG), *GVS* 62441 (MH); Anantagiri (VSKP), *GVS* 21781 (MH).

INDIA: Throughout India.

WORLD: Australia, Bhutan, Brazil, China, Indo-china, Japan, Malesia, Mexico, Nepal, Philippines, Sri Lanka, Thailand.

Ficus mollis Vahl, Symb. Bot. 1:82. 1790; Sudhakar *et al.*, Figs Eastern Ghats 67. 2017. *F. tomentosa* Roxb. ex Willd., Sp. Pl. 4: 1136. 1806; FBI 5: 501.1888; Fischer 3: 1361. 1928.

Evergreen tree, up to 15 m high; bark greenish-white, smooth, scaly; wood white; young branches densely, yellowish tomentose; aerial roots several. Leaves crowded towards the ends of branches, coriaceous, ovate-elliptic or obovate-elliptic, 4-12 × 3-7 cm, glabrous or nearly so and dotted above, densely rusty-grey tomentose beneath, base cordate with a gland at the base of the midrib beneath, margin entire, apex obtuse to sub-acute, main lateral nerves 5-6 pairs; petiole to 3 cm. Figs subglobose, sessile, axillary in pairs with rusty tomentum, green, 1 cm in dia.

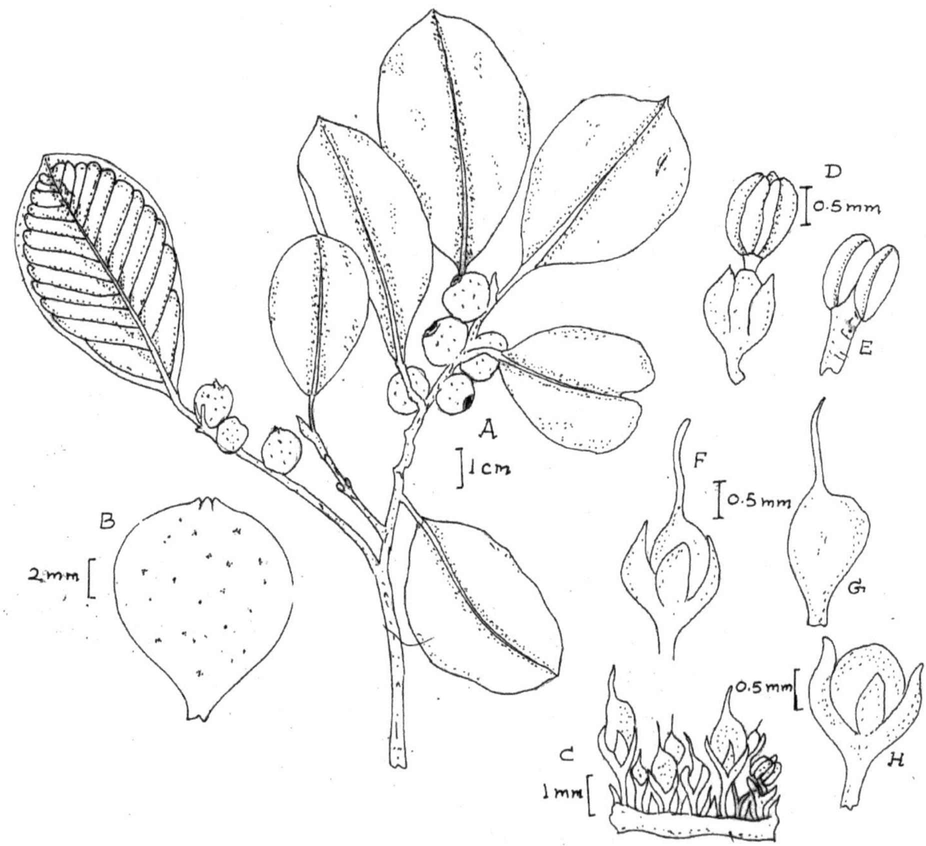

Figure 123. **Ficus microcarpa** L.f.
A. Twig, B. Fruit, C. L.s. of receptacle part enlarged, D. Male flower, E. Stamen, F. Female flower, G. Pistil, H. Achene.

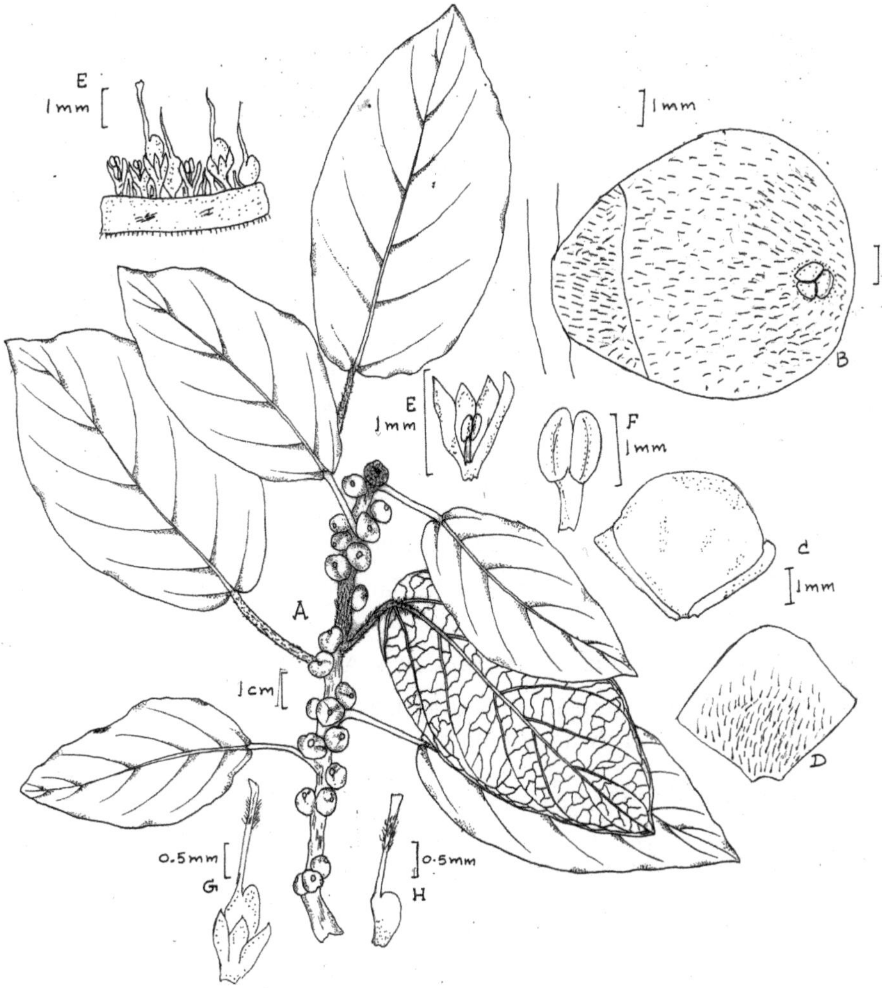

Figure 124. **Ficus mollis** Vahl
A. Twig, B. Fruit, C and D. Bracts, E. L.s. of receptacle part enlarged, F. Stamen, G. Female flower, H. Pistil, I. Male flower.

Common, in forests and villages in dry localities. Fl. & Fr.: Almost throughout the year. Vern.: Tel.: *Jivi, Juvvi, Kal juvi*; Ori.: *Jari*; Tam.: *Kal-ichchi, Kattarasu, Ichchi.*

Gundumala (ATP), *TP* & *NY* 1234; Madhavaram RF (KNL), *TP* & *PVR* 1384; Rachakuntapalli (CDP), *KRKS* 39015; Tirumala (CTR), *SSR* 13742; Kalyani dam (CTR), *KJ* & *SSR* 12553; Nagarjunakonda (GNT), *VRK* 6707; B. Velamala kota (EG), *TP* & *KJ* 12536; Sangam (NLR), *MSR* 1148 (CAL); Rampachodavaram to Maredumilli (EG), *MM* 102686; Kakarapadu (VSKP), *GVS* 65645 (MH); Buguda (GJM), *SX* & *MB* 2777 (RRL-B); Mahendragiri (GJM), *SX* & *MB* 3386 (RRL-B).

INDIA: Peninsular India, Bihar, Jharkhand, Rajasthan, Uttar Pradesh.

WORLD: Sri Lanka.

Ficus nervosa B. Heyne ex Roth, Nov. Pl. Sp. 388. 1821; FBI 5: 512. 1888 *p.p.*; Fischer 3: 1364. 1928; Sudhakar *et al.*, Figs Eastern Ghats 70. 2017.

Evergreen, medium to large tree; aerial roots absent; bark brown, mottled white; wood white, young branches minutely, appressedly pubescent; latex white turning yellow. Leaves thinly coriaceous, elliptic, oblong-lanceolate or oblanceolate, 10-16 × 5-8 cm, glabrous, shining, base narrowed, subequilateral, margin entire to slightly undulate, apex abruptly and narrowly acuminate, 3-nerved, not stronger, secondary nerves 9-11 pairs; petiole 0.8-25 cm. Receptacles subglobose, smooth or slightly verrucate, peduncle to 1.5 cm, yellow to red when ripe; achene ovoid, acuminate, smooth.

Occasional, moist deciduous forests above 1000 m. Fl. & Fr.: March-December. Vern.: Tel.: *Vonjar;*Tam.: *Neer-aal, Niral, Yeeppamaram.*

Vaiyar (VSKP), *PSPB* 1181; Doljharan, Gandhamardhan (Brgarh), *SX* & *MB* 6474 (RRL-B).

INDIA: Peninsular India, Bihar, Jharkhand, Assam, Arunachal Pradesh.

WORLD: Australia, Bangladesh Bhutan, China, Indo-china, Myanmar, Nepal, New Guinea, Sri Lanka, Taiwan, Vietnam.

Ficus palmata Forssk., Fl. Aegypt.-Arab. 179. 1775 subsp. **virgata** (Roxb.) P. Browicz in Rechinger, Fl. Iranica 153. 12. 1982; Abdulkader, J. Econ. Taxon. Bot. 33(2) 339. 2009; Sudhakar *et al.*, Figs Eastern Ghats 73. 2017. *F. virgata* Roxb., Fl. Ind. 3: 530. 1832. *F. palmata* sensu King in FBI 5: 530. 1888; Fischer 3: 1364. 1928. .

Shrubs or small trees, without aerial roots; bark grey, smooth; branchlets, petioles, young shoots soft-tomentose. Leaves 6.5-11 × 6-9.5 cm, broadly ovate, alternate, base broad truncate or cordate, margin dentate, at times deeply lobed, apex acute, lower surface soft tomentose, upper surface pubescent. Figs monoecious, axillary, more or less pear-shaped or obovoid-globose, 1.4 cm across; densely pubescent, yellow, edible when ripe; male and gall flowers in one receptacle, female flowers on separate receptacles. Scales near the osteole up to 0.12 cm long, deltoid. Male flowers: in 2 series, osteolar, pedicellate; tepals generally 4-5, ovate; stamens

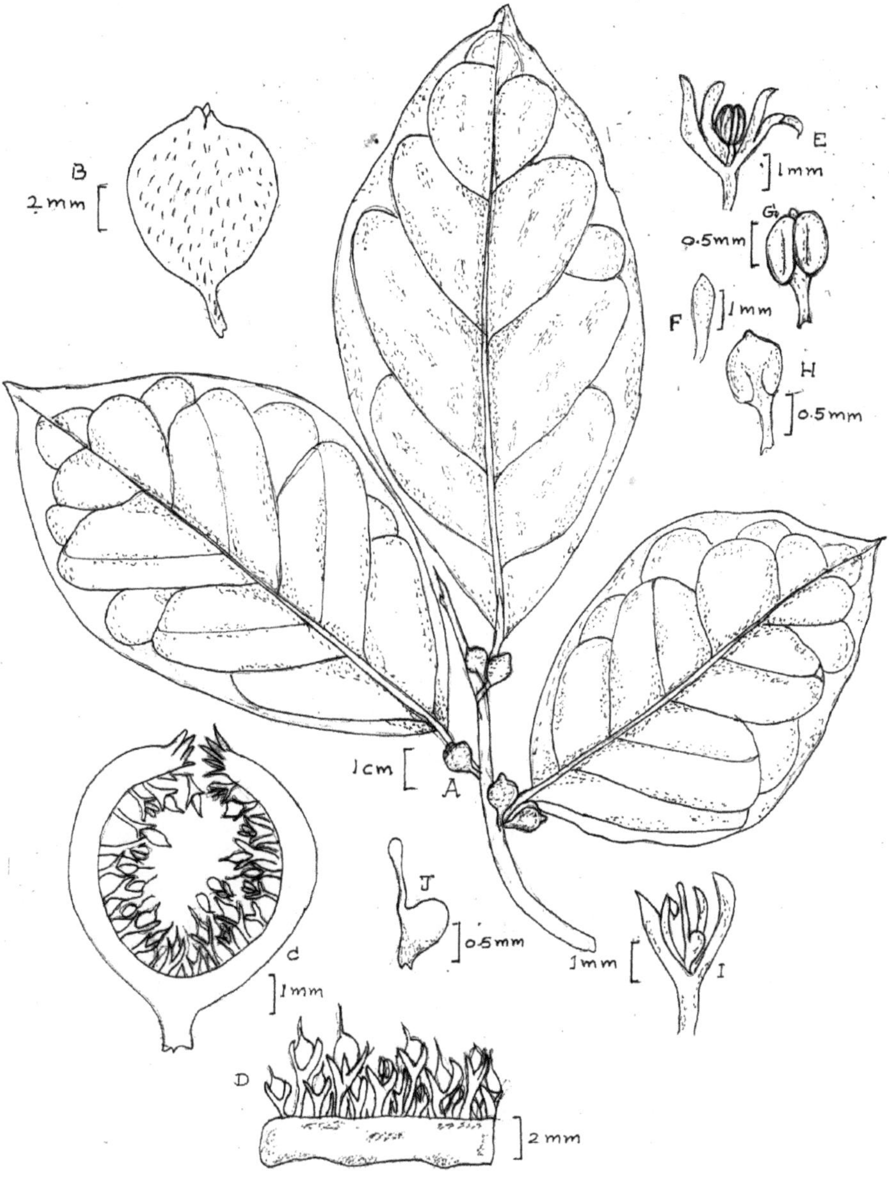

Figure 125. **Ficus nervosa** B. Heyne ex Roth
A. Twig, B. Fruit, C and D. Receptacle L.s. and part enlarged, E. Male flower, F. Perianth lobe, G and H. Anthers, I. Female flower, J. Pistil.

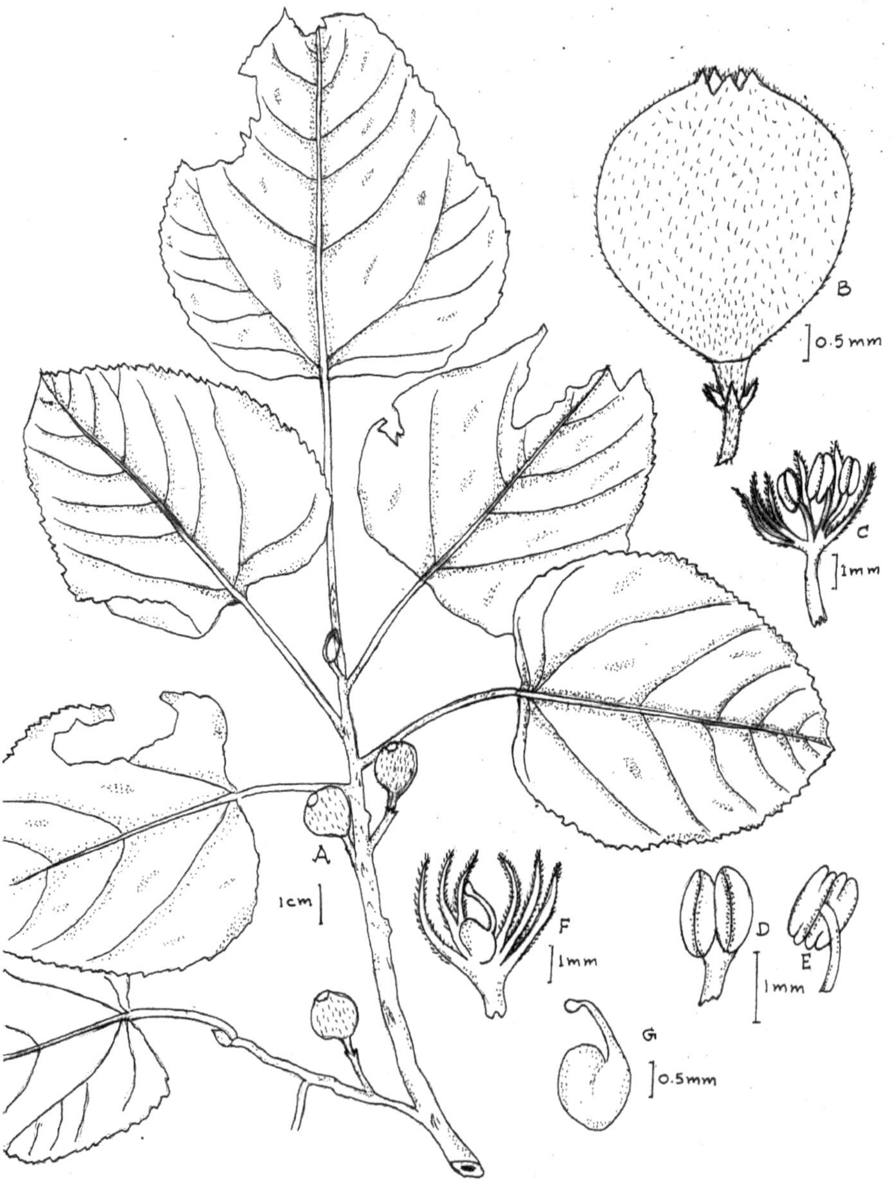

Figure 126. **Ficus palmata** Forssk.
A. Twig, B. Fruit, C. Male flower, D and E. Anthers, F. Female flower, G. Pistil.

4-5. Gall flowers: sessile or pedicellate; tepals 4-5, linear-lanceolate, pubescent without; ovary ovoid, style lateral. Female flowers: sessile, tepals up to 7, similar to those of gall flowers; ovary obliquely ovoid, style subterminal, stigma pointed. Achenes smooth.

Not common, usually found in the forests of Northern Eastern Ghats. Fl. & Fr.: January – April. Vern.: Tel.: *Manchi-medi, Medi.*

Balikonda (VSKP), *KJ* & *SSR* 12585; Bispuram (VSKP), *GVS* 44288 (MH); Sunkarimetta (VSKP), *GVS* 10901 (MH); Araku valley (VSKP), *GVS* 19665 (MH); Kasipur (KHD), *HFM* 2544 (DD).

Note: Quite common and variable in height of the plants and leaf shape and size. In majority of the Indian floras this species has been identified as *F. palmata* Forssk., which occurs only in East Africa and Saudi Arabia. The Indian materials belong to subsp. *virgata* (Roxb.) Browicz.

INDIA: Andhra Pradesh, Odisha, Bihar, Chhattisgarh, Delhi, Gujarat, Haryana, Himachal Pradesh, Jammu & Kashmir, Madhya Pradesh, Maharashtra, Rajsthan, Uttar Pradesh, West Bengal.

WORLD: Afghanistan, Pakistan, Nepal.

Ficus racemosa L., Sp. 1060. 1753. *F. glomerata* Roxb., Pl. Coromandel 2: 13. t. 123. 1799; FBI 5: 535. 1888; Fischer 3: 1364. 1928; Sudhakar *et al.*, Figs Eastern Ghats 75. 2017.

Deciduous tree, up to 15 m high with few or no aerial roots; bark smooth, grey; wood greyish brown, soft grained or shallowly or irregularly interlocked grained, very coarse and uneven textured; branchlets minutely pubescent, reddish-brown. Leaves coriaceous, elliptic-oblong, 4-13.5 × 2.5-6 cm, glabrous, base oblique, margin entire, apex acute or obtuse, 3-nerved; secondary nerves 6-8 pairs; petiole to 4 cm. Receptacles monoecious, globose, 2.5-4 cm in dia., reddish, white pilose on tubercled leafless branchlets from trunk; orifice slightly sunken. Achene tuberculate.

Common, along river banks, roadsides, village surrounds, up to 1500 m. Syconia: Throughout the year. Vern.: Tel.: *Medi, Atti, Moida, Bodda, Paidu;* Ori.: *Dimiri, Dimuri, Dumri;* Tam.: *Attimaram, Nalla atthi, Vellai atthi.*

Itukalapalli (ATP), *NY* 175; Ghani RF (KNL), *TP* & *RVR* 1794; Palakonda hills (CDP), *CS* 9321; Mabbuchintalapalle (CDP), *TP* & *SSR* 33659; Siddulaiah kona (NLR), *SSR* 19383; Macherla (GNT), *VRK* 3944; Maredumilli (EG), *TP* & *KJ* 12503; Puttagondhilanka, near Tadepalli (EG), *DN* 85532 (BSID); Vangara (VZN), *NRR* & *RAN* 79468 (BSID).

INDIA: Throughout India.

WORLD: Australia, Bangladesh, China, Indo-china, Malesia, Myanmar, Nepal, Pakistan, Sri Lanka.

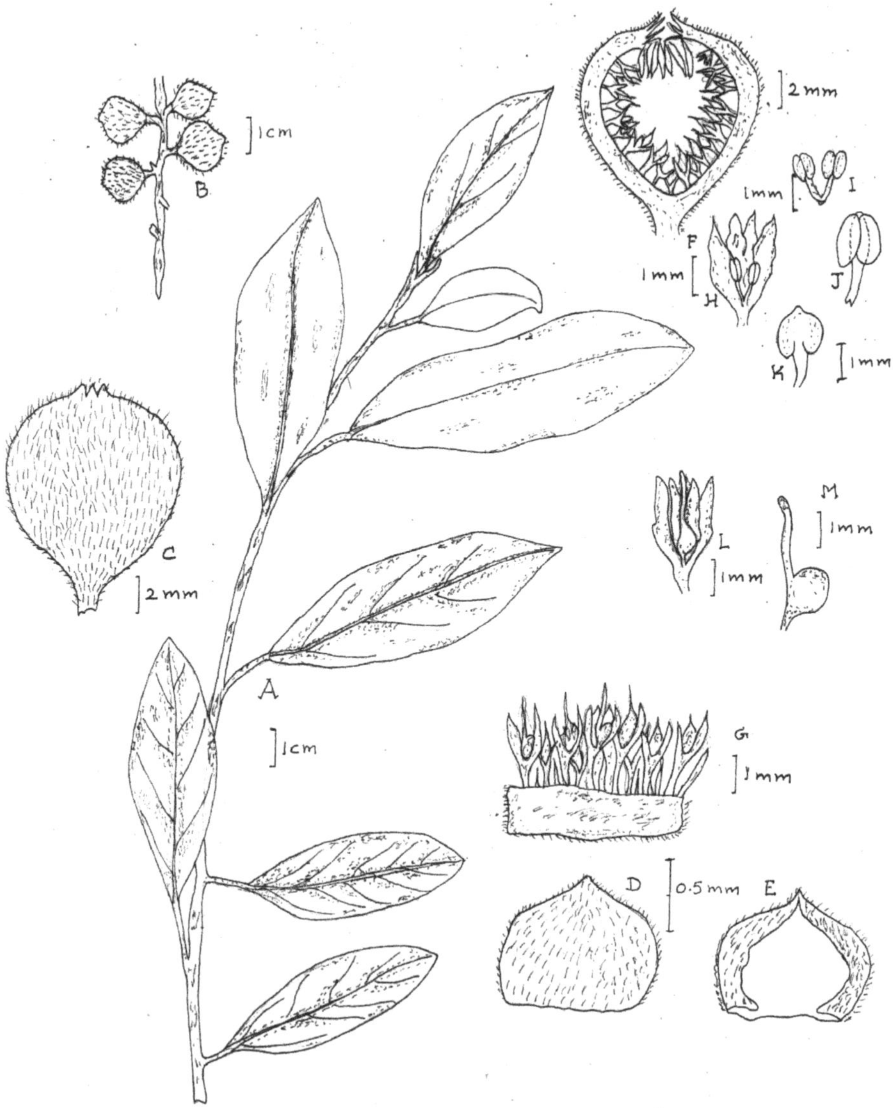

Figure 127. **Ficus racemosa** L.
A. Twig, B. Fruiting branch, C. Fruit, D and E. Bracts, F and G. Receptacle L.s. and part enlarged, H. Male flowr, I. Stamen, J and K. Anthers, L. Female flower, M. Pistil.

Ficus religiosa L., Sp. Pl. 1059. 1753; FBI 5: 513. 1888; Fischer 3; 1363. 1928; Sudhakar *et al.*, Figs Eastern Ghats 78. 2017.

Deciduous, fast growing tree, up to 20 m high; often epiphytic when young; aerial roots absent; bark greenish, smooth, wood greyish-white or yellowish white, fibrous, broadly and irregularly interlocked grained, coarse and uneven textured; branchlets yellowish grey, glabrous. Leaves coriaceous, ovate, 10-18 × 5-13 cm, glabrous, base truncate or sometimes cordate, margin wavy, apex long caudate-acuminate, secondary nerves ca. 8 pairs, inter-secondaries present, petiole to 12 cm. Receptacles sessile, paired, smooth, globose, 1.2 cm dia., dark purple when ripe.

Common, along roadsides, near temples, wild in village surrounding, river banks up to 1000 m; it is a sacred tree of Hindus, and Buddhists. It is also planted for shade. Fl. & Fr.: May-August. Vern.: Tel.: *Ravi, Ragi, Raigu, Ashwarthamu*; Or.: *Aswatha, Jori, Pipalo, Usto;* Tam.: *Arasu, Arasumaram*. Eng.: The peepal tree.

Talakona (CTR), *KJ* & *SSR* 12550; Owk (KNL), *RVR* 3052.

INDIA: Throughout India.

WORLD: Bangladesh, China, Indo-china, Malesia, Pakistan, Myanmar, Napal, Sri Lanka.

Ficus rumphi Blume, Bijdr. 437. 1825; FBI 5: 512. 1888; Sudhakar *et al.*, Figs Eastern Ghats 81. 2017.

Evergreen tree, up to 20 m high; without aerial roots, at first usually epiphytic; bark greyish-black, smooth; branchlets glabrous. Leaves subcoriaceous, broadly ovate, 10-15 × 6-8.5 cm, glabrous, shining, base subcordate, margin undulate, apex acuminate, 3-5-ribbed; lateral nerves 6-8 pairs above basal ones, conspicuous; petiole to 9.2 cm, lenticel like structures are in the base of the petiole; stipules ovate-lanceolate. Receptacles globular, 1.3 cm dia., smooth, cream with reddish spots, sessile in axillary pairs on old leaf scars, black when ripe, basal bracts 3, rotund; orifice circular, slightly raised, covered by 3 apical bracts. Achene minute.

Occasional, along river banks and moist deciduous forests in Visakhapatnam district at 800-1000 m. Syconia: January-Apr. Ripen in May-June. Vern.: Tel.: *Bodhi;* Ori.: *Jhida.*

Anantagiri (VSKP), *SSR* 15934, *KJ* & *SSR* 12586; Bodalur, Maredumilli (EG), *MM* 100767 (BSID); Satwa near Araku (VSKP), *NPBK* 10858 (MH).

INDIA: Andhra Pradesh, Odisha, Bihar, Gujarat, Himachal Pradesh, Jammu & Kashmir, Madhya Pradesh, Maharashtra, Rajasthan, Uttar Pradesh, West Bengal.

WORLD: Bhutan, China, Indo-china, Malesia, Myanmar, Nepal, Thailand.

Ficus semicordata Buch.-Ham. ex J.E. Smith, Rees Cyclop. 14: *Ficus* n. 71. 1810; Sudhakar *et al.*, Figs Eastern Ghats 84. 2017. *F. cunia* Buch.-Ham. ex Roxb., Fl. Ind. 3: 561. 1832; FBI 5; 523. 1888; Fischer 3: 1366. 1928.

Evergreen tree, up to 12 m high; without aerial roots, bark rough, reddish brown; wood greyish-brown; branchlets pubescent. Leaves coriaceous, semisagittate, one

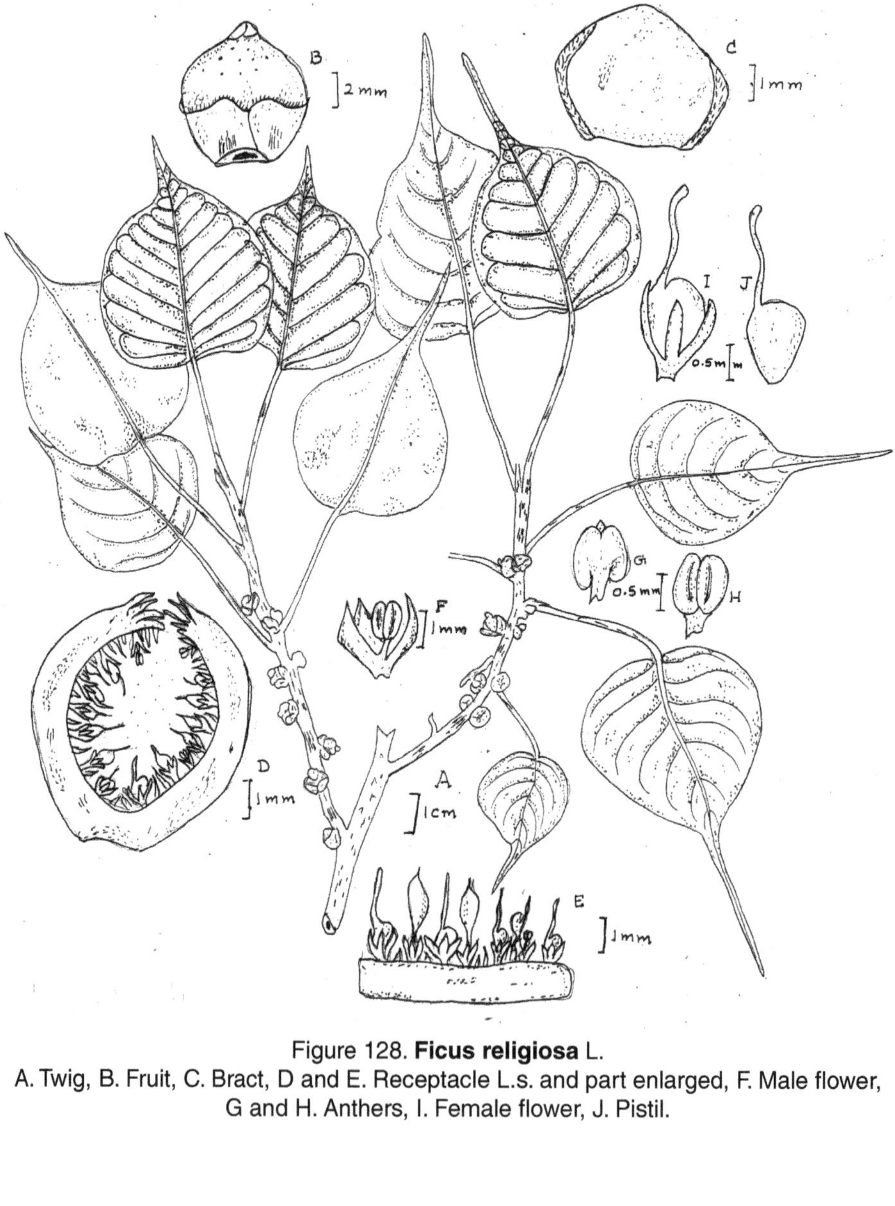

Figure 128. **Ficus religiosa** L.
A. Twig, B. Fruit, C. Bract, D and E. Receptacle L.s. and part enlarged, F. Male flower, G and H. Anthers, I. Female flower, J. Pistil.

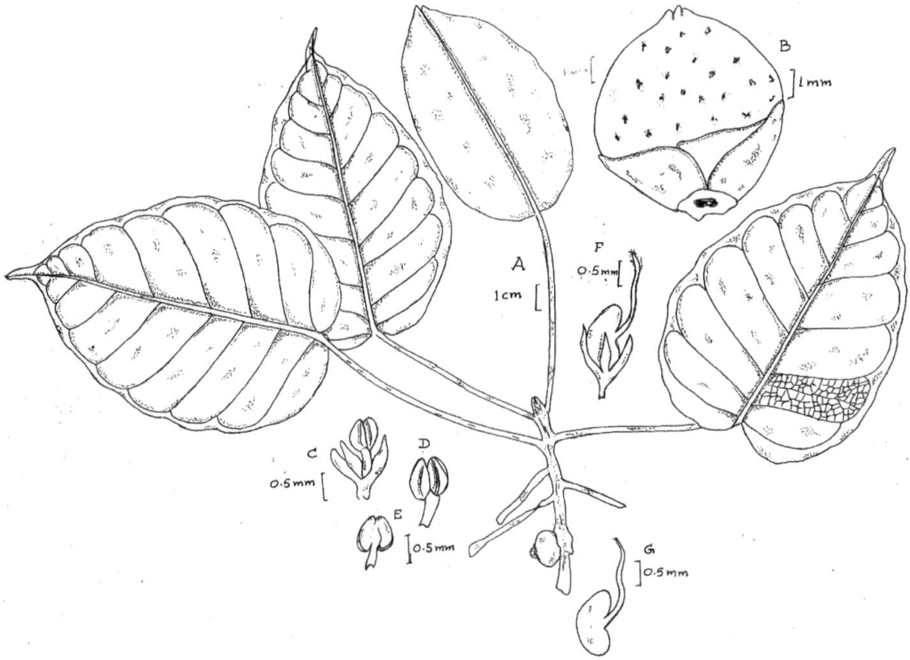

Figure 129. **Ficus rumphii** Blume
A. Twig, B. Fruit, C. Male flower, D and E. Anthers, F. Female flower, G. Pistil.

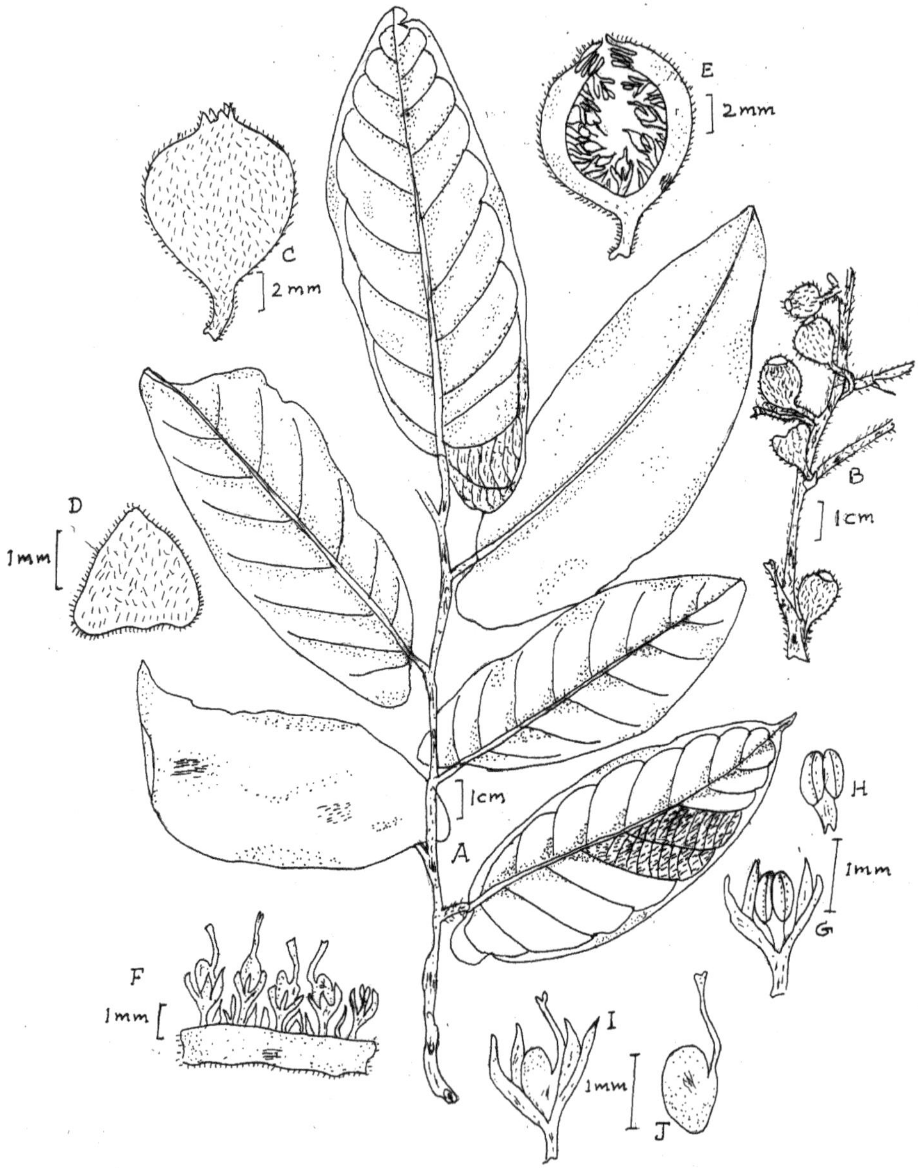

Figure 131. **Ficus semicordata** Buch.-Ham. ex J.E.Smith
A. Twig, B. Fruiting branch, C. Fruit, D. Bract, E and F. Receptacle L.s and part enlarged, G. Male flower, H. Anther, I. Female flower, J. Pistil.

side with a large 3-4 ribbed lobe, the other cuneate, elliptic, oblong-lanceolate, 4-22 × 4-7 cm, scabrid above, more or less pubescent beneath, base semicordate, margin entire or slightly serrate, apex shortly acuminate, lateral nerves 9-14 pairs, raised above, flattened below; petiole to 2.5 cm. Receptacles 2.5 cm in dia., paired, pear-shaped, peduncled on long leafless scaly branchlets from the trunk near base, reddish-brown when ripe; basal bracts 3, ovate, pubescent; orifice circular, covered by 3 apical bracts.

Common near streams and on cool, damp slopes of hills above 900 m in Northern Eastern Ghats. Syconia: Most of the year and usually ripe by May-June, and Geocarpic. Vern.: Tel.: *Erubodda, Bomma marri*; Ori.: *Bhui dumri, Godma, Kharsal, Podia dumri, Potkuli, Purhei.*

Pamuleru (EG), *TP* & *KJ* 12524; Galikonda (VSKP), *SSR* 15915; Valamuru to Maredumilli (EG), *MM* 100888; Anantagiri (VSKP), *GVS* 19557 (MH); Gamperkot (SKLM), *KCJ* 1732 (MH).

INDIA: Andhra Pradesh, Odisha, Bihar, Himachal Pradesh, Jammu & Kashmir, Jharkhand, Madhya Pradesh, Maharashtra, Uttar Pradesh, West Bengal, Northeast.

WORLD: Bangladesh, Bhutan, Pakistan, China, Malesia, Myanmar, Nepal, Thailand, Vietnam.

Ficus talbotii King, Ann. Roy. Bot. Gard. (Calcutta), 1: t. 63. 84Q. 1887 (*talboti*) and in FBI 5: 512. 1888; Fischer 3: 1363. 1928; Sudhakar *et al.*, Figs Eastern Ghats 87. 2017.

Evergreen tree, 10-20 m high; aerial roots absent; bark green, smooth; wood very white, turning yellow. Leaves subcoriaceous, narrow elliptic to broad ovate, 7-11 × 4-6 cm, glabrous, base rounded, margin entire, apex shortly, acutely caudate, 3-nerved; secondary nerves 8-10 pairs, raised on both surfaces; petiole to 3 cm, glandular at apex below. Receptacles obovoid, sessile, ripening yellowish, spotted with red colour dots, in axillary pairs; achenes smooth.

Occasional in forests. Fl. & Fr.: December-February. Vern.: Tam.: *Erali, Kalitthi, Ichchimaram, Ottal.*

Gundumala RF (ATP), *NY* 1236; Bairluty (KNL). *SSR* & *KJ* 15517.

INDIA: Peninsular India, Madhya Pradesh.

WORLD: Cambodia, Laos, Malaysia, Myanmar, S.China, Sri Lanka, Thailand, Vietnam.

Ficus tinctoria Forster f., Fl. Ins. Austr. 76. 1786 subsp. **gibbosa** (Blume) Corner, Gard. Bull. Singapore 17: 476. 1960; Sudhakar *et al.*, Figs Eastern Ghats 89. 2017. *F. gibbosa* Blume, Bijdr. Fl. Ned. Ind. 9: 466. 1825; King, Ann. Roy. Bot. Gard. (Calcutta) 1: 4. t. 2. 1887 & in FBI 5: 497. 1888; Fischer 3: 1365. 1928. *F. parasitica* Koenig ex Willd. in Mem. Acad. Roy. Soc. Hist. 2: 102. t. 3. 1798. *F. tinctoria* Forster f. subsp. *parasitica* (Koenig ex Willd.) Corner, Gard. Bull. Straits Settlem. 17: 476. 1960.

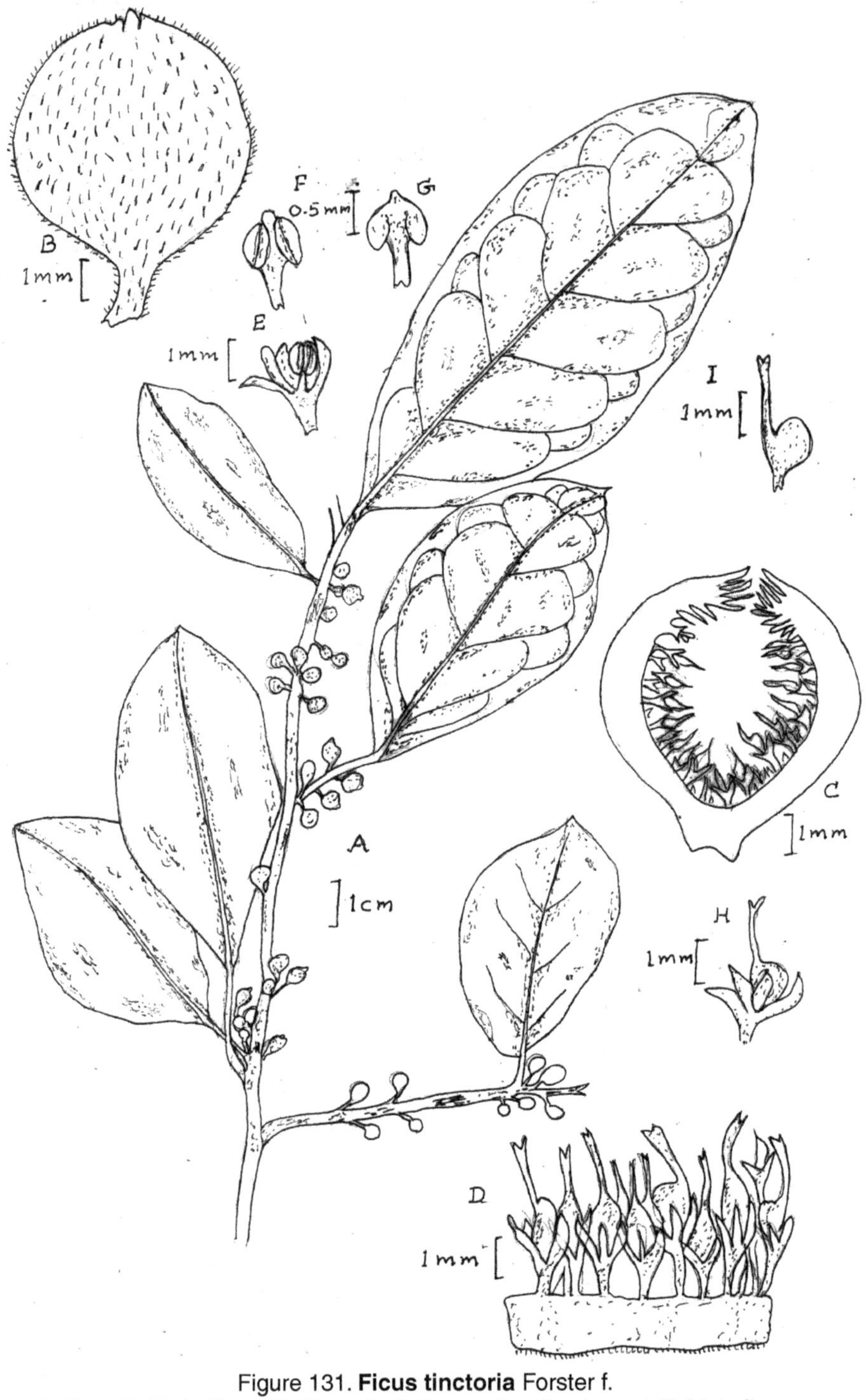

Figure 131. **Ficus tinctoria** Forster f.
A. Twig, B. Fruit, C and D. Receptacle L.s. and part enlarged, E. Male flower, F and G. Anthers, H. Female flower, I. Pistil.

Tree, up to 10 m high; often epiphytic, enclosing the trunks of tree in a perfect network of branches and creeping along the wells and on the side of wells, without aerial roots, bark greyish, smooth; wood brownish grey. Leaves thinly coriaceous, rhomboid or sub-rhomboid or elliptic ovate, 4-12 × 3-6 cm, hispid and dark green above, scabrid and minutely hispid and paler beneath; apex acute, margin entire, base oblique, 5-7 mm across, paired, globose, peduncled, hairy; male receptacles yellow; female receptacles red when ripe.

Common in open places 1200 m. Syconia: February-April. Vern.: Tel.: *Konda juvvi, Pakki;* Ori.: *Kharsara, Kortosani, Jhirang, Jhori, Udambar;* Tam.: *Kallathi, Kal-itthi, Kal-perukam.*

Bommalingam (PKM), *SKB* & *BSS* 32688; Y. Ramavaram (EG), *TP* & *KJ* 12526; Araku (VSKP), *KJ* 12584; Eethakayalamandapam (CTR), *GVS* 46899 (NH); Bheematheertham - Tirumala hills (CTR), *DRC* 1173 (MH); Rampachodavaram (EG), *GVS* 24473 (MH); Duggeru forest (SKLM), *GVS* 62153 (MH); Cherukonda (VSKP), *GVS* 30204 (MH); Cuddapah (CDP), *JSG* 10823 (DD); Similipahar (MBJ), *SX* 363 (RRL-B), *SX* & *MB* 4420, 4591 (RRL-B); Sirumalai hills (DGL), *SKS* 689 (SGH).

INDIA: Throughout India.

WORLD: Bangladesh,Bhutan, Cambodia, China, Laos, Indo-china, Indonesia, Malesia, Myanmar, Nepal, Sri Lanka, Thailand, Vietnam.

Ficus tsjahela Rheede ex Burm. f., Fl. Ind. 227. 1768; King, Ann. Roy Bot. Gard. Calcutta, 1:57, t. 70 & 84. 1887(*tsjakela*) & in FBI 5; 14. 1888; Fischer 3: 1862.1928; Corner, Gard. Bull. Singapore 21: 7. 1965; Sudhakar *et al.,* Figs Eastern Ghats 92. 2017.

Deciduous tree, up to 18 m high; aerial roots absent; bark coloured, rough; wood brown; branchlets glabrous, dark brown. Leaves coriaceous, oblong-elliptic, or elliptic-ovate, 6-19 × 3-10.5 cm, glabrous, dark green above, pale beneath, ashy grey when dry, base truncate or rounded, margin entire, apex shortly acuminate, 3-ribbed; 7-12 pairs above basal ones, conspicuous, to 2.5 cm apart from basal ones to just above lateral nerves; petiole to 8 cm. Receptacle monoecious, depressed-globose, 6 mm across, greyish-purple, sessile, 2-6 together on tubercles in fallen leaf branches, basal bracts 3, bifid, orifice plane, circular; apical bracts 3.

In moist regions of moist deciduous forests above 600 m. Syconia: Ripen by May; Vern.: Tam.: *Kal-aal, Karibasari, Kattal, Kurugu, Itti.*

G. Madugula (VSKP), *KJ* & *SSR* 12589; Tirumala (CTR), *SSR* 16184.

Note: The correct spelling of this species is *'tsjahela'* as given by Corner (*loc.cit.*); Fischer gives it as *tjakela* (*loc.cit.*).

INDIA: Peninsular India.

WORLD: Sri Lanka.

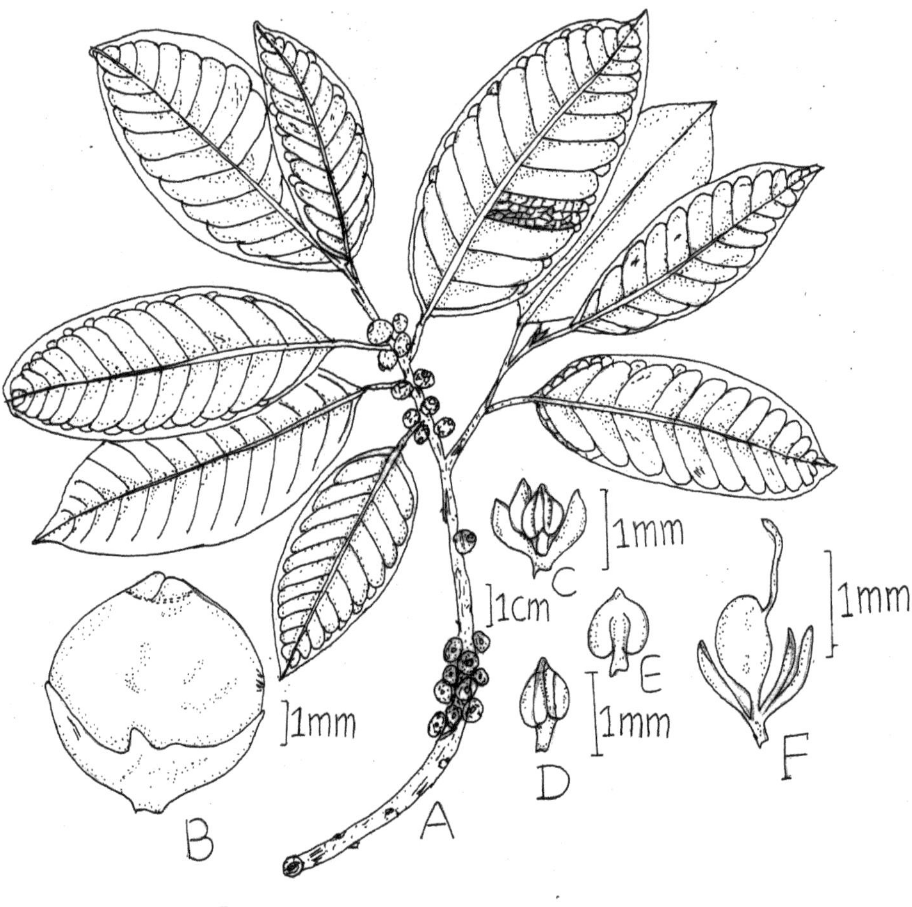

Figure 132. **Ficus tsjahela** Rheede ex Burm.f.
A. Twig, B. Fruit, C. Male flower, D and E. Anthers, F. Female flower.

Ficus variegata Blume, Bijdr. Fl. Ned. Ind. 9: 459. 1825; FBI 5: 535. 1888; Sudhakar & Murthy, Rheedea 22: 62. 2012; Sudhakar *et al.*, Figs Eastern Ghats 95. 2017.

Trees, up to 30 m high, deciduous; trunk buttressed, straight; branches numerous on top, spreading in all directions; aerial roots absent; branchlets glabrous, pubescent or glabrous when young; bark smooth, pale brown with reddish spots. Leaves simple, alternate, symmetric, broadly ovate to ovate-elliptic, sometimes cordate, 10 – 30 × 2.5 –11 cm, base rounded or cordate to subcordate, margin subrepand or coarsely denticulate, apex acuminate to subacute at apex with *ca* 1 cm long acumen, coriaceous with cystoliths only on lower surface, glabrous, puberulous below when young, brown with ash colour patches above, dark brown below when dry; basal nerves 3 – 5 (2 inconspicuous), waxy glands at slit-shaped extensions of axils of main basal nerves; midrib prominent below; lateral nerves 3 – 5(– 8) pairs, prominent below, looping intramarginally; tertiary nerves transverse, minutely reticulate; petioles stout, 2.5 – 6.5 × 0.25 cm, grooved above, glabrous or appressed -puberulous, brown with flaking off epidermis; stipules terminal, ovate-lanceolate, 0.5– 2 cm long, caducous, glabrous or puberulous. Receptacles in fascicles from woody tubercles, ramiflorous to cauliflorous, dioecious, drooping, long-pedunculate, globose or subglobose to pyriform, 2 – 3.5 cm across, smooth, glabrous, yellowish green, pink to red with white streaks and spots when ripe, sometimes with a short stipe (up to 8 mm), fig wall up to 6 mm thick; peduncles moderately stout, 2 – 5 × 0.1 – 0.2 mm, glabrous; basal bracts 3, basally connate, ovate-acute, 1.5 – 2.5 mm long, deciduous or persistent, leaving annular scar, pubescent; ostiole slightly depressed, 4 – 5 mm across, 1.5 – 2 mm depth; ostiolar bracts 3 – 5, free, ovate-acute or ovate-obtuse, 1 – 2 mm long, pubescent; internal bristles obtuse, 1 – 1.5 mm long, glabrous, hairy at margin; interfloral bracts usually absent. Male flowers: sessile, found near ostiole, 2 – 2.5 mm long; perianth more or less polyphyllous; tepals 3 or 4, linear-lanceolate or spathulate, 1.5 – 2 mm long; stamens 2; anthers broadly ovoid; filaments short. Gall flowers: sessile or shortly pedicellate, 2 – 3 mm long; pedicels 0.6 – 0.8 mm long; perianth gamophyllous; tubular, 4 or 5-lobed, 1.5 – 2.5 mm long, inflated at apex; ovary ovoid 1 – 1.2 mm long, red-brown to pale; style lateral, 0.5 – 0.8 mm long, white; stigma infundibuliform. Female flowers: sessile or shortly pedicellate, 2.5 – 3 mm long; pedicels 0.7 – 1 mm long; perianth gamophyllous or basally connate; tepals 3 or 4, lanceolate or spathulate, *ca* 2 mm long, inflated at apex, reddish or brownish; ovary ovoid, 1 – 1.3 mm, reddish or pale brown; style lateral, cylindric, 1.3 – 1.5 mm long, thick, white; stigma infundibuliform. Achenes ovoid or obovoid, 1 – 1.5 mm long, minutely tuberculate.

Dry deciduous and semi evergreen forests, to 700. Fl. & Fr.: January – September.

On the way Valamuru toTiger Camp, near Maredumilli (EG), *J.V. Sudhakar* 126741, 126743 (MH); Maredumilli to Valamuru way (EG), *J.V. Sudhakar* 126742 (MH; Gudem to Sapparla way, near Sileru (VSKP), *J.V. Sudhakar* 126 (MH).

INDIA: Andhra Pradesh, Assam, Meghalaya, Andaman & Nicobar Islands.

WORLD: Australia, Bangladesh, China, Indo-china, Japan, Malesia, Myanmar, Ryukyu Islands, Solomon Islands, Thailand, Taiwan.

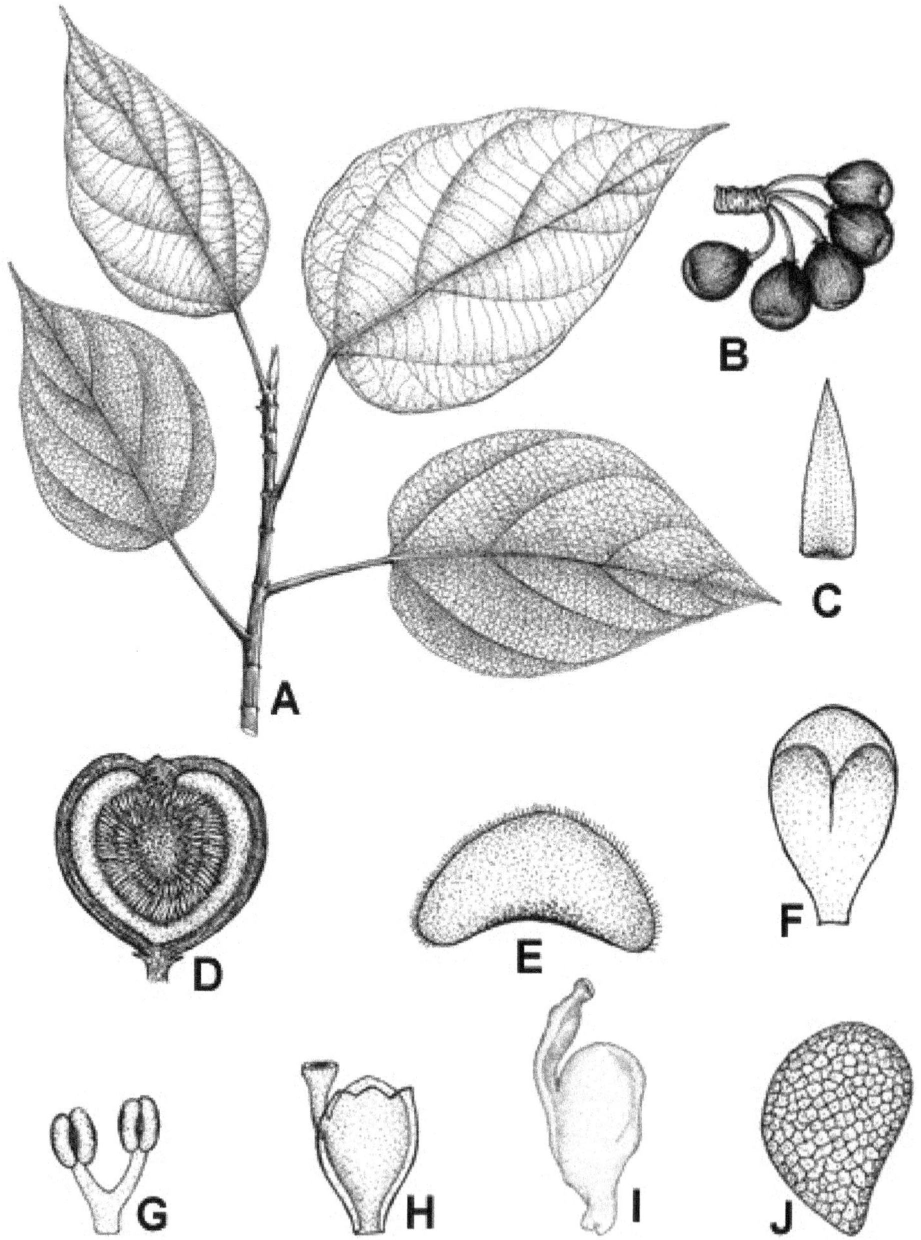

Figure 133. **Ficus variegata** Blume
A. Habit; B. Receptacles; C. Stipule; D. L.S. of Receptacle; E. Ostiolar bract; F. Male flower; G. Stamens; H. Gall flower; I. Female flower; J. Achene.
(after Sudhakar and Murthy, 2012).

Ficus virens Aiton, Hort, Kew. 3: 451. 1789; Sudhakar *et al.*, Figs Eastern Ghats 98. 2017. *F. infectoria* Roxb., Fl. Ind. 3: 551. 1832 (non Willd); 1806; FBI 5; 515. 1888; Fischer 3: 1362. 1928. *F. infectoria* Roxb. var. *lambertiana* (Miq.) King, Ann. Roy. Bot. Gard. (Calcutta) 1: 63. t. 76. 1887; Fischer 3: 1363. 1928. *F. virens* Aiton var. *lambertiana* (Miq.) Raizada, Indian For. 94: 443. 1968; C. S. Reddy *et al.*, Suppl. Fl. Andhra Pradesh 109. 2008. *Urostigma lambertiana* Miq. London J. Bot. 6: 565. 1847.

Large, spreading, deciduous tree, epiphytic in early stages, with as few aerial roots; bark greenish-grey, smooth; wood grey. Leaves subcoriaceous, elliptic, oblong, oblong-ovate, 7-18 × 2.6 - 5.7 cm, glabrous, apex abruptly acuminate with blunt tip, margin entire to slightly undulate, base acute, truncate or subcordate; secondary nerves 7-12 pairs, lopped near margin, basal pair shorter, intercostals zig zag; petiole 2.5-7 cm, articulate, glandular at the apex below. Receptacles axillary, paired, shortly peduncled, globose, whitish with red dots.

Common, in plains to hills up to 1200 m. Fl. & Fr.: October-May. Vern.: Tel.: *Banda juvvi, Jati juvvi*; Ori.: *Pakodo, Kua-Jari*; Tam.: *Kalla, Kalithi, Kurugatti, Ichchimaram, Jovi, Malai ichchi.*

Anantagiri (VSKP), *SSR* 15935, *GVS* 21716 (MH); Balapalle (CDP), *JLE* 14327 (MH); Talakona temple stream (CTR), *GVS* 46010 (MH); Tirumala (CTR), *SSR* 16186; Gogarbham, Tirumala hills (CTR), *DRC* 1499 (MH); Valamura (EG), *MM* 101363 (BSID).

INDIA: Throughout India.

WORLD: Australia, Bangladesh, Bhutan, Cambodia, China, Indo-china, Japan, Laos, Malesia, Myanmar, Nepal, Pakistan, Solomon Islands, Sri Lanka, Taiwan, Thailand, Vietnam.

MACLURA Nuttall *nom. cons.*

Maclura cochinchinensis (Lour.) Corner, Gard. Bull. Straits. Settlem. 19: 239. 1962. *Vaniera cochinchinensis* Lour., Fl. Cochinch. 564. 1790. *Cudrania cochinchinensis* (Lour.) Yakuro Kudo & Masam., Ann. Rep. Taihoku Bot. Gard. 2: 27. 1932. *C. javanensis* Trecul, Ann. Sci. Nat., Bot. Ser. 3. 8: 123 1847; FBI 5: 538. 1888.

A straggling or subscandent shrub, or small tree with drooping branchlets; spines up to 3.2 cm long, fulvous hairy, ultimately glabrous. Leaves 2.5-7.5 × 2-3.5 cm, oblong or obovate to oblanceolate, alternate, margin entire, apex acute, or acuminate, glabrous; lateral nerves up to 10 pairs, bifurcating near the margin and uniting with the branches of the next to form an intramarginal nerve; petiole up to 1.2 cm long, fulvous-pubescent; stipules up to 2 mm long, deltoid, fulvous-tomentose, caducous. Flowers dioecious in small, naked globose heads, solitary, shortly peduncled, pubescent. Male flower heads: cream coloured; tepals 4, free, obovate-oblong, pubescent without in the upper half; stamens 4, included. Syncarp up to 2.5 cm across, peduncles up to 5 mm long, tomentose. Achenes enclosed in the fleshy bracts and perianth forming a globose fleshy head, ovoid.

Rare in forests of Northern Eastern Ghats. Fl. & Fr.: April- July.

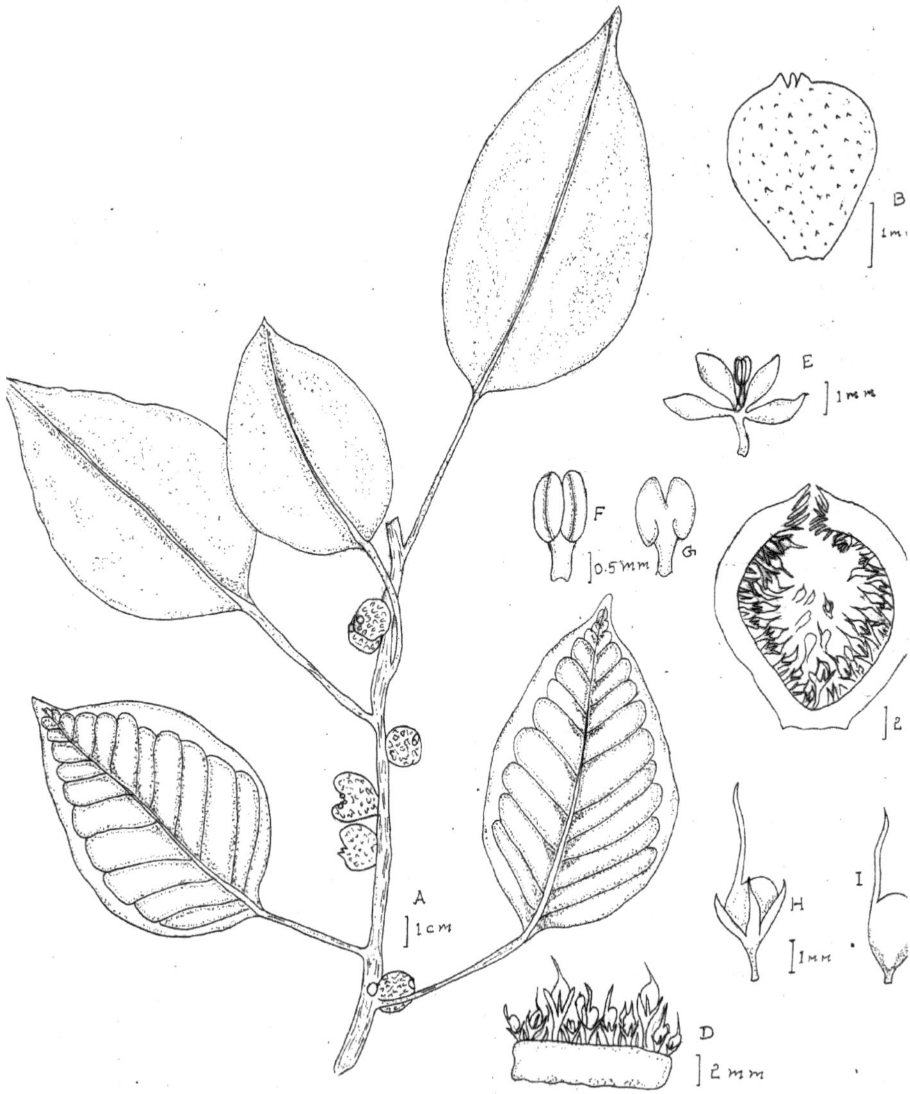

Figure 134. **Ficus virens** Aiton
A. Twig, B. Fruit, C and D. Receptacle L.s. and part enlarged, E. Male flower, F and G. Anthers, H. Female flower, I. Pistil.

Sunkarimetta (VSKP), *NPBK* 745 (MH); Minumuluru (VSKP), *GVS* 28253 (MH); Bhalupodar valley, Pottangi (KPT), *HFM* 3864 (DD).

INDIA: Tropical Himalayas, Sikkim, Khasia Hills, Bihar, Odisha, Andhra Pradesh.

WORLD: Bangladesh, Sri Lanka, Myanmar, Malay Peninsula, China, E. Africa, Australia.

MORUS Linnaeus

1. Syncarp less than 2.5 cm long:
    2. Bracts conspiculous; peduncle more than 1 cm long ....**M. australis**
    2. Bracts minute; peduncle less than 1 cm long ....**M. alba**
1. Syncarp 3-6 cm .... **M. macroura**

Morus alba L., Sp. Pl. 986. 1753; FBI 5: 492. 1888; Fischer 3: 1370. 1928.

Deciduous, medium-sized tree; crown round; branches grey or greyish yellow; young shoots pubescent. Leaves thin, sub-coriaceous, broadly ovate, 7-12 × 4-7 cm, glabrous and shining above, slightly pubescent along nerves beneath, base subcordate, margin dentate or coarsely toothed, apex acute, basal nerves 3-5; secondary nerves 3-5 pairs; petiole 1-2.5 cm. Flowers monoecious, greenish, aggregated into spike. Fruit red, then purple, sweet.

Usually cultivated, for rearing the silk worms. Fl. & Fr.: Throughout the year. Vern.: Tel.: *Resham chettu*; Tam.: *Pattu*. Eng.: Mulberry.

S.K. University campus (ATP), *TP & NY* 1250; Eswaraamala (ATP), *R.Thirupel Reddy* 374 (BSID).

INDIA: Cultivated in almost all the states of India.

WORLD: Native of northern China, cultivated elsewhere in USA, Australia, Mexico, Argentina, Madagascar, Africa and Europe.

Morus australis Poir., in Lam., Encycl. 4: 380. 1797. *M. indica* auct. non L., 1753; FBI 5: 492. 1888; Fischer 3: 1370. 1928.

Moderate sized trees. Leaves ovate, up to 7.5 × 4.5 cm, base cordate, margin serrate, apex acuminate, glabrescent above, pubescent at least on nerves beneath, 3-nerved at base; petioles up to 2 cm long, pubescent. Flowers in axillary spikes. Female spikes: up to 1.8 cm long, rachis pubescent; tepals 4, ciliate, outer 2 obovate, inner 2 much broader than the outer ones; ovary obovate, styles 2, connate at base. Achenes ovoid or globose.

In deciduous forests in Visakhapatnam district, probably escape from cultivation (Subba Rao and Kumari, 2008). Fl. & Fr.: May.

INDIA: Himalaya, up to 2100 m, Sikkim, widely cultivated throughout Idia.

WORLD: Myanmar, W. China.

Morus macroura Miq., Pl. Jungh. 42. 1851. *M. laevigata* Wall. ex Brandis, For. Fl. 409. 1874; FBI 5: 492. 1888; Fischer 3: 1370. 1928.

Medium sized trees, young parts pubescent, branches lenticellate. Leaves ovate, up to 13 × 7 cm, ovate, base rounded, often oblique, margin finely serrate, apex shortly acuminate, 3-nerved at base; petiole up to 4.4 cm long, glabrous. Flowers dioecious, axillary in pendulous spikes. Female spikes: up to 9.3 cm long, pubescent or glabrous; tepals 4, outer pair of tepals up to 2.2 mm long, keeled, ciliate, inner pair broadly obovate, flattened up to 2.2 mm across; ovary ovoid, compressed, styls 2, papillose, connate at base.

Occasional in deciduous forests in Visakhapatnam district, probably escape from cultivation (Subba Rao and Kumari, 2008). Fl. & Fr.: May.

INDIA: Himalayas. Cultivated elsewhere for rearing silkworms.

WORLD: Indo-China.

PLECOSPERMUM Trecul

Plecospermum spinosum Trecul, Ann. Sci. Nat. Bot. (Ser.3) 8: 124. 1847; FBI 5: 491. 1888; Fischer 3: 1353. 1928. *Trophis spinosa* Heyne ex Willd., Sp. Pl. 4: 734. 1806.

Large dioecious thorny shrubs or straggler or small trees, branchlets drooping, minutely pubescent; spines stout, axillary, straight. Leaves 4-6.5 × 2.5-4 cm, obovate or ovate-elliptic, base acute, or narrowed into petiole, alternate, margin entire, apex obtuse or acute, sub-coriaceous, glabrous. Flowers unisexual, aggregated in groups of 1-3, axillary capitable clusters. Syncarps angled.

Occasional in forests. Fl. & Fr.: April – August. Vern.: Tel.: *Kokinta, Potti murda chettu;* Ori.: *Berakanta, Banabanka, Banabana.*

Erramalai hills (KNL), *TP* & *RVR* 1329; Pecheruvu (KNL), *BR* & *SKB* 30160; Namixal RF (CTR), *DRC* 321 (BSID); Mamandur (CTR), *GVS* 31954 (MH); Nattavaram, *CAB* 1904 (MH); Tanjavanam RF (VSKP), *GVS* 47294 (MH).

INDIA: Sub-Himalayan tract from Punjab eastwards, Peninsular India.

WORLD: Sri Lanka.

STREBLUS Loureiro

1. Branchlets thorny; leaves smooth, margin distantly serrate towards the very upper half; male flowers subsessile; fruit obliquely ovoid .... **S. taxoides**
1. Branchlets not thorny; leaves rough on both surfaces, margin irregularly dentate; male flowers pedicellate; fruit globose .... **S. asper**

Streblus asper Lour., Fl. Cochinch. 1: 615. 1790; FBI 5: 489.1888; Fischer 3: 1353. 1928.

Evergreen scraggy or bushy tree, up to 7 m high; latex milky, bark smooth, grey; wood yellowish or brownwish-white; branchlets hairy. Leaves thick-coriaceous, elliptic-obovate or rhomboid, 1.7-6.8 × 1-3 cm, rough on both surfaces, base acute or cuneate, margin irregularly dentate, apex acute, lateral nerves 4-6 pairs, raised on under surface; petiole to 2 mm. Stipules linear, caducous. Flowers dioecious, male flowers in globose heads, cream-coloured, axillary, peduncle to 5 mm; female flowers solitary on axillary, usually fascicled peduncles. Fruit globose, 6 mm across, yellow, enclosed in the enlarged fleshy sepals, style persistent; seed 1, globose, smooth, 3 mm across.

Common, in shrubberies, hedges in dry deciduous forests up to 1000 m. Fl.: January- March, Fr.: April. Vern.: Tel.: *Stianike, Barranki, Barniki, Pakki, Bamigi.*

Way to Upper Ahobilam (KNL), *SSR* 15292; Kusimil (SKLM), *SSR* 15600; Nallamalai (KNL), *JSG* 10934 (DD); Balithogu forest, Mumumuru (KNL), *TRS* 85175 (MH); Balapalli (CDP), *JLE* 15745 (MH); Putula (KSN), *PVK* 5714 (MH); Koruturu (WG), *DN* 84238 (BSID); Gokavaram (EG), *JSG* 1216 (DD); Ragupalem (VSKP), *CAB* 1589 (MH); Punyagiri (VZN), *GVS* 23327 (MH).

INDIA: Widely distributed throughout drier parts of India.

WORLD: Sri Lanka, Myanmar, S. China, Hainan, Indo-China, Thailand, Malesia.

Streblus taxoides (Heyne ex Roth) Kurz, Forest Fl. Burma 2; 465. 1877. *Trophis taxoides* Heyne ex Roth, Nov. Pl. Sp. 368. 1821. *Phyllochlamys spinosa* Bureau in DC., Prodr. 17: 218. 1873; FBI 5: 488. 1888; Fischer 3: 1352. 1928.

Evergreen, small tree; bark peeling off in thin flakes; branchlets glabrous, minutely puberulous, ending with a spine. Leaves obovate or oblanceolate, 3-8.5 × 1-3 cm cm, glabrous, base cuneate, or slightly cordate, margin distantly serrate towards the very upper half, apex acuminate, secondary nerves 8-10 pairs, intersecondaries numerous, united near the margin and form an intra-marginal vein; petiole 3-4 mm. Male flowers brownish, 5 mm across sessile to subsessile, aggregated in clusters; female flowers solitary, pedicellate, sometimes two together with bracts at the base. Drupe obliquely ovoid, 5 mm, enclosed by enlarged foliaceous perianth.

Ocassional, in semievergreen forests, especially banks of streams at 700-1200 m. Fl.: February-March, Fr.: April-June. Vern.: Tel.: *Barivenka, Sukali.*

Talakona (CTR), *SSR* 19680; Maredumilli (EG), *GVS* 24274; Chintapalli (VSKP), *JLE* 51194; Krishnapuram (VSKP), *CAB* 1987 (CAL); Tirumala (CTR), *DRC* 2417 (MH).

INDIA: Bihar, Orisa, Deccan Peninsula, Andaman.

WORLD: Sri Lanka, Myanmar, Indo-China, Thailand, Malesia.

CASUARINACEAE

CASUARINA Adanson

Casuarina equisetifolia L., Amoen. Acad. 4: 143. 1759; FBI 5: 598. 1888; Fischer 3: 1389. 1928.

Tall leafless trees, up to 15 m high; branchlets arising from the axils of scales. Leaves scaly, to 1 mm long, alternating with ribs of the upper node. Male flowers in terminal catkins, to 6 mm long, tepals 2, 1 mm, scarious, thinly ciliate, acute, stamens 2-3 mm long. Female flowers in globose-ovoid cone like heads, bracts and bracteoles persistent and woody; tepals absent, ovary ovoid, 1 mm, style filiform, to 8 mm. Nutlets compressed, winged.

Widely planted along the coast. Fl. & Fr.: February – August. Vern.: Tel.: *Sarugudu;* Tam.: *Chavukku.*

Ethipothala Falls, Macherla RF (GNT), *VRK* 3566.

INDIA: Widely planted along the sandy sea-coast and sometimes also in the interior.

WORLD: Indigenous on the sandy sea-coasts of Chittagong and Myanamar, Malay Isls, Australia, Pacific.

SALICACEAE

SALIX Linnaeus

Salix tetrasperma Roxb., Pl. Coromandel t. 97. 1795; Fischer 3: 1389. 1928; Beddome, Fl. Sylv. S. India t. 302. 1873.

Deciduous, small or medium-sized tree; bark greyish-brown, rough with deep, vertical fissures; wood red, straight or somewhat interlocked grained, medium-fine and mostly even textured; branchlets silky-pubescent. Leaves membranous, lanceolate or oblanceolate to ovate-lanceolate, 5-16 × 1.5-6 cm, glabrous above, sparsely hairy or glabrescent and pale glaucous beneath, base narrowed or rounded, margin glandular-serrate, apex subacuminate, petiole to 2.5 cm. Flowers cream coloured in lax or dense-flowered catkins; one flower to each bract; perianth absent, disc cupular, stamens 6-9, unequal, exserted, 5.5 mm long, basifixed, ovary conical. Capsule 3-4 mm long, 2-valved, pubescent; valves recurved; seeds ovate-lanceolate, flattened, 2 mm across with brownish hairs.

Rare along the banks of rivers, streams and in wet swampy places, above 1000 m. Fl.: October-February, Fr.: December-May. Vern.: Tel.: *Iroda, Etipala, Konda ganneru, Vodime.*

INDIA: More or less throughout India, up to 1800 m.

WORLD: Afghanistan, Indo-China, Malesia, S. China.

CERATOPHYLLACEAE

CERATOPHYLLUM Linnaeus

Ceratophyllym demersum L., Sp. Pl. 992. 1753; FBI 5: 639. 1888; Fischer 3: 1391. 1928.

Submerged, slender, root-less aquatic herbs, branchlets brittle, green, reddish, glabrous. Leaves whorled, 7-12 in close verticels, 1.5-2.5 cm long, sometimes sparsely sericeous, margin denticulate, apex acute. Flowers minute, monoecious, axillary, solitary; tepals 12-14, subequal, oblanceolate, 1 mm, scarious, stamens in male flowers 16-24, anther 1.5 mm, basifixed, extrorse, ovary globose, 1.5 mm, style 2.5 mm, subulate, stigmatic on one side. Fruit 4 cm long with 3 spines.

Occasional in ponds, ditches and pools. Fl. & Fr.: September – December.

INDIA:Madhya Pradesh, Odisha, Andhra Pradesh, Tamil Nadu.

WORLD: Cosmopolitan.

References

Anitha, K. 2011. *Zeylanidium lichenoides*. The IUCN Red List of Threatened Species 2011: e.T177349A7419247. http://dx.doi.org/10.2305/IUCN.UK.2011-1.RLTS.T177349A7419247.en. Downloaded on 16 August 2017.

Balachandran, N. and K.Ravikumar. 2013. Notes on new distributional plant records from Southern India. Phtytaxonomy **13:** 92-95.

Balakrishnan, N.P. 2007. The Family Euphorbiaceae in India. A synopsis of its profile, taxonomy and Bibliography. Bishen Singh Mahendra Pal Singh, Dehra Dun.

Benjamin, J. H. F. and G. V. S. Murthy. 2013. Flora of Sri Venkateswara National Park, Andhra Pradesh. Botanical Survey of India, Kolkata.

Binojkumar, M.S. and N.P.Balakrishnan. 2010. The genus *Euphorbia* L. (Euphorbiaceae) in India: A taxonomic revision. Bishen Singh Mahendra Pal Singh, Dehra Dun.

Bahadur, B., T.Pullaiah and G.V.S.Murthy. 2013. Systematics of Indian Jatrophas. I: B.Bahadur et al. (eds.) *Jatropha,* Challenges for a New Energy Crop. Volume 2. pp. 153-186. New York, Springer Science.

Chakrabarty, T. and P. G. Diwakar. 2009. An undescribed species of *Neolitsea* (Benth.) Merr. (Lauraceae) from Andhra Pradesh. J. Econ. Taxon. Bot. 33(2): 345-346.

Chandra, R. and P. A. Azeez. 2011. The flora of Araku valley, Visakhapatnam, Andhra Pradesh, India. J. Econ. Taxon. Bot. 35(4): 816-836.

Chaudhary, L. B., J. V. Sudhakar,A. Kumar, O. Bajpai, R. Tiwari, and G. V. S. Murthy. 2012. Synopsis of the genus *Ficus* L. (Moraceae) in India. Taiwania **57(2):** 193-216.

Cook, C.D.K. 1996. Aquatic and Wetland Plants of India. Oxford University Press, Oxford.

Debasmita, D.P., G.G.Maiti and N.S.Mondal. 2015. A taxonomic revision of the genus *Boerhavia* L. (Nyctaginaceae) in India. J. Global Biosci. 4(3): 1704-1707.

Haridasan, K., K. Ravikumar, D. Saha and D. Ved. 2015. *Myristica dactyloides*. The IUCN Red List of Threatened Species 2015: e.T33526A50131225. http://dx.doi.org/10.2305/IUCN.UK.2015-2.RLTS.T33526A50131225.en. Downloaded on **17 August 2017**

Kalpana, B., R. Devilal, M. Sharath Goud, B. Chandra Shekar, B. Sadasivaiah and N.Babu Rao. 2016. *Laportea interrupta* (L.) Chew (Urticaceae) – a new distributional record from central Eastern Ghats, India. J. Biol. Records 9: 81-85.

Karuppusamy, S. and T. Pullaiah. 2006. Occurrence of Podostemaceae – a new report to the flora of Andhra Pradesh. J. Econ. Taxon. Bot. **30**: 397-398.

Kottaimuthu, R. 2012. *Cleidion nitidum* (Muell.-Arg.) Thw. ex Kurz. (Euphorbiaceae): a new record for Peninsular India. Life Sciences Leaflets 4: 31-35.

Krishna Prasad, V. 1999. Studies on the biodiversity and forest cover dynamics of Rampa revenue division (Eastern Ghats) using remote sensing technique. Ph.D. thesis, Osmania University.

Kullayiswamy, K.R., S.Sandhya Rani, B.Sadasivaiah and T.Pullaiah. 2017. Flora of Thummalapalle Uranium Mining Area, Andhra Pradesh, India. Regency Publications, New delhi.

Lalithamba, A. 2014. The new distributional record of *Phyllanthus scabrifolius* Hook. f. (Phyllanthaceae) from Velugonda hills of Nellore district, Andhra Pradesh, India. J. Indian Bot. Soc. 93: 284-285

Matthew, K.M. 1983. Flora of Tamil Nadu Carnatic. Rapinat Herbarium, Tiruchirapalli.

Matthew, K.M. 1991. An excursion flora of Central Tamil Nadu, India. Oxford and IBH Publishing Co. Pvt. Ltd., New Delhi.

McNeill, J. *et al.* (eds.) 2012. International Code of Nomenclature for algae, fungi, and plants (Melbourne Code). International Association for Plant Taxonomy. Bratislava.

Meerabai, G. and B. Padmavathi. 2011. Plant diversity in protected area of Nallamalai forest at Velugodu, Kurnool, Andhra Pradeh, India. Indian Forester 137(4): 512-520.

Murugesan *et al.*, 2009. J. Econ Taxon. Bot. 33: 553.

Naidu, M. T., J.P.Rao, D.S Rao, P.P. Murthy, and M.Venkaiah. 2012. Plant wealth of Kambalakonda wild life sanctuary, Visakhapatnam, Andhr Pradesh. Indian Forester 138(7): 593-699.

Narasimhan, D., N. Rama Rao and T. Ravisankar. 1989. Two rare interesting taxa of Euphorbiaceae from Andhra Pradesh, India. J. Econ. Taxon. Bot. 13: 56-59.

Panigrahi, G., S. Chowdhury, D.C.S.Raju and G.K. Deka G. 1964. A contribution to the Botany of Orissa. Bull. Bot. Surv. India. 6(2-4): 237-266.

Prakasa Rao, J. and K. V. Satish. 2016. *Persicaria perfoliata* (L.) H. Gross (Polygonaceae): A species new to Eastern Ghats of India. Trop. Plant Res. 3(2): 249-252.

Prakasa Rao, J., K. V. Satish and O. A. Kumar. 2016. *Achyranthes longifolia* (Makino) Makino (Amaranthaceae) – An angiosperm new to India. National Acad. Sci. Letters DOI 10. 1007/s40009-016-0515-4

Prakasa Rao, J., K. V. Satish, B. Siva Sankar, C. S. Reddy and O. Aniel Kumar. 2015. On the occurrence of parasitic plant *Balanophora fungosa* J. R. Forster & G. Forster (Balanophoraceeae) in Andhra Pradesh, India. J. Threatened Taxa 7(2): 6943-6946.

Prasad, K. and P.V. Prasanna. 2016a. *Euphorbia seshachalamensis* (Euphorbiaceae), a new species from Andhra Pradesh, India. Ann. Bot. Fennici 53: 73-76.

Pullaiah, T., D.Ali Moulali and S. Sandhya Rani. 2018. Flora of Andhra Pradesh, Volume 4, 2nd Edition., Scientific Publishers, Jodhpur.

Raju, V.S., S.Suthari and A. Ragan. 2011. The identity and occurrence of *Phyllanthus hookeri* Muell-Arg. and *P. nozeranii* Rossignol & Haicour (Euphorbiaceae) in India. Bangladesh J. Plant Taxon. 18(1): 57-63.

Rasingam, L., A. R. Chorghe, P. V. Prasanna and M. Sankara Rao. 2014. *Glochidion tirupathiense* (Phyllanthaceae) – A new species from Seshachalam Biosphere reserve of Andhra Pradesh. Taiwania 59(1): 9-12.

Ravikumar, K., N. Balachandran, S.N. Begum, P. Patchaimal, M. Bhanja and K. Lohitasyudu. 2014. Interesting plant records from Visakhapatnam District, Andhra Pradesh, India. J. Threatened Taxa 6(8): 6108–6121; http://dx. doi. org/10. 11609/JoTT. o3144. 6108-21

Reddy, C. S. and V. S. Raju 2002. Additions to the weed flora of Andhra Pradesh. J. Econ. Taxon. Bot. **26**: 195-198.

Rehel, S. 2011. *Polypleurum filifolium*. The IUCN Red List of Threatened Species 2011: e.T177034A7362830. http://dx.doi.org/10.2305/IUCN.UK.2011-1.RLTS. T177034A7362830.en. Downloaded on **16 August 2017**.

Sankara Rao, M., J. Swamy, S Nagaraju, S.B. Padal, M. Tarakeswara Naidu, K.Chandramohan and T.Thulasaiah. 2016. *Glochidion talakonense sp. nov.* (Phyllanthaceae) from Seshachalam Biosphere Reserve, Andhra Pradesh, India. Bangladesh J. Plant Taxon. 23(1): 59-63.

Sarojinidevi, N. and R. R. Venkataraju. 2014. *Euphorbia kadapensis* (Euphorbiaceae), a new species from southern India. Phytotaxa 181 (3): 179-183.

Sarojinidevi, N. and R.R. Venkataraju. 2017. *Euphorbia royleana* Boiss. (Euphorbiaceae) a new record for the Eastern Ghats, India. J. Threatened Taxa 9(5): 10233-10235.

Shivakumar, S. and M. Prabhakar. 2013. *Amaranthus polygonoides* Linn.: A new record to Andhra Pradesh. J. Econ. Taxon. Bot. 37(2): 414.

Subramanyam, K. 1962. Aquatic Angiosperms.

Subramanyam, K. and C. P. Sreemadhavan. 1969. A conspectus of the families Podostemaceae and Tristichaceae. Bull. Bot. Surv. India **11**: 161-168.

Sudhakar, J.V., N. Chandra Mohan Reddy and G.V.S. Murthy. 2017. Figs of Eastern Ghats. Pragathi offset, Hyderabad.

Viswanatha, R. Murugan and R. R. Rao. 2011. Extended distribution of two little known herbaceous *Phyllanthus* species in India. J. Econ. Taxon. Bot. **35**: 206-209.

Colour Plates

Boerhavia chinensis

Boerhavia diffusa

Boerhavia erecta

Mirabilis jalapa

Bougainvillea glabra

Pisonia aculeata

Images by A. Lalithamba

Aerva javanica

Aerva lanata

Achyranthes aspera* var. *pubescens

Allmania nodiflora

Alternanthera ficoidea

Alternanthera pungens

Images by A. Lalithamba

Amaranthus dubius

Amaranthus viridis

Amaranthus polygamus

Amaranthus spinosus

Celosia argentea var. *cristata*

Celosia polygonoides

Images by A. Lalithamba

Digera muricata

Gomphrena serrata

Nothosaerva brachiata

Puppalia lappacia

Puppalia lappacia* var. *lappacia

Trichuriella monsoniae

Chenopodium murale

Images by A. Lalithamba

Basella alba

Antigonon leptopus

Persicaria chinensis

Persicaria glabra

Polygonum plebium

Rumex vesicarius

Images by A. Lalithamba

Aristolochia bracteolata

Aristolochia indica

Piper longum

Piper nigrum

Peparomia tetraphylla

Actinodaphne madraspatana

Images by A. Lalithamba

Litsea glutinosa

Litsea deccanensis

Persea macrantha

Cassytha filiformis

Gyrocarpus americanus

Grevillea robusta

Images by A. Lalithamba

Dendrophthoe falcata* var. *pubescens

Dendrophthoe falcata* var. *falcata

Dendrophthoe glabrascens

Taxillus heyneanus

Viscum ramosissimum

Santalum album

Images by A. Lalithamba

Acalypha alnifolia

Antidesma ghaesembilla

Baliospermum solanifolium

Bridelia glauca

Bridelia montana

Bridelia retusa

Images by A. Lalithamba

Breynia compressa

Breynia vitis-idaea

Cleistanthus collinus

Cleistanthus patulus

Chrozophoa rotlleri

Croton persimilis

Images by A. Lalithamba

Croton scabiosus

Dalechampia indica

Dimorphocayx glabellus

Drypetes sepiaria

Euphorbia caudicifolia

Euphorbia clarkeana

Images by A. Lalithamba

Euphorbia indica

Euphorbia lacei

Euphorbia thymifolia

Euphorbia deccanensis var. *nalllamalyana*

Fluggea leucopyrus

Fluggea virosa

Images by A. Lalithamba

Glochidion ellipticum

Givotia moluccana

***Homonoia riparia* (male)**

Jatropha curcas

Macaranga peltata

Margaritaria indica

Images by A. Lalithamba

Mallotus philippinensis

Mallotus repandus

Microstachys chamaelea

Micrococcoa mercurialis

Meineckia parviflora

Putranjiva roxburghii

Images by A. Lalithamba

Phyllanthus emblica

Phyllanthus indofischeri

Phyllanthus leschenaultii

Phyllanthus maderaspatensis

Phyllanthus polyphyllus

Phyllanthus lawi seed

Images by A. Lalithamba

Phyllanthus rheedei

Phyllanthus reticulates

Phyllanthus rotundifolius

Phyllanthus virgatus

Phyllanthus fraternus

Phyllanthus urinaria

Phyllanthus scabrifolius

Images by A. Lalithamba

Ricinus communis

Sauropus androgynous

Suregada lanceolata

Tragia involucrate

Artocarpus heterophyllus

Artocarpus lacoocha

Images by A. Lalithamba

Ficus amplissima

Ficus arnottina

Ficus benghalensis

Ficus benjamina

Ficus microcarpa

Ficus mollis

Images by A. Lalithamba

Ficus racemosa

Ficus religiosa

Ficus rumphi

Ficus tinctoria

Ficus tsjakela

Ficus virens

Images by A. Lalithamba

Morus alba

Plecospermum spinarum

Trema orientalis

Holoptelia integrifolia

Cannabis sativa

***Celtis philippinensis* with fruit**

Images by A. Lalithamba

Celtis philippinensis

Celtis timorensis

Pouzolzia pentandra

Pouzolzia zeylanica

Streblus asper

Salix tetrasperma

Images by A. Lalithamba

Index to Families, Genera and Species
(Italicized names indicate Synonyms)

Index to Vernacular Names

Telugu

TAMIL

ORIYA

www.ingramcontent.com/pod-product-compliance
Ingram Content Group UK Ltd.
Pitfield, Milton Keynes, MK11 3LW, UK
UKHW021322190726
13845UKWH00003B/25

9 789386 071996